Invasive Species Reviews: 2018–2024

Invasive Species Reviews: 2018–2024

Edited by

David Hemming

CABI

CABI is a trading name of CAB International

CABI
Nosworthy Way
Wallingford
Oxfordshire OX10 8DE
UK

Tel: +44 (0)1491 832111
E-mail: info@cabi.org
Website: www.cabi.org

CABI
200 Portland Street
Boston
MA 02114
USA

T: +1 (617)682-9015
E-mail: cabi-nao@cabi.org

A catalogue record for this book is available from the British Library, London, UK.

ISBN-13: 9781800628908 (hardback)
9781800628915 (ePDF)
9781800628922 (ePub)

DOI: 10.1079/9781800628922.0000

Commissioning Editor: Alexandra Lainsbury
Editorial Assistant: Emma McCann
Production Editor: Shankari Wilford

Typeset by Exeter Premedia Services Pvt Ltd, Chennai, India
Printed and bound in the UK by CPI Group (UK) Ltd, Croydon, CRO 4YY

Contents

Contributors

Nahuel Aizen: Grupo de Ecología de la Polinización, INIBIOMA, Universidad Nacional del Comahue, CONICET, Quintral 1250, Bariloche 8400, Argentina

B.M.A. Abdel-Banat: Date Palm Research Centre of Excellence, King Faisal University, P.O. Box 400, Al-Hassa-31982, Saudi Arabia

A. Alba: Laboratorio de Anticuerpos Monoclonales, Departamento de Parasitología, Instituto de Medicina Tropical -Pedro Kourí", Ave Novia del Mediodía km 6 1/2, AP 601, Havana, Cuba

P. Alda: Centro de Estudios Parasitológicos y Vectores, Consejo Nacional del Investigaciones Científicas y Técnicas, Universidad Nacional de La Plata (CCT-La Plata-CONICET-UNLP), Calle 2 No. 584, 1900 La Plata, Buenos Aires, Argentina

Marina P. Arbetman: Departamento de Zoologia, Universidade de Brasilia, Campus Darcy Ribeiro, Brasilia, DF, Brazil

C. Bergamino: Department of Veterinary Sciences, University of Turin, Largo Paolo Braccini 2, 10095 Grugliasco, Turin, Italy

R.C. Blackman: School of Environmental Sciences, University of Hull, Cottingham Road, Hull, HU6 7RX, UK

M.C. Boukouvala: Laboratory of Agricultural Zoology and Entomology, Department of Crop Science, Agricultural University of Athens, 75 Iera Odos str., 11855, Athens, Attica, Greece

Matthew L. Buffington: USDA-ARS Systematic Entomology Laboratory, Washington D.C., USA

S. Burela: Laboratorio de Ecologia, Departamento de Biologia, Bioquimica y Farmacia, Universidad Nacional del Sur, San Juan 670 (8000), Bahia Blanca, Argentina

Nicolás R. Cecchetto: Universidad Nacional del Comahue (INIBIOMA-CONICET) Río Negro, Argentina

M.J.W. Cock: CABI, Bakeham Lane, Egham, Surrey TW20 9TY, UK

Kent M. Daane: University of California Berkeley, Berkeley, California, USA

M.D. Day: Department of Agriculture and Fisheries, Ecosciences Precinct, GPO Box 267, Brisbane, Queensland 4001, Australia

H.A.F. El-Shafie: Date Palm Research Centre of Excellence, King Faisal University, P.O. Box 400, Al-Hassa-31982, Saudi Arabia

R. Enderle: Institute for Plant Protection in Horticulture and Forests, Julius Kuehn Institute, 38104 Braunschweig, Germany

H.C. Evans: CABI, Bakeham Lane, Egham, Surrey TW20 9TY, UK

Zhang FuDou: Agricultural Environment & Resource Research Institute, Yunnan Academy of Agricultural Sciences, Ku

Xu GaoFeng: Agricultural Environment and Resource Research Institute, Yunnan Academy of Agricultural Sciences, Kunming Yunnan 650205, China

S. García-Lara: Unidad de Estrés Biótico, Programa Global de Maíz, CIMMYT, Apartado Postal 6-641, 06600 México D. F., Mexico

E.W. Githae: Department of Biological Sciences, Chuka University, P. O. Box 109-60400, Chuka, Kenya

Jennifer Grenz: Department of Forest Resources Management, Faculty of Forestry, University of British Columbia, 2424 Main Mall, Vancouver, BC V6T 1Z4, Canada

V.C. Griess: Institute of Forest Management, Center of Life and Food Sciences Weihenstephan, Technische Universität München, Hans-Carl-von-Carlowitz-Platz 2, 85354 Freising, Germany

B. Hänfling: Department of Biological Sciences, University of Hull, Cottingham Road, Hull HU6 7RX, UK

Kim A. Hoelmer: USDA-ARS Beneficial Insects Introduction Research Unit, Newark, Delaware, USA

A.J. Hruska: Food and Agriculture Organization of the United Nations, Viale delle Terme di Caracalla, Rome 00153, Italy

S. Hurtrez-Boussès: Département Biologie des Populations, CEFE-CNRS, BP 5051, 34033 Montpellier Cedex 1, France

Vanessa L. Jones: Faculty of Land and Food Systems, 248-2357 Main Mall, University of British Columbia, Vancouver, BC V6T 1Z4, Canada

Brooks A. Kaiser: Department of Sociology, Environmental and Business Economics, University of Southern Denmark, 6705, Esbjerg Ø, Denmark

N.G. Kavallieratos: Lab. of Agricultural Zoology and Entomology, Agricultural University of Athens, 11855 Athens, Greece

Melina Kourantidou: Department of Sociology, Environmental and Business Economics, University of Southern Denmark, 6705, Esbjerg Ø, Denmark; Hellenic Centre for Marine Research, Institute of Marine Biological Resources and Inland Waters, 16452, Athens, Greece; Woods Hole Oceanographic Institution, Marine Policy Center, Woods Hole, MA, 02543, USA

V. Lafond: Cemagref-EMGR, 2 rue de la Papeterie, BP 76, 38402 Saint-Martin-d'Hères Cedex, France

L. Lawson-Handley: School of Natural Sciences, University of Hull, Cottingham Road, Hull, HU6 7RX, UK

Jana C. Lee: USDA-ARS Horticultural Crops Research Unit, Corvallis, Oregon, USA

S. Lioy: Department of Agricultural, Forest and Food Sciences, University of Turin, Largo Paolo Braccini 2, 10095 Grugliasco, Turin, Italy

Josefina Lohrmann: Grupo de Ecología de la Polinización, INIBIOMA, Universidad Nacional del Comahue, CONICET, Quintral 1250, Bariloche 8400, Argentina

L. M. López-Castillo: Laboratory of Biochemical and Instrumental Analysis, Department of Biotechnology and Biochemistry, Cinvestav Unidad Irapuato, Km. 9.6 Libramiento Norte, Carretera Irapuato-León, 36821 Irapuato, Guanajuato, Mexico

M. Lounnas: MIVEGEC, UMR IRD 224 CNRS 5290 UM1-UM2, 911 Avenue Agropolis, BP 64501, 34394 Montpellier Cedex 5, France

P.R. Martín: Laboratorio de Zoología de Invertebrados I, Departamento de Biología, Bioquímica y Farmacia, Universidad Nacional del Sur, San Juan 670, (8000) Bahía Blanca, Argentina

B.L. Muatinte: Department of Biological Sciences, Faculty of Sciences, Eduardo Mondlane University, P.O. Box 257, Maputo, Mozambique

B.M. Mvumi: Department of Soil Science and Agricultural Engineering, University of Zimbabwe, P.O Box MP167, Mt Pleasant, Harare, Zimbabwe

J.P. Pointier: Laboratoire de Zoologie de l'INRA, Petit-Bourg, 97170 Guadeloupe, French West Indies

M. Porporato: Department of Agricultural, Forest and Food Sciences, University of Turin, Largo Paolo Braccini 2, 10095 Grugliasco, Turin, Italy

M. Rusdy: Department of Forage Science and Grassland Management, Faculty of Animal Science, Hasanuddin University, Makassar, Indonesia

E. Sabourin: Centro de Cooperação Internacional em Pesquisa Agronômica para o Desenvolvimento (CIRAD), Unidade de Pesquisa "Actions collectives, politiques publiques et marchés", Departamento de Sociologia da Universidade de Brasília (UnB), Brasília, DF, Brazil

L. Saveanu: Laboratorio de Ecología, Departamento de Biología, Bioquímica y Farmacia, Universidad Nacional del Sur. San Juan 670, 8000, Bahía Blanca, Argentina

M.E. Seuffert: Departamento de Biología, Bioquímica y Farmacia, Universidad Nacional del Sur, San Juan 670, 8000, Bahía Blanca, Argentina

R.H. Shaw: CABI, Bakeham Lane, Egham, Surrey TW20 9TY, UK

Shen ShiCai: Center for Biodiversity and Indigenous Knowledge, 3rd Floor Building A, Zhonghuandasha, Yanjiadi, China

V. Srivastava: Laboratory of Green Chemistry, School of Engineering Science, Lappeenranta University of Technology, Sammonkatu 12, FI-50130, Mikkeli, Finland

John P. Stanga: 1501 Mercer University Dr., Macon, GA 31211, USA

J. Stenlid: Swedish University of Agricultural Sciences, S-750 07 Uppsala, Sweden

N.E. Tamburi: INBIOSUR (UNS-CONICET), Universidad Nacional del Sur, Departamento de Biología, Bioquímica y Farmacia, San Juan 670, 8000 Bahía Blanca, Argentina

I. Unlu: Center for Vector Biology, Department of Entomology, Rutgers University, New Brunswick, NJ 08901, USA

Osariyekemwen Uyi: Department of Animal and Environmental Biology, University of Benin, P.M.B. 1154, Benin City, Nigeria; Department of Entomology, University of Fort Hare, Private Bag X1314, 5700 Alice, South Africa; Department of Entomology, Pennsylvania State University, University Park, Pennsylvania, 16802, USA

R. Vasaitis: Department of Forest Mycology & Pathology, Swedish University of Agricultural Sciences, P.O. Box 7026, SE-75007 Uppsala, Sweden

A.A. Vázquez: Laboratorio de Malacología, Instituto de Medicina Tropical Pedro Kourí, Apartado Postal 601, Marianao 13, La Habana, Cuba

Xingeng Wang: USDA-ARS Beneficial Insects Introduction Research Unit, Newark, Delaware, USA

R.L. Winston: MIA Consulting, 316 N. Hansen Ave., Shelley, ID 83274, USA

Eduardo E. Zattara: Grupo de Ecología de la Polinización, INIBIOMA, Universidad Nacional del Comahue, CONICET, Quintral 1250, Bariloche 8400, Argentina

Introduction

A recent study by Bradshaw *et al.* (Ecological Economics, 2024) estimated the damage costs associated with invasive species at US$1.578 trillion and management expenditure at US$147 billion. Invasive species affect agriculture, forestry and fishing, and can even damage buildings and infrastructure. This harms local and national economies and hinders sustainable economic growth. The economic impacts of invasive species include lower crop productivity, harm to livestock, and soil degradation. The value of land is often decreased, and producers have to spend more on controlling pests, weeds and diseases. Invasive species also reduce biodiversity, impacting the functioning of ecosystems, and affecting native plant and animal species.

This collection of articles looks at a wide range of invasive species and their impacts, including insects, plants, snails, and fungal diseases, including: *Mimosa diplotricha*, *Chromolaena odorata*, privet, *Opuntia*, fall armyworm, *Aedes albopictus*, *Prostephanus truncatus*, *Pomacea* and ash dieback. The articles examine mechanisms for detecting the spread of invasive species, and models for understanding the mechanisms of invasion alongside control and management approaches with a particular focus on biological control. The articles have been specially selected from contributions to the international peer-reviewed journal *CABI Reviews*.

Reference

Bradshaw, C.J.A., Hulme, P.E., Hudgins, E.J., Leung, B., Kourantidou, M. *et al.* (2024) Damage costs from invasive species exceed management expenditure in nations experiencing lower economic activity. *Ecological Economics*, 220, 108166.

1 A review of the impacts and management of invasive plants in forestry

Vanessa L. Jones and Jennifer Grenz*
Department of Forest Resources Management, Faculty of Forestry, University of British Columbia, 2424 Main Mall, Vancouver, BC V6T 1Z4, Canada

Abstract
While the impacts of invasive plants are commonly researched and described within the context of agriculture and restoration ecology, they receive less attention within the specific context of forestry. Forestry operations are both vulnerable to and could exacerbate the spread of invasive plants through all aspects of silviculture, all of which can lead to reduced profitability and negative impacts to the sustainability and resiliency of the ecosystems they operate within. The purpose of this review article was to synthesize the current academic and gray literature pertaining to invasive plants and forestry to inform prevention and management approaches and identify gaps in the research. We incorporated a case study from interviews with major forestry company professionals in British Columbia, Canada managing invasive species within their operations to provide critical and often overlooked perspectives and experiences. Our review provides key insights into the risks invasive plants pose to forest community and tree health, operations, economic value, and ecosystem health and identifies the need for research specific to the impacts of invasive plants and their management strategies within the context of forestry operations.

Keywords: invasive plants, invasion biology, forestry, invasive plant management, restoration ecology, noxious weeds

Introduction

Invasive plants have become a recognized threat to forestry worldwide, causing both direct and indirect negative impacts to forest health, human livelihoods, and surrounding ecosystems (Fine, 2002; Rai and Singh, 2020). Reductions in forest productivity and regeneration are causing economic losses through the resulting decrease in timber values and increased costs associated with control measures (Holmes *et al.*, 2009). Invasive plants have been known to negatively impact forest ecosystems by altering nutrient cycling, changing soil structure and aggregation, modifying fire regimes, reducing water availability, and altering community composition and structure (Langmaier and Lapin, 2020; Pyšek *et al.*, 2020; Rai and Singh, 2020). Impacts that have the potential to be compounded by climate change as it is anticipated that changes such as hotter summers, more growing degree days, fewer frosts, and increased atmospheric CO_2 will favor invasive plant presence and decrease the efficacy of management approaches (IPCC, 2014; Birthisel *et al.*, 2021). Invasive plants employ strategies such as rapid reproduction and evolution and can adapt to these changes better than native forest tree

*Corresponding author: jennifer.grenz@ubc.ca

 Invasive Species Reviews: 2018–2024 (ed. D. Hemming)

species (Dukes *et al.*, 2009; Birthisel *et al.*, 2021; Shephard *et al.*, 2022).

The generally accepted definition of an invasive species is a nonnative species to a particular ecosystem that is causing some degree of harm to humans and/or the environment (Richardson and Pysek, 2004; NISC, 2005). This definition is not without controversy, however, as there are varying viewpoints among researchers and practitioners regarding the limitation of the categorization of an invasive species to include only non-native species (Colautti and MacIsaac, 2004). Many examples exist of species native to a region that exhibit invasive qualities (Kay, 1994; Bai *et al.*, 2004; Valéry *et al.*, 2008, 2009; Dunwiddie *et al.*, 2011), including those that have negative impacts on forest operations such as the Mountain pine beetle (*Dendroctonus ponderosae* Hopkins), a species native to western Canada which infects and often leads to the death of Lodgepole pine (*Pinus contorta* Dougl.) (Carroll *et al.*, 2004; Safranyik and Wilson, 2006). Another example from Canada includes several native plant species that are often described within forestry operations as behaving invasively, including wild raspberry (*Rubus strigosus* Michx.), thimbleberry (*Rubus parviflorus* Nutt.), and salmonberry (*Rubus spectabilis* Pursh); these species form extensive, dense monoclonal stands, rapidly invading disturbed areas (Oleskevich *et al.*, 1996). For the purposes of this review, we will define invasive plants using the generally accepted definition provided above while including consideration of their economic harms. Native plant species which behave invasively will not be included in our definition of invasive plants.

While widespread recognition of the potential risks of invasive plants has catalyzed a growing body of research over the past two decades (Watson *et al.*, 2019; European Commission, 2021), in addition to the challenges of achieving consensus on what defines an invasive species, there is the related issue of confirmation bias toward those labeled as such. This may be evidenced by common citations repeated in research articles and education materials (Davis, 2009) that often describe invasive species impacts in generalized terms that lack quantifiable, empirical data to measure them, especially when determining ecological rather than economic or other human impacts (Parker *et al.*, 1999; Valéry *et al.*, 2008; Crystal-Ornelas and Lockwood, 2020). Given the potential significance of the impacts of invasive plants to forestry and the ecosystems forestry operations are part of, and the criticism that invasion biology applies generalized research results to inappropriate contexts (Davis, 2009), we focus our review on specific and quantifiable impacts of invasive plants within forestry operations. This review aims to assist with informing the development of invasive plant prevention strategies and management plans through a comprehensive synthesis of the current academic and gray literature, shared insights and lessons learned by industry professionals, and to identify research gaps and opportunities. Through this synthesis, four distinct themes pertaining to impacts emerged that will form the review structure: forest community and tree health; operations; economic value; and ecosystem health.

Review methodology

This review was compiled by searching for literature using open access journals from MDPI and CABI, and additional paid-access journals were searched through the University of British Columbia OpenAthens Online Library. The following terms or combination of terms were used when searching databases: invasive plants, forestry, forestry operations, invasive weeds, weed management, invasive plant management, invasive plant monitoring, invasive plant prevention, economic impacts, environmental impacts, timber production, timber value, herbicides, forest recreation, climate change.

Relevant gray literature was sought from a variety of global government agencies including economic reports, legislation and policies, invasive plant management practices, and forestry operations practices; these were accessed from the respective government agency websites. Additional information was obtained through three open-ended, semi-structured interviews with representatives from major forestry companies in British Columbia responsible for the management of invasive plants and by speaking to research colleagues to check for any upcoming studies not yet published.

Review of impacts of invasive plant species to forestry

People and invasive plants share an extensive co-evolutionary history, with the emergence of invasive plants being traced to the development of agriculture. With more recent developments in silviculture, including tree propagation and timber harvesting, it follows that invasive plant species would impact these operations (Sitzia, 2014).

Similar to agriculture, silviculture describes the activities used to grow and manage trees including harvesting, regeneration/replanting, fertilizing, pruning, thinning, and more (Government of BC, 2023e). Many of these practices allow for the introduction and proliferation of invasive plants, especially if preventative measures are not in place (Friday *et al.*, 2008; Peter and Harrington, 2018). Additionally, invasive plants may impact other aspects of forestry operations, such as the building and maintenance of logging roads and camps, as invasive plants may cause safety issues by blocking sightlines or compromising infrastructure (Veldman and Putz, 2010). As a result, strategies and management plans within forestry operations often include methods for preventing, managing, and monitoring invasive plants to minimize impacts to various aspects of operations, timber value (Shephard *et al.*, 2022), and to ensure environmental stewardship within their operational areas and the ecosystems they are within.

Forest community and tree health

Within the context of forest ecosystems, invasive plants can have significant effects on forest composition and structure and tree health (Sokol *et al.*, 2017; Wang *et al.*, 2022). As composition of managed forests is directly tied to productivity and timber value, these impacts can, in turn, affect profitability (Holmes *et al.*, 2009).

Temperate deciduous forests, often associated with human activities, including forestry operations, are examples of systems whose composition and structure are vulnerable to the introduction of invasive plants. A study of 38 sites across the eastern United States by Trammell *et al.* (2020) revealed that although canopy trees were 98% native species, the understory was composed of 65% non-native woody plants. Of those, 64% of shrubs and 76% of vines were invasive species. Four dominant invasive plant species across the sites accounted for 64% of all total understory stems resulting in a lack of variation and decreased evenness in understory composition across the sites (Trammell *et al.*, 2020; Jobin *et al.*, 2023). It is important to recognize, however, that plant invasion is not a linear trajectory of ever-increasing abundance over time (Rasp, 2019) in such contexts. Invasive plant species abundance can vary with stages of canopy closure with areas with more recent canopy closure being more abundant than those that have been closed for longer (Trammell *et al.*, 2020). An important consideration for prioritizing invasive plant management. The impacts of invasive plants on tree health may be caused by a variety of mechanisms such as competition for resources and changes to the physical environment (Willoughby *et al.*, 2006; Maguire *et al.*, 2009). Invasive plants use mechanisms to impact both the abiotic and biotic environment, which in turn impacts the forest community composition (Bellerose *et al.*, 2022). Invaded forest sites are vulnerable to competition by invasive plants that can outcompete woody tree seedlings for resources such as bare soil for seedling establishment, and water, nutrient, and light availability (Burnham and Lee, 2010; Carter *et al.*, 2019; Harrington, 2020). For instance, the invasive species tree of heaven (*Ailanthus altissima* (Mill.) Swingle), an invader of European forests, competes with the native sweet chestnut tree (*Castanea sativa* Mill.) for water during droughts. Historically, it has been shown that tree of heaven maintains higher levels of drought resistance than sweet chestnut, demonstrated by decreased *C. sativa* radial growth during periods of drought (Knüsel *et al.*, 2015). With more frequent and severe droughts expected in the future due to climate change, invasive species that compete with trees for water are especially concerning (IPCC, 2014; Kelso *et al.*, 2020; Orbán *et al.*, 2021). Analysis of Alpine and Mediterranean ecosystems, both of which support forestry operations, found that invasive plants have higher evolutionary potential to adapt to climate change (Gianoli and Molina-Montenegro, 2021). The photosynthetic rate was measured as a physiological indicator of evolutionary potential and was significantly higher in invasive species compared to their native counterparts (Gianoli and Molina-Montenegro, 2021). In another example in European forests, silver birch (*Betula pendula* Roth) is subject to competition for water by herbaceous invasive plants such as spear thistle (*Cirsium vulgare* L.) as well as competition for bare soil for seedling establishment with invasive grasses such as perennial ryegrass (*Lolium perenne* L.). Similarly, Scotch broom (*Cytisus scoparius* (L.) Link (ITIS)), a common invader of forests in the Pacific Northwest of North America, uses more water than native plants during the early and late-growing season, which are essential times for stem wood production in plantation trees (Carter *et al.*, 2019; Slesak *et al.*, 2022). As nitrogen (N)-fixing leguminous species, Scotch broom also alters the soil nutrient profile by increasing N as well as other nutrients like magnesium (Mg), calcium (Ca), and potassium (K), which can in turn create positive feedbacks for other invasive plants that prefer those conditions, such as sweet vernalgrass (*Anthoxanthum odoratum* L.) (Carter *et al.*, 2019). N-fixing plants like Scotch broom, which significantly change the soil chemical makeup, pose a unique problem for management as removal of the plant may result in a corresponding decrease in available soil nutrients for planted trees, which may in turn require mitigation with fertilizers (Welham *et al.*, 2007; Slesak *et al.*, 2022).

Invasive plants may also release allelochemical compounds, which are chemicals released into the soil or air that directly impact the growth, germination, or establishment of nearby plants or alter the composition of fungi and bacteria in the soil (Qu *et al.*, 2021; Xu *et al.*, 2023). One such plant is Himalayan balsam (*Impatiens glandulifera* Royle), an invader of deciduous and coniferous forests in both Europe and North America (Gaggini *et al.*, 2018). This species releases allelopathic compounds including naphthoquinones which are associated with an altered soil bacterial and fungal community and subsequently altered plant species composition, with less native coniferous and deciduous species associated with Himalayan balsam invasion (Rusterholz *et al.*, 2014; Gaggini *et al.*, 2018). At the soil microbial community level, the release of allelochemicals by invasive plants can also cause changes to the abundance and diversity of fungi and bacteria. This is especially important when considering that many coniferous trees are reliant upon mycorrhizal fungal networks for sharing resources like carbon and nitrogen (Pickles *et al.*, 2017; Qu *et al.*, 2021) which may in turn affect tree community health (Beck *et al.*, 2020).

Generally, short-lived plant species with smaller growth forms, such as herbs or shrubs, demonstrate higher rates of reproductive output and allocation of energy toward reproduction than longerlived species like deciduous and coniferous trees (Wenk and Falster, 2015). Accordingly, invasive plants often possess such abilities to reproduce, hybridize, and evolve more rapidly than tree species used in forest plantations, again contributing to their competitiveness (Clements and Jones, 2021). It has been demonstrated that competition from invasive plants for space and resources significantly decreases stem wood production and biomass in trees such as poplar and birch (Willoughby *et al.*, 2006; Welham *et al.*, 2007). Opening of the tree canopy through logging practices, while considered beneficial for the rapid growth of tree species, may also act as an opportunity for more aggressive invasive plants to outcompete trees for available light (Balandier *et al.*, 2006; Villegas *et al.*, 2009). Competition for light is an essential step in the establishment of tree seedlings, and as a result, light limitation during early growth periods can result in significantly suppressed height and stem wood production; for example, European forests invaded by the light-demanding black cherry (*Prunus serotina* Ehrh.) experience significant reductions in regeneration of timber species such as English oak (*Quercus robur* L.) (Closset-Kopp *et al.*, 2007; Terwei *et al.*, 2013).

Invasive plants that form vines can also cause direct damage to the bark or stem wood of timber trees as their attachment structures penetrate the outer layer of bark (Melzer *et al.*, 2010). For example, the invasive vine kudzu (*Pueraria* montana var. lobata (Willd.) Maesen & S. M. Almeida ex Sanjappa & Predeep (ITIS)), reduces available light for the tree and causes damage to the bark layers as it forms attachment roots around the trunk and limbs; if left untreated, this can result in tree death in as little as 2 years (Grebner *et al.*, 2011). In forest plantations, this can result directly in loss of profit due to unusable trees (Grebner *et al.*, 2011; Padmanaba and Corlett, 2014). This phenomenon is also observed in other climbing vines, such as English ivy (*Hedera helix* L.) and Oriental bittersweet (*Celastrus orbiculatus* Thunb.), both invasive species on multiple continents including Europe and North America that can cause significant damage to trees (McNab and Meeker, 1987; Melzer *et al.*, 2010; Gudžinskas *et al.*, 2020).

Forest operations

The presence of invasive plants can have an impact on forestry operations, including harvesting of timber, debris removal, building and maintenance of logging roads, and regeneration of new trees (Dobert *et al.*, 2018).

Logging is known to facilitate the establishment and growth of certain invasive plants, including Scotch broom, alder buckthorn (*Frangula alnus* Mill.), and bellucia (*Bellucia pentamera* Naud.), as the opening of the canopy can be enough for invasive plant seeds present in the seed bank to break dormancy and germinate (Burnham and Lee, 2010; Carter *et al.*, 2019). Seed dormancy, which can be defined as, "the internal condition of the seed that impedes its germination under otherwise adequate hydric, thermal and gaseous conditions," has been observed in many different invasive plant species (Benech-Arnold *et al.*, 2000). One of the key factors that can contribute to the termination of seed dormancy is the availability of light. A fraction of a second of light exposure can be sufficient for allowing seeds the potential to germinate (Finch-Savage and Leubner-Metzger, 2006; Batlla and Benech-Arnold, 2014). Therefore, opening of the tree canopy from harvesting practices, or causing any level of soil disturbance through other forest operations, can facilitate the breaking of dormancy and subsequent germination (Dillis *et al.*, 2017).

Harvesting of timber can also be delayed or disrupted due to infestations of invasive plants. For instance, logging in the Northwestern USA is often timed seasonally to avoid the growing periods of understory plants, including invasive species, with winter logging often being preferential (Wolf *et al.*, 2007; Nash *et al.*, 2020). Forest operations in British Columbia have been halted due to infestations of Japanese knotweed (*Reynoutria japonica* (Houtt.) Ronse Decr. (ITIS)) being found to mitigate spread until the plants can be treated with herbicide. Forest Service Roads in BC have also experienced closures so that government agencies can prevent the spread of and work to control knotweed species (*Reynoutria* spp.) (see Fig. 1.1); operations could not resume until spread prevention and control plans were in place (Jennifer Grenz, 2023, personal communication).

Logging debris left behind after timber harvest is often piled and burned or removed, practices which can be impacted by the presence of invasive plants and conversely, impact the presence of invasive plant populations (see Fig. 1.2). For example, species that are more flammable, such as Scotch broom, which has high oil content (Sheppard and Hosking, 2000; Leblanc, 2001), could influence fire behavior, increasing risk during controlled burns (DiTomaso *et al.*, 2006). Previous studies have predicted the wildfire risk of invasive plants based on their flammability including ignitability, sustainability, combustibility, and consumability (Underwood *et al.*, 2019; Faccenda and Daehler, 2022; Tomat-Kelly and Flory, 2022). Invasive plants that are considered a high wildfire risk, such as Scotch broom, can increase the frequency of wildfires as they provide fuel that can be both ignited and sustained easily, and may provide fuel during different growing periods/seasons than native plants, allowing for a longer window of opportunity and environmental conditions for wildfires (Downey, 2000; Brooks *et al.*, 2004). There are many invasive plant species that are categorized as high fire risk in different regions (Grice *et al.*, 2011; Marshall *et al.*, 2012; Rossi *et al.*, 2014). Saltcedar (*Tamarix* spp.) and Giant reed (*Arundo donax* L.) are such examples in southern California where despite ecosystems being adapted to frequent wildfires, where native sage, scrubland, and grassland vegetation returns within several years, these invasive plants cause wildfires to burn more intensely and enable spread to nearby forests where it can damage the canopy of native trees such as Sycamore (*Platanus occidentalis* L.) and Cottonwood (*Populus* spp.) (Bell *et al.*, 2009). While fire can cause seed mortality in some invasive species, in other species such as *Acacia melanoxylon* L., *Conyza canadensis* (L.) Cronq., and *Eucalyptus globulus* Labill., which are some of the most widely distributed invasive species of forest ecosystems globally, it may stimulate germination of their seeds in the seed bank (Haysom and Murphy, 2004; Aran *et al.*, 2013). It has been demonstrated that the removal of logging debris is associated with increased soil temperatures, lower soil moisture, and lower nutrient levels, all of which impact the growth and establishment of native species and favor establishment of microclimates that favor invasive plants (Harrington *et al.*, 2012). As a result, forestry companies that operate in areas with high densities of invasive plants may consider altering typical debris removal practices by leaving more debris on site, rather than leaving disturbed, exposed soil. This could facilitate the growth of native species (Harrington and Schoenholtz, 2010) and benefit newly planted trees.

Building and maintenance of logging roads are other forestry operations that could be impacted by invasive plants. It has long been established that logging roads act as a facilitator in their spread; most invasive plants found in managed forests are within close proximity to logging roads (Mortensen *et al.*, 2009; Padmanaba and Sheil, 2014) (See Figs 1.3 and 1.4). Disturbance to the soil and seed bank as well as opening of the tree canopy allows for the establishment of invasive plants as roads are built, and these roads can become dispersal channels themselves (Mortensen *et al.*, 2009; Meunier and Lavoie, 2012). Equipment and heavy machinery are known to harbor invasive plant propagules which can spread during construction and maintenance; these propagules may be vegetative plant fragments as well as fruits or seeds (Christen and Matlack, 2008). Species such as *R. japonica*, the roots and stems of which can be fragmented into pieces as small as 1 cm in length and 0.7 g in weight and still regenerate into new plants (Bailey *et al.*, 2009) make it easy for logging vehicles themselves to act as dispersal vectors (Veldman

Fig. 1.1. Forest service road closure due to invasive plants in British Columbia.

and Putz, 2010). This exemplifies the need for preventative measures when building, maintaining, and operating logging roads; vehicles and equipment must be cleaned to remove any plant pieces from previous work, and the gravel used to build roads also needs to be free from invasive species seeds or propagules (Rauschert *et al.*, 2017) (see Fig. 1.5). This can impact the cost and time needed to build logging roads if clean materials must be sourced. It is also recommended that roads are graded in directions that avoid spreading plant parts and seeds from known infestations of invasive plants to uninvaded areas (ISCBC, 2019a).

Prevention of invasive plants is often said to be the cheapest method of control. Methods to inhibit invasion are essential to multiple aspects of operations. Prevention practices can include seeding bare mineral soil with native plant species, surveying for and reporting invasive plants regularly, and ensuring clean equipment is used by implementing wash-down and inspection stations (Baer and Gray, 2022). Utilization of GIS and predictive modeling methods to determine areas of high priority for invasive species monitoring or management, as well as to communicate essential information about potential threats both rapidly and early enough to inform better management decisions (Hamelin and Roe, 2019; Martynenko *et al.*, 2019) are essential at all production stages. Species distribution models use abiotic and biotic environmental requirements as well as the current distribution of certain invasive plants to predict geographic areas where they are most likely to invade; this is a relatively new method that will likely increase in accuracy over time, but

Fig. 1.2. Invasive plant Bohemian knotweed growing around a slash pile.

nonetheless is a useful tool for determining areas to prioritize monitoring and managing specific invasive plants (Peeler and Smithwick, 2018; Hamelin and Roe, 2019).

Invasive plants can also negatively impact forest regeneration, as competition from invasive plant species has been shown to decrease the abundance, diversity, and richness of tree seedlings and saplings, especially in economically viable coniferous tree species like pine (*Pinus* spp.) (Lázaro-Lobo *et al.*, 2021). For example, the invasive glossy buckthorn, uses allelopathy to inhibit the seedling growth of desired trees (Fagan and Peart, 2004; Frappier *et al.*, 2004). In the case of glossy buckthorn, native juvenile tree species such as sugar maple (*Acer saccharum* Marsh), red maple, (*Acer rubrum* L.), white ash (*Fraxinus americana* L.), and white pine (*Pinus strobus* L.) all experience reductions in growth, survival, and abundance under buckthorn competition, leading to reductions in overall recruitment and a widening age gap in these tree species (Fagan and Peart, 2004; Lee and Thompson, 2012).

Economic

In general, the economic impacts of invasive species have been well documented; however, there is little information available regarding the economic costs of invasive plants on the forestry industry specifically. A summary of the economic costs caused by invasive plant infestations and their management found in the literature can be seen in Table 1.1. In Canada, the combined impact of the 16 worst invasive species results in an estimated annual revenue loss of CAD$7.5 billion to agriculture and forestry industries (Canadian Council on Invasive Species, 2023). In BC specifically, it was estimated that the damage caused by invasive species would reach CAD$139 million to control by 2020 (ISCBC, 2019b). In the USA, annual costs associated with damage caused by invasive plants are estimated at US$21 billion annually or US$1.2 trillion from 1960 to 2020, with costs to the forestry sector estimated at US$42 billion from 1960 to 2020 (Crystal-Ornelas *et al.*, 2021; Fantle-Lepczyk *et al.*, 2022). Looking specifically at the costs incurred as a result of managing invasive plants, it was determined that plants were the second costliest invaders in the USA, costing US$190 billion from 1960 to 2020 after mammals when also considering insects, birds, mollusks, fungi, reptiles, fish, and amphibians (Fantle-Lepczyk *et al.*, 2022).

Economic costs incurred by invasive plants may occur as a result of loss of profits through reduction in timber value, or more indirectly through the management costs associated with prevention and control. Preventative measures including washing stations for vehicles and equipment and seeding bare ground, and management activities like the spraying of herbicides and follow-up monitoring by field crews are essential for maintaining productive forests but also contribute to the economic costs of invasive plants (ISCBC, 2019a).

Beyond economic impacts to forestry companies through management costs and reduction in timber value, these impacts are also felt by private landowners (Ma *et al.*, 2018; Snyder *et al.*, 2021). For private landowners, infestations of invasive plants on their forested properties may result in decreased property values, agricultural productivity, outdoor recreation, and can also result in unforeseen costs for control and management (Zenni *et al.*, 2021).

In productive forest operations, the invasive plants that are of high management priority are often those with the greatest economic impacts. Aggressive clonal invaders in North America such as Kudzu in the south-eastern USA and knotweed in the Pacific north-west can be just as costly as dealing with fungal or insect pathogens in forestry plantations (Grebner *et al.*, 2011; Hocking *et al.*, 2023). Kudzu has rapidly elongating, viny stems, large leaves, and frequent rooting at nodes, all of which allow it to compete with native trees in forest ecosystems and plantations (Forseth and Innis, 2004). As a result of these negative impacts, economic productivity of Kudzu-infested forests is greatly decreased; on average, it is reported that the cost of Kudzu control is US$200 per acre per year, which is a large proportion of the average US$650 per acre per year value of an average 25-year-old white pine stand (Van Driesche *et al.*, 2002). Although not all invasive plants present the extreme negative impacts to trees and their value that Kudzu does, this species serves as an example of how high the economic costs of invasive plant control can reach.

Fig. 1.3. Forest service road infested with Bohemian knotweed due to road maintenance activities.

Impacts to ecosystems

Beyond the direct harm that invasive plants may cause to forestry operations, forestry companies should also be concerned with engaging in and maintaining good environmental stewardship practices in the areas they operate as invasive plants can have negative impacts on the ecosystems where forestry operations take place (Wang *et al.*, 2022). Such stewardship practices include preventing the introduction of invasive plants through forestry operations and managing current invasive plant infestations to prevent their spread. As previously mentioned, several aspects of forestry operations can act as vectors for spread. Movement of animals and workers over large areas can spread seeds or other propagules stuck to fur, clothes, and shoes; vehicles and equipment used in forestry operations can also harbor plant propagules that can be transported to new, uninvaded sites (Baer and Gray, 2022). Tree health in forest ecosystems is directly tied to environmental conditions and productivity; as a result, forestry companies should also have a vested interest in maintaining overall ecosystem health.

In addition to influencing native plant community assemblages and tree health, invasive plants may outcompete native organisms for limited resources, alter critical and sensitive habitat, and lower biodiversity of an area (NOAA, 2021). They can alter forest ecosystem processes including carbon and nitrogen cycling, as well as hydrologic cycles which can in turn impact water availability and quality (Lepš *et al.*, 2002). In general, invasive plants use more water than natives, resulting in lowered soil moisture (Cavaleri and Sack, 2010; Pyšek *et al.*, 2012). In some cases, invasive plants can alter carbon cycling in forest ecosystems by causing soil carbon stock to decrease (von Lützow *et al.*, 2007; Liao *et al.*, 2008). Additions of organic matter to the soil through decomposition of litter, root exudates, alterations in soil microbiome, changes in plant functional types, and different nutrient requirements of invasive plants can all contribute to alteration of nutrient cycles of an ecosystem after invasion (Gill and Burke, 1999; Hickman *et al.*, 2013). Invasive plants have been known to increase N and C availability to other plants and produce litter that decomposes at higher rates compared to native counterparts; however, the exact opposite patterns have been observed in other invasivenative co-occurring plant relationships (Ehrenfeld, 2003; Hickman *et al.*, 2013). Some invasive plant species can alter nutrient cycles differently depending on the environment they are introduced to. For example, the invasive Cheatgrass (*Bromus tectorum* L.) is known to increase N mineralization and nitrification rates in the cool, desert ecosystems of eastern Washington state, USA, whereas in the more arid, warm grassland ecosystems in Colorado state, USA, it decreases N mineralization and nitrification rates (Bolton *et al.*, 1990; Evans *et al.*, 2001). This suggests that the ecological impacts of invasion are context-specific, rather than just species-specific, and thus will have varying impacts depending on a variety of ecological factors. Impacts which generally lower the environmental suitability for preferential native tree and understory species in favor of invaders.

The impacts of invasive species on local populations of native species and community composition are more commonly documented than more generalized ecosystem impacts, such as hydrological cycling (Flood *et al.*, 2020). In terms of plant community composition, many studies agree that plant invasions significantly decrease native species richness and diversity (Powell *et al.*, 2011; Dornelas *et al.*, 2014; Rivas-Torres *et al.*, 2017). Regarding the aquatic ecosystem impacts, woody invasive plants such as *Tamarix* spp., can alter flow regimes and flood patterns due to their deep root systems that can better access alluvial groundwater compared to native counterparts (Stromberg *et al.*, 2007). Rhizomatous invasive plants such as knotweed contribute to streambank erosion by outcompeting and displacing native riparian species and their roots, resulting in a loss of soil stability and therefore increased erosion and water turbidity (Colleran *et al.*, 2020).

Control methods used to manage invasive plants can also impact ecosystem function;

Fig. 1.4. Forest service road infested by Bohemian knotweed due to road maintenance activities.

for example, it has been documented that the removal of invasive plants along stream banks results in higher water temperatures that in turn can impact local fish populations (Trimmel *et al.*, 2018). Chemical methods are also one of the most common methods for controlling invasive plants, and despite herbicides such as glyphosate being classified by regulatory bodies as having low environmental impact, it is still a controversial topic that is subject to ongoing research (Hagner *et al.*, 2019).

Invasive plants can alter soil structural properties such as the amount and size of soil aggregates, rates of soil erosion, and water table depth (Mboukou-Kimbatsa *et al.*, 2007; Corbin and D'Antonio, 2012). For example, compared to native grassland species, invasions of Spotted knapweed (*Centaurea stoebe* ssp. *micranthos* (Gugler) Hayek) significantly increased surface runoff and erosion, in addition to lowering C:N ratios; the legacy of these changes were recorded even after removal of the knapweed infestations (Corbin and D'Antonio, 2012; Singh *et al.*, 2022). Invasive plants are also known to cause changes to the soil microbiome, they can alter the biomass, abundance, and diversity of bacteria and fungi (Zhang *et al.*, 2019). This may occur due to allelopathic chemicals, hormones, and other exudates from the roots, or from the accumulation of plant litter from the invasive population (Farrer and Goldberg, 2009). Litter produced by invasive plants may require specific bacteria or fungi for decomposition and mineralization which can in turn result in changes to the microbial community composition (Farrer and Goldberg, 2009). A meta-analysis of 211 cases of plant invasion effects on soil bacteria found that invasive plants significantly increase the bacterial diversity of soil, again due to the specific bacteria required for cycling of these compounds (Torres *et al.*, 2021) the implications of which, regarding forest community health, are not yet understood.

Fig. 1.5. Gravel for maintenance activities infested with Bohemian knotweed.

Table 1.1. Economic impact assessments of costs associated with invasive species to foresty.

Impact analysis	Total cost incurred	Annual cost	Source
16 worst invasive species in forestry and agriculture in Canada	N/A	CAD$7.5 billion/year	Canadian Council on Invasive Species (2023)
Invasive plant damage to all sectors of industry in the United States	US$1.2 trillion from 1960–2020	US$21 billion/year	Crystal-Ornelas *et al.* (2021)
Invasive plant damage to forestry operations in the United States	US$42 billion from 1960–2020	US$700 million/year	Fantle-Lepczyk *et al.* (2022)
Invasive plant management in the United States	US$190 billion from 1960–2020	US$3.2 billion/year	Fantle-Lepczyk *et al.* (2022)

BC as a case study

Across Canada, the need for fast-growing, high-yielding forest plantations has risen to meet the increasing global demands for wood products. As a result, expansive plantation and breeding programs have been established in provinces of Canada, including British Columbia (BC). As valuable practical experience is often missing from reviews, we conducted interviews with three foresters responsible for the management of invasive plants from differing major forestry companies in BC to learn more about the impacts of invasive plants to their operations. Interviewees were asked about potential impacts and disruptions that they have seen invasive plants cause to forestry operations; the costs, personnel, and equipment involved in prevention, management, and monitoring of invasive plant populations; as well as the legislation and information available regarding the roles of forestry companies in invasive species management.

Examples of invasive plants impacting forestry operations

There were three plant species of concern that were mentioned by all interviewees: knotweed species, giant hogweed (*Heracleum mantegazzianum* Sommier & Levier (ITIS)), and Scotch broom. These species have reportedly halted forestry operations in some capacity. In the case of knotweeds, it was reported that large infestations of Bohemian knotweed (*Reynoutria* x *bohemica* Chrtek & Chrtkova) have impacted forest operations in multiple areas of the province, sometimes for several years in a row. Knotweeds are known to form large, monoclonal stands that can cover hundreds of square meters, and the underground rhizomes can extend up to 20 m laterally and 3 m deep (Beerling *et al.*, 1994; Martin *et al.*, 2020). These large patches are usually found in open, sunny areas, but knotweeds are also found in shaded forest environments, usually due to lateral expansion of nearby populations (Beerling *et al.*, 1994). Due to the large size and rapid expansion of *Reynoutria* spp., they are usually a top priority for management with forestry companies. Infestations of *Reynoutria* spp. have been found around forestry working camps and can even invade newly built logging roads. These types of infestations must be managed for operations to continue, as they can cause damage to infrastructure and pose safety issues by blocking lines of sight, which can slow down or stop operations (Bashtanova *et al.*, 2009).

Infestations of Giant hogweed can persist for many years due to prolific seed production; one flowering plant can produce upwards of 20,000 seeds which can spread by wind and persist in the seed bank for 5 years, causing further problems if left untreated (Nielsen *et al.*, 2005). Not only does Giant hogweed impose ecological and economic impacts like other invasive plants, but also poses a human health risk due to the phytotoxic sap produced by the plant which can cause severe chemical burns and blindness (Nielsen *et al.*, 2005). Removal of Giant hogweed can be complicated due to its toxicity but is essential for protecting those working in the area.

Scotch broom, which is not considered a provincially noxious weed in BC, has also presented significant challenges for invasive plant managers (see Fig. 1.6). Not only does Scotch broom possess qualities similar to other invasive species, including prolific production of seeds that can persist up to 30 years in the seed bank, but it is also highly flammable and is known to increase wildfire severity (Prasad, 1999). Over the past several years, BC has experienced longer, hotter summers, warmer annual temperatures, and more frequent and severe droughts and these trends are forecasted to continue in the coming years, which in turn increases the risk for more frequent and severe wildfires (Metro Vancouver, 2016; Government of BC, 2023c). One company reported that it will completely shut down operations when summer temperatures are too high, beginning with areas infested with Scotch broom, in order to prevent any accidental wildfires caused by equipment working in these dry, flammable areas. Additionally, in the case of some species such as cheatgrass (*Bromus tectorum* L.), fire is known to be a vector of invasive plant spread within forest ecosystems (Veldman *et al.*, 2009; Peeler and Smithwick, 2018).

Strategies employed to manage invasive plant species in BC

The costs associated with the control of invasive plant species can be high, especially when it comes to management of clonal or perennial species like knotweeds and Scotch broom. In one

Fig. 1.6. Scotch broom infestation on secondary forest service road on Vancouver Island, British Columbia.

interview, an invasive plant manager reported that their company spends at least CAD$1 million per year on management of Scotch broom alone. Another company reported that they would consider "average" maintenance costs of a forest plantation to be CAD$5–6 per cubic meter; this cost would include preventative measures such as risk assessments and seeding of mineral soils, regular brush clearing and timber harvesting, and replanting. An area infested with problematic invasives costs more like CAD$16–18 per cubic meter to manage, accounting for regular maintenance in addition to the costs of managing the invasive species, with herbicides unless the infestation is deemed small enough for mechanical removal, monitoring later in the season, and additional seeding any bare soil to help prevent regrowth.

In BC, invasive plant management in the forestry sector is a complex task due to the various jurisdictions, legislations, policies, and guidelines involved. Currently, forestry operations in BC are regulated under the Forest and Range Practices Act (Government of BC, 2023d). Division 1, section 47 outlines the requirements for dealing with invasive plants and specifies that any person or company carrying out forestry practices must have measures in place to prevent the introduction or further spread of an invasive species into an area. The Invasive Plants Regulation section of the Act lists the invasive plant species that are subject to section 47. Furthermore, the Integrated Pest Management Act (Government of BC, 2023a) regulates the use of herbicides to manage invasive plant species. However, this legislation does not directly address the responsibilities of forest companies to actively manage invasive species through some control methods, or to continually monitor and report on the status of invasive species after treatment. Ultimately, the development and implementation of strategies and best management practices used to manage each invasive plant species is the responsibility of the forestry company itself, although the BC government and the Invasive Species Council of BC (ISCBC), a non-profit society, have published various management strategies and best management practices for common invasive species in BC that are available online. Forestry companies often have to decide which invasive species to manage using *The Invasive Plant Core Ranking Process*, a tool provided by the BC government to assist land managers in making invasive plant management decisions. The intent of the tool is to assist land managers in the identification of invasive plants that should be actively managed, and invasive plants that should be monitored (Government of BC, 2023b).

The forestry companies interviewed all concluded that the use of herbicides was generally the preferred method for invasive plant management. In BC, the use of herbicides is restricted near waterways, these areas are known as pesticide-free zones (PFZs) (Gaskin *et al.*, 2014). According to BC's Integrated Pest Management Act (2016) (Government of BC, 2023a), PFZs consist of a 10-m buffer maintained around bodies of water, measured from the high water mark, where herbicides cannot be used. Depending on the type of herbicide used, this distance can sometimes be reduced to 1 m. This means that infestations of invasive plants deemed too large for mechanical removal may be left untreated or only partially treated with herbicides near waterways, which is especially concerning as waterways can act as a major vector of spread for seeds or other plant propagules (Rouifed *et al.*, 2011). Again, the forestry companies interviewed agreed that these PFZ regulations are strict, and result in untreated infestations that can disperse and create new infestations nearby, creating more problems over time.

The invasive plant managers interviewed for this article agreed that they would like to see more research involving the long term, residual impacts of herbicides on the forest ecosystems of BC, particularly areas that are subjected to the repeated herbicide treatment over many years, as only one paper could be found addressing this specific issue (Botten *et al.*, 2021). Some of the biggest challenges that forestry companies face when managing invasive species is the pushback from members of the public. There are often misunderstandings about the impacts and use of herbicides in areas that people may use for recreation or food harvesting. As a result, getting local residents to be completely on board with the best management practices used by forestry companies can be difficult, and can even halt management operations. The invasive plant managers we spoke to emphasized the need for further research in this area, but also for improved public outreach and education on how these practices impact their local environment.

Conclusion

Within the context of forestry, the available literature makes a compelling argument that invasive plants result in significant economic losses through both the reduction of timber values and increased costs associated with control measures while having negative impacts on forest ecosystem health. Our review revealed a considerable gap in the literature regarding impacts associated solely with managing invasive plants in the forestry sector. Most often, impacts were generalized and/or characterized in combination with other sectors across the globe, such as agriculture. Of the 121 reviewed research articles, 29 were directly discussing the impacts of invasive plants to forestry operations. While it may be appropriate for literature regarding the impacts of invasive plants to forest ecosystems or agricultural settings to be applied to forestry, operational activities and timelines are distinct and management goals, such as the tolerance for different species or management approaches, may differ. We also identified an existing gap in the literature regarding the impacts of invasive plant management strategies on forestry operations and tree health. Invasive plant managers interviewed agreed that the long-term use of herbicides, for instance, should be studied within the context of forest health. Furthermore, little information is available regarding the responsibilities and legal obligations of forestry companies to manage and mitigate the spread of invasive plants globally; it seems that in many countries, there exists only voluntary, suggested strategies for the prevention and management of invasive plants leaving the engagement in such responsibility up to industry. More detailed information and stricter guidance and/or regulations are often available and applicable to specific invasive insects or fungal pathogens, such that it seems as though invasive plants and their impacts are diminished or overlooked completely.

Our case study, the management of invasive plant species in BC's forestry industry, as part of this literature review, illuminated the often-noted gap between practitioners and academic researchers as well as the complexities of forest operations that make bridging that gap even more imperative. Moving forward it is clear that research specific to the impacts of invasive plants and their management at each stage of forestry operations is needed across a variety of global contexts. A worthy investment to ensure more profitable, productive forests while upholding sound environmental stewardship.

Conflict of Interest

The authors declare no conflict of interest.

Author Contributions

J.G. conceptualized the data; V.L.J. carried out the initial drafts of Sections 1–4; editing and literature review by J.G. and V.L.J.; Figures provided from the personal photo collections of J.G. Both authors have read and agreed to the published version of the manuscript.

Funding Statement

We acknowledge the support of the Natural Sciences and Engineering Research Council of Canada (NSERC), [funding reference number RGPIN-2023-03308].

References

Aran, D., Garcia-Duro, J., Reyes, O. and Casal, M. (2013) Fire and invasive species: Modifications in the germination potential of *Acacia melanoxylon, Conyza canadensis* and *Eucalyptus globulus*. *Forest Ecology and Management* 302, 7–13.

Baer, K.C. and Gray, A.N. (2022) Biotic predictors improve species distribution models for invasive plants in Western U.S. Forests at high but not low spatial resolutions. *Forest Ecology and Management* 518, 120249.

Bai, Y., Thompson, D. and Broersma, K. (2004) Douglas fir and ponderosa pine seed dormancy as regulated by grassland seedbed conditions. *Journal of Range Management* 57, 661–667.

Bailey, J.P., Bimova, K. and Mandak, B. (2009) Asexual spread versus sexual reproduction and evolution in Japanese Knotweed s.l. sets the stage for the "Battle of the Clones". *Biological Invasions* 11, 1189–1203.

Balandier, P., Collet, C., Miller, J.H., Reynolds, P.E. and Zedacker, S.M. (2006) Designing forest vegetation management strategies based on the mechanisms and dynamics of crop tree competition by neighbouring vegetation. *Forestry* 79, 3–27.

Barney, J.N., Tekiela, D.R., Dollete, E.S. and Tomasek, B.J. (2013) What is the "real" impact of invasive plant species? *Frontiers in Ecology and the Environment* 11(6), 322–329.

Bashtanova, U.B., Beckett, K.P. and Flowers, T.J. (2009) Review: Physiological approaches to the improvement of chemical control of Japanese knotweed (*Fallopia japonica*). *Weed Science* 57, 584.

Batlla, D. and Benech-Arnold, R.L. (2014) Weed seed germination and the light environment: Implications for weed management. *Weed Biology and Management* 14(2), 77–87.

Beck, J., Cale, J., Rodriguez-Ramos, J., Kanekar, S., Karst, J., Cahill, J., Simard, S. and Erbilgin, N. (2020) Changes in soil fungal communities following anthropogenic disturbance are linked to decreased lodgepole pine seedling performance. *Journal of Applied Ecology* 57, 1292–1302.

Beerling, D.J., Bailey, J.P. and Conolly, A.P. (1994) Fallopia Japonica (Houtt.) Ronse Decraene. *Journal of Ecology* 82(4), 959–979.

Bell, C.E., DiTomaso, J.M. and Brooks, M.L. (2009) Invasive Plants and Wildfires in Southern California. UC Agriculture & Natural Resources Peer-Reviewed Publications Available at: https://escholarship.org/uc/item/3tk834s7#main (accessed 25 October 2023).

Bellerose, J. Dupuch, A. and Aubin, I. (2022) Changes in understory composition of rural North American temperate forests after a 14-year period with focus on exotic and sensitive plant species. *Forests* 13(5), 678.

Benech-Arnold, R.L., Sanchez, R.A., Forcella, F., Kruk, B.C. and Ghersa, C.M. (2000) Environmental control of dormancy in weed seed soil banks. *Field Crop Research* 67, 105–122.

Birthisel, S.K., Clements, R.S. and Gallandt, E.R. (2021) Review: How will climate change impact the 'many little hammers' of ecological weed management? *Weed Research* 61(5), 327–341.

Bolton, H., Smith, J.L. and Wilding, R.E. (1990) Nitrogen mineralization potentials of shrub-steppe soils with different disturbance histories. *Soil Science Society of America* 54, 887–891.

Botten, N., Wood, L.J. and Werner, J.R. (2021) Glyphosate remains in forest plant tissues for a decade or more. *Forest Ecology and Management* 493, 119259.

Brooks, M.L., D'Antonio, C.M., Richardson, D.M. *et al.* (2004) Effects of invasive alien plants on fire regimes. *BioScience* 54, 677–688.

Burnham, K.M. and Lee, T.D. (2010) Canopy gaps facilitate establishment, growth, and reproduction of invasive *Frangula alnus* in a *Tsuga canadensis* dominated forest. *Biological Invasions* 12(6), 1509–1520.

Canadian Council on Invasive Species (2023) Invasive Species: Economic Impacts. Available at: https://canadainvasives.ca/invasive-species/ (accessed 2 October 2023).

Carroll, A.L., Taylor, S.W., Regniere, J. and Safranyik, L. (2004) Effects of Climate Change on Range Expansion by the Mountain Pine Beetle in British Columbia. Available at: http://cfs.nrcan.gc.ca/publications?id=25051 (accessed 15 September 2023).

Carter, D.R., Slesak, R.A., Harrington, T.B., Peter, D.H. and D'Amato, A.W. (2019) Scotch broom (*Cytisus scoparius*) modifies microenvironment to promote nonnative plant communities. *Biological Invasions* 21(4), 1055–1073.

Cavaleri, M.A. and Sack, L. (2010) Comparative water use of native and invasive plants at multiple scales: A global meta-analysis. *Ecology* 91, 2705–2715.

Christen, D.C. and Matlack, G.R. (2008) The habitat and conduit functions of roads in the spread of three invasive plant species. *Biological Invasions* 11(2), 453–465. DOI: 10.1007/s10530-008-9262-x.
Clements, D.R. and Jones, V.L. (2021) Ten ways that weed evolution defies human management efforts amidst a changing climate. *Agronomy* 11(2), 284.
Closset-Kopp, D., Chabrerie, O., Valentin, B., Delachapelle, H. and Decocq, G. (2007) When oskar meets alice: Does a lack of trade-off in r/K-strategies make *Prunus Serotina* a successful invader of European forests? *Forest Ecology and Management* 247(1), 120–130.
Colautti, R.I. and MacIsaac, H.J. (2004) A neutral terminology to define 'invasive' species: Defining invasive species. *Diversity and Distributions* 10(2), 135–141.
Colleran, B., Lacy, S.N. and Retamal, M.R. (2020) Invasive Japanese knotweed (*Reynoutria japonica* Houtt.) and related knotweeds as catalysts for streambank erosion. *River Research and Applications* 36, 1962–1969.
Corbin, J.D. and D'Antonio, C.M. (2012) Gone but Not Forgotten? Invasive Plants' Legacies on Community and Ecosystem Properties. *Invasive Plant Science and Management* 5(1), 117–124.
Crystal-Ornelas, R. and Lockwood, J.L. (2020) The 'known unknowns' of invasive species impact measurement. *Biological Invasions* 22(4), 1513–1525.
Crystal-Ornelas, R., Hudgins, E.J., Cuthbert, R.N. *et al.* (2021) Economic costs of biological invasions within North America. *NeoBiota* 67, 485–510.
Davis, M.A. (2009) *Invasion Biology*. Oxford University Press, Oxford, UK. Available at: https://books.google.ca/books?id=41jS4xjy3P8C&lpg=PR7&ots=5atgyFxaCX&lr&pg=PR7#v=onepage&q&f=false (accessed 20 November 2023).
Dillis, C., Marshall, A.J. and Rejmanek, M. (2017) Change in disturbance regime facilitates invasion by *Bellucia pentamera* Naudin (Melastomataceae) at Gunung Palung National Park, Indonesia. *Biological Invasions* 19(4), 1329–1337.
DiTomaso, J.M., Brooks, M.L., Allen, E.B. *et al.* (2006) Control of invasive weeds with prescribed burning. *Weed Technology* 20, 535–545.
Dobert, T.F., Webber, B.L., Sugau, J.B., Dickinson, K.J.M. and Didham, R.K. (2018) Logging, exotic plant invasions, and native plant reassembly in a lowland tropical rain forest. *Biotropica* 50(2), 254–265.
Dornelas, M., Gotelli, N.J., McGill, B. *et al.* (2014) Assemblage time series reveal biodiversity change but not systematic loss. *Science* 344(6181), 296–299.
Downey, P.O. (2000) Broom (*Cytisus scoparius* (L.) Link) and fire: Management implications. *Plant Protection Quarterly* 15(4), 178–183.
Dukes, J.S., Pontius, J., Orwig, D. *et al.* (2009) Responses of insect pests, pathogens, and invasive plant species to climate change in the forests of northeastern North America: What can we predict? *Canadian Journal of Forest Research* 30(2), 231–240.
Dunwiddie, P.W., Bakker, J.D., Almaguer-Bay, M. and Sprenger, C.B. (2011) Environmental history of a Garry Oak/Douglas-Fir Woodland on Waldron Island, Washington. *Northwest Science* 85(2), 130–140.
Ehrenfeld, J.G. (2003) Effects of exotic plant invasions on soil nutrient cycling processes. *Ecosystems* 6, 503–523.
European Commission (2021) Report from the Commission to the European Parliament and Council on the review of the application of Regulation (EU) No 1143/2014 of the European Parliament and of the Council of 22 October 2014 on the prevention and management of the introduction and spread of invasive alien species.
Evans, R.D., Rimer, R., Sperry, L. and Belnap, J. (2001) Exotic plant invasion alters nitrogen dynamics in an arid grassland. *Ecological Applications* 11, 1301–1310.
Faccenda, K. and Daehler, C.C. (2022) A screening system to predict wildfire risk of invasive plants. *Biological Invasions* 24, 575–589.
Fagan, M.E. and Peart, D.R. (2004) Impact of the invasive shrub glossy buckthorn (*Rhamnus frangula* L.) on juvenile recruitment by canopy trees. *Forest Ecology and Management* 194(1–3), 95–107.
Fantle-Lepczyk, J.E., Haubrock, P.J., Kramer, A.M. *et al.* (2022) Economic costs of biological invasions in the United States. *Science of the Total Environment* 806, 151318.
Farrer, E.C. and Goldberg, D.E. (2009) Litter drives ecosystem and plant community changes in cattail invasion. *Ecological Applications* 19(2), 398–412.
Finch-Savage, W.E. and Leubner-Metzger, G. (2006) Seed dormancy and the control of germination. *New Phytologist* 171, 501–523.
Fine, P.V.A. (2002) The invasibility of tropical forests by exotic plants. *Journal of Tropical Ecology* 18(5), 687–705.

Flood, P.J., Duran, A., Barton, M., Mercado-Molina, A.E. and Trexler, J.C. (2020) Invasion impacts on functions and services of aquatic ecosystems. *Hydrobiologia* 847(7), 1571–1586.

Forseth, I.N. and Innis, A.I. (2004) Kudzu (Pueraria montana): History, physiology, and ecology combine to make a major ecosystem threat. *Critical Reviews in Plant Sciences* 23(5), 401–413.

Frappier, B., Eckert, R.T. and Lee, T.D. (2004) Experimental removal of the non-indigenous shrub *Rhamnus frangula* (Glossy Buckthorn): Effects on native herbs and woody seedlings. *Northeastern Naturalist* 11(3), 333–342.

Friday, J.B., Scowcroft, P.G. and Ares, A. (2008) Responses of native and invasive plant species to selective logging in an acacia koa-Metrosideros polymorpha forest in Hawai'i. *Applied Vegetation Science* 11(4), 471–482.

Gaggini, L., Rusterholz, H.P.P. and Baur, B. (2018) The invasive plant impatiens glandulifera affects soil fungal diversity and the bacterial community in forests. Applied *Soil Ecology* 124, 335–343.

Gaskin, J.F., Schwarzlander, M., Grevstad, F.S., Haverhals, M.A., Bourchier, R.S. and Miller, T.W. (2014) Extreme differences in population structure and genetic diversity for three invasive congeners: Knotweeds in western North America. *Biological Invasions* 16(10), 2127–2136.

Gianoli, E. and Molina-Montenegro, M.A. (2021) Evolution of physiological performance in invasive plants under climate change. *Evolution* 75(12), 3181–3190.

Gill, R.A. and Burke, I.C. (1999) Ecosystem consequences of plant life form changes at three sites in the semiarid United States. *Oecologia* 121, 551–563.

Government of BC (2023a) B.C. Integrated Pest Management Act. Available at: https://www.bclaws.gov.bc.ca/civix/document/id/complete/statreg/00_03058_01 (accessed 12 November 2023).

Government of BC (2023b) B.C. Invasive Plant Core Ranking Process Tool. Available at: https://www.for.gov.bc.ca/hra/invasive-species/calculatorIndex.htm (accessed 8 September 2023).

Government of BC (2023c) CleanBC: Wildfires. Available at: https://cleanbc.gov.bc.ca/about-climate-change/impacts/wildfires/ (accessed 20 October 2023).

Government of BC (2023d) Forest & Range Practices Act, Amended 2021. Available at: https://www.bclaws.gov.bc.ca/civix/document/id/complete/statreg/00_02069_01 (accessed 8 September 2023).

Government of BC (2023e) Forest Stewardship: Silviculture. Available at: https://www2.gov.bc.ca/gov/content/industry/forestry/managing-our-forestresources/silviculture (accessed 8 September 2023).

Grebner, D.L., Ezell, A.W., Prevost, J.D. and Gaddis, D.A. (2011) Kudzu control and impact on monetary returns to non-industrial private forest landowners in Mississippi. *Journal of Sustainable Forestry* 30(3), 204–223.

Grice, A.C., Friedel, M.H., Marshall, N. and Van Klinken, R.D. (2011) Tackling contentious invasive plant species: A case study of buffel grass in Australia. *Environmental Management* 49(2), 285–294.

Gudžinskas, Z., Petrulaitis, L. and Žalneravičius, E. (2020) Emerging invasion threat of the liana *Celastrus orbiculatus* (Celastraceae) in Europe. *NeoBiota* 56, 1–25.

Hagner, M., Mikola, J., Saloniemi, I., Saikkonen, K. and Helander, M. (2019) Effects of a glyphosate-based herbicide on soil animal trophic groups and associated ecosystem functioning in a northern agricultural field. *Scientific Reports* 9, 8540.

Hamelin, R.C. and Roe, A.D. (2019) Genomic biosurveillance of forest invasive alien enemies: A story written in code. *Evolutionary Applications* 13(1), 95–115.

Harrington, T.B. (2020) Scotch broom (*Cytisus scoparius*) germination and growth responses to light: Implications for logging debris retention after forest harvesting. *Weed Science* 68(5), 517–526.

Harrington, T.B. and Schoenholtz, S.H. (2010) Effects of logging debris treatments on five-year development of competing vegetation and planted Douglas-fir. *Canadian Journal of Forest Research* 40, 500–510.

Harrington, T. *et al.* (2012) Logging debris matters: Better soil, fewer invasive plants. *Science Findings* 145, 1–6.

Haysom, K. and Murphy, S. (2004) The status of invasiveness of forest tree species outside their natural habitat: A global review and discussion paper. *Forest Genetic Resources* 31, 5–8.

Hickman, J.E., Ashton, I.W., Howe, K.M. and Lerdau, M.T. (2013) The native–invasive balance: Implications for nutrient cycling in ecosystems. *Oecologia* 173, 319–328.

Hocking, S., Toop, T., Jones, D., Graham, I. and Eastwood, D. (2023) Assessing the relative impacts and economic costs of Japanese knotweed management methods. *Scientific Reports* 13, 3872.

Holmes, T.P., Aukema, J.E., Von Holle, B., Liebhold, A. and Sills, E. (2009) Economic impacts of invasive species in forests. *Annals of the New York Academy of Sciences* 1162(1), 18–38.

IPCC (2014) Summary for policymakers. In: Field, C.B., Barros, V.R., Dokken, D.J. *et al.* (eds) *Climate Change 2014: Impacts, Adaptation, and Vulnerability. Part A: Global and Sectoral Aspects. Contribution of Working Group II to the Fifth Assessment Report of the Intergovernmental Panel on Climate Change*. IPCC, Cambridge, NY, pp. 1–32.

ISCBC (2019a) Forestry Operations Factsheet. Available at: https://bcinvasives.ca/wp-content/uploads/2021/01/Forestry_OperationsFINAL_ 04_24_2019.pdf (accessed 8 September 2023).

ISCBC (2019b) Invasive Species Council of BC: Annual Report: Economic Impacts of Invasive Plants in British Columbia. Available at: https://bcinvasives.ca/wp-content/uploads/2021/01/ISCBC_Annual_Report_2019-web-spreads.pdf (accessed 12 November 2023).

Jobin, V., Das, A., Harikrishnan, C.P., Chanda, R., Lawrence, S. and Robin, V.V. (2023) Patterns of understory invasion in invasive timber stands of a tropical sky island. *Ecology and Evolution* 13(4), e9995.

Kay, M. (1994) Biological control for invasive tree species. *N Z Forestry* 39, 35–37.

Kelso, M.A., Wigginton, R.D. and Grosholz, E.D. (2020) Nutrients mitigate the impacts of extreme drought on plant invasions. *Ecology* 101(4), e02980.

Knüsel, S., Conedera, M., Rigling, A., Fonti, P. and Wunder, J. (2015) A TreeRing perspective on the invasion of ailanthus altissima in protection forests. *Forest Ecology and Management* 354, 334–343.

Langmaier, M. and Lapin, K. (2020) A systematic review of the impact of invasive alien plants on forest regeneration in European temperate forests. *Frontiers in Plant Science* 11, 524969.

Lázaro-Lobo, A., Lucardi, R.D., Ramirez-Reyes, C. and Ervin, G.N. (2021) Region-wide assessment of fine-scale associations between invasive plants and forest regeneration. *Forest Ecology and Management* 483, 118930.

Leblanc, J.W. (2001) Getting a Handle on Broom: Scotch, French, Spanish, and Portuguese Brooms in California. Available at: https://escholarship.org/uc/item/7bt479qt (accessed 8 September 2023).

Lee, T.D. and Thompson, J.H. (2012) Effects of logging history on invasion of eastern white pine forests by exotic glossy buckthorn (*Frangula alnus* P. Mill.). *Forest Ecology and Management* 265(1), 201–210.

Lepš, J., Novotný, V., Čižek, L. *et al.* (2002) Successful invasion of the neotropical species *Piper aduncum* in rain forests in Papua New Guinea. *Applied Vegetation Science* 5(2), 255–262.

Liao, C., Peng, R., Luo, Y. *et al.* (2008) Altered ecosystem carbon and nitrogen cycles by plant invasion: A meta-analysis. *New Phytologist* 177, 706–714.

Ma, Z., Clarke, M. and Church, S.P. (2018) Insights into individual and cooperative invasive plant management on family forestlands. *Land Use Policy* 75, 682–693.

Maguire, D.A., Mainwaring, D.B., Rose, R., Garber, S.M. and Dinger, E.J. (2009) Response of coastal Douglas-fir and competing vegetation to repeated and delayed weed control treatments during early plantation development. *Canadian Journal of Forest Research* 39(6), 1208–1219.

Marshall, V.M., Lewis, M. and Ostendorf, B. (2012) Buffel grass (Cenchrus ciliaris) as an invader and threat to biodiversity in arid environments: A review. *Journal of Arid Environments* 78, 1–12.

Martin, F., Dommanget, F., Lavallee, F. and Evette, A. (2020) Clonal growth strategies of *Reynoutria japonica* in response to light, shade, and mowing, and perspectives for management. *Neobiota* 56(56), 89–110.

Martynenko, O.V., Karminov, V.N., Yugaj, E.S., Mutygullin, I.R. and Ontikov, P.V. (2019) The use of GIS for agrochemical soil characteristics and weed infestation of Grebnevskiy nursery of Shchelkovskiy teaching and experimental forestry. *Бюллетень Почвенного института им ВВ Докучаева* 0(99), 5–20.

Mboukou-Kimbatsa, I., Bernhard-Reversat, F., Loumeto, J.J., Ngao, J. and Lavelle, P. (2007) Understory vegetation, soil structure and soil invertebrates in Congolese eucalypt plantations, with special reference to the invasive plant Chromolaena odorata and earthworm populations. *European Journal of Soil Biology* 43(1), 48–56.

McNab, W.H. and Meeker, M. (1987) Oriental bittersweet: A growing threat to hardwood silviculture in the appalachians. *Northern Journal of Applied Forestry* 4(4), 174–177.

Melzer, B., Steinbrecher, T., Seidel, R., Kraft, O., Schwaiger, R. and Speck, T. (2010) The attachment strategy of English ivy: A complex mechanism acting on several hierarchical levels. *Journal of The Royal Society Interface* 7(50), 1383–1389.

Metro Vancouver (2016) Climate Change Projections for Metro Vancouver Report. Available at: http://www.metrovancouver.org/services/air-quality/AirQualityPublications/ClimateProjectionsForMetroVancouver.pdf (accessed 10 August 2023).

Meunier, G. and Lavoie, C. (2012) Roads as corridors for invasive plant species: New evidence from smooth bedstraw (*Galium mollugo*). *Invasive Plant Science and Management* 5(1), 92–100.

Mortensen, D.A., Rauschert, E.S.J., Nord, A.N. and Jones, B.P. (2009) Forest roads facilitate the spread of invasive plants. *Invasive Plant Science and Management* 2(3), 191–199.

Nash, M., Napper, C., Page-Dumroese, D., Alger, R., Wagenbrenner, J. *et al*. (2020) Winter Logging for Mechanical Harvesting and Fuel Treatment Operations. Available at: https://www.fs.usda.gov/rm/pubs_journals/2020/rmrs_2020_nash_m001.pdf (accessed 10 August 2023).

Nielsen, C., Ravn, H.P., Nentwig, W. and Wade, M. (2005) The giant hogweed best practice manual: Guidelines for the management and control of invasive weeds in Europe. *Forest & Landscape Denmark, Hoersholm, Denmark*, 44p. Available at: http://labgis.ibot.cas.cz/wpcontent/uploads/pdf/Booy%20et%20al.%20-%202005%20-%20The%20 giant%20hogweed%20best%20practice%20 manual%20guidelines%20for%20the%20management%20and%20control%20of%20invasive %20 weeds%20in%20Europe.pdf (accessed 20 November 2023).

NISC (National Invasive Species Council) (2005) *Five-Year Review of Executive Order 13112 on Invasive Species*. NISC, Washington, DC.

NOAA (2021) What is an Invasive Species? Available at: https://oceanservice.noaa.gov/facts/invasive.html (accessed 20 November 2023).

Oleskevich, C., Punja, Z.K. and Shamoun, S.F. (1996) The biology of Canadian weeds. 105. Rubus strigosus Michx., Rubus parviflorus Nutt., and Rubus spectabilis Pursh. *Canadian Journal of Plant Science* 76(1), 187–201.

Orbán, I., Szitár, K., Kalapos, T. and Körel-Dulay, G. (2021) The role of disturbance in invasive plant establishment in a changing climate: Insights from a drought experiment. *Biological Invasions* 23(6), 1877–1890.

Padmanaba, M. and Corlett, R.T. (2014) Minimizing risks of invasive alien plant species in tropical production forest management. *Forests* 5(8), 1982–1998.

Padmanaba, M. and Sheil, D. (2014) Spread of the invasive alien species *Piper Aduncum* via logging roads in Borneo. *Tropical Conservation Science* 7(1), 35–44.

Parker, I.M., Simberloff, D., Lonsdale, W.M., Goodell, K., Wonham, M. *et al*. (1999) Impact: Toward a framework for understanding ecological effects of invaders. *Biological Invasions* 1, 3–19.

Peeler, J.L. and Smithwick, E.A.H. (2018) Exploring invasibility with species distribution modeling: How does fire promote cheatgrass (*Bromus tectorum*) invasion within lower montane forests? *Diversity and Distributions* 24(9), 1308–1320.

Peter, D.H. and Harrington, T.B. (2018) Effects of forest harvesting, logging debris, and herbicides on the composition, diversity and assembly of a western Washington, USA plant community. *Forest Ecology and Management* 417, 18–30.

Pickles, B.J., Wilhelm, R., Asay, A.K., Hahn, A.S., Simard, S.W. and Mohn, W.W. (2017) Transfer of 13C between paired Douglas-fir seedlings reveals plant kinship effects and uptake of exudates by ectomycorrhizas. *New Phytologist* 214(1), 400–411.

Powell, K.I., Chase, J.M. and Knight, T.M. (2011) A synthesis of plant invasion effects on biodiversity across spatial scales. *American Journal of Botany* 98, 539–548.

Prasad, R. (1999) Scotch Broom, Cytisus scoparius L. in British Columbia. Available at: https://cfs.nrcan.gc.ca/pubwarehouse/pdfs/31653.pdf (accessed 12 November 2023).

Pyšek, P., Jarošík, V., Hulme, P.E. *et al*. (2012) A global assessment of invasive plant impacts on resident species, communities and ecosystems: The interaction of impact measures, invading species' traits and environment. *Global Change Biology* 18, 1725–1737.

Pyšek, P., Hulme, P.E., Simberloff, D., Bacher, S., Blackburn, T.M. *et al*. (2020) Scientists' warning on invasive alien species. *Biological Reviews* 95(6), 1511–1534.

Qu, T., Du, X., Peng, Y., Guo, W., Zhao, C. and Losapio, G. (2021) Invasive species allelopathy decreases plant growth and soil microbial activity. *Plos One* 16(2), e0246685.

Rai, K.P. and Singh, J.S. (2020) Invasive alien plant species: Their impact on environment, ecosystem services and human health. *Ecological Indicators* 111, 106020.

Rasp, B.C. (2019) Spatial and Temporal Patterns of Invasive Exotic Plant Species in Response to Timber Harvesting in a Mixed Mesophytic Forest of Eastern Kentucky. Available at: https://uknowledge.uky.edu/forestry_ etds/45/ (accessed 10 August 2023).

Rauschert, E.S.J., Mortensen, D.A. and Bloser, S.M. (2017) Humanmediated dispersal via rural road maintenance can move invasive propagules. *Biological Invasions* 19(7), 2047–2058.

Richardson, D.M. and Pysek, P. (2004) What is an Invasive Species? Available at: http://www.cabicompendium.org/cpc/library/invasiveplants/ (accessed 20 October 2023).

Rivas-Torres, G., Flory, S.L. and Loiselle, B. (2017) Plant community composition and structural characteristics of an invaded forest in the Galapagos. *Biodiversity and Conservation* 27, 329–344.

Rossi, R.D., Martins, C.R., Viana, P.L., Rodrigues, E.L. and Figueira, J.E.C. (2014) Impact of invasion by molasses grass (*Melinis minutiflora* P. Beauv.) on native species and on fires in areas of campo-cerrado in Brazil. *Acta Botanica Brasilica* 28(4), 631–637.

Rouifed, S., Puijalon, S., Viricel, M. and Piola, F. (2011) Achene Buoyancy and germinability of the terrestrial invasive Fallopia × Bohemica in aquatic environment: A new vector of dispersion? *Ecoscience* 18(1), 79–84.

Rusterholz, H.P., Salamon, J.A., Ruckli, R. and Baur, B. (2014) Effects of the annual invasive plant impatiens glandulifera on the collembola and acari communities in a deciduous forest. *Pedobiologia* 57(4–6), 285–291.

Safranyik, L. and Wilson, B. (2006) *The Mountain Pine Beetle: A Synthesis of Biology, Management and Impacts on Lodgepole Pine*. Natural Resources Canada, Canadian Forest Service, Pacific Forestry Centre Victoria, BC, Canada. Available at: https://www.for.gov.bc.ca/hfd/library/documents/bib96122.pdf (accessed 20 August 2023).

Shephard, N.T., Narine, L., Peng, Y. and Maggard, A. (2022) Climate smart forestry in the Southern United States. *Forests* 13(9), 1460.

Sheppard, A.W. and Hosking, J.R. (2000) Broom management. *Plant Protection Quarterly* 15, 134–186.

Singh, J.P., Kuang, Y., Plough, L., Coghill, M. and Fraser, L.H. (2022) Spotted knapweed (Centaurea stoebe) creates a soil legacy effect by modulating soil elemental composition in a semi-arid grassland ecosystem. *Journal of Environmental Management* 317, 115391.

Sitzia, T. (2014) A call to silviculturists for a new field of science: The forestry of invasive alien species. *The Forestry Chronicle* 90(04), 486–488.

Slesak, R.A., Harrington, T.B., D'Amato, A.W. and Peter, D.H. (2022) Removal of invasive Scotch broom increases its negative effects on soil chemistry and plant communities. *Oecologia* 198(1), 243–254.

Snyder, S.A., Blinn, C.R. and Peterson, R.R. (2021) Examining loggers' attitudes and behaviors toward invasive forest plants: A minnesota case study. *Journal of Sustainable Forestry* 40(3), 300–318.

Sokol, N.W., Kuebbing, S.E. and Bradford, M.A. (2017) Impacts of an invasive plant are fundamentally altered by a co-occurring forest disturbance. *Ecology* 98(8), 2133–2144.

Stromberg, J.C., Lite, S.J., Marler, R., Paradzick, C., Shafroth, P.B. *et al.* (2007) Altered stream-flow regimes and invasive plant species: The Tamarix case. *Global Ecology and Biogeography* 16, 381–393.

Terwei, A., Zerbe, S., Zeileis, A., Annighofer, P., Kawaletz, H. *et al.* (2013) Which are the factors controlling tree seedling establishment in North Italian floodplain forests invaded by non-native tree species? *Forest Ecology and Management* 304, 192–203.

Tomat-Kelly, G. and Flory, S.L. (2022) Research gaps limit understanding of invasion-fire cycles. *Biological Invasions* 25, 693–711.

Torres, N., Herrera, I., Fajardo, L. and Bustamante, R.O. (2021) Metaanalysis of the impact of plant invasions on soil microbial communities. *BMC Ecology and Evolution* 21(1), 172.

Trammell, T.L.E., D'Amico, V., Avolio, M.L., Mitchell, J.C. and Moore, E. (2020) Temperate deciduous forests embedded across developed landscapes: Younger forests harbour invasive plants and urban forests maintain native plants. *Journal of Ecology* 108(6), 2366–2375.

Trimmel, H., Weihs, P., Leidinger, D., Formayer, H., Kalny, G. and Melcher, A. (2018) Can riparian vegetation shade mitigate the expected rise in stream temperatures due to climate change during heat waves in a human-impacted pre-alpine river? *Hydrology and Earth System Sciences* 22(1), 437–461.

Underwood, E.C., Klinger, R.C. and Brooks, M.L. (2019) Effects of invasive plants on fire regimes and postfire vegetation diversity in an arid ecosystem. *Ecological Evolution* 9(22), 12421–12435.

Valéry, L., Fritz, H., Lefeuvre, J.-C. and Simberloff, D. (2008) In search of a real definition of the biological invasion phenomenon itself. *Biological Invasions* 10, 1345–1351.

Valéry, L., Fritz, H., Lefeuvre, J.-C. and Simberloff, D. (2009) Invasive species can also be native. *Trends in Ecology & Evolution* 24(11), 585.

Van Driesche, R. *et al.* (2002) Biological Control of Invasive Plants in the Eastern United States, USDA Forest Service Publication FHTET-2002-04, 41. Available at: https://www.invasive.org/biocontrol/25Kudzu.cfm#:~:text=Kudzu%20control%20costs%20exceed%20%24200,not%20economically%20feasible.%E2%80%9D%20Dr (accessed 29 September 2023).

Veldman, J.W. and Putz, F.E. (2010) Long-distance dispersal of invasive grasses by logging vehicles in a tropical dry forest. *Biotropica* 42(6), 697–703.

Veldman, J.W., Mostacedo, B., Peña-Claros, M. and Putz, F.E. (2009) Selective logging and fire as drivers of alien grass invasion in a Bolivian tropical dry forest. *Forest Ecology and Management* 258(7), 1643–1649.

Villegas, Z., Peña-Claros, M., Mostacedo, B. *et al*. (2009) Silvicultural treatments enhance growth rates of future crop trees in a tropical dry forest. *Forest Ecology and Management* 258(6), 971–977.

von Lützow, M., Kögel-Knabner, I., Ekschmitt, K., Flessa, H., Guggenberger, G., Matzner, E. and Marschner, B. (2007) SOM fractionation methods: Relevance to functional pools and to stabilization mechanisms. *Soil Biology and Biochemistry* 39, 2183–2207.

Wang, Z., Wu, J., Zhang, D., Gong, C., Wang, Q., Zhai, C. *et al*. (2022) Effects of *Rhus typhina* invasion on soil physicochemical properties and carbon emissions in Urban green spaces. *Forests* 13(11), 1827.

Watson, R.T., Baste, I.A., Larigauderie, A., Leadley, P., Pascual, U. and Baptiste, B. (2019) The global assessment report on BIODIVERSITY AND ECOSYSTEM SERVICES: Summary for Policymakers. Intergovernmental Science-Policy Platform on Biodiversity and Ecosystem Services (IPBES).

Welham, C., Van Rees, K., Seely, B. and Kimmins, H. (2007) Projected long-term productivity in Saskatchewan hybrid poplar plantations: Weed competition and fertilizer effects. *Canadian Journal of Forest Research* 37(2), 356–370.

Willoughby, I., Clay, D.V., Dixon, F.L. and Morgan, G.W. (2006) The effect of competition from different weed species on the growth of *Betula pendula* seedlings. *Canadian Journal of Forest Research* 36(8), 1900–1912.

Wenk, E.H. and Falster, D.S. (2015) Quantifying and understanding reproductive allocation schedules in plants. *Ecology and Evolution* 5, 5521–5538.

Wolf, A.T., Parker, L., Fewless, G., Corio, K., Sundance, J. *et al*. (2007) Impacts of summer versus winter logging on understory vegetation in the Chequamegon-Nicolet National Forest. *Forest Ecology and Management* 254(1), 35–45.

Xu, Y., Chen, X., Ding, L. and Kong, C.-H. (2023) Allelopathy and allelochemicals in grasslands and forests. *Forests* 14(3), 562.

Zenni, R.D., Essl, F., Garcia-Berthou, E. and McDermott, S.M. (2021) The economic costs of biological invasions around the world. *NeoBiota* 67, 1–9.

Zhang, P., Li, B., Wu, J. and Hu, S. (2019) Invasive plants differentially affect soil biota through litter and rhizosphere pathways: A meta-analysis. *Ecology Letters* 22(1), 200–210.

2 Recent developments in parasitic plant biology

John P. Stanga
1501 Mercer University Dr., Macon, GA 31207, USA

Abstract
Parasitic plants are taxonomically diverse, globally distributed, and capable of parasitizing a wide array of host plants, often with preferences for specific hosts. Some parasitic plants are damaging pests of numerous crops; their effects can be especially detrimental in low-input agriculture. Effective breeding and management strategies rely on comprehensive information about the life history, evolution, genetics, biochemistry, physiology, communication, and ecology of parasitic plants and their hosts. Highthroughput sequencing approaches are used to advance research in genome evolution, to identify genetic loci for breeding, to characterize microbial communities associated with parasitic plants, and to discover molecular mechanisms of parasitism and host responses. Molecular and biochemical investigations are elucidating the mechanisms by which parasitic seeds germinate in response to host phytohormones in the rhizosphere. Structural information guides synthesis- and screening-based methods to identify chemicals that could be used to promote or inhibit parasitic seed germination. Microbial communities in the rhizosphere influence parasitic plants and hosts as a system; fungal and bacterial species themselves, and products derived from them, have the potential for agronomic application. Proteins and nucleic acids exchanged through the haustorium reveal the complex interplay between parasitic plant pathogenesis and host defense responses. This review focuses on the latest developments published within the last two years.

Keywords: parasitic plants, parasitic weeds, strigolactones, germination, haustorium, weed control

Introduction

Parasitic plants exist in nearly every biome globally and parasitize a wide array of host plants. Life histories, diversity, hosts, and distributions of parasitic plants are broadly reviewed in [1]. Parasitic angiosperms are characterized by a haustorium, a specialized organ that attaches to a host plant, invading its tissues, and forming a connection with the host vasculature [2]. Parasitic plants are often described as obligate or facultative; stem parasitic or root parasitic; and holoparasitic or hemiparasitic (hemiparasites are plants that retain any degree of autotrophy, and are the predominant form of parasitic plants). Endoparasites spend most of their life cycle within a host plant; like other types of parasitic plants, their seeds germinate and are distributed outside of the host [3].

Phylogenomic analyses demonstrate that parasitism in angiosperms has evolved independently at least twelve times, resulting in parasitic plants in several different families

*Corresponding author: stanga_jp@mercer.edu

[2]. Of the approximately 292 genera and 4750 species of parasitic plants, fewer than 10% of these genera are crop-damaging pathogens [2]. Despite representing a relatively small fraction of all parasitic plants, pathogenic plants are estimated to cause billions of dollars of losses annually. Although most attention to parasitic plants is because of their negative impact as agricultural weeds, many parasitic plants are beneficial. For example, parasitic plants can function as keystone species, may help control invasive species, can contribute to ecotourism, and synthesize several phytochemicals with biological activity in animals and animal pathogens [4–8].

This review focuses on advances in parasitic plant biology with implications for breeding and management, with an emphasis on reports and reviews published within the last two years. First reviewed are recent findings on parasitic plant diversity and genomics, then a summary of progress toward understanding the role of strigolactones as endogenous phytohormones with a major role in stimulating seed germination. Finally, chemotaxis toward the host plant, haustorium formation, and direct interactions between host and parasitic plants are discussed.

Review Methodology

PubMed and CAB Reviews were searched using the keyword search term "parasitic plant." For selected articles, I also explored references within, and articles citing them, to identify notable sources excluded by the initial search parameters.

Genomics

Sequencing data is allowing new evaluations of population diversity and genome structure and evolution; reviewed in [9]. Recent contributions to the growing collection of genomics databases include an assembly at the chromosome level of *Taxillus chinensis* (Loranthaceae) constructed using chromosome conformation capture data [10]. Three new Orobanchaceae chromosome-level genomes were assembled for holoparasites *Orobanche cumana* and *Phelipanche aegyptiaca*, and an autotroph, *Lindenbergia luchunensis* [11]. Together with five other Orobanchaceae genomes [12–16], the authors conducted a comparative genomics analysis, revealing that although whole genome duplication did occur prior to Orobanchaceae divergence, holoparasitism in Orobanchaceae is not associated with whole genome duplications. The authors also associated parasitism with genomic expansions that may have facilitated haustorium development [11]. Analysis of gene ontology descriptions for genes lost in parasitic lineages reveals convergent patterns among two species within Orobanchaceae, similar to patterns observed in Cuscuta australis (Convolvulaceae) [11]. Loss of photosynthesis and floral regulation genes in many parasitic lineages was also identified by the authors of transcriptome studies and mitochondrial and plastid genome analyses [17–21]. Wholegenome analysis of endoparasitic *Sapria* (Rafflesiaceae) also demonstrates convergent patterns of gene loss across different genera, where the degree of gene loss appears to be correlated with the degree of parasitism [22]. Patterns of gene loss seen in parasitic plants appear to be conserved even when comparisons are extended to include mycoheterotrophs (plants that rely on fungal hosts for nutrition) [23]. Cytogenetic methods to understand karyotypic evolution are complementary to sequencing-based approaches and have been used to identify evolution driven by duplication or dysploidy within different *Cuscuta* species [24]. Whole-genome sequencing alternatives are also used to evaluate population structure, taxonomy, and evolution. For example, the traditional method of ribosomal RNA ITS sequencing has recently been applied to describe a new species of *Cuscuta* [25], and a new ecotype of *Orobanche* [26]. RAPD-PCR was used to identify high genetic diversity among *Orobanche* in sunflower fields from eleven locations, and single-gene analysis of RuBisCO subunit *rbcL* was used to resolve differences between different species of *Orobanche* [27]. Together, diverse methods allow for the characterization of genome evolution.

Genome-wide sequencing methods are valuable for advances in breeding resistance into agronomically important crops that are impacted by parasites. Different host strategies for resistance to parasitic plants, reviewed in

[28], may affect pre-attachment processes like host-specific seed germination, chemotropism, and early haustorium development, and post-attachment processes, like host immune responses and parasitic effector signaling. Recent reports that apply diverse genomics methods add to our understanding of host resistance described in the aforementioned review. For example, transcriptomics methods applied to *Cuscuta* resistant cultivars of tomato allowed researchers to identify four regulatory genes involved in lignification and signaling [29]. A multi-locus genome-wide association study (GWAS) for maize resistance to *Striga* (Orobanchaceae) identified several resistance loci that had not been previously identified. Candidate loci include transcription factors and defense-related genes [30]. Similarly, GWAS of sorghum resistance to *Striga* (Orobanchaceae) has identified numerous candidate loci for both pre-attachment [31] and post-attachment strategies [32]. Quantitative trait locus (QTL) mapping for faba bean resistance to *Orobanche* has identified several loci; although there are numerous candidate genes within the loci, QTL studies can be applied to marker-assisted selection [33]. Introgression of resistance loci is useful to develop new varieties, and also to provide materials for research on the mechanisms of parasitism. For example, researchers bred a locus conferring improved *Orobanche* resistance into a sunflower and demonstrated inhibition of parasite development at the cortex prior to establishing vascular connections [34]. In contrast to GWAS and QTL studies that require considerable phenotyping and genotyping, and may be limited by the genotypes used, researchers are increasingly applying genomic methods in an effort to identify local adaption occurring in both hosts and parasites. Genotyping by sequencing (GBS) was used for Orobanche crenata populations associated with different hosts to identify population genetic structure associated with hosts, and also influenced by geographic location [35]. A genome-environment association (GEA) approach was used to leverage whole genome sequence data from several sorghum landraces to identify loci associated with local *Striga* populations [36]. The researchers identified multiple alleles of a gene, *LGS1*, for which a role in strigolactone biosynthesis had previously been described [37]. A comprehensive understanding of the developmental and molecular events underpinning the parasitic-host interactions should lead to greater precision in breeding efforts. Genome editing technology will be an important method for engineering resistant hosts, although cultural, economic, ecological, and regulatory factors may affect their adoption [38–40].

Host-derived lactones stimulate germination

Strigolactones (SLs) are a diverse class of butenolide phytohormones, named for their ability to stimulate germination in *Striga* [41, 42]. SLs are involved in numerous biological processes in plants, including root growth, seedling photomorphogenesis, shoot architecture and branching, and senescence (reviewed in [43, 44]). SLs also moderate responses to biotic and abiotic factors [45]. Recently, roles for SLs have been reported in salicylic acid-mediated disease responses [46], abscisic acid-mediated alkaline tolerance in seedlings [47], cytochrome p450-mediated developmental timing of leaf emergence [48], and carboxylesterasemediated axillary bud outgrowth [49]. Considering the numerous physiological processes involving SLs, it is important to evaluate hosts holistically for other growth, development, and response phenotypes when altering host SL metabolism to target for preemergent breeding efforts [50].

Holoparasites across ten families produce dust seeds, which are small seeds with undifferentiated embryos. With the exception of Orobanchaceae, germination is poorly described for many of these holoparasites (reviewed in [51]). SLs are well known for their role in promoting seed germination of Orobanchaceae, among other weedy plants. Over 30 structurally diverse SLs have been identified, including the canonical SLs strigol and orobanchol, among other non-canonical types [52, 53]. In the absence of suitable germination signals, *Striga* seeds may remain viable and dormant in the soil for years or decades [54]. Numerous studies, especially in *Arabidopsis* and rice, have described molecular details of SL signaling in seed germination [55, 56]. Briefly, SLs interact with receptor proteins to initiate the degradation of transcriptional repressors. Many of these signaling proteins have conserved roles across plants, even in mosses [57]. Insight into these mechanisms has come from karrikins, chemically similar molecules found in smoke that stimulate germination of fire-following and other species, including *Arabidopsis* [58]. Indeed, only three mutations are sufficient to convert a karrikin receptor into an SL receptor [59].

Structural and stereochemical differences among SLs and SL receptors contribute to host recognition; because SLs differ across species, it is important to characterize the biosynthesis of SLs in diverse plants. For example, the characterization of SL biosynthesis in Sorghum bicolor (Poaceae) identified a mechanism apparently unique to this species [60]. Furthermore, SLs are not the only class of bioactive chemicals in root exudates: analysis of root exudates from *Orobanche*-resistant and susceptible sunflowers identified different chemical profiles that may be correlated with resistance, including sesquiterpene lactones [61]. When isolating and characterizing root exudates [62], it is also important to evaluate SL capacity for stimulating germination. SL bioactivity differs across species and includes the ability to induce germination in non-parasitic plants, notably *Arabidopsis*. On the other hand, not all parasitic weeds may be responsive to a given SL, and some holoparasites may not require a host chemical signal at all [51]. Inquiry into the activity and sensitivity of SLs in plants is aided by the design of chemical probes that report a fluorescent signal upon enzymatic interaction with SL receptors [63, 64]. Numerous recent reports and reviews describe relationships between SL-selectivity and host detection in a variety of species, including Pisum sativum (Fabaceae) [65] and predominantly in parasitic plants of Orobanchaceae: *Phelipanche spp.* [66–68] and Striga hermonthica [69–71]. Understanding roles and mechanisms for SLs and SL analog action in host detection and germination presents intriguing possibilities for parasite-specific management. Using structural information and computational methods, effective germination agonists have been designed [72]. Chemical syntheses have also produced several promising germination stimulants [73, 74]; methods for synthesizing SLs and SL analog molecules are reviewed in [75]. Field and greenhouse tests of SL formulations show promising results [76] for suicidal germination as a practical method for parasitic weed control. As an alternative or complement to suicidal germination, parasite-specific strigolactone receptor inhibitors are also under investigation [77, 78]. Seed germination molecules that function in germination downstream of SL reception may also represent viable targets for the management of parasitic plants [79].

Rhizosphere influences

Host plants exuding SLs into the rhizosphere come at the potential cost of inviting parasitic plant colonization. SLs can also promote beneficial associations with arbuscular mycorrhizal fungi and root nodulation [80, 81]. Root-associated microorganisms may prime plants for parasitic attacks, and also affect parasitic plants directly [82]. One research group used a model system composed of combinations of a tobacco host, holoparasitic *Phelipanche spp.*, and an arbuscular mycorrhizal fungus, to identify differences among endogenous and exuded levels of four major phytohormones and orobanchol SLs [83]. They discovered that each condition has a different hormone profile, and the presence of the parasite reduces arbuscular mycorrhizal colonization and resembles tobacco and *Phelipanche* grown without the fungus [83]. In contrast, another group used a pot study to grow a root hemiparasite, *Pedicularis kansuensis* (Orobanchaceae), and an arbuscular mycorrhizal fungus on a grass host or a legume host. They observed improved growth performance associated with the fungus, especially for the legume [84]. Both of these studies underscore the need to evaluate parasite, host, and community diversity when evaluating possible beneficial effects of rhizosphere microorganisms in parasitic plant systems. Metagenomic methods are a powerful way to assess the spectrum of microbiota in rhizosphere communities. Associations between Xanthomonadaceae bacteria and parasitism were identified using metagenomic analysis of sunflower field soils with naturally differing levels of holoparasitic *Orobanche cumana* infection [85]. These findings led to germination and pot studies, which were used to identify both inhibitory and stimulatory roles for different bacteria and their metabolites. Natural bacterial and fungal metabolites are an attractive set of candidates to screen for effects on parasitic seed germination and colonization. Hundreds of fungal isolates were screened for growth effects on holoparasitic *Phelipanche ramosa*; a *Fusarium* species emerged as a promising mycoherbicide that inhibited germination, necrotized stems, and was not pathogenic to the tobacco host [86]. *Fusarium* metabolites have also been demonstrated as effective in germination assays against *Striga hermonthica* [87]. Bacterial metabolites have likewise been demonstrated as viable herbicidal candidates [88].

The haustorium and plant–plant exchange

Haustoriogenesis, the developmental process in which a parasitic plant attaches, penetrates, and connects to a host, is reviewed in [89]. Histological and microscopic methods have long been used to describe the structure of the haustorium, and continue to contribute to our understanding of the haustorium structure. For example, because the early development of endophytic parasites occurs entirely within the host plant, observations of life stages before bud emergence have been limited. By microtome sectioning numerous samples of *Rafflesia consueloae* (Rafflesiaceae) and its host, researchers were able to describe early developmental events [90], including the possibility that small groups of parasitic cells may persist innocuously prior to flower formation. By applying a similar approach to an endophytic plant, *Lathrophytum peckoltii* (Balanophoraceae), researchers observed late cell differentiation, followed by changes in host root cambium and establishment of vascular connections [91]. While these methods can lead to important insights, they typically do not describe spatial relationships. An exception to this can be seen in [92], where the authors overcame the challenges posed by thick tissues by collecting serial thin sections, then reconstructing three-dimensional models. Applying this approach to two species of parasites within Orobanchaceae, one obligate (*Striga hermonthica*) and one facultative (*Phtheirospermum japonicum*), allowed them to identify structural differences, especially at the xylem connection. To visualize the complex haustorium-host structures without disrupting their physical associations, high-resolution X-ray microtomography may be used [93, 94]. Endoparasites were long considered to be monophyletic, but analysis of development using microtomography and histology supports endoparasitism as an example of convergent evolution [95]. Visualizing early parasitic colonization under controlled conditions may be useful for screening populations for resistance [96]. Once connected through the vasculature, the different species become symplastic with one another, allowing for the transport of nutrients and signaling molecules [97, 98]. Metabolomic studies demonstrate that Cuscuta campestris haustoria have more sugars and polyols than other organs, and their metabolic profiles depend on host identity [99]. Parasite presence affects metabolism in host plants [100], and host plant responses may even be influenced by the sex of a dioecious parasite [101]. Hydrogen isotopes allow for estimates of how much carbon is transferred to parasites versus gained through photosynthesis, in parasites with varying degrees of heterotrophy [102]. This relatively new approach was taken because carbon isotope composition is too similar between C3 hemiparasites associated with C3 host plants, which allowed the researchers to associate carbon uptake with haustoria complexity. The haustorial transfer of metals has physiological and ecological implications. While the transfer of cadmium between *Cuscuta chinensis* and soybean hosts may be selective and not result in the bioaccumulation of cadmium [103], other species demonstrate varying degrees of metal transfer. For example, *Orobanche laxissima* grown on pomegranate and ash hosts resulted in lower heavy metals in the hosts [104].

Evidence from diverse parasites and their hosts demonstrates the importance of the symplastic exchange of RNA and protein signaling molecules responsible for parasitic virulence and host defenses [105]. Investigating interspecific communication should benefit conservation efforts for threatened species, such as endoparasitic *Rafflesia* species, by elucidating species-specific hormonal and nutritional requirements for propagation, an important tool in conservation efforts [106]. Noteworthy signaling molecules transmitted from parasite to host include phytohormones, such as cytokinin [107], and effector-like molecules [108]. Both of these examples are believed to suppress host responses. RNA is of particular interest as a stable, transmittable signaling molecule capable of altering gene expression. In a system composed of *Cuscuta australis*, cucumbers, and aphids, mRNAs were transferred among three organisms; the researchers noted that mRNA abundance and length correlated with movement across species [109]. The latter observation was in contrast to earlier findings [110]; more research will clarify the mechanisms of interspecific mRNA transport. For example, structures that resemble transfer-RNAs may facilitate the transport of mRNAs, increasing mRNA expression in other species [111]. Noncoding RNAs like lncRNA [112] and miRNA [113] also play roles in inter-specific communication and are secreted into growth substrates [114]. Other evidence of nucleic acid transfer between species comes from genomic analyses, which support a history of mitochondrial and nuclear horizontal gene transfer (HGT), reviewed in [9]. An example comes from *Lophophytum mirabile* (Balanophoraceae), in which mitochondrial oxidative phosphorylation complexes appear to be composed of both foreign and native subunits [115]. Analysis of mitochondrial genomes from 26 *Cuscuta* species suggests that HGT of coding genes is rare [116], although there is evidence that limited mitochondrial gene transfer does occur in at least one *Cuscuta* species [117].

Haustorium formation in the hemiparasite *Phtheirospermum japonicum* is sensitive to environmental conditions, as nitrogen availability affects abscisic acid signaling, resulting in reduced haustoria formation [118]. Despite the biological importance of parasitic seedlings to orient their growth toward suitable host roots in preparation for attachment, many of the molecular details of the chemotropic process have not been deeply described. Root chamber assays provide an accessible model to identify factors affecting germ tube growth relative to host roots. Such an assay was used to identify a role for costunolide, a sesquiterpene lactone, but not for SLs in *Orobanche cumana* chemotropism toward a sunflower host [119]. In contrast, SLs were identified as chemical signals for chemotropism in other Orobanchaceae, *Phtheirospermum japonicum* and *Striga hermonthica* [120].

Early haustorium development from a *Striga* radicle appears to be coordinated by auxin [121]. Haustoriogenesis in Cuscuta japonica and *Cassytha filiformis* (Lauraceae) stems also features roles for auxin, as well as cytokinin [122]. This study also recognizes the importance of blue light and brassinolide for twining, and low farred- to-red ratios for haustoriogenesis. RNA sequencing has been used to define markers for developmental stages of haustorigenesis in Cuscuta campestris [123], *Orobanche* [89], and *Psittacanthus schiedeanus* (Loranthaceae; mistletoe) [124]. Similar methods demonstrated that *Phelipanche aegyptiaca*, despite its status as a non-photosynthetic holoparasite, has retained the photosynthesis-related electron carrier phylloquinone [125]. Transcriptomics have even been applied at the tissue level using laser-capture microdissection to identify candidate regulators of haustorium development in *Cuscuta campestris* on tomato hosts [126, 127]. Candidate-gene expression was manipulated using host-derived gene silencing, resulting in impaired parasitism. These studies demonstrate a plausible strategy for genetically engineered parasite resistance.

Conclusions

Parasitic plants are diverse taxonomically, physiologically, biochemically, and in their host requirements. Parasitic plants are members of complex inter-species systems that include microbial life and are responsive to environmental conditions. Advances in genomics, structural biology and biochemistry, development and physiology, and ecology all contribute greatly to understanding parasitic plants in their natural contexts. Because of the variability and complexity of parasitic plant systems, it is important for management strategies to consider as many factors as appropriate, including soil composition, breeding for host resistance, agronomic practices such as intercropping and pushpull, microbial formulations, germination agents, early detection under field conditions, post-attachment herbicides, identification of management-resistant varieties, and mechanical removal [128–133]. Multifaceted management may help to limit the emergence of parasitic weeds that may evolve to evade treatment. Moreover, parasitic plant systems are dynamic and responsive to the environment, an important consideration as climate models predict alterations to growing regions [134–136].

Acknowledgments

I am grateful to Dr. Amy Klocko at the University of Colorado Colorado Springs, for introducing me to CABI; to Dr. David Hemming at CABI, for welcoming this review; and to Dr. Caitlin Conn at Berry College, for reviewing critically.

References

1. Teixeira-Costa L, Davis CC. Life history, diversity, and distribution in parasitic flowering plants. Plant Physiology 2021;187(1):32–51.
2. Nickrent DL. Parasitic angiosperms: How often and how many? Taxon 2020;69(1):5–27.
3. Thorogood CJ, Teixeira-Costa L, Ceccantini G, Davis C, Hiscock SJ. Endoparasitic plants and fungi show evolutionary convergence across phylogenetic divisions. The New Phytologist 2021;232(3):1159–67.
4. Dell'Oste V, Spyrakis F, Prandi C. Strigolactones, from plants to human health: Achievements and challenges. Molecules 2021;26(15):4579.
5. Těšitel J, Li AR, Knotkova K, McLellan R, Bandaranayake PCG, Watson DM. The bright side of parasitic plants: What are they good for? Plant Physiology 2021;185(4):1309–24.
6. Berillo D, Kozhahmetova M, Lebedeva L. Overview of the biological activity of anthraquinons and flavanoids of the plant rumex species. Molecules 2022;27(4):1204.
7. Genovese C, D'Angeli F, Attanasio F, Caserta G, Scarpaci KS, Nicolosi D. Phytochemical composition and biological activities of Orobanche crenata Forssk.: A review. Natural Product Research 2021;35(22):4579–95.
8. D'Angeli F, Guadagni F, Genovese C, Nicolosi D, Trovato Salinaro A, Spampinato M, et al. Anticandidal activity of the parasitic plant orobanche crenata forssk. Antibiotics 2021;10(11):1373.
9. Yoshida S, Kee YJ. Large-scale sequencing paves the way for genomic and genetic analyses in parasitic plants. Current Opinion in Biotechnology 2021;70:248–54.
10. Fu J, Wan L, Song L, He L, Jiang N, Long H, et al. Chromosome-level genome assembly of the hemiparasitic Taxillus chinensis (DC.) Danser Genome Biology and Evolution 2022;14(5):evac060.
11. Xu Y, Zhang J, Ma C, Lei Y, Shen G, Jin J, et al. Comparative genomics of orobanchaceous species with different parasitic lifestyles reveals the origin and stepwise evolution of plant parasitism. Molecular Plant 2022;15(8):1384–99.
12. Ma L, Dong C, Song C, Wang X, Zheng X, Niu Y, et al. De novo genome assembly of the potent medicinal plant Rehmannia glutinosa using nanopore technology. Computational and Structural Biotechnology Journal 2021;19:3954–63.
13. Yoshida S, Kim S, Wafula EK, Tanskanen J, Kim YM, Honaas L, et al. Genome sequence of *Striga asiatica* provides insight into the evolution of plant parasitism. Current Biology 2019;29(18):3041–3052.e4.
14. Becher H, Brown MR, Powell G, Metherell C, Riddiford NJ, Twyford AD. Maintenance of species differences in closely related tetraploid parasitic euphrasia (Orobanchaceae) on an isolated island. Plant Communications 2020;1(6):100105.

15. Cui S, Kubota T, Nishiyama T, Ishida JK, Shigenobu S, Shibata TF, et al. Ethylene signaling mediates host invasion by parasitic plants. Science Advances 2020;6(44):eabc2385.
16. *Cuscuta japonica* (ID 805569) - BioProject - NCBI [Internet]. Available from: https://www.ncbi.nlm.nih.gov/bioproject/805569
17. Jin DP, Kim JS, Ku YB, Lim CE. The complete chloroplast genome sequence of Melampyrum koreanum (Orobanchaceae), an endemic and hemi-parasitic herb in Korea. Mitochondrial DNA Part B. 2021;6(11):3122–4.
18. Yu R, Sun C, Zhong Y, Liu Y, Sanchez-Puerta MV, Mower JP, et al. The minicircular and extremely heteroplasmic mitogenome of the holoparasitic plant Rhopalocnemis phalloides. Current Biology 2022;32(2):470–479.e5.
19. Lin Y, Li P, Zhang Y, Akhter D, Pan R, Fu Z, et al. Unprecedented organelle genomic variations in morning glories reveal independent evolutionary scenarios of parasitic plants and the diversification of plant mitochondrial complexes. BMC Biology 2022;20(1):49.
20. Schelkunov MI, Nuraliev MS, Logacheva MD. Genomic comparison of non-photosynthetic plants from the family Balanophoraceae with their photosynthetic relatives. PeerJ 2021;9:e12106.
21. Duan K, Fu H, Fang D, Wang K, Zhang W, Liu H, et al. Genome-wide analysis of the MADS-box gene family in holoparasitic plants (*Balanophora subcupularis* and *Balanophora fungosa* var. *globosa*). Frontiers in Plant Science 2022;13:846697. Available from: https://www.frontiersin.org/articles/10.3389/fpls.2022.846697
22. Cai L, Arnold BJ, Xi Z, Khost DE, Patel N, Hartmann CB, et al. Deeply altered genome architecture in the endoparasitic flowering plant Sapria himalayana Griff. (Rafflesiaceae). Current Biology 2021;31(5):1002–1011.e9.
23. Xu Y, Lei Y, Su Z, Zhao M, Zhang J, Shen G, et al. A chromosomescale Gastrodia elata genome and large-scale comparative genomic analysis indicate convergent evolution by gene loss in mycoheterotrophic and parasitic plants. The Plant Journal 2021;108(6):1609–23.
24. Ibiapino A, Garcia MA, Amorim B, Baez M, Costea M, Stefanović S, et al. The evolution of cytogenetic traits in Cuscuta (Convolvulaceae), the genus with the most diverse chromosomes in angiosperms. Frontiers in Plant Science 2022;13:842260. Available from: https://www.frontiersin.org/articles/10.3389/fpls.2022.842260
25. Costea M, Soares da Silva S, Simao-Bianchini R, Simoes ARG, Stefanovic S. Notes on the systematics of Cuscuta sect. Subulatae (subg. Grammica) with the description of Cuscuta mantiqueirana, a new species from Brazil. PhytoKeys. 2021;184:27–44.
26. Yousefi AR, Ahmadikhah A, Fotovat R, Rohani L, Soheily F, Uberti DL, et al. Molecular characterization of a new ecotype of holoparasitic plant Orobanche L. on host weed *Xanthium spinosum* L. Plants. 2022;11(11):1406.
27. Ivanović Ž, Marisavljević D, Marinković R, Mitrović P, Blagojević J, Nikolić I, et al. Genetic diversity of Orobanche cumana populations in Serbia. Plant Pathology Journal 2021;37(6):512–20.
28. Fishman MR, Shirasu K. How to resist parasitic plants: Pre- and postattachment strategies. Current Opinion in Plant Biology 2021;62:102004.
29. Jhu MY, Farhi M, Wang L, Philbrook RN, Belcher MS, Nakayama H, et al. Heinz-resistant tomato cultivars exhibit a lignin-based resistance to field dodder (*Cuscuta campestris*) parasitism. Plant Physiology 2022;189(1):129–51.
30. Stanley AE, Menkir A, Ifie B, Paterne AA, Unachukwu NN, Meseka S, et al. Association analysis for resistance to Striga hermonthica in diverse tropical maize inbred lines. Scientific Reports 2021;11(1):24193.
31. Mallu TS, Irafasha G, Mutinda S, Owuor E, Githiri SM, Odeny DA, et al. Mechanisms of pre-attachment Striga resistance in sorghum through genome-wide association studies. Molecular Genetics and Genomics 2022;297(3):751–62.
32. Kavuluko J, Kibe M, Sugut I, Kibet W, Masanga J, Mutinda S, et al. GWAS provides biological insights into mechanisms of the parasitic plant (Striga) resistance in sorghum. BMC Plant Biology 2021;21(1):392.
33. Gutierrez N, Torres AM. QTL dissection and mining of candidate genes for Ascochyta fabae and Orobanche crenata resistance in faba bean (*Vicia faba* L.). BMC Plant Biology 2021;21(1):551.
34. Fernandez-Aparicio M, del Moral L, Munos S, Velasco L, Perez-Vich B. Genetic and physiological characterization of sunflower resistance provided by the wild-derived OrDeb2 gene against highly virulent races of Orobanche cumana Wallr. Theoretical and Applied Genetics 2022;135(2):501–25.

35. Bendaoud F, Kim G, Larose H, Westwood JH, Zermane N, Haak DC. Genotyping-by-sequencing analysis of Orobanche crenata populations in Algeria reveals genetic differentiation. Ecology and Evolution 2022;12(3):e8750.
36. Bellis ES, Kelly EA, Lorts CM, Gao H, DeLeo VL, Rouhan G, et al. Genomics of sorghum local adaptation to a parasitic plant. Proceedings of the National Academy of Sciences 2020;117(8):4243–51.
37. Gobena D, Shimels M, Rich PJ, Ruyter-Spira C, Bouwmeester H, Kanuganti S, et al. Mutation in sorghum LOW GERMINATION STIMULANT 1 alters strigolactones and causes Striga resistance. Proceedings of the National Academy of Sciences 2017;114(17):4471–6.
38. Tripathi L, Dhugga KS, Ntui VO, Runo S, Syombua ED, Muiruri S, et al. Genome editing for sustainable agriculture in Africa. Frontiers in Genome Editing 2022;4:876697. Available from: https://www.frontiersin.org/articles/10.3389/fgeed.2022.876697
39. Klocko AL. Strategies to facilitate containment of genetically engineered crops. CAB Reviews Perspectives in Agriculture Veterinary Science Nutrition and Natural Resources. 2020;15(055):1–11. Available from: http://www.cabi.org/cabreviews/review/20203552556
40. Rato C, Carvalho MF, Azevedo C, Oblessuc PR. Genome editing for resistance against plant pests and pathogens. Transgenic Research 2021;30(4):427–59.
41. Yoneyama K, Xie X, Yoneyama K, Takeuchi Y. Strigolactones: Structures and biological activities. Pest Management Science 2009;65(5):467–70.
42. Xie X, Yoneyama K, Yoneyama K. The Strigolactone Story. Annual Review of Phytopathology 2010;48(1):93–117.
43. Bunsick M, McCullough R, McCourt P, Lumba S. Plant hormone signaling: Is upside down right side up? Current Opinion in Plant Biology 2021;63:102070.
44. Rehman NU, Li X, Zeng P, Guo S, Jan S, Liu Y, et al. Harmony but not uniformity: Role of strigolactone in plants. Biomolecules 2021;11(11):1616.
45. Kyozuka J, Nomura T, Shimamura M. Origins and evolution of the dual functions of strigolactones as rhizosphere signaling molecules and plant hormones. Current Opinion in Plant Biology 2022;65:102154.
46. Kusajima M, Fujita M, Soudthedlath K, Nakamura H, Yoneyama K, Nomura T, et al. Strigolactones modulate salicylic acid-mediated disease resistance in *Arabidopsis thaliana*. International Journal of Molecular Sciences 2022;23(9):5246.
47. Mi X, Anwar S, Chen C, Jin X, Yu L, Ali N, et al. Strigolactone enhances alkaline tolerance in soybean seeds germination by altering expression profiles of ABA biosynthetic and signaling genes. Journal of Plant Biology 2022. DOI. 10.1007/s12374-022-09357-2
48. Cornet F, Pillot JP, Le Bris P, Pouvreau JB, Arnaud N, de Saint Germain A, et al. Strigolactones (SLs) modulate the plastochron by regulating KLUH (KLU) transcript abundance in Arabidopsis. The New Phytologist 2021;232(5):1909–16.
49. Xu E, Chai L, Zhang S, Yu R, Zhang X, Xu C, et al. Catabolism of strigolactones by a carboxylesterase. Nature Plants 2021;7(11):1495–504.
50. Kawa D, Taylor T, Thiombiano B, Musa Z, Vahldick HE, Walmsley A, et al. Characterization of growth and development of sorghum genotypes with differential susceptibility to Striga hermonthica. Journal of Experimental Botany 2021;72(22):7970–83.
51. Baskin JM, Baskin CC. Germination and seed/embryo size in holoparasitic flowering plants with "dust seeds" and an undifferentiated embryo. The Botanical Review 2022;88(1):1–49.
52. Yoneyama K, Brewer PB. Strigolactones, how are they synthesized to regulate plant growth and development? Cell Signaling and Gene Expression 2021;63:102072.
53. Wakabayashi T, Ueno K, Sugimoto Y. Structure elucidation and biosynthesis of orobanchol. Frontiers in Plant Science 2022;13:835160. Available from: https://www.frontiersin.org/articles/10.3389/fpls.2022.835160
54. Runo S, Kuria EK. Habits of a highly successful cereal killer, Striga. PLoS Pathogens 2018;14(1):e1006731.
55. Burger M, Chory J. The many models of strigolactone signaling. Trends in Plant Science 2020;25(4):395–405.
56. Nelson DC. The mechanism of host-induced germination in root parasitic plants. Plant Physiology 2021;185(4):1353–73.
57. Lopez-Obando M, Guillory A, Boyer FD, Cornu D, Hoffmann B, Le Bris P, et al. The Physcomitrium (Physcomitrella) patens PpKAI2L receptors for strigolactones and related compounds function via MAX2-dependent and -independent pathways. The Plant Cell 2021;33(11):3487–512.

58. Yao J, Waters MT. Perception of karrikins by plants: A continuing enigma. Journal of Experimental Botany 2020;71(6):1774–81.
59. Arellano-Saab A, Bunsick M, Al Galib H, Zhao W, Schuetz S, Bradley JM, et al. Three mutations repurpose a plant karrikin receptor to a strigolactone receptor. Proceedings of the National Academy of Sciences 2021;118(30):e2103175118.
60. Yoda A, Mori N, Akiyama K, Kikuchi M, Xie X, Miura K, et al. Strigolactone biosynthesis catalyzed by cytochrome P450 and sulfotransferase in sorghum. The New Phytologist 2021;232(5):1999–2010.
61. Wu W, Huang H, Su J, Yun X, Zhang Y, Wei S, et al. Dynamics of germination stimulants dehydrocostus lactone and costunolide in the root exudates and extracts of sunflower. Plant Signaling & Behavior 2022;17(1):2025669.
62. Wang JY, Chen GTE, Jamil M, Braguy J, Sioud S, Liew KX, et al. Protocol for characterizing strigolactones released by plant roots. STAR Protocols 2022;3(2):101352.
63. de Saint Germain A, Clave G, Schouveiler P, Pillot JP, Singh AV, Chevalier A, et al. Expansion of the strigolactone profluorescent probes repertory: The right probe for the right application. Frontiers in Plant Science 2022;13:887347. Available from: https://www.frontiersin.org/articles/10.3389/fpls.2022.887347
64. Wang DW, Yu SY, Pang ZL, Ma DJ, Liang L, Wang X, et al. Discovery of a broad-spectrum fluorogenic agonist for strigolactone receptors through a computational approach. Journal of Agricultural and Food Chemistry 2021;69(36):10486–95.
65. Guercio AM, Torabi S, Cornu D, Dalmais M, Bendahmane A, Le Signor C, et al. Structural and functional analyses explain Pea KAI2 receptor diversity and reveal stereoselective catalysis during signal perception. Communications Biology 2022;5(1):1–16.
66. Fornier SD, de Saint Germain A, Retailleau P, Pillot JP, Taulera Q, Andna L, et al. Noncanonical strigolactone analogues highlight selectivity for stimulating germination in two Phelipanche ramosa populations. Journal of Natural Products. 2022;85(8):1976–92. DOI: 10.1021/acs. jnatprod.2c00282
67. de Saint Germain A, Jacobs A, Brun G, Pouvreau JB, Braem L, Cornu D, et al. A Phelipanche ramosa KAI2 protein perceives strigolactones and isothiocyanates enzymatically. Plant Communications 2021;2(5):100166.
68. Miura H, Ochi R, Nishiwaki H, Yamauchi S, Xie X, Nakamura H, et al. Germination stimulant activity of isothiocyanates on Phelipanche spp. Plants 2022;11(5):606.
69. Chen J, White A, Nelson DC, Shukla D. Role of substrate recognition in modulating strigolactone receptor selectivity in witchweed. Journal of Biological Chemistry 2021;297(4) . Available from: https://www.jbc.org/article/S0021-9258(21)00895-4/abstract
70. Chen J, Nelson DC, Shukla D. Activation mechanism of strigolactone receptors and its impact on ligand selectivity between host and parasitic plants. Journal of Chemical Information and Modeling 2022;62(7):1712–22.
71. Wang Y, Yao R, Du X, Guo L, Chen L, Xie D, et al. Molecular basis for high ligand sensitivity and selectivity of strigolactone receptors in Striga. Plant Physiology 2021;185(4):1411–28.
72. Wang D, Pang Z, Yu H, Thiombiano B, Walmsley A, Yu S, et al. Probing strigolactone perception mechanisms with rationally designed small-molecule agonists stimulating germination of root parasitic weeds. Nature Communications 2022;13(1):3987.
73. Chen Y, Kuang Y, Shi L, Wang X, Fu H, Yang S, et al. Synthesis and evaluation of new halogenated GR24 analogs as germination promotors for Orobanche cumana. Frontiers in Plant Science 2021;12:725949. Available from: https://www.frontiersin.org/articles/10.3389/fpls.2021.725949
74. Suzuki T, Kuruma M, Seto Y. A new series of strigolactone analogs derived from cinnamic acids as germination inducers for root parasitic plants. Frontiers in Plant Science 2022;13:843362. Available from: https://www.frontiersin.org/articles/10.3389/fpls.2022.843362
75. Zorrilla JG, Rial C, Varela RM, Molinillo JMG, Macias FA. Strategies for the synthesis of canonical, non-canonical and analogues of strigolactones, and evaluation of their parasitic weed germination activity. Phytochemistry Reviews. 2022:1–33. DOI: 10.1007/s11101-022-09801-8
76. Jamil M, Wang JY, Yonli D, Patil RH, Riyazaddin M, Gangashetty P, et al. A new formulation for strigolactone suicidal germination agents, towards successful Striga management. Plants 2022;11(6):808.
77. Zarban RA, Hameed UFS, Jamil M, Ota T, Wang JY, Arold ST, et al. Rational design of Striga hermonthica-specific seed germination inhibitors. Plant Physiology 2022;188(2):1369–84.
78. Arellano-Saab A, McErlean CSP, Lumba S, Savchenko A, Stogios PJ, McCourt P. A novel strigolactone receptor antagonist provides insights into the structural inhibition, conditioning, and germination

of the crop parasite Striga. Journal of Biological Chemistry. 2022;298(4):101734. Available from: https://www.jbc.org/article/S0021-9258(22)00174-0/abstract
79. Okazawa A, Baba A, Okano H, Tokunaga T, Nakaue T, Ogawa T, et al. Involvement of α-galactosidase OmAGAL2 in planteose hydrolysis during seed germination of Orobanche minor. Journal of Experimental Botany. 2022;73(7):1992–2004.
80. Bennett AE, Groten K. The costs and benefits of plant–arbuscular mycorrhizal fungal interactions. Annual Review of Plant Biology 2022;73(1):649–72.
81. Mitra D, Rad KV, Chaudhary P, Ruparelia J, Sagarika MS, Boutaj H, et al. Involvement of strigolactone hormone in root development, influence and interaction with mycorrhizal fungi in plant: Mini-review. Current Research in Microbial Sciences 2021;2:100026.
82. Mounde LG. Tripartite interaction between Striga spp., cereals, and plant root-associated microorganisms: A review. CAB Reviews Perspectives in Agriculture Veterinary Science Nutrition and Natural Resources. 2020;15(005):1–17. Available from: http://www.cabi.org/cabreviews/review/20203185495
83. Mishev K, Dobrev PI, Lacek J, Filepova R, Yuperlieva-Mateeva B, Kostadinova A, et al. Hormonomic changes driving the negative impact of broomrape on plant host interactions with arbuscular mycorrhizal fungi. International Journal of Molecular Sciences 2021;22(24):13677.
84. Sui X, Guan K, Chen Y, Xue R, Li A. A legume host benefits more from arbuscular mycorrhizal fungi than a grass host in the presence of a root hemiparasitic plant. Microorganisms. 2022;10(2):440.
85. Xi J, Lei B, Liu YX, Ding Z, Liu J, Xu T, et al. Microbial community roles and chemical mechanisms in the parasitic development of Orobanche cumana. iMeta. 2022;1(3):e31.
86. Gibot-Leclerc S, Guinchard L, Edel-Hermann V, Dessaint F, Cartry D, Reibel C, et al. Screening for potential mycoherbicides within the endophyte community of Phelipanche ramosa parasitizing tobacco. FEMS Microbiology Ecology 2022;98(3):fiac024.
87. Anteyi WO, Klaiber I, Rasche F. Diacetoxyscirpenol, a Fusarium exometabolite, prevents efficiently the incidence of the parasitic weed Striga hermonthica. BMC Plant Biology 2022;22(1):84.
88. He W, Li Y, Luo W, Zhou J, Zhao S, Xu J. Herbicidal secondary metabolites from Bacillus velezensis JTB8-2 against Orobanche aegyptiaca. AMB Express 2022;12(1):52.
89. Brun G, Spallek T, Simier P, Delavault P. Molecular actors of seed germination and haustoriogenesis in parasitic weeds. Plant Physiology. 2021;185(4):1270–81.
90. Bascos EMA, Fernando ES, Duya MV, Rodriguez LJV. Beginnings of a plant parasite: Early development of Rafflesia consueloae inside its Tetrastigma host. Planta 2021;254(3):61.
91. Pellissari LC, Teixeira-Costa L, Ceccantini G, Tamaio N, Cardoso LJ, Braga JM, et al. Parasitic plant, from inside out: Endophytic development in Lathrophytum peckoltii (Balanophoraceae) in host liana roots from tribe Paullineae (Sapindaceae). Annals of Botany. 2022;129(3):331–42.
92. Masumoto N, Suzuki Y, Cui S, Wakazaki M, Sato M, Kumaishi K, et al. Three-dimensional reconstructions of haustoria in two parasitic plant species in the Orobanchaceae. Plant Physiology 2021;185(4):1429–42.
93. Teixeira-Costa L. Leveraging micro-CT scanning to analyze parasitic plant-host interactions. JoVE Journal of Visualized Experiments 2022;(179):e63423.
94. Mylo MD, Hofmann M, Delp A, Scholz R, Walther F, Speck T, et al. Advances on the visualization of the internal structures of the european mistletoe: 3D reconstruction using microtomography. Frontiers in Plant Science 2021;12:715711. Available from: https://www.frontiersin.org/articles/10.3389/fpls.2021.715711
95. Teixeira-Costa L, Davis CC, Ceccantini G. Striking developmental convergence in angiosperm endoparasites. American Journal of Botany 2021;108(5):756–68.
96. Le Ru A, Ibarcq G, Boniface MC, Baussart A, Munos S, Chabaud M. Image analysis for the automatic phenotyping of Orobanche cumana tubercles on sunflower roots. Plant Methods 2021;17(1):80.
97. Fischer K, Lachner LAM, Olsen S, Mulisch M, Krause K. The enigma of interspecific plasmodesmata: Insight from parasitic plants. Frontiers in Plant Science 2021;12:641924. Available from: https://www.frontiersin.org/articles/10.3389/fpls.2021.641924
98. Banerjee A. Inter-plant communication via parasitic bridging. Journal of Experimental Botany 2020;71(3):749–50.
99. Kumar K, Amir R. The effect of a host on the primary metabolic profiling of cuscuta campestris' main organs, haustoria, stem and flower. Plants 2021;10(10):2098.

100. Lázaro-Gonzalez A, Gargallo-Garriga A, Hodar JA, Sardans J, Oravec M, Urban O, et al. Implications of mistletoe parasitism for the host metabolome: A new plant identity in the forest canopy. Plant, Cell & Environment 2021;44(11):3655–66.
101. Burns KC. Gender dimorphism in the virulence of a dioecious mistletoe. International Journal for Parasitology 2021;51(12):985–7.
102. Giesemann P, Gebauer G. Distinguishing carbon gains from photosynthesis and heterotrophy in C3-hemiparasite–C3-host pairs. Annals of Botany. 2022;129(6):647–56.
103. Chen BJW, Xu J, Wang X. Trophic transfer without biomagnification of cadmium in a soybean-dodder parasitic system. Plants. 2021;10(12):2690.
104. Piwowarczyk R, Ochmian I, Lachowicz S, Kapusta I, Malinowska K, Ruraż K. Correlational nutritional relationships and interactions between expansive holoparasite Orobanche laxissima and woody hosts on metalrich soils. Phytochemistry 2021;190:112844.
105. Albert M, Axtell MJ, Timko MP. Mechanisms of resistance and virulence in parasitic plant–host interactions. Plant Physiology 2021;185(4):1282–91.
106. Molina J, Nikolic D, Jeevarathanam JR, Abzalimov R, Park EJ, Pedales R, et al. Living with a giant, flowering parasite: Metabolic differences between Tetrastigma loheri Gagnep. (Vitaceae) shoots uninfected and infected with Rafflesia (Rafflesiaceae) and potential applications for propagation. Planta. 2021;255(1):4.
107. Greifenhagen A, Braunstein I, Pfannstiel J, Yoshida S, Shirasu K, Schaller A, et al. The Phtheirospermum japonicum isopentenyltransferase PjIPT1a regulates host cytokinin responses in Arabidopsis. The New Phytologist 2021;232(4):1582–90.
108. Su C, Liu H, Wafula EK, Honaas L, de Pamphilis CW, Timko MP. SHR4z, a novel decoy effector from the haustorium of the parasitic weed Striga gesnerioides, suppresses host plant immunity. The New Phytologist 2020;226(3):891–908.
109. Song J, Bian J, Xue N, Xu Y, Wu J. Inter-species mRNA transfer among green peach aphids, dodder parasites, and cucumber host plants. Plant Divers. 2022;44(1):1–10.
110. Calderwood A, Kopriva S, Morris RJ. Transcript abundance explains mRNA mobility data in Arabidopsis thaliana. The Plant Cell 2016;28(3):610–5.
111. Park SY, Shimizu K, Brown J, Aoki K, Westwood JH. Mobile host mRNAs are translated to protein in the associated parasitic plant Cuscuta campestris. Plants 2022;11(1):93.
112. Wu Y, Luo D, Fang L, Zhou Q, Liu W, Liu Z. Bidirectional lncRNA transfer between Cuscuta parasites and their host plant. International Journal of Molecular Sciences 2022;23(1):561.
113. Betti F, Ladera-Carmona MJ, Weits DA, Ferri G, Iacopino S, Novi G, et al. Exogenous miRNAs induce post-transcriptional gene silencing in plants. Nature Plants 2021;7(10):1379–88.
114. Loreti E, Perata P. Mobile plant microRNAs allow communication within and between organisms. New Phytologist. 2022;235(6):2176–82.
115. Gatica-Soria LM, Ceriotti LF, Garcia LE, Virginia Sanchez-Puerta M. Native and foreign mitochondrial and nuclear encoded proteins conform the OXPHOS complexes of a holoparasitic plant. Gene 2022;817:146176.
116. Anderson BM, Krause K, Petersen G. Mitochondrial genomes of two parasitic Cuscuta species lack clear evidence of horizontal gene transfer and retain unusually fragmented ccmFC genes. BMC Genomics 2021;22(1):816.
117. Lin Q, Banerjee A, Stefanović S. Mitochondrial phylogenomics of Cuscuta (Convolvulaceae) reveals a potentially functional horizontal gene transfer from the host. Genome Biology and Evolution 2022;14(6):evac091.
118. Kokla A, Leso M, Zhang X, Simura J, Serivichyaswat PT, Cui S, et al. Nitrogen represses haustoria formation through abscisic acid in the parasitic plant *Phtheirospermum japonicum*. Nature Communications 2022;13(1):2976.
119. Krupp A, Bertsch B, Spring O. Costunolide influences germ tube orientation in sunflower broomrape—A first step toward understanding chemotropism. Frontiers in Plant Science 2021;12:699068. Available from: https://www.frontiersin.org/articles/10.3389/fpls.2021.699068
120. Ogawa S, Cui S, White ARF, Nelson DC, Yoshida S, Shirasu K. Strigolactones are chemoattractants for host tropism in Orobanchaceae parasitic plants. Nature Communications 2022;13(1):4653.
121. Xiao TT, Kirschner GK, Kountche BA, Jamil M, Savina M, Lube V, et al. A PLETHORA/PIN-FORMED/auxin network mediates prehaustorium formation in the parasitic plant Striga hermonthica. Plant Physiology. 2022;189(4):2281–97.

122. Furuhashi K, Iwase K, Furuhashi T. Role of light and plant hormones in stem parasitic plant (Cuscuta and Cassytha) twining and haustoria induction. Photochemistry and Photobiology 2021;97(5):1054–62.
123. Bawin T, Bruckmuller J, Olsen S, Krause K. A host-free transcriptome for haustoriogenesis in Cuscuta campestris: Signature gene expression identifies markers of successive development stages. Physiologia Plantarum 2022;174(2):e13628.
124. Ibarra-Laclette E, Venancio-Rodriguez CA, Vasquez-Aguilar AA, Alonso-Sanchez AG, Perez-Torres CA, Villafan E, et al. Transcriptional basis for haustorium formation and host establishment in hemiparasitic Psittacanthus schiedeanus Mistletoes. Frontiers in Genetics 2022;13:929490. Available from: https://www.frontiersin.org/articles/10.3389/fgene.2022.929490
125. Gu X, Chen IG, Tsai CJ. How do holoparasitic plants exploit vitamin K1? Plant Signaling & Behavior 2021;16(11):1976546.
126. Jhu MY, Ichihashi Y, Farhi M, Wong C, Sinha NR. LATERAL ORGAN BOUNDARIES DOMAIN 25 functions as a key regulator of haustorium development in dodders. Plant Physiology. 2021;186(4):2093–110.
127. Jhu MY, Farhi M, Wang L, Zumstein K, Sinha NR. Investigating host and parasitic plant interaction by tissue-specific gene analyses on tomato and cuscuta campestris interface at three haustorial developmental stages. Frontiers in Plant Science 2022;12:764843. Available from: https://www.frontiersin.org/articles/10.3389/fpls.2021.764843
128. Parker C. A personal history in parasitic weeds and their control. Plants. 2021;10(11):2249.
129. Masteling R, Lombard L, de Boer W, Raaijmakers JM, Dini-Andreote F. Harnessing the microbiome to control plant parasitic weeds. Current Opinion in Microbiology 2019;49:26–33.
130. Atsmon G, Nehurai O, Kizel F, Eizenberg H, Nisim Lati R. Hyperspectral imaging facilitates early detection of Orobanche cumana below-ground parasitism on sunflower under field conditions. Computers and Electronics in Agriculture 2022;196:106881.
131. Fatino MJ, Hanson BD. Evaluating branched broomrape (*Phelipanche ramosa*) management strategies in california processing tomato (*Solanum lycopersicum*). Plants. 2022;11(3):438.
132. Bhat KA, Akhtar S, Dar NA, Bhat MI, Bhat FA, Rizwan R, et al. Mistletoe eradicator—A novel tool for simultaneous mechanical and chemical control of mistletoe. Journal of Visualized Experiments 2022;(181):e63455.
133. Scott D, Scholes JD, Randrianjafizanaka MT, Randriamampianina JA, Autfray P, Freckleton RP. Identifying existing management practices in the control of Striga asiatica within rice–maize systems in mid-west Madagascar. Ecology and Evolution 2021;11(19):13579–92.
134. Zagorchev L, Stoggl W, Teofanova D, Li J, Kranner I. Plant parasites under pressure: Effects of abiotic stress on the interactions between parasitic plants and their hosts. International Journal of Molecular Sciences 2021;22(14):7418.
135. 135. Ossa CG, Aros-Mualin D, Mujica MI, Perez F. The physiological effect of a holoparasite over a cactus along an environmental gradient. Frontiers in Plant Science 2021;12:763446. Available from: https://www.frontiersin. org/articles/10.3389/fpls.2021.763446
136. Zhang L, Cao X, Yao Z, Dong X, Chen M, Xiao L, et al. Identification of risk areas for Orobanche cumana and Phelipanche aegyptiaca in China, based on the major host plant and CMIP6 climate scenarios. Ecology and Evolution 2022;12(4):e8824.

3 The invasive hornet *Vespa velutina*: distribution, impacts and management options

S. Lioy[1,*], C. Bergamino[2] and M. Porporato[1]

[1]Department of Agricultural, Forest and Food Sciences, University of Turin, Largo Paolo Braccini 2, 10095 Grugliasco, Turin, Italy; [2]Department of Veterinary Sciences, University of Turin, Largo Paolo Braccini 2, 10095 Grugliasco, Turin, Italy

Abstract

The Asian yellow-legged hornet *Vespa velutina* is an invasive alien species introduced and widespread in several countries of Europe and Asia. Its diffusion generates relevant environmental and socio-economic impacts. Environmental impacts include threats to the native insect biodiversity and the pollination ecosystem services. Socio-economic impacts include threats to the apiculture sector, economic consequences for the adoption of management strategies, social concern and health issues. Different options were developed and adopted for (i) preventing the introduction of *V. velutina*, (ii) early detecting its presence, (iii) eradicating populations at the initial stage of invasion or (iv) controlling populations for limiting and mitigating its impacts. The aim of this review was to provide an updated overview about the distribution, impacts and options for managing *V. velutina* populations, through a literature review of the published academic documents. Moreover, this study highlights that some topics received little attention (impacts of *V. velutina* on the biodiversity, on the pollination ecosystem services, on the economy) or require further research efforts (effective control methods for *V. velutina*); therefore, future research should be directed towards filling these gaps of knowledge.

Keywords: yellow-legged hornet, wasps, spread, monitoring, eradication, control, pests

Introduction

Invasive alien species are a serious threat to biodiversity and human activities [1] whose accumulation in the last centuries shows no sign of saturation for most of the introduced taxa [2]. Among insects, social wasps and hornets proved to be successful invaders, with 34 species introduced worldwide until 2010 and, among these, at least seven recorded as invasive species [3]. One of the introduced invasive species is *Vespa velutina* Lepeletier, 1836 (Hymenoptera: Vespidae), a hornet introduced in Europe, South Korea and Japan whose spread poses concern

*Corresponding author: simone.lioy@unito.it

for the conservation of insect biodiversity, pollination ecosystem services as well as human activities, economies and health [4, 5]. From its introduction onwards, the knowledge about the species continually increased, with yearly new information on the biology, ecology, population dynamics and impacts of the species as well as on the techniques for monitoring its presence or controlling its spread (Fig. 3.1) [6, 7]. Therefore, the aim of this review is to provide an up-to-date summary of the knowledge about the distribution of *V. velutina*, its impacts, and the options for limiting its spread, with a brief insight into its biology and life cycle.

Review Methodology

The literature review was performed by searching a defined combination of words in the title, abstract and keywords of the documents indexed in three online databases (Scopus, Web of Science and CAB Direct). The searching string adopted for this purpose considered the following keywords: *Vespa velutina*, *Vespa velutina nigrithorax*, Asian yellow-legged hornet, Asian hornet. An overall number of 328 unique documents were found in the period 1975–2021, both considering indexed documents and references cited within them. Of these, 28 documents were excluded since unconnected to the species of interest. The remaining 300 documents were analysed for evaluating the covered research topics and their spatial distribution at the global level. For this purpose, each record was classified on the basis of the continent in which the research was conducted and the covered topics: (1) distribution, (2) Species Distribution Models or spread modalities, (3) impacts on biodiversity, (4) impacts on ecosystem services, (5) impacts on apiculture, (6) impacts on human health, (7) impacts on the economy, (8) positive effects, (9) monitoring, (10) eradication, (11) control, (12) genetic, (13) viruses and parasites, (14) biology, (15) taxonomy, (16) reviews on the species and (17) other topics not covered by the previous categories (Fig. 3.1). Of the analysed documents, 19 were in languages other than English and 30 did not have accessible full text. Finally, based on their relevance, 140 documents were selected for the preparation of this review.

Biology and life cycle

Morphological identification

The subspecies introduced in Europe and other countries of Asia where the species is not native, *Vespa velutina nigrithorax* (du Buysson, 1905) [8–11], is characterised by a dark-brown almost black thorax. The abdomen of the females, which is composed of six apparent tergites that correspond to the II-VII gastral terga (the first morphological segment, the propodeum, is fused with the thorax), is characterised by three tergites almost black with a yellow margin at the end (II-IV) and a tergum (V) almost yellow. The legs are dark, except for the tarsi that are yellow. The dorsal and ventral parts of the antennae are respectively black and dark-brown (Fig. 3.2a) [12, 13]. The absence of the sting and longer antennae allows to discriminate males from workers and queens. Although individual weight increases throughout the year, body weight provides an indication for caste differentiation between workers (mean ± SD of wet weight ranges from 188.8 ± 44.9 mg in June to 386.4 ± 88.3 mg in November) from queens (from 624 ± 15.2 mg in September to 721.3 ± 73.2 mg in November) [14].

Life cycle

As for the other social wasps, the life cycle of a colony of *V. velutina* could be divided into four different phases: the foundation, the worker, the reproductive and the intermediate phase [13]. During spring, mated queens that survive the winter begin the foundation of the initial nest, of small size, composed of a single comb and used by the queen to layer the eggs of the first workers [4, 5]. The length of the development stages was estimated in 48.1 days (13 days for the eggs, 15.8 days for the larvae and 19.3 days for the sealed brood) [13]. During the worker phase (Fig. 3.2b), the emerged workers will then contribute to the development of the colony, by enlarging the nest, recovering proteins for feeding the brood, protecting the colony from enemies, and contributing to cleaning and ventilation [4, 5, 15]. The reproductive phase starts with the emergence of males that generally

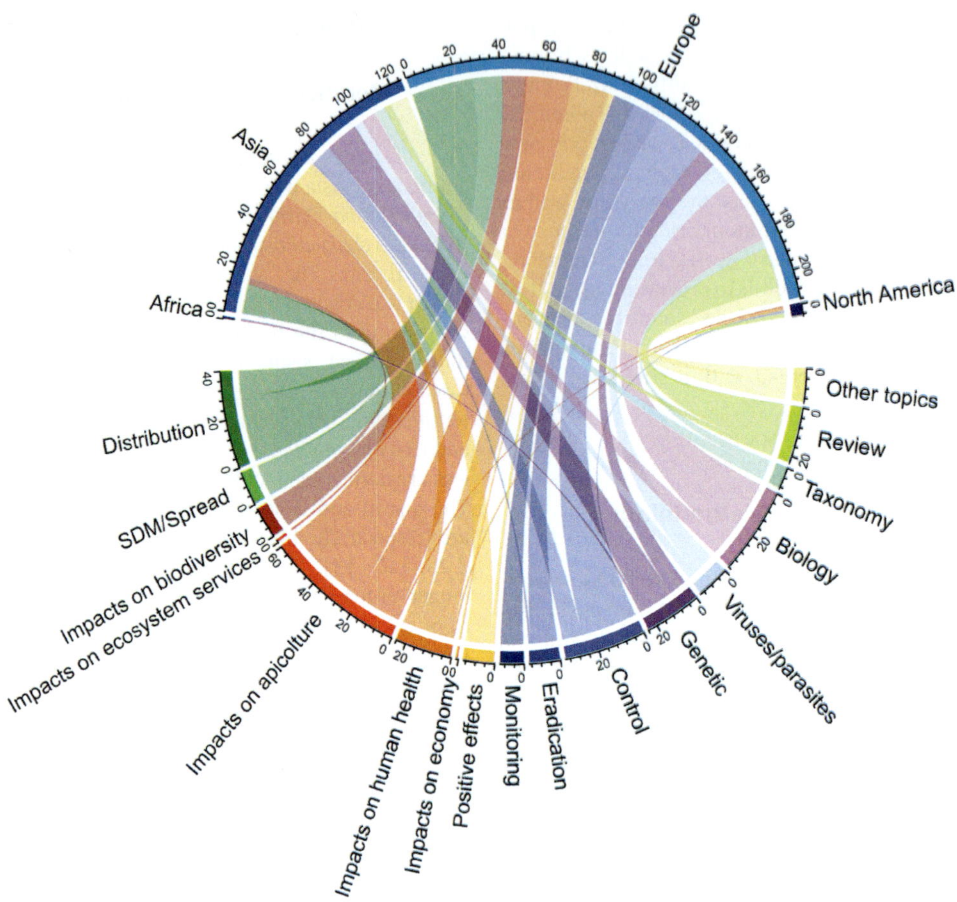

Fig. 3.1. Research topics concerning *V. velutina* covered in the literature in the period 1975–2021 in relation to the continent in which the research was conducted.

occurs from early September on, about 15 days before the emergence of gynes (potential reproductive females). Reproductive castes reached their maximum numbers in November (mean number of gynes: 191 ± 195.7), and most of them left the colony before the end of this month [14]. During winter, the colony gradually collapses, although in Spain nests recovered in January still contained live specimens including gynes [16]. Winter active colonies were also observed in Italy (Lioy S., pers. comm.).

The dimensions of the colony vary throughout the year: the number of combs increases up to a mean of eight combs in December; and the total number of individuals produced in a year by the colonies increases up to a mean of 6151 individuals in November, with a maximum value of 13,340 individuals. The initial nest built by the queen could be relocated by the workers during the annual life cycle of the colony. A study conducted in France observed the shape and dimension of the first comb in 49 nests of *V. velutina*, and about 70% were considered relocated colonies [14]. When a colony is relocated, it is commonly named secondary nest, to distinguish it from the primary nest built by the queen and enlarged by the workers. The distinction between primary and secondary nest could only be performed by the internal morphological analysis of the structure of the nest, and not from the size of the nest or the time of the year. Indeed, primary nests are characterised by a first

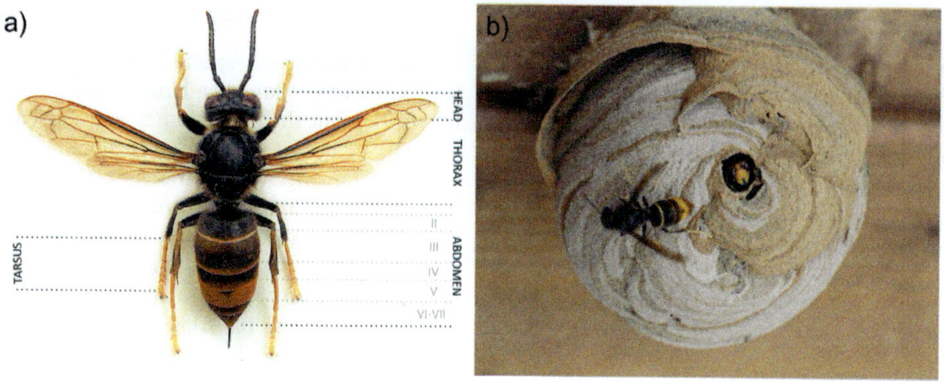

Fig. 3.2. (a) Female specimen of *V. velutina nigrithorax* from Italy, with an indication of the main morphological parts that allow species identification (photo by S. Lioy). (b) Nest of the species observed in June (photo by M. Porporato).

comb of irregular structure, with the embryo comb still visible at its centre, while secondary nests by a first comb of regular shape [14].

Distribution

Native range

Vespa velutina is native to south-east Asia and occurs in Afghanistan, Bangladesh, Bhutan, China, Hong Kong, India, Indonesia, Laos, Malaysia, Myanmar, Nepal, Pakistan, Taiwan, Thailand and Vietnam (Fig. 3.3) [13, 18–23]. The species presents 12 colour subspecies across its native range, four on the mainland (*auraria*, *variana*, *divergens* and *nigrithorax*, the latter introduced in Europe, South Korea and Japan) and eight on islands (*flavitarsus*, *karyni*, *velutina*, *ardens*, *sumbana*, *floresiana*, *timorensis* and *celebensis*) [13]. Some authors [19, 24] recognised the subspecies *pruthii* as distinct subspecies, while others suggested to treat it as a synonym of *auraria* due to their minor colour differences and their range overlap [13].

Alien range in Europe

V. velutina was accidentally introduced to Europe in 2004, probably through boat transport of horticultural products from an area between the Chinese provinces of Zhejiang and Jiangsu [4, 9, 25]. Following the introduction, which occurred in south-west France, the species spread across various European countries, with the French population acting as a source of diffusion or a source for new introductions [26]. The observed spread of *V. velutina* in Europe was impressively fast, probably due to a diffusion process that encompassed natural dispersal and human-mediated transportation [27–30], which in some cases (i.e. Portugal) has been correlated with the motorway traffic [31]. Hereafter, a summary of the evolution of *V. velutina* colonisation across European countries.

France

Detected for the first time in the department of Lot-et-Garonne in 2004 [4, 32], in the following years the species started to spread in the neighbouring departments [33] colonising an area of 190,000 km^2 by 2010 [34], 360,000 km^2 by 2012 [35] and mostly all the country by 2017 [29]. The estimated spread rate observed in France was 78 km/year [29]. In an area of western France (Andernos-les-Bains), nest density increased in the period 2007–2014 up to 4.81 nests/km^2 (considering both urban and non-urban areas) [36], and up to 10.23–12.26 nests/km^2 considering only urban areas [36, 37].

Spain

Detected for the first time in 2010 in an area close to the border with France, in Guipuzcoa and Navarra provinces [38–40], the species continued to spread westwards and eastwards until reaching in 2012 Galicia [41] and Catalonia, respectively [42]. *V. velutina* was also observed in the Balearic Island of Mallorca in 2015 [43], but here the species has been declared eradicated after some years of management [44].

Portugal

Detected in the Minho province in 2011 [45], in the following years *V. velutina* expanded its range southwards at an estimated spread rate of 37–45 km/year [31, 46], and by 2020, 62% of the mainland Portugal was colonised [46]. Urban nest density was estimated at the local scale in 5.4 ± 3.3 nests/km^2 [46].

Belgium

The species was firstly observed in 2011 in Wallonia [47] but the first nests were reported only in 2016 and 2017 in Wallonia and Flanders, respectively. In the following years, the population spread rapidly from west to east [48].

Italy

V. velutina was firstly spotted in 2012 in the Liguria region, in an area close to the motorway [49], while the first colonies were detected in 2013 in both Liguria and Piedmont [50]. In the following years, the species spread along the coastline of Liguria, increasing the colonised area from 205 km^2 in 2013 to 930 km^2 in 2015 at an estimated spread rate of 18.3 ± 3.3 km/year, reaching densities of 2.9–3.5 nests/km^2 (considering both urban and woodland areas) [27]. In 2016 and 2017, the species was observed in two areas at several hundred kilometres from the main invaded range, respectively in Veneto and in an area between Tuscany and eastern Liguria, but only in the latter case *V. velutina* established viable populations [5]. In years 2020–2021 a new outbreak was also reported in Lombardy (Lioy S., pers. comm.). Genetic studies confirmed that Italian populations most likely derived from the eastern spread of the species from France [51].

Germany

The species started to colonise the south-western part of Germany in 2014 [52]. In 2019, *V. velutina* was observed in Hamburg, several hundred kilometres of distance from other colonised areas [53]. The introduction in

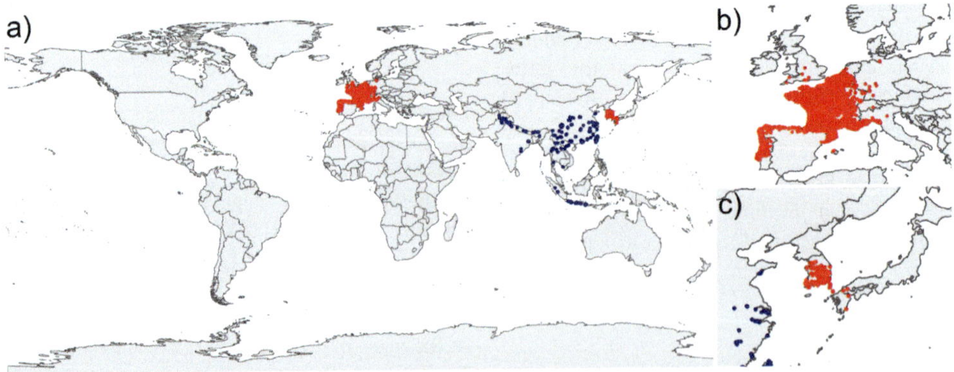

Fig. 3.3. (a) Worldwide observations of *V. velutina* in its native range (blue) and in the areas of introduction (red). (b) Details of the alien range in Europe. (c) Details of the alien range in Asia. Records of *V. velutina* were obtained by aggregating data collected by the authors of this review with observations derived from GBIF [17] and from the literature.

northern Germany probably occurred by the long-distance transportation of the species from the European population rather than a new introduction from Asia [54].

UK

V. velutina was firstly observed in 2016 in Tetbury [55], and in the following years, few nests were periodically located in various localities in southern England [56]. Based on their geographic distribution and on microsatellite analyses, it is supposed that all UK colonies found up to 2019 were derived from separate introductions from the European population, since none of the discovered nests were directly descended from nests of the previous years found in the UK [56]. The species has also been present in the Channel Islands since 2016, due to the natural spread or the introduction of the species from France [11].

Other countries in Europe

Recently, *V. velutina* continued to spread in Europe and it was observed in other countries such as the Netherlands, Switzerland [5], and Luxemburg [57]. Moreover, a single specimen was reported from Ireland [58].

Alien range in Asia

Besides the European alien range, *V. velutina* has established expanding population also in two other Asian countries.

South Korea

The species was firstly discovered in 2003 in the city of Busan [12]. From the site of introduction, the species spread northwards at a rate of 10–20 km per year [59]. Independent introductions probably helped *V. velutina* to spread into South Korea [60].

Japan

V. velutina was spotted for the first time in 2012 on Tsushima Island [61, 62], and afterwards on Kyushu Island in 2015 [63] and Iki Island in 2017 [11, 64]. The genetic similarity of the individuals from the Japanese and South Korean populations suggests that *V. velutina* in Japan originated from a second event of introduction from the population in South Korea [10, 11].

Risk of invasion in other countries

After its introduction and initial spread, several studies investigated the potential invasion risk of the species, mostly based on the assessment of the climatic suitability [25, 65] or by a combination of multiple factors (e.g. climatic and habitat variables) [66, 67]. Most of the continents are characterised by temperate and subtropical areas with suitable climatic conditions for *V. velutina*, including the eastern part of North America, Europe, East Asia, and small areas in South America, South Africa and Australia [25, 65]. In Europe, many countries along the northern Mediterranean and Atlantic coasts, including wide areas of the UK [68] as well as coastal areas of the Balkan Peninsula and Turkey, exhibited a high probability of being invaded [69], and the observed spread in years 2011–2015 confirms the accuracy of the predictions made with data from the early stage of the invasion [70]. Furthermore, climate change could contribute to further increase the invasion risk, particularly in the northern hemisphere [71].

Impacts

Environmental impacts

The presence of *Vespa velutina* outside its native range is posing a concern due to the multiple impacts that the species could generate on native biodiversity. This could occur by the combination of different mechanisms: predation, competition and alteration of the pollination ecosystem service [4, 5].

The species preys upon other insects for feeding the brood, acting as a generalist predator [15, 34, 72, 73]. The analysis of its diet in an invaded area of France revealed that a single colony could prey upon at least 159 different species (11 orders and 43 families of insects, three families of spiders and four families of vertebrates), consuming about 11.32 kg of insect biomass in one season. Among insects, the mostly consumed preys are Hymenoptera (60.1%), particularly honey bees (38.1%) and social wasps (19.7%), followed by Diptera (29.9%) and other less frequent groups such as Lepidoptera, Hemiptera, Coleoptera, Mecoptera and Orthoptera [72]. The food spectrum could differ in relation to the habitat that surround *V. velutina* colonies, with a higher proportion of social wasps and flies in woodlands (respectively 28.3% and 31.5%) than in urban areas (7.8% and 17.2%), and a higher presence of bees in the latter (66.6% in urban areas and 33.4% in woodlands) [34].

Another impact mechanism to be considered is the competition, which may occur between *V. velutina* and those species with a similar ecological niche, namely, other wasps and hornets. Different mechanisms may generate competition, such as the overlap in the trophic niche between the invasive and the native species [74], the overlap in the seasonal phenology [75], or higher performance of the invasive species in terms of reproductive potential [76], boldness/exploration scores [77] and aggressiveness scores over smaller species [78]. Reproductive interference has also been observed in Japan between *V. velutina* and the native *V. simillima*, with 43% of native queen species inseminated by males of the invasive species and, of these, 28% having only sperms of *V. velutina* [79]. Moreover, a negative correlation in the abundances of *V. velutina* and three other native *Vespa* species of Japan (*V. simillima xanthoptera*, *V. analis* and *V. mandarinia japonica*) suggest a possible negative effect posed by the presence of the invasive species due to competition and/or predation [80].

Despite its importance, the impact of *V. velutina* on pollination has rarely been analysed. An observational study conducted in Spain demonstrated, for the first time, that the presence of the invasive hornet could induce changes in the foraging behaviour of several groups of pollinators (honey bees, bumblebees, small hymenopterans and syrphids) due to its preying behaviour on flowers, thus affecting the abundance of pollinators and their flower visitation frequency. This could generate an indirect effect on the quantity of pollen deposited on stigmas of a wild plant, *Mentha suaveolens* [81].

Socio-economic impacts

Besides the environmental issues, the presence of *V. velutina* generates several not neglectable socio-economic impacts such as consequences for the beekeeping sector, management costs and health issues.

Honey bees are among the most insects preyed by *V. velutina* [15, 34, 72, 73]. This could be related to the olfactory attraction produced by honey bee colony odours (pollen and honey) and pheromones (geraniol, the honey bee aggregation pheromone) that may signal a high density of prey [82] or due to a combination of visual and olfactory cues that attract the species [83]. *V. velutina* prey on honey bees mostly by hovering in front of the hive for catching the returning bees [84]. Once a honey bee is caught, the hornet removes all the less protein portions of the insect, keeping only the thorax that is brought back to the colony for feeding the larvae [15]. Individuals of *V. velutina* are able to revisit the apiary more time during the day, suggesting a learning behaviour that could benefit the foraging performance [85]. Predation pressure on honey bees increases during the season (in France from July to October) according to the increase in the population of *V. velutina* colonies [84]. Moreover, predation pressure could vary in relation to environmental conditions such as temperature, humidity or wind speed [86]. The predation success increases in relation to the number of chasing hornets, with peaks around nine hornets per hive [84]. The overall observed predation success in South Korea was 80%, with peaks up to 96% in September [87]. This intensive predation activity is a source of oxidative stress in honey bees [88] and generates foraging paralysis that, in turn, could increase the homing failure of returning honey bees due to higher predation success of the hornets [89]. With time, the predation pressure could generate the collapse of honey bee colonies due to a reduction of the population and of the stocks of honey and pollen used by the honey bees for surviving the winter [5].

Unlike colonies of *Apis cerana*, which occur in sympatry with *V. velutina*, colonies of *Apis mellifera* are more susceptible to its predation pressure due to (i) a lower performance or absence of defensive behaviour, such as the heat balling

[90, 91] and the wing shimmering [92, 93], (ii) the inability to eavesdrop the alarm pheromone of the hornet [94, 95], (iii) the lower number of guard bees at hive entrance [92] and (iv) the lower speed rate adopted by *A. mellifera* for entering inside the hive [92]. In fact, the hawking success of *V. velutina* is three times higher for *A. mellifera* foragers than that for *A. cerana* [92]. Concerning defensive behaviour, several studies have investigated the heat balling behaviour, in which honey bee workers respond to hornets' attacks by forming a ball of bees around the hornet with which they manage to kill the threat owing to the increase in both core-temperature and carbon dioxide concentration [90, 91]. Workers of *A. cerana* are better reacting with the heat balling behaviour to *V. velutina* attacks, due to a greater number of individuals engaged in this behaviour (32.2 ± 3.2 workers for *A. cerana* and 22.7 ± 3.1 workers for *A. mellifera*) and a higher core temperature reached by the ball (45.6 ± 0.1°C for *A. cerana* and 44.3 ± 0.2°C for *A. mellifera*) [90].

Besides the direct impacts on the beekeeping sector, the presence of *V. velutina* in invaded countries also generates massive economic costs for developing management practices, particularly those related to the implementation of nest destruction activities. It has been estimated that control activities performed in France between the years 2006–2015 cost about €23 million and, if the species continues to spread colonising all the areas predicted to be suitable, yearly cost could reach €29.5 million in three invaded countries of Europe (France, Italy and UK) and €31.4 million in two invaded countries of Asia (South Korea and Japan) [96].

As in the other eusocial wasps, workers of *V. velutina* tend to defend their colony in case the insects feel threatened, becoming a potential source of danger for people. Defensive behaviours could occur when people approach within three metres from the nest and could last as long as the distance is not increased to a few tens of metres, despite in some cases, few hornets could still attack up to 300 m [97]. Defensive behaviours are more frequent towards aggressors dressed in dark colours [98]. After stings, a series of reactions of different severity could occur, from allergic to toxic reactions that could lead to anaphylaxis and multiple organ failures [99–101] that in some cases could lead to fatal events [102, 103]. Moreover, ocular lesions could occur due to the projection of venom into the eyes of people: 29 cases were reported in France until 2019, and 80% of these were associated with occupational exposure due to nest destruction activities [104]. However, determining whether the presence of *V. velutina* causes a general increase in Hymenoptera stings is not always straightforward, also because in most cases, it is impossible to determinate the responsible species. A review of the data registered in the French Poison Control Centers highlighted the absence of a correlation between the increase in *V. velutina* populations and the number of Hymenoptera stings during the initial phase of invasion [105]. However, a study conducted in Spain reported that *V. velutina* has become one of the common cause of anaphylaxis [103], while an analysis of cases and emergency calls in South Korea demonstrated that *V. velutina* became one of the most prevalent sources of issues as the species was more widespread in the country and abundant in urban areas [106]. Furthermore, the global increase in temperature due to climate change could increase the areas of suitability for *V. velutina* [71], increasing thus the risks for humans in relation to a greater diffusion and abundance of Hymenoptera in the environment [107].

Potential positive effects

Despite the many negative impacts associated with the presence of *V. velutina*, some studies highlighted that the species might generate benefits or positive effects (Fig. 3.1). For example, studies conducted in the native range showed that the species could contribute to pollination [108, 109] and seed dispersal [110, 111], or could be exploited as a food resource by humans [112, 113]. Recent studies from the invaded range highlighted that the abundance of *V. velutina* might be considered as a promising alternative source of chitin [114], while the antimicrobial proprieties of their nests might be used against pathogenic bacteria [115]. Moreover, *V. velutina* venom could be extracted and used for several purposes. For example, it can be used for treating allergy and/or anaphylaxis through immunotherapy [116], for obtaining peptides for the development of new antioxidants and antimicrobial agents [117, 118], or for obtaining anti-inflammatory compounds [119].

Management options

Tackling the spread of invasive alien species requires the adoption of different strategies based on the stage of invasion and the objectives to be achieved. Generally, the options to be considered for this purpose could be divided into four categories: (i) prevention, (ii) early detection, (iii) eradication and (iv) control [120]. These options are also suggested for managing *Vespa velutina* introductions and spread [5].

Prevention

Preventing new introductions of *V. velutina* queens is hardly feasible due to the number of drivers and pathways with which the species could be transported, such as contaminant of plants, wood, nursery or habitat material, or as hitchhikers on containers, boats or vehicles [121, 122]. However, including *V. velutina* in regulation's "black lists" as occurred in Europe (where the species is designated as invasive alien species of Union concern in all European countries; EU No 1143/2014, Reg. EU No 1141/2016) or in Japan (designated under the invasive alien species act) [64] could help to prevent new introductions of the species besides to adopt eradication/control measures. In fact, listing *V. velutina* as a species capable of generating massive impacts foresees the adoption of transportation restrictions, and action plans on the pathways of introductions and other related measures for raising awareness in people and stakeholders that may contribute to prevent new introductions.

Early Detection

In countries or areas not yet colonised by *V. velutina*, surveillance strategies could contribute to early detecting the arrival of the species for attempt an eradication of the populations at early stages of invasions. For example, a monitoring strategy was established in Italy in 2007, several years earlier than the appearance of the species in the country, allowing to intercept a first specimen in 2012 [49] and detecting the first hornets and nests in the following year [50]. Surveillance strategies mostly adopted for monitoring the presence of *V. velutina* encompass (i) citizen science monitoring schemes and (ii) the engagement of relevant stakeholders such as the beekeepers, for monitoring the presence of *V. velutina* through monitoring traps and observations in apiaries [5, 7].

Citizen science projects were specifically developed for monitoring *V. velutina* in several countries worldwide [7] including, as nonexhaustive examples, Belgium [48], France [67], Italy [30], Luxembourg [57], Portugal [46, 123], Spain [44] and the UK [124]. The collection of reports through citizens could be based on website solutions, smartphone applications or hybrid systems [7]. Moreover, citizen science monitoring schemes could focus on detecting several wasp species simultaneously, such as the system developed in Sardinia (Italy) for monitoring the spread of *V. crabro* and detecting the arrival of *V. velutina* [125]. Additionally, the engagement of non-experts does not negatively affect the reliability of the collected data [123].

Beekeepers and beekeeper associations have been engaged in monitoring the presence of *V. velutina* mostly by organising local networks by using monitoring traps [49, 126, 127] or by reporting observations of nests and adults in the environment and near apiaries [5], the latter in relation to the attractiveness exerted by honey bees [82, 83]. Monitoring traps, either commercial or handmade, are nowadays unselective for *V. velutina* since the target species represent about 1% of all trapped insects [126, 127]. Although automatic detection systems are not widely adopted, the direct observation of hornets in apiaries could eventually be supported by automatic systems for the recognition of the species [128], particularly at the early stage of invasion when the number of hornets preying on honey bees is restrained, which implies a low probability of observing the species by the beekeepers.

Finally, rapid molecular methods have been developed for the in-field and laboratory identification of the species [129]. These tools would be useful for the determination of the species from damaged/incomplete specimens or immature life stages when morphological analysis is not feasible.

Eradication

Once locally established in a new area, *V. velutina* could be eradicated by detecting and removing all the present nests, ideally before the reproductive phase of the colonies. The observation of multiple hornets preying in apiaries or feeding on flowers is an indicator of the presence of a colony, most likely within 1 km of distance [130]. Once discovered, colonies of *V. velutina* could be dissected and analysed for assessing their reproductive output and understanding if the next generation of queens was produced and dispersed into the surrounding environment [56]. Follow-up monitoring in the subsequent years should then be performed for confirm eradication, by establishing monitoring areas in the environment around the removed colonies with an extension that considers the dispersal capability of the species [30]. Successful examples of eradication are the cases from Mallorca Island (Spain) [44] and the different incursions managed in the UK [56].

Different techniques for improving colony detectability could be adopted for eradication. Techniques could be divided into (i) methods for tracking the flight of *V. velutina* foragers back to their colonies and (ii) methods for spotting *V. velutina* nests. Effective tracking methods currently adopted for tracking *V. velutina* encompass the following techniques.

- Harmonic radar tracking: hornets are trapped in apiaries, equipped with a passive (without battery) lightweight transponder (~15 mg) and tracked with a radar system [131–133] either in flat or complex environments characterised by a different degree of obstacles and slopes [134].
- Radio tracking: hornets are trapped, equipped with an active transmitter (~150–312 mg) and tracked manually in complex environments with a tracking receiver [135].
- Visual tracking and triangulation of flying directions: hornets are attracted by multiple baited food stations, followed on sight, and finally, the colonies are discovered by triangulating the flying directions [44].

A network of unmanned aerial vehicles (UAVs) has also been proposed as a tracking method for tagged hornets [136] but, to the best of our knowledge, experimental confirmations are currently lacking.

In addition to the tracking methods, other techniques for improving the probabilities of spotting *V. velutina* colonies could be adopted, particularly when nests are covered or masked by leaves. These techniques rely on the use of thermal imaging cameras to detect nest position, either with an operator acting from the ground [137] or with cameras mounted on drones for an aerial survey [138].

Control

Since no control method, taken individually, allows an effective control of the species, the control of *V. velutina* populations currently relies on a series of measures to limit or mitigate its environmental and socio-economic impacts [5, 64]. Due to their eusociality, an effective control of social insects should mainly target the colony rather than individuals [64]. For this reason, despite the associated costs [96], long-term nest destruction strategies are one of the most adopted techniques for both limiting the impacts of *V. velutina* and providing an answer to the requests of citizens asking for nests' removal [5, 37, 139]. Nest destruction is commonly performed by introducing insecticides, such as pyrethroids, directly inside the nests [5, 64, 139]. Steam injection has also been proposed as an alternative green control method for destroying *V. velutina* colonies [140].

A complementary method to nest destruction, often adopted in the invaded areas for reducing the number of individuals, is the use of traps for trapping queens or other castes [139]. Spring queen trapping is one of the most popular adopted techniques; however, its effectiveness is questionable since (i) few surviving queens can establish colonies able to produce hundreds of new reproductive individuals [37] and (ii) the available traps are not selective (on average, the number of *V. velutina* trapped is 1% of the overall number of insects caught in the traps). Trapping workers also appear ineffective since colonies are composed by several hundreds of individuals that could compensate for the losses [64]. Moreover, several studies have suggested possible side effects of trapping on native entomofauna, also toward those species or groups that are less affected by *V. velutina* predation, such as *V. crabro*, Lepidoptera or Coleoptera. Therefore, further research is requested for improving trapping methods [37, 126, 127, 141, 142].

Toxic baits are among the methods examined for the control of *V. velutina* [64, 139]. The approach relies on the behaviour of workers to collect proteins to feed the larvae, a behaviour that can be exploited to convey toxic substances to the colonies. However, the use of insecticides or insect growth regulators (IGRs) severely limits their use due to the potential negative effects on non-target species, the potential dissemination of substances in the environment (since nest position is unknown), together with legislation restrictions and the lack of specific information on their application to *V. velutina* [64, 139].

Different methods have also been developed by the beekeepers and adopted to mitigate the impact on honey bees, such as hive muzzles [139, 143, 144] and electric harps [139]. Muzzles prevent the hornets from getting close to the hives, reducing the stress of bees and their foraging paralysis [144, 145]. Electric harps are capable of electrocuting hornets without affecting the bees [139].

Other potential control methods have been suggested in the literature, such as the biocontrol through predators or parasites. These methods currently appear unsuitable for a large-scale control of *V. velutina* and therefore not covered within this review.

Conclusions

This review provides an update on the distribution, impacts and options currently available for managing *V. velutina* populations in the invaded countries. The observed spread from the early 2000s demonstrates that the species could rapidly increase its range, helped by multiple introductions due to human-mediated transportation. In the invaded range, *V. velutina* could generate massive impacts on the environment (biodiversity and pollination service) and the society (beekeepers, public administrations, public health). Several options and strategies could be adopted to prevent or mitigate *V. velutina* effects: including *V. velutina* in actionable lists (as already performed in Europe and other invaded countries of Asia); early detect new introductions by engaging citizens and stakeholders (i.e. beekeepers) in monitoring *V. velutina* presence; attempt eradication at the early stage of invasion, thanks also to the new technologies developed for locating nests; control *V. velutina* populations preferring strategies for targeting the colonies (e.g. nest removal). Additionally, the literature review highlighted that some topics received little attention or required further research efforts. For example, the impacts of *V. velutina* on the native entomofauna, on the pollination ecosystem services, and on the economy were investigated to a lesser extent than the impacts towards honey bees. Moreover, no control method, taken individually, allows to effectively interrupt the spread and impact of the species; therefore, further research is still needed to fill these gaps of knowledge.

References

1. Early R, Bradley BA, Dukes JS, Lawler JJ, Olden JD, Blumenthal DM, et al. Global threats from invasive alien species in the twenty-first century and national response capacities. Nature Communications 2016;7(1):12485.
2. Seebens H, Blackburn TM, Dyer EE, Genovesi P, Hulme PE, Jeschke JM, et al. No saturation in the accumulation of alien species worldwide. Nature Communications 2017;8(1):14435.
3. Beggs JR, Brockerhoff EG, Corley JC, Kenis M, Masciocchi M, Muller F, et al. Ecological effects and management of invasive alien Vespidae. BioControl 2011;56(4):505–26.
4. Monceau K, Bonnard O, Thiéry D. *Vespa velutina*: A new invasive predator of honeybees in Europe. Journal of Pest Science 2014;87(1):1–16.
5. Laurino D, Lioy S, Carisio L, Manino A, Porporato M. *Vespa velutina*: An Alien driver of honey bee colony losses. Diversity 2020;12(1):5.

6. Do Y, Kim JB, Shim J, Kim C, Kwon O, Choi MB. Quantitative analysis of research topics and public concern on *V. velutina* as invasive species in Asian and European countries. Entomological Research 2019;49(10):456–61.
7. Laurino D, Gajger IT, Lioy S, Porporato M. COLOSS task force to investigate and *reduce Vespa velutina* impacts and spread. Bee World 2022;99(1):26–28.
8. Choi MB, Lee SA, Suk HY, Lee JW. Microsatellite variation in colonizing populations of yellow-legged Asian hornet, *Vespa velutina nigrithorax* in South Korea. Entomological Research 2013;43(4):208–14.
9. Arca M, Mougel F, Guillemaud T, Dupas S, Rome Q, Perrard A, et al. Reconstructing the invasion and the demographic history of the yellow-legged hornet, *Vespa velutina*, in Europe. Biological Invasions 2015;17(8):2357–71.
10. Takeuchi T, Takahashi R, Kiyoshi T, Nakamura M, Minoshima YN, Takahashi J. The origin and genetic diversity of the yellow-legged hornet, *Vespa velutina* introduced in Japan. Insect Sociaux 2017;64(3):313–20.
11. Takahashi J, Okuyama H, Kiyoshi T, Takeuchi T, Martin SJ. Origins of *Vespa velutina* hornets that recently invaded Iki Island, Japan and Jersey Island, UK. Mitochondrial DNA Part A 2019;30(3):434–9.
12. Kim JK, Choi M, Moon TY. Occurrence of *Vespa velutina* Lepeletier from Korea, and a revised key for Korean Vespa species (Hymenoptera: Vespidae). Entomological Research 2006;36(2):112–5.
13. Archer M. Vespine wasps of the world: behaviour, ecology & taxonomy of the Vespinae. Siri Scientific Press Monograph Series 4. Rochdale, Lancashire, UK: Siri Scientific Press; 2012.
14. Rome Q, Muller FJ, Touret-Alby A, Darrouzet E, Perrard A, Villemant C. Caste differentiation and seasonal changes in *Vespa velutina* (Hym.: Vespidae) colonies in its introduced range. Journal of Applied Entomology 2015;139(10):771–82.
15. Perrard A, Haxaire J, Rortais A, Villemant C. Observations on the colony activity of the Asian hornet *Vespa velutina* Lepeletier 1836 (Hymenoptera: Vespidae: Vespinae) in France. Annales de la Société entomologique de France 2009;45(1):119–27.
16. Feás Sánchez X, Charles RJ. Notes on the Nest Architecture and Colony Composition in Winter of the Yellow-Legged Asian Hornet, *Vespa velutina* Lepeletier 1836 (Hym.: Vespidae), in Its Introduced Habitat in Galicia (NW Spain). Insects 2019;10(8):237.
17. GBIF.org. GBIF Occurrence Download. 2022. Available from: https://doi.org/10.15468/dl.6jf3ts
18. Archer ME. Taxonomy, distribution and nesting biology of the Vespa Bicolor group (Hym., Vespinae). Entomologist's Monthly Magazine 1994;130:149–158.
19. Perrard A, Arca M, Rome Q, Muller F, Tan J, Bista S, et al. Geographic variation of melanisation patterns in a hornet species: Genetic differences, climatic pressures or aposematic constraints? PLoS One 2014;9(4):e94162.
20. Kumar PG, Mazumdar PC, Phong HP. New records of hornet wasps of the genus Vespa Linnaeus (Hymenoptera: Vespidae) from Indian States and Bangladesh. Records of the Zoological Survey of India 2015;115:215–9.
21. Siddiqui JA, Bodlah I, Carpenter JM, Naeem M, Ahmad M, Bodlah MA. Vespidae (Hymenoptera) of the Pothwar region of Punjab, Pakistan. Zootaxa 2015;3914(5):501.
22. Dorji P, Gyeltshen T, Klein W, Nidup T. New records of social wasps (Hymenoptera: Vespinae: Vespa and Provespa) from Bhutan. Journal of Threatened Taxa 2017;9(4):10102–8.
23. Handru A, Nugroho H, Saito-Morooka F, Ubaidillah R, Kojima JI. Eusocial wasp fauna of Sulawesi Island, the central island of Wallacea (Hymenoptera: Vespidae; Polistinae, Vespinae). Zootaxa 2020;4885(4):541–59.
24. van der Vecht J. Notes on Oriental Vespidae, including some species from China and Japan. Zoologische Mededelingen 1959;36:205–32.
25. Villemant C, Barbet-Massin M, Perrard A, Muller F, Gargominy O, Jiguet F, et al. Predicting the invasion risk by the alien bee-hawking Yellow-legged hornet *Vespa velutina nigrithorax* across Europe and other continents with niche models. Biological Conservation 2011;144(9):2142–50.
26. Quaresma A, Henriques D, Godinho J, Gmaside X, Bortolotti L, Pinto MA. Invasion genetics of the Asian hornet *Vespa velutina nigrithorax* in Southern Europe. Biological Invasions 2022;24:1479–99.
27. Bertolino S, Lioy S, Laurino D, Manino A, Porporato M. Spread of the invasive yellow-legged hornet *Vespa velutina* (Hymenoptera: Vespidae) in Italy. Applied Entomology and Zoology 2016;51(4):589–97.
28. Robinet C, Suppo C, Darrouzet E. Rapid spread of the invasive yellow-legged hornet in France: The role of human-mediated dispersal and the effects of control measures. Journal of Applied Ecology 2017;54(1):205–15.

29. Robinet C, Darrouzet E, Suppo C. Spread modelling: A suitable tool to explore the role of human-mediated dispersal in the range expansion of the yellow-legged hornet in Europe. International Journal of Pest Management 2019;65(3):258–67.
30. Lioy S, Manino A, Porporato M, Laurino D, Romano A, Capello M, et al. Establishing surveillance areas for tackling the invasion of *Vespa velutina* in outbreaks and over the border of its expanding range. NeoBiota 2019;46:51–69.
31. Verdasca MJ, Rebelo H, Carvalheiro LG, Rebelo R. Invasive hornets on the road: Motorway-driven dispersal must be considered in management plans of *Vespa velutina*. NeoBiota 2021;69:177–98.
32. Haxaire J, Tamisier JP, Bouguet JP. *Vespa velutina* Lepeletier, 1836, une redoutable nouveauté pour la faune de France (Hym., Vespidae). Bulletin de la Société entomologique de France 2006;111(2):194.
33. Rome Q, Muller F, Gargominy O, Villemant C. Bilan 2008 de l'invasion de *Vespa velutina* Lepeletier en France (Hymenoptera, Vespidae). Bulletin de la Société Entomologique de France 2009;114(3):297–302.
34. Villemant C, Muller F, Haubois S, Perrard A, Darrouzet E, Rome Q. Bilan des travaux (MNHN et IRBI) sur l'invasion en france de *Vespa velutina*, le frelon asiatique prédateur d'abeilles. Proceedings of the Journée Scientifique Apicole; 2011 February 11; Arles, ONIRIS-FNOSAD, Nantes, France; 2011. p. 3–12.
35. Rome Q, Dambrine L, Onate C, Muller F, Villemant C, García-Pérez A, et al. Spread of the invasive hornet *Vespa velutina* Lepeletier, 1836, in Europe in 2012 (Hym., Vespidae). Bulletin de la Société Entomologique de France 2013;118(1):21–2.
36. Franklin DN, Brown MA, Datta S, Cuthbertson AGS, Budge GE, Keeling MJ. Invasion dynamics of Asian hornet, *Vespa velutina* (Hymenoptera: Vespidae): a case study of a commune in south-west France. Applied Entomology and Zoology 2017;52(2):221–9.
37. Monceau K, Thiéry D. *Vespa velutina* nest distribution at a local scale: An 8-year survey of the invasive honeybee predator. Insect Science 2017;24(4):663–74.
38. Castro L, Pagola-Carte S. *Vespa velutina* Lepeletier, 1836 (Hymenoptera: Vespidae), recolectada en la Península Ibérica. Heteropterus Revista de Entomología 2010;10(2):193–6.
39. López S, González M, Goldarazena A. *Vespa velutina* Lepeletier, 1836 (Hymenoptera: Vespidae): first records in Iberian Peninsula. EPPO Bulletin 2011;41(3):439–41.
40. Goldarazena A, de Heredia IP, Romon P, Iturrondobeitia JC, Gonzalez M, Lopez S. Spread of the yellow-legged hornet *Vespa velutina nigrithorax* du Buysson (Hymenoptera: Vespidae) across Northern Spain. EPPO Bulletin 2015;45(1):133–8.
41. Rodríguez-Flores MS, Seijo-Rodríguez A, Escuredo O, Seijo M. Spreading of *Vespa velutina* in northwestern Spain: influence of elevation and meteorological factors and effect of bait trapping on target and nontarget living organisms. Journal of Pest Science 2019;92(2):557–65.
42. Pujade-Villar J, Torrell A, Rojo M. Confirmada la presència a Catalunya d'una vespa originària d'Àsia molt perillosa per als ruscs. Butlletí de la Institució Catalana d'Historia Natural 2012;77:173–6.
43. Leza M, Miranda MÁ, Colomar V. First detection of *Vespa velutina nigrithorax* (Hymenoptera: Vespidae) in the Balearic Islands (Western Mediterranean): a challenging study case. Biological Invasions 2018;20(7):1643–9.
44. Leza M, Herrera C, Picó G, Morro T, Colomar V. Six years of controlling the invasive species *Vespa velutina* in a Mediterranean island: The promising results of an eradication plan. Pest Management Science 2021;77(5):2375–84.
45. Grosso-Silva JM, Maia M. *Vespa velutina* Lepeletier, 1836 (Hymenoptera, Vespidae), new species for Portugal. Arquivos Entomolóxicos 2012;6:53–4.
46. Carvalho J, Hipólito D, Santarém F, Martins R, Gomes A, Carmo P, et al. Patterns of *Vespa velutina* invasion in Portugal using crowdsourced data. Insect Conservation and Diversity 2020;13(5):501–7.
47. Rome Q, Muller F, Villemant C. Expansion en 2011 de *Vespa velutina* Lepeletier en Europe (Hym., Vespidae). Bulletin de la Société entomologique de France 2012;117(1):114.
48. Schoonvaere K, Laget D, Adriaens T. Vespa-watch-invasion monitoring of the Asian hornet *Vespa velutina* with beekeepers and citizens. Poster Presentation BEES Market. Brussels, Belgium: Belgian Ecosystems and Society community; 2019.
49. Demichelis S, Manino A, Minuto G, Mariotti M, Porporato M. Social wasp trapping in north west Italy: Comparison of different bait-traps and first detection of *Vespa velutina*. Bulletin of Insectology 2014;67(2):307–17.
50. Porporato M, Manino A, Laurino D, Demichelis S. *Vespa velutina* Lepeletier (Hymenoptera Vespidae): A first assessment two years after its arrival in Italy. Redia 2014;97:189–94.

51. Granato A, Negrisolo E, Bonomi J, Zulian L, Cappa F, Bortolotti L, et al. Recent confirmation of a single haplotype in the Italian population of *Vespa velutina*. Biological Invasions 2019;21(9):2811–7.
52. Witt R. Erstfund eines Nestes der Asiatischen Hornisse *Vespa velutina* Lepeletier,1838 in Deutschland und Details zum Nestbau (Hymenoptera, Vespinae). Ampulex 2015;7:42–53.
53. Husemann M, Sterr A, Mack S, Abraham R. The northernmost record of the Asian hornet *Vespa velutina nigrithorax* (Hymenoptera, Vespidae). Evolutionary Systematics 2020;4(1):1–4.
54. Husemann M, Dey LS, Hawlitschek O. *Vespa velutina nigrithorax* Lepeltier, 1836 from Hamburg (Northern Germany) shares the same COI haplotype with other European populations. Journal of Hymenoptera Research 2020;79:111–5.
55. Budge GE, Hodgetts J, Jones EP, Ostojá-Starzewski JC, Hall J, Tomkies V, et al. The invasion, provenance and diversity of *Vespa velutina* Lepeletier (Hymenoptera: Vespidae) in Great Britain. PLoS One 2017;12(9):e0185172.
56. Jones EP, Conyers C, Tomkies V, Semmence N, Fouracre D, Wakefield M, et al. Managing incursions of *Vespa velutina nigrithorax* in the UK: An emerging threat to apiculture. Scientific Reports 2020;10(1):19553.
57. Ries C, Schneider N, Vitali F, Weigand A. First records and distribution of the invasive alien hornet *Vespa velutina nigrithorax* du Buysson, 1905 (Hymenoptera: Vespidae) in Luxembourg. Bulletin de la Société des Naturalistes Luxembourgeois 2021;123:181–93.
58. Government of Ireland. NPWS confirms first discovery of a live specimen Asian Hornet in Ireland. 2021. Available from: https://www.gov.ie/en/press-release/63eaf-npws-confirms-first-discovery-of-a-live-specimenasian-hornet-in-ireland/#
59. Choi MB, Martin SJ, Lee JW. Distribution, spread, and impact of the invasive hornet *Vespa velutina* in South Korea. Journal of Asia-Pacific Entomology 2012;15(3):473–7.
60. Jeong JS, Kim MJ, Park JS, Lee KH, Jo YH, Takahashi JI, et al. Tracing the invasion characteristics of the yellow-legged hornet, *Vespa velutina nigrithorax* (Hymenoptera: Vespidae), in Korea using newly detected variable mitochondrial DNA sequences. Journal of Asia-Pacific Entomology 2021;24(2):135–47.
61. Sakai Y, Takahashi J. Discovery of a worker of *Vespa velutina* (Hymenoptera: Vespidae) from Tsushima Island, Japan. Japanese Journal of Entomology 2014;17(1):32–6.
62. Ueno T. Establishment of the invasive hornet *Vespa velutina* (Hymenoptera: Vespidae) in Japan. International Journal of Chemical, Environmental & Biological Sciences 2014;2(4):220–2.
63. Minoshima YN, Yamane S, Ueno T. An invasive alien hornet, *Vespa velutina nigrithorax* du Buysson (Hymenoptera, Vespidae), found in Kitakyushu, Kyushu Island: A first record of the species from mainland Japan. Japanese Journal of Systematic Entomology 2015;21(2):259–61.
64. Kishi S, Goka K. Review of the invasive yellow-legged hornet, *Vespa velutina nigrithorax* (Hymenoptera: Vespidae), in Japan and its possible chemical control. Applied Entomology and Zoology 2017;52(3):361–8.
65. Kim SH, Kim DE, Lee H, Jung S, Lee WH. Ensemble evaluation of the potential risk areas of yellow-legged hornet distribution. Environmental Monitoring and Assessment 2021;193(9):601.
66. Bessa AS, Carvalho J, Gomes A, Santarém F. Climate and landuse drivers of invasion: Predicting the expansion of *Vespa velutina nigrithorax* into the Iberian Peninsula. Insect Conservation and Diversity 2016;9(1):27–37.
67. Fournier A, Barbet-Massin M, Rome Q, Courchamp F. Predicting species distribution combining multi-scale drivers. Global Ecology and Conservation 2017;12:215–26.
68. Keeling MJ, Franklin DN, Datta S, Brown MA, Budge GE. Predicting the spread of the Asian hornet (*Vespa velutina*) following its incursion into Great Britain. Scientific Reports 2017;7(1):6240.
69. Villemant C, Muller F, Rome Q, Perrard A, Barbet-Massin M, Jiguet F. Estimating the Potential Range Expansion and Environmental Impact of the Invasive Bee-Hawking Hornet. In Silico Bees 2014;269–87.
70. Barbet-Massin M, Rome Q, Villemant C, Courchamp F. Can species distribution models really predict the expansion of invasive species? PLoS One 2018;13(3):e0193085.
71. Barbet-Massin M, Rome Q, Muller F, Perrard A, Villemant C, Jiguet F. Climate change increases the risk of invasion by the Yellow-legged hornet. Biological Conservation 2013;157:4–10.
72. Rome Q, Perrard A, Muller F, Fontaine C, Quilès A, Zuccon D, et al. Not just honeybees: Predatory habits of *Vespa velutina* (Hymenoptera: Vespidae) in France. Annales de la Société Entomologique de France 2021;57(1):1–11.

73. Verdasca MJ, Godinho R, Rocha RG, Portocarrero M, Carvalheiro LG, Rebelo R, et al. A metabarcoding tool to detect predation of the honeybee *Apis mellifera* and other wild insects by the invasive *Vespa velutina*. Journal of Pest Science 2021;95:997–1007.
74. Cini A, Cappa F, Petrocelli I, Pepiciello I, Bortolotti L, Cervo R. Competition between the native and the introduced hornets *Vespa crabro* and *Vespa velutina*: A comparison of potentially relevant life-history traits. Ecological Entomology 2018;43(3):351–62.
75. Monceau K, Maher N, Bonnard O, Thiéry D. Evaluation of competition between a native and an invasive hornet species: Do seasonal phenologies overlap? Bulletin of Entomological Research 2015;105(4):462–9.
76. Poidatz J, Bressac C, Bonnard O, Thiéry D. Comparison of reproductive traits of foundresses in a native and an invasive hornet in Europe. Journal of Insect Physiology 2018;109:93–9.
77. Monceau K, Moreau J, Poidatz J, Bonnard O, Thiéry D. Behavioral syndrome in a native and an invasive hymenoptera species. Insect Science 2015;22(4):541–8.
78. Kwon O, Choi MB. Interspecific hierarchies from aggressiveness and body size among the invasive alien hornet, *Vespa velutina nigrithorax*, and five native hornets in South Korea. PLoS One 2020;15(7):e0226934.
79. Yamasaki K, Takahashi R, Harada R, Matsuo Y, Nakamura M, Takahashi JI. Reproductive interference by alien hornet *Vespa velutina* threatens the native populations of *Vespa simillima* in Japan. The Science of Nature 2019;106(5–6):15.
80. Ikegami M, Tsujii K, Ishizuka A, Nakagawa N, Kishi S, Sakamoto Y, et al. Environments, spatial structures, and species competitions: Determining the impact of yellow-legged hornets, *Vespa velutina*, on native wasps and bees on Tsushima Island, Japan. Biological Invasions 2020;22:3131–43.
81. Rojas-Nossa SV, Calviño-Cancela M. The invasive hornet *Vespa velutina* affects pollination of a wild plant through changes in abundance and behaviour of floral visitors. Biological Invasions 2020;22:2609–18.
82. Couto A, Monceau K, Bonnard O, Thiéry D, Sandoz JC. Olfactory attraction of the hornet *Vespa velutina* to honeybee colony odors and pheromones. PLoS One 2014;9(12):e115943.
83. Wang ZW, Chen G, Tan K. Both olfactory and visual cues promote the hornet *Vespa velutina* to locate its honeybee prey *Apis cerana*. Insectes Sociaux 2014;61(1):67–70.
84. Monceau K, Arca M, Leprêtre L, Mougel F, Bonnard O, Silvain JF, et al. Native prey and invasive predator patterns of foraging activity: The case of the yellow-legged hornet predation at European honeybee hives. PLoS One 2013;8(6):e66492.
85. Monceau K, Bonnard O, Moreau J, Thiéry D. Spatial distribution of *Vespa velutina* individuals hunting at domestic honeybee hives: Heterogeneity at a local scale. Insect Science 2014;21(6):765–74.
86. Monceau K, Maher N, Bonnard O, Thiéry D. Predation pressure dynamics study of the recently introduced honeybee killer *Vespa velutina*: Learning from the enemy. Apidologie 2013;44(2):209–21.
87. Choi MB. Foraging behavior of an invasive alien hornet (*Vespa velutina*) at *Apis mellifera* hives in Korea: Foraging duration and success rate. Entomological Research 2021;51(3):143–8.
88. Leza M, Herrera C, Marques A, Roca P, Sastre-Serra J, Pons DG. The impact of the invasive species *Vespa velutina* on honeybees: A new approach based on oxidative stress. Science of the Total Environment 2019;689:709–15.
89. Requier F, Rome Q, Chiron G, Decante D, Marion S, Menard M, et al. Predation of the invasive Asian hornet affects foraging activity and survival probability of honey bees in Western Europe. Journal of Pest Science 2019;92(2):567–78.
90. Ken T, Hepburn HR, Radloff SE, Yusheng Y, Yiqiu L, Danyin Z, et al. Heat-balling wasps by honeybees. Naturwissenschaften 2005;92(10):492–5.
91. Arca M, Papachristoforou A, Mougel F, Rortais A, Monceau K, Bonnard O, et al. Defensive behaviour of *Apis mellifera* against *Vespa velutina* in France: Testing whether European honeybees can develop an effective collective defence against a new predator. Behavioural Processes 2014;106:122–9.
92. Tan K, Radloff SE, Li JJ, Hepburn HR, Yang MX, Zhang LJ, et al. Beehawking by the wasp, *Vespa velutina*, on the honeybees *Apis cerana* and *A. mellifera*. Naturwissenschaften 2007;94(6):469–72.
93. Tan K, Wang Z, Li H, Yang S, Hu Z, Kastberger G, et al. An 'I see you' prey–predator signal between the Asian honeybee, *Apis cerana*, and the hornet, *Vespa velutina*. Animal Behaviour 2012;83(4):879–82.
94. Wang Z, Qu Y, Dong S, Wen P, Li J, Tan K, et al. Honey bees modulate their olfactory learning in the presence of hornet predators and alarm component. PLoS One 2016;11(2):e0150399.
95. Dong S, Wen P, Zhang Q, Wang Y, Cheng Y, Tan K, et al. Olfactory eavesdropping of predator alarm pheromone by sympatric but not allopatric prey. Animal Behaviour 2018;141:115–25.

96. Barbet-Massin M, Salles JM, Courchamp F. The economic cost of control of the invasive yellow-legged Asian hornet. NeoBiota 2020;55:11–25.
97. Choi MB. Defensive behavior of the invasive alien hornet *Vespa velutina nigrithorax* against potential human aggressors. Entomological Research 2021;51(4):186–95.
98. Choi MB, Hong EJ, Kwon O. Defensive behavior of the invasive alien hornet, *Vespa velutina*, against color, hair and auditory stimuli of potential aggressors. PeerJ 2021;9:e11249.
99. Liu Z, Chen S, Zhou Y, Xie C, Zhu B, Zhu H, et al. Deciphering the Venomic Transcriptome of Killer-Wasp *Vespa velutina*. Scientific Reports 2015;5(1):9454.
100. Tabar AI, Chugo S, Joral A, Lizaso MT, Lizarza S, Alvarez-Puebla MJ, et al. *Vespa velutina nigritorax*: A new causative agent for anaphylaxis. Clinical and Translational Allergy 2015;5(S3):P43.
101. Vidal C. The Asian wasp *Vespa velutina nigrithorax*: Entomological and allergological characteristics. Clinical & Experimental Allergy 2022;52(4):489–98.
102. Feás X. Human fatalities caused by hornet, wasp and bee stings in Spain: Epidemiology at state and sub-state level from 1999 to 2018. Biology 2021;10(2):73.
103. Vidal C, Armisén M, Monsalve R, González-Vidal T, Lojo S, López-Freire S, et al. Anaphylaxis to *Vespa velutina nigrithorax*: Pattern of sensitization for an emerging problem in Western countries. Journal of Investigational Allergology and Clinical Immunology 2021;31(3):228–35.
104. Laborde-Castérot H, Darrouzet E, Le Roux G, Labadie M, Delcourt N, de Haro L, et al. Ocular lesions other than stings following yellow-legged hornet (*Vespa velutina nigrithorax*) projections, as reported to french poison control centers. JAMA Ophthalmology 2021;139(1):105–8.
105. De Haro L, Labadie M, Chanseau P, Cabot C, Blanc-Brisset I, Penouil F. Medical consequences of the Asian black hornet (*Vespa velutina*) invasion in Southwestern France. Toxicon 2010;55(2–3):650–2.
106. Choi MB, Kim TG, Kwon O. Recent trends in wasp nest removal and hymenoptera stings in South Korea. Journal of Medical Entomology 2019;56(1):254–60.
107. Demain JG. Hymenoptera allergy and anaphylaxis: Are warmer temperatures changing the impact? Current Opinion in Allergy and Clinical Immunology 2020;20(5):438–44.
108. Ren ZX, Wang H, Bernhardt P, Li DZ. Insect pollination and self-incompatibility in edible and/or medicinal crops in southwestern China, a global hotspot of biodiversity. American Journal of Botany 2014;101(10):1700–1710.
109. Wei W, Wu H, Li X, Wei X, Lu W, Zheng X. Diversity, daily activity patterns, and pollination effectiveness of the insects visiting *Camellia osmantha*, *C. vietnamensis*, and *C. oleifera* in South China. Insects 2019;10(4):98.
110. Chen G, Wang ZW, Wen P, Wei W, Chen Y, Ai H, Sun WB. Hydrocarbons mediate seed dispersal: A new mechanism of vespicochory. New Phytologist 2018;220(3):714–725.
111. Chen G, Zhang Z, Chomicki G, Sun W. The flip side of the coin: Ecological function of the bee-hawking Asian hornet. Integrative Zoology 2020;15(2):156–159.
112. Jeong H, Kim JM, Kim B, Nam JO, Hahn D, Choi MB. Nutritional value of the larvae of the alien invasive wasp *Vespa velutina nigrithorax* and amino acid composition of the larval saliva. Food 2020;9(7):885.
113. Ghosh S, Namin SM, Meyer-Rochow VB, Jung C. Chemical composition and nutritional value of different species of *Vespa* hornets. Food 2021;10(2):418.
114. Feás X, Vázquez-Tato MP, Seijas JA, Nikalje APG, Fraga-López F. Extraction and physicochemical characterization of chitin derived from the Asian hornet, *Vespa velutina* Lepeletier 1836 (Hym.: Vespidae). Molecules 2020;25(2):384.
115. Kim J, Kim M, Lee M, Lee YJ, Kim HR, Nam JO, et al. Antibacterial potential of *Nidus vespae* built by invasive alien hornet, *Vespa velutina nigrithorax*, against food-borne pathogenic bacteria. Entomological Research 2020;50(1):28–33.
116. Feás X, Vidal C, Vázquez-Tato MP, Seijas JA. Asian hornet, *Vespa velutina* Lepeletier 1836 (Hym.: Vespidae), venom obtention based on an electric stimulation protocol. Molecules 2022;27(1):138.
117. Le TN, Da Silva D, Colas C, Darrouzet É, Baril P, Leseurre L, Maunit B. Asian hornet *Vespa velutina nigrithorax* venom: Evaluation and identification of the bioactive compound responsible for human keratinocyte protection against oxidative stress. Toxicon 2020;176:1–9.
118. Meng YC, Mo XG, He TT, Wen XX, Nieh JC, Yang XW, Tan K. New bioactive peptides from the venom gland of a social hornet *Vespa velutina*. Toxicon 2021;199:94–100.
119. Yun HS, Oh J, Lim JS, Kim HJ, Kim JS. Anti-inflammatory effect of wasp venom in BV-2 microglial cells in comparison with bee venom. Insects 2021;12(4):297.

120. Wittenberg R, Cock MJW. Invasive alien species: A toolkit of best prevention and management practices. Wallingford: CABI; 2001.
121. Marris G, Brown M, Cuthbertson AG. GB Non-native Organism Risk Assessment for *Vespa velutina nigrithorax*. 2011. Available from: www. nonnativespecies.org
122. Lioy S, Kenis M. The threat posed by invasive alien wasps (Vespidae) to the EU, their pathways of importation and the potential measures to be applied for preventing introductions. Technical note prepared by IUCN for the European Commission. Gland, Switzerland: International Union for Conservation of Nature; 2021.
123. de Medeiros CM, Hernández-Lambraño RE, Sánchez Agudo JÁ. How reliable is the untrained eye in the identification of an invasive species? The case of alien bee-hawking yellow-legged hornet in Iberian Peninsula. Contemporary Problems of Ecology 2018;11(6):666–81.
124. Roy HE, Rorke SL, Beckmann B, Booy O, Botham MS, Brown PMJ, et al. The contribution of volunteer recorders to our understanding of biological invasions. Biological Journal of the Linnean Society 2015;115(3):678–89.
125. Pusceddu M, Floris I, Mannu R, Cocco A, Satta A. Using verified citizen science as a tool for monitoring the European hornet (*Vespa crabro*) in the island of Sardinia (Italy). NeoBiota 2019;50:97–108.
126. Rojas-Nossa SV, Novoa N, Serrano A, Calviño-Cancela M. Performance of baited traps used as control tools for the invasive hornet *Vespa velutina* and their impact on non-target insects. Apidologie 2018;49(6):872–85.
127. Lioy S, Laurino D, Capello M, Romano A, Manino A, Porporato M. Effectiveness and selectiveness of traps and baits for catching the invasive hornet *Vespa velutina*. Insects 2020;11(10):706.
128. Braga D, Madureira A. Towards a decision support system for the automatic detection of Asian hornets and removal planning. International Journal of Computer Information Systems and Industrial Management Applications 2020;12:239–47.
129. Stainton K, Hall J, Budge GE, Boonham N, Hodgetts J. Rapid molecular methods for in-field and laboratory identification of the yellowlegged Asian hornet (*Vespa velutina nigrithorax*). Journal of Applied Entomology 2018;142(6):610–6.
130. Poidatz J, Monceau K, Bonnard O, Thiéry D. Activity rhythm and action range of workers of the invasive hornet predator of honeybees *Vespa velutina*, measured by radio frequency identification tags. Ecology and Evolution 2018;8(15):7588–98.
131. Milanesio D, Saccani M, Maggiora R, Laurino D, Porporato M. Design of an harmonic radar for the tracking of the Asian yellow-legged hornet. Ecology and Evolution 2016;6(7):2170–8.
132. Milanesio D, Saccani M, Maggiora R, Laurino D, Porporato M. Recent upgrades of the harmonic radar for the tracking of the Asian yellow-legged hornet. Ecology and Evolution 2017;7(13):4599–606.
133. Maggiora R, Saccani M, Milanesio D, Porporato M. An innovative harmonic radar to track flying insects: the case of *Vespa velutina*. Scientific Reports 2019;9(1):11964.
134. Lioy S, Laurino D, Maggiora R, Milanesio D, Saccani M, Mazzoglio PJ, et al. Tracking the invasive hornet *Vespa velutina* in complex environments by means of a harmonic radar. Scientific Reports 2021;11(1):12143.
135. Kennedy PJ, Ford SM, Poidatz J, Thiéry D, Osborne JL. Searching for nests of the invasive Asian hornet (*Vespa velutina*) using radiotelemetry. Communications Biology 2018;1(1):88.
136. Reynaud L, Guérin-Lassous I. Design of a force-based controlled mobility on aerial vehicles for pest management. Ad Hoc Networks 2016;53:41–52.
137. Lioy S, Bianchi E, Biglia A, Bessone M, Laurino D, Porporato M. Viability of thermal imaging in detecting nests of the invasive hornet *Vespa velutina*. Insect Science 2021;28(1):271–7.
138. Shams T, Desbarats P. Detection of Asian hornet's nest on drone acquired FLIR and color images using deep learning methods. In: Proceedings of the Tenth International Conference on Image Processing Theory, Tools and Applications (IPTA); 2020 November 9-12; Paris, France. Institute of Electrical and Electronics Engineers; 2020. p. 1–6.
139. Turchi L, Derijard B. Options for the biological and physical control of *Vespa velutina nigrithorax* (Hym.: Vespidae) in Europe: A review. Journal of Applied Entomology 2018;142(6):553–62.
140. Ruiz-Cristi I, Berville L, Darrouzet E. Characterizing thermal tolerance in the invasive yellow-legged hornet (*Vespa velutina nigrithorax*): The first step toward a green control method. PLoS One 2020;15(10):e0239742.
141. Monceau K, Bonnard O, Thiéry D. Chasing the queens of the alien predator of honeybees: A water drop in the invasiveness ocean. Open Journal of Ecology 2012;2(4):183–91.

142. Sánchez O, Arias A. All that glitters is not gold: The other insects that fall into the Asian Yellow-Legged Hornet *Vespa velutina* 'specific' traps. Biology 2021;10(5):448.
143. Blanco CA, Corona M, Hernández G, Smith-Pardo AH, Dively GP. Metal screen at the entrance of a hive does not affect production and reproduction of honey bees and can prevent predation by hornets. Southwestern Entomologist 2020;45(4):843–51.
144. Bonnefond L, Paute S, Andalo C. Testing muzzle and ploy devices to reduce predation of bees by Asian hornets. Journal of Applied Entomology 2021;145:145–57.
145. Requier F, Rome Q, Villemant C, Henry M. A biodiversity-friendly method to mitigate the invasive Asian hornet's impact on European honey bees. Journal of Pest Science 2020;93(1):1–9.

4 When bio is not green: the impacts of bumblebee translocation and invasion on native ecosystems

Josefina Lohrmann, Nicolás R. Cecchetto, Nahuel Aizen, Marina P. Arbetman, and Eduardo E. Zattara*

Grupo de Ecología de la Polinización, INIBIOMA, Universidad Nacional del Comahue, CONICET, Quintral 1250, Bariloche 8400, Argentina

Abstract

Adequate pollination is fundamental to optimize reproduction and yield of most flowering plants, including many staple food crops. Plants depending on insect pollination rely heavily on many wild species of solitary and social bees, and declines or absence of bees often hampers crop productivity, prompting supplementation of pollination services with managed bees. Though honeybees are the most widely deployed managed pollinators, many high-value crops are pollinated more efficiently by bumblebees (*Bombus* spp.), prompting domestication and commercial rearing of several species. This led to a blooming international trade that translocated species outside their native range, where they escaped management and invaded the ecosystems around their deployment sites. Here, we briefly review the history of bumblebee invasions and their main impacts on invaded ecosystems, and close by discussing alternatives to the use of commercially reared bumblebees to enhance crop pollination. As evidence of widespread negative effects on local ecosystems of bumblebee invasions builds up, bumblebee trade adds to the list of examples of "biological" strategies devised to solve agricultural problems that ended up being far from the "green," eco-friendly solutions they were expected to be.

Keywords: biological invasions, pollination services, bee pathogens, bumblebee trade

Review methodology

To write this review, we drafted the main sections based on the different aspects of bumblebee trade and translocation, loosely using [1] as a starting point that needed updating. Then, each author was assigned a section/topic, using Google Scholar as a search engine with keywords related to the topic to complement the list of literature already known to us, and incorporating relevant articles. We also checked the reference lists of all consulted articles to look for relevant works not revealed by our search strategy.

*Corresponding author: ezattara@comahue-conicet.gob.ar

Introduction

"...(The) history of evolution is that life
escapes all barriers.
Life breaks free.
Life expands to new territories.
Painfully, perhaps even dangerously."
Michael Crichton, Jurassic Park [2]

Insufficient or inadequate pollination services translate into lower seed set and poorer fruit yields. Although this fact has been known since humans started translocating crop species beyond their native ranges or growing crops at densities or conditions with which their natural pollinators could not cope, commercial pollinator supplementation is a relatively new agricultural strategy. By far, the most common managed pollinator is the European honeybee *Apis mellifera*, which has been moved all over the world from its native African and Eurasian range, becoming naturalized (and sometimes invasive, see below) in all areas it was introduced to [3, 4]; however, honeybees do not work well in enclosed spaces (like greenhouses) and are less efficient pollinating specialized flowers with deeper corollas or requiring buzz pollination. This prompted attempts to domesticate bumblebees (genus *Bombus*); it took over 60 years to achieve this goal, but in 1987, commercial rearing began, and since then, production and sale of millions of colonies of several species have been the basis for a blooming business [5].

Marketed as a "biological" solution to an agricultural need (i.e., an ecosystem-friendly, "green" alternative to technological solutions), farmers across the world embraced the use of commercial bumblebee colonies and their breeding and trade became a profitable market [6]. Even though the risks of translocating species have always been well known, not all species of *Bombus* are equally easy or profitable to domesticate for commercial breeding; this resulted in both pressures from farmers of regions lacking commercial rearing of native species to import colonies of foreign species and an eagerness from companies producing them to sell them [5]. Despite some early assertions that commercial colonies were unlikely to genetically pollute their wild conspecifics or become invasive (based on commercial colonies being "out of sync" with natural populations, outside their climatic niches or prevented from releasing reproductive individuals through devices like queen excluders), as Jurassic Park's Dr. Ian Malcolm wisely said, "life finds a way": commercial bumblebees quickly escaped, established feral populations, and became invasive (defined as explosively expanding in abundance and geographic range and negatively interfering with ecosystems) almost everywhere they were deployed, leading to disruption of natural habitats and driving some native bee species near extinction [1, 3, 7]. In this article, we review evidence for the impacts on native ecosystems of commercial bumblebee deployment, escape, establishment, and invasion. We discuss impacts associated with competition for resources, pathogen co-introduction, genetic introgression and hybridization, and changes in pollination networks by the three species with best documented invasions (*Bombus impatiens*, *Bombus ruderatus*, and *Bombus terrestris*). We then briefly discuss the main drivers of invasion and finalize by discussing the ecological cost/benefit of supplementing pollination services with commercial bumblebee colonies and the main alternatives to their use.

Invasive bumblebee species and invasion events

Several of the approximately 250 described species of *Bombus* bumblebees [8] have been translocated through history, but only three have shown enough range expansion outside of their native distribution to be considered invasive [1]: the North American *Bombus impatiens* and the European *Bombus ruderatus* and *Bombus terrestris*. While *B. ruderatus* was translocated only through wild-caught fertilized queens released to form feral colonies, most invasion events for *B. impatiens* and *B. terrestris* were driven by trading of commercially reared queens and colonies.

Bombus (Pyrobombus) impatiens is a species native to eastern North America, where it is the most frequently found bumblebee. After commercial breeding of captive *B. terrestris* started in Europe in the late 1980s, both Canada and the US banned its import, prompting the development of the native *B. impatiens* instead

[1]. Commercial production of *B. impatiens* from wild queens started in eastern Canada in 1990; later, queens of *B. impatiens* were translocated to Europe between 1992 and 1994 to develop this species for commercial rearing in the same facilities used to develop *B. terrestris*, and the resulting colonies were brought back to the US [9]. During the 1990's, deployment of commercial colonies of this species was restricted to areas within its native distribution; in western North America, *Bombus (Bombus) occidentalis*, a species native to that region, had been developed instead. However, commercial colonies of this species crashed in 1999, likely due to *Nosema bombi* infections; in response, importation of *B. impatiens* colonies was permitted to western Canada first and soon to western US states [10] (Fig. 4.1A). Presumptively, feral individuals of *B. impatiens* were found in British Columbia in 2003 [12] and have since then been expanding in the Pacific Northwest [10]. Starting in the mid-90s, commercial colonies have also been imported to Mexico in large numbers for exclusive greenhouse use [5, 13]; escaped individuals have been observed from Baja California to Querétaro [11] (Fig. 4.1A). Habitat suitability modeling for the species suggests it has a low chance of establishing feral populations in northern México (including Baja California) [10]; however, some invasive species have been known to expand their ecological niches upon arrival to new habitats [14–16].

Bombus (Megabombus) ruderatus is a species native to Western and Central Europe, and is declining in parts of its native range, but abundant where it was introduced [1, 17]. Although *B. ruderatus* has never been commercially reared, it was translocated from its European range to two separate continents in the Southern Hemisphere (Fig. 4.1B) to enhance the reproduction of another introduced European species, the red clover *Trifolium repens*: first to New Zealand over 130 years ago and more recently to southern South America. In both cases, invasion required just a few introductory events [5]. In New Zealand, UK-caught *B. ruderatus* queens were introduced in 1885 and 1906, along with at least six other bumblebee species [18, 19]. By 1960, *B. ruderatus* had already spread over large areas of the Southern Island—concurrently with *B. terrestris* and *B. hortorum* [20]—and by 1995, it had become, along with *B. terrestris*, ubiquitous throughout New Zealand [21]. In South America, about 300 *B. ruderatus* queens brought from New Zealand were released in December 1982 and November 1983 in two regions of south-central Chile, representing the first introduction of European bumblebees into South America [22]. *Bombus ruderatus* rapidly spread and invaded southward at both sides of the Andes. Although in some areas of New Zealand and Argentina *B. ruderatus* seems to have declined compared to the initial invasion stage (especially after *B. terrestris* introduction, see below), this species is still abundant in its non-native ranges [23, 24].

Bombus (Bombus) terrestris is a species native to temperate Eurasia [25] that is a matter of concern worldwide due to its invasiveness and impacts [26] (Fig. 4.1C). Besides its rearing within its native range, it has been exported to Israel, New Zealand, Tasmania, China, South Korea, Japan, México and Chile [27]. In Europe, commercial rearing of *B. terrestris* began in the late 80's for pollination of many crops (particularly tomato), replacing manual pollination and faring even better than honeybee pollination [28]. The species comprises several subspecies with distinct distribution across the continent, and commercial rearing is somewhat customized to different regions [5]: while in continental Europe mostly *B. terrestris terrestris*, *B. t. sassaricus* and *B. t. dalmatinus* are used for pollination efforts, only the local subspecies *B. terrestris audax* in the United Kingdom and *B. terrestris canariensis* in the Canary Islands are allowed for outdoor pollination. Despite these practices, among European countries, only Norway, Turkey, and the Canary Islands (Spain) have actually imposed restrictions on the import of exotic bumblebees [5]. *B. terrestris* has also shown invasive behavior in Israel, but whether this resulted from early introduction of commercial colonies [27] or range expansion is unclear, since this region is at the margins of the natural distribution of the species [29]: although a local subspecies was present as early as 1930 at northern Galilee, its distribution has been expanding south since then, probably

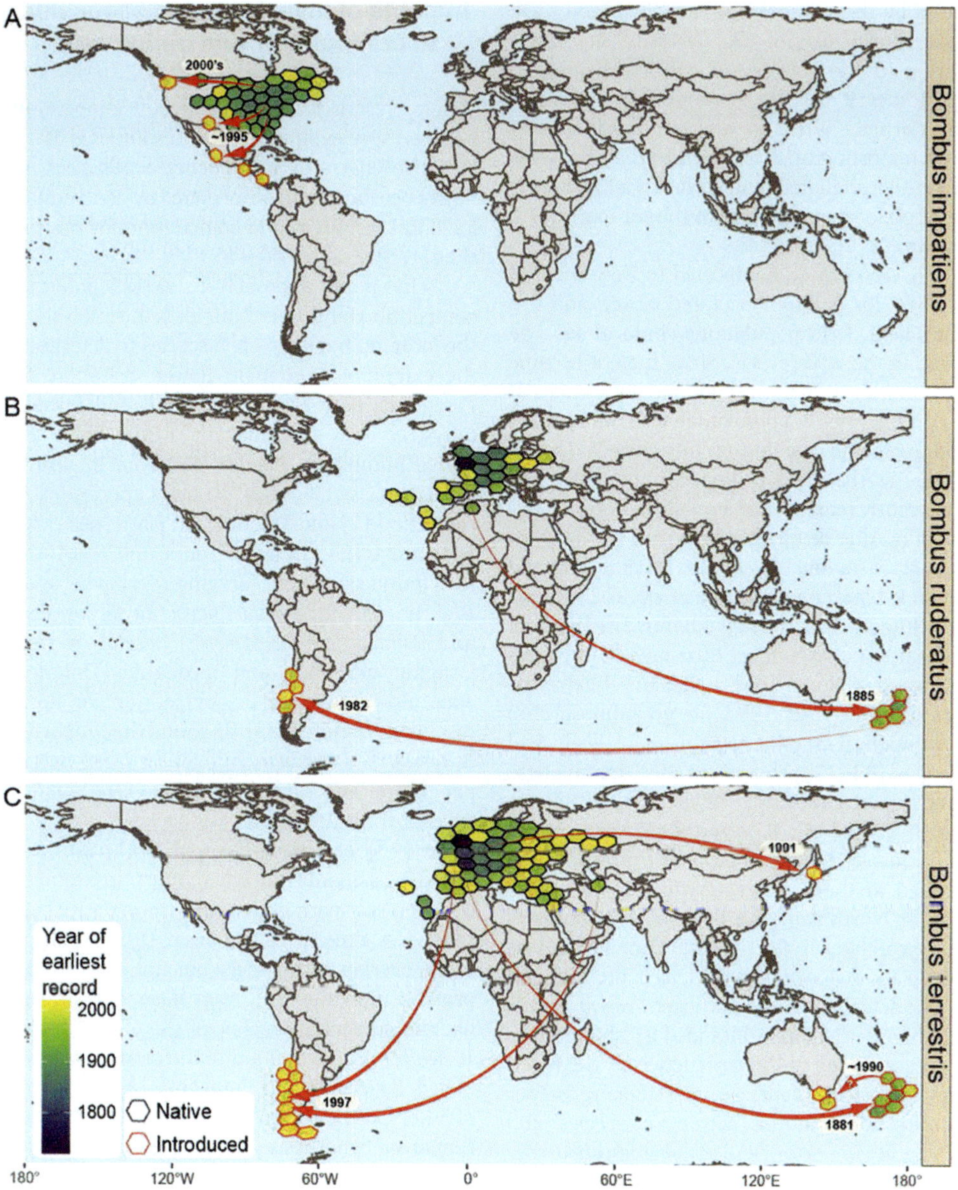

Fig. 4.1. Global distribution of the three main invasive bumblebee species. Occurrences of each species were downloaded from GBIF [11], projected on a global grid of approximately 520-km radius hexagonal cells, and each cell was color-coded according to the year of the earliest reported occurrence within the cell; cells within invaded ranges have red outlines. Red arrows indicate the approximate origin and date of the initial translocation event (the "?" indicates a non-confirmed translocation of *Bombus terrestris* to Tasmania from New Zealand). Only cells with more than five records were retained; some reports of *B. terrestris* in Canada were removed since they actually refer to *Bombus terricola* records. (A) *Bombus impatiens*. (B) *Bombus ruderatus*. (C) *Bombus terrestris*.

fueled by the expansion of irrigated land [30]. Agricultural use of *B. terrestris dalmatinus* started in the country in 1991 [5], and 10 years later it was the predominant species in Mt. Carmel, where it was previously absent [30]. Importing of *B. terrestris* to the country is presently prohibited, and commercial rearing is based on local populations of the subspecies *B. t. dalmatinus* [5].

B. terrestris was imported to New Zealand in 1881 for pollination of red clover, and five years later, feral populations could already be found in the wild [5, 31]. Since then, it became naturalized, rapidly expanding its range at maximum rates of approximately 90 km/yr [18] and replacing previously introduced *Bombus* species as the main pollinating bumblebees. It is currently reared commercially for tomato pollination. After being introduced in 1990 likely in Hobart, Tasmania, *B. terrestris* became feral and expanded its range at a rate of up to 25 km/yr [32]. Import of *B. terrestris* is banned in Australia and alternative native buzz-pollinating bees are being explored [26]. In Japan, *B. terrestris* was introduced in 1992 for greenhouse crop pollination with almost immediate ecological impacts [33–37]. Therefore, after the Japanese government enacted its Invasive Alien Species Act in 2004 [38], *B. terrestris* was categorized as an invasive species and its introduction was banned in the country nearly 20 years later [39]. In North America, *B. terrestris* import has been prohibited in the USA and Canada, but was briefly allowed into Jalisco, Mexico, in 1995 and 1996; detection of the pathogen *Nosema bombi* in a batch of colonies intended for shipment to that locality led to the destruction of the batch and removal of import permits, stopping further introductions of this European species and prompting importation of *Bombus impatiens* (see above) instead [40]. Feral colonies of *B. terrestris* have not been reported in Mexico to date. In South America, import of colonies of *B. terrestris* from Belgium and Israel began in central Chile in 1997; by 2006, they had been already detected across the Andes on the Argentinian side, where its range continues to expand South and East [7]. Although in neighboring countries import of *B. terrestris* is forbidden, it remains legal in Chile in spite of efforts to impose a ban [41]; consequently, thousands of queens are imported yearly to this day.

Impacts of invasive bumblebees due to competition with native bees

When resources are limited, competition between species sharing these resources is likely. Negative effects of introducing exotic bees on native bee fauna can be mediated by competition for floral resources and competition for nesting sites [1, 3].

The main approach to assess potential competition for floral resources is the analysis of the foraging habits of each species to determine the extent of overlap in flower visitation. In Patagonia, both *Bombus ruderatus* and *Bombus terrestris* show a significant overlap with the native bumblebee *Bombus dahlbomii* in visited plant species: 17 out of 20 plant species for *B. ruderatus* [42] and 15 out of 20 plant species for *B. terrestris* [24]. In Japan, Inoue and Yokoyama [37] found significant foraging overlap between *B. terrestris* with *Bombus hypocrita sapporoensis* and *Bombus pseudobaicalensis*. Besides overlap, foraging efficiency and mechanical interference also contribute to competition for floral resources: a study in the UK found that imported *B. terrestris dalmatinus* can collect more nectar per hour and achieve higher reproductive success than the native subspecies *B. terrestris audax* [43], and Hingston and McQuillan [44] found in Tasmania that when *B. terrestris* foraged concurrently with native bees, these native bees spent less time at each flower. Demonstrating niche overlap is necessary but not sufficient to prove competition [1]; even in extreme cases, like the precipitous decline of the native species *B. dahlbomii* concomitant to the arrival of introduced *B. terrestris* to Patagonia [24], the role of interspecific competition for floral resources remains a hypothesis.

Competition for nesting sites with native bees is even harder to demonstrate, since *Bombus* nests are notoriously hard to locate [3]; thus, less evidence for it is currently available. Studies in Japan—where native bumblebees have declined coincidently with *B. terrestris* invasion—found overlap in nest site preferences between *B. terrestris* and Japan's native bumblebee species [34, 36]. Evidence of more direct competition for sites came from laboratory experiments that found that queens of *B. terrestris* succeeded in usurping the nests of native bumblebees within artificial nest boxes [45], and reports of dead queens of native bumblebee species found in and around *B. terrestris* nests in the field in Japan [37]. Differences in phenology can also impose an asymmetry to competition: *Bombus terrestris* not only benefits from an earlier emergence from hibernation in comparison with its native competition [34], but is also more flexible regarding nesting selection, thriving on artificial nests, rodent cavities, and even anthropic debris [34, 46].

Impacts of invasive bumblebees due to pathogen co-introduction

There is strong evidence that introduced bumblebees bring with them a significant array of pathogenic parasites with large potential to spill over to native bee species [1, 47]. On the one hand, introduced bumblebees can harbor pathogenic species or strains that are novel to native species populations and for which they lack adequate immune defenses; on the other hand, the high demographic densities implied by commercial scale rearing often promote both high levels of parasite infection and promotes selection of highly parasite-tolerant genotypes [48, 49]. Native populations can be exposed to novel pathogens as a result of contact with introduced exotic species [13, 50–52], or by contact with conspecific colonies that have been translocated to rearing facilities outside their native range, infected there by pathogens, and taken back to their native range [5, 9, 49, 53]. The susceptibility to decline is phylogenetically structured within the *Bombus* genus, with some clades more prone to decline than others, with non-declining species harboring internal parasites with rather small fitness costs, probably because highly lethal parasites act as strong ecological filters [54]. The impact of each pathogen on different species is still under investigation, but it is clear that pathogens can have synergistic impacts on bumblebee health [6].

Pathogens have been shown to regulate, to a large extent, the fate of bumblebee colonies [55]. To understand their impact on colony development, it is important to understand their roles, their routes of infection, how infection affects each individual bee, and which stages in the colony cycle are most affected by the pathogen. Most species of *Bombus* native to cold and temperate regions are univoltine (i.e., produce a single colony cycle per season), although recent observations suggest that climate change-driven warmer and longer summers might be resulting in bivoltine populations [56, 57]. The colony cycle begins as fertilized queens that have been overwintering emerge, feed, and start looking for suitable nesting sites to found a new colony. Once a site is found, the queen lays the eggs for the first workers and looks after them until they emerge from the pupal case. Until this point, the queen is the only foraging member of the new colony, but after the first workers emerge, they take over foraging first and then start helping with brooding the next batch of eggs and larvae. As the season proceeds, the number of workers increases, until the colony begins to generate reproductive individuals: gynes (future queens) and males, which will leave the nest to mate. Past this point, the colony enters into a senescent phase, in which the dominance of the queen over the workers debilitates as the founder queen ages and eventually dies, and the colony degenerates quickly as parasites and commensals take over the nest [1].

Emergence from hibernation and nest initiation is probably the most critical stage of the bumblebee life cycle: finding a suitable place for colony foundation, including wax production and egg laying, finding flowers to feed, and gathering pollen to supply brood [1, 25]. Several pathogens strongly impact queens at this stage. The nematode *Sphaerularia bombi* induces complex changes in bumblebee's gene expression and behavior, described as mixed and disorientated, with characteristic flight behavior and sound production [58, 59]. Infected new emerging queens search for overwintering sites instead of seeking for a suitable place for nesting, after which the nematode reproduces, killing the host and leaving infective progeny waiting for the next generation of new fertilized queens that are looking for hibernation sites [58]. The apicomplexan *Apicystis bombi* and the microsporidian *Nosema bombi* also can preclude infected queens from founding a colony [60]. The effects of infection by the trypanosomatid *Crithidia bombi* on hibernation are less clear: studies showed that hibernating infected queens lost more body mass and were less likely to initiate a colony than control ones, but there was no difference in survival of each group; survival varies strongly among families, and evidence of reduction in hibernation success is only indirect [61–63].

After finding a nesting site, founder queens lay the first batch of eggs and brood them. *Nosema bombi* can travel through ovaries to the offspring and cause their death; healthy *B. terrestris* colonies can also be infected from outside sources, with younger bees showing higher susceptibility to infection [60, 64, 65]. *Crithidia bombi* has lethal and sublethal effects on different *Bombus* species. They are not vertically transmitted, but acquired from spore-contaminated flowers by foragers, which can pass spores to larvae and in turn infect brooding workers. The impact of this parasite is very context-dependent, particularly with diet, and includes behavioral and cognitive impairments (e.g., reduced ability to learn the color of rewarding flowers) [62, 66, 67]. Single infections with the neogregarine *Apicystis bombi* cause moderate mortality (<20% of queens), but reduce fat reserves, which might hamper overwintering [68]. Viruses also affect colony startup: infection with Kashmir bee virus (KBV) and Israeli acute paralysis virus (IAPV) causes slower offspring production and/or delayed colony startup [69]. In contrast to the previous pathogens, viral infections by KBV and IAPV seem to be sourced from honeybee colonies, with little evidence for secondary propagation within the bumblebee community [70, 71].

A final point for impact is at the reproductive stage of the colony; most pathogens affect reproductive output indirectly, by causing the colony to produce lower-quality reproductive individuals or fewer of them. *Nosema bombi* lowers colony fitness mainly through lower male life span and fertility [55, 72]. Colonies infected with *Crithidia bombi* produced fewer males and thus had lower fitness than healthy colonies [62, 63]. Viral infections by KBV and IAPV also reduce the production of males [69].

Impacts of invasive bumblebees due to introgression and hybridization

Bumblebee males are prone to rather indiscriminate mating behaviors and in laboratory settings will readily mate females of other subspecies or even distantly related species [3]; although natural settings might alleviate these tendencies, there are numerous anecdotal reports of interspecific mating attempts in the wild. Thus, bumblebee trade and translocation that create new areas of contact between geographically distant and genetically different populations imply a risk of hybridization and genetic introgression.

Even though nine *Bombus terrestris* subspecies, differing in morphology, genetics, behavior, resistance to pathogens, and/or phenology, have been described throughout Europe, commercial lines traded within and outside the continent are mainly derived from two: *B. terrestris terrestris* and *B. terrestris dalmatinus* [5, 73–75]. This reduced representation promotes genetic homogenization and introgression of non-local (and probably non-adaptive) alleles into endemic populations [76]. Evidence of interbreeding between the native *B. terrestris lusitanicus* and the commercial *B. terrestris* strains has been reported throughout the entire Iberian Peninsula [77–79]. In Spain, only 19% of wild-caught bumblebees were identified as genetically pure individuals, while more than 45% showed signs of interbreeding with commercial bumblebees; introgressing alleles were found at distances greater than 60 km from managed colonies [79]. In Poland, commercial bumblebees frequently escape and mate with wild *B. terrestris* populations [80]. In contrast, in England no signs of introgression from commercial *B. terrestris lusitanicus* into native *B. terrestris audax* were found [81]. Interestingly, population genetic studies on eastern North American *Bombus impatiens* (for which no subspecies are described) found no widespread introgression from commercial to wild individuals [82].

While the consequences of genetic introgression within species are difficult to measure and predict, the impact of hybridization among species is more evident. Both in the field and in the laboratory, reproductive individuals from commercial *Bombus terrestris* and the native Japanese and Korean species *Bombus ignitus* and *Bombus hypocrita* (*hypocrita* and *sapporoensis* subspecies) mate readily [35, 83, 84]. About 20% of *B. h. hypocrita* and 30% of *B. h. sapporoensis* wild queens had sperm from *B. terrestris* males in their spermathecas [84], and 70% of the *B.*

ignitus queens copulated with *B. terrestris* males in laboratory conditions [35]. Viability of diploid (worker) hybrid eggs between native bumblebee queens and *B. terrestris* males after experimental mating gave variable results, and hybrid genetic background of the workers was not confirmed for apparently successful crosses [35, 83, 85, 86]. However, reduced fecundity due to the high frequency of nonviable diploid hybrid eggs could be a significant threat to these native bumblebee species [35, 84, 87]. Although bumblebees are generally monandrous (queens mate once in their life) [88–90], *B. terrestris* males in some regions of Japan outnumber native males by a factor of more than 50 to 1 [36] and might have evolved a higher mating frequency [87]. Thus, the mating opportunities of local bumblebees with their conspecifics are substantially diminished. Interestingly, recent studies found that *B. h. sapporoensis* and *B. terrestris* queens were more likely to show polyandry in regions of Japan where the latter are more abundant (*unpublished data* in [35]). Polyandry in both bumblebee species (which increases the probability of mating with their conspecifics) could thus be an adaptive response to among-species reproductive interference [35, 87].

Impacts of invasive bumblebees due to changes in pollination networks

Plant-pollinator interactions established during or after a bee invasion can range from beneficial to detrimental to native plant communities and even for cultivated plants [91]. For example, although the native Patagonian bumblebee *Bombus dahlbomii* is the most efficient pollinator of *Alstroemeria aurea* in terms of pollen quantity and quality deposited per visit, the high-visit frequency of invasive *Bombus ruderatus* compensates for this lack and outperforms *B. dahlbomii* when cumulative visits per flower are considered [92, 93]. Similarly, after the drastic decline of *B. dahlbomii*, invasive *B. terrestris* and *B. ruderatus* became the main pollinators of two Andean orchids [94]. In an experimental setup, when seven Japanese native plant species were exposed only to *B. terrestris* pollination, five species whose flowers the invasive bumblebees could only access through nectar robbing showed decreased fruit set or fruit quality or both, but when the same plants were exposed to a mix of *B. terrestris* and native bumblebees, pollination success varied unpredictably [95–97]. This difference in pollination success suggests that native plants are relatively specialized to native bumblebees, and therefore, invasive bumblebees could be poor substitutes for local ones, and that even at medium invasion densities, the fertility of native plants is compromised. Invasive bumblebees also show patterns of pollen transport different to those of native pollinators, potentially altering the genetic structure of plant populations—as it was reported in pollination networks invaded by honeybees [98–100]. For raspberry crops in northwestern Patagonia, disproportionate abundance of invasive *B. terrestris* relative to flowers altered the historical cost-benefit balance of pollination interactions and drove this mutualism toward antagonism by decreasing fruit yields [101, 102]. Overall, current data show that impacts of invasive pollinators on plants are predominantly negative for native species, mixed for crops, and positive for invasive species [91].

Drivers and facilitators of bumblebee invasions

While most arthropod biological invasions are usually related to undesirable species, like pests or disease vectors, that are accidentally translocated out of their home range, there are no known reports of unintentional bumblebee translocations across continents. Instead, the main driver of *Bombus* invasions has been the provision of pollination services—whether needed or unneeded—to augment yield of plant species of economic value, like crops. Climate change might also be driving invasions within continents through ecological niche range shifts of native species, and facilitating establishment and range expansion of intentionally translocated species. Introduction of species happen most often as a single translocation event in which only a few members of a given species are brought into a novel environment. Feral populations resulting from such events will inevitably have a reduced genetic pool and will be particularly vulnerable to stochastic events [103, 104].

Thus, most introduction events fail to establish naturalized populations, and the relatively few that achieve critical numbers can take decades to become invasive. Despite this difficulty, even a single introduction event can result in successful invasion, as shown by the point release of fertilized queens that primed the invasion of Patagonia by *Bombus ruderatus* [22]. However, in most recent pollinator invasions, driven by commercial provision of pollination services, the foreign species is introduced in great numbers over a prolonged period of time. Each year, hundreds of thousands of bumblebee hives are brought into new environments to provide pollination services for crops [5, 7]. Because of this, alien bumblebees are mostly safe from stochastic forces, harbor more genetic diversity, and are also actively catered by human care [105, 106]. Such constant subsidies ensure that if a species has any invasive potential at all, then it will eventually become established. Even without invasive potential, introduced pollinators can impact native ecosystems if sustained through human care. For example, honeybees can outcompete local pollinators within environments in which they would not thrive on its own [107]. As long as the pollinator is domestically bred, it might not matter if a natural ecosystem acts as a population sink: the foreign species could exploit resources, displace local species, and spread diseases, as an invasive species would.

Climate change is a driving force of many ecological invasions since climate is one of the main factors defining species distribution [108, 109]. As the thermal landscape of the world is altered, species can venture into previously inaccessible ecosystems where they might become invasive. Climate change can also displace species from their original habitats, liberating niches that foreign species can exploit and, even when not displaced, species can experience a loss of fitness as a result of suboptimal climatic conditions, which can allow low-fitness foreign species to outcompete the native ones, leading to invasion events. The Brazilian bumblebee *Bombus bellicosus*, for example, is slowly being displaced toward colder latitudes to the southernmost regions of Brazil and further into Argentina as a consequence of heat stress suffered in its historical range [110]; this southern shift could potentially make it invasive within the novel range, while facilitating invasions from other species after abandoning parts of its historical range. Other examples of potential invasion driven by climate change are the spread of *Bombus terrestris* and *Bombus lapidarius* above the Arctic Circle [111], the expansion of *B. terrestris* over northern Israel [30], or the arrival to the British islands of *Bombus hypnorum* from mainland Europe [112, 113]. Climate change can also worsen ongoing invasion events, by introducing additional stress on native species and enabling invaders to advance further and faster.

Since organisms are not passive subjects to their environment, but instead regulate their development and behavior in response to climate change and other external challenges [114], it is fundamental to incorporate perspectives that take into account the differential ability of bumblebees to adapt and respond through both phenotypic plasticity and adaptive evolution [115]. This can result in complex interactions with other drivers, like changes in land use: in North America, average bumblebee worker size, measured from museum specimens of four *Bombus* species, decreased from historical (1897–1926) to recent (1978–2016) collections, a change associated with increasingly unfavorable environmental conditions [116]. In contrast, studies in Belgium and Germany have found trends toward an increasing body size of queens and workers [117, 118]; such increase in size does not seem to be a direct response to warmer temperatures, but is instead associated with changes in land use such as agricultural intensification and urbanization. The direction and intensity of these documented trends of size change are species-specific [116–118] and thus can potentially impact overall pollination networks by altering size-dependent plant-pollinator interactions and changing the composition of local pollinator communities. Additionally, invasive species have been shown to shift and expand their climatic niches within their invaded ranges [119], as described for example for the Asian giant resin bee *Megachile sculpturalis* [120, 121] or the bull-headed dung beetle *Onthophagus taurus* [15]. Such examples demonstrate the plasticity and adaptability of invasive species in the face of environmental challenges, and the complex network of interactions between environmental drivers and biological capabilities that need to be traced to be able to predict species invasive potential and introduction and invasion outcomes. In other words, our current state of knowledge is often still insufficient to fully and confidently ascertain the risks of species translocation.

Concluding remarks: do we really need to keep importing bumblebees?

Deployment of commercially reared bumblebee colonies is often portrayed as a biological, or "green," eco-friendly solution to missing or deficient pollination services for crops. However, worldwide implementation has repeatedly resulted in biological invasions that are now near impossible to revert. As we summarized above, these invasions are taking a heavy toll on native ecosystems. Just as it happened with other examples of "biological" solutions (e.g., the worldwide importation of the Asian harlequin ladybeetle *Harmonia axyridis* for pest control [122]), translocation of bumblebees has proven to be far from environmentally friendly. Some countries, like Australia, have thus chosen to ban the importation of bumblebees, despite discontent from the agricultural sector [26]. But what are the alternatives?

A first answer would be the local rearing of native species. Commercial rearing of *Bombus impatiens*, *Bombus vosnesenskii*, and *Bombus huntii* in North America, and *Bombus ignitus* in Japan is ongoing [5, 123, 124]. In South America, commercial rearing of *Bombus pauloensis* (= *B. atratus*) and/or *Bombus bellicosus* is ongoing or being developed in Colombia, Argentina, and Uruguay [125–130]. In Pakistan, *Bombus haemorrhoidalis* is being explored as an option to replace *B. terrestris* [131]. However, rearing conditions favor the growth of *Bombus* pathogens and select for a small fraction of genotypes, leading to potential erosion of genetic diversity of wild populations and pathogen spillover, as mentioned above. Other buzz-pollinating bee species have been explored as viable options in crops that require this specific type of pollination (like tomatoes). For example, a number of native species such as *Examalopsis* spp. and *Augochloropsis* spp. have been explored in Mexico to avoid import of *B. impatiens* [132]. In Brazil, breeding of two stingless bee species (*Tetragonisca angustula* and *Melipona fasciculata*) is widely established but not at the scale needed for crop pollination [133]. Import of *Melipona quadrifasciata* from Brazil to Japan has been suggested, given the very low risk of this tropical species becoming invasive in temperate and cold environments [134]. In Australia, potential native crop pollinators such as *Xylocopa lestis*, *Amegilla* spp., *Austroplebeia australis*, *Tetragonula carbonaria*, and even the introduced syrphid hoverfly *Eristalis tenax* have been explored [135–141]. However, in many cases, the biology of native pollinator species does not allow for cost-efficient, profitable rearing at commercial scale [5].

The well-proven risks of commercial-scale deployment of managed pollinators have prompted a search for technological rather than biological alternatives [142]. Asian engineers developed a proof-of-concept pollination system based on flying drones blowing pollen-laden soap bubbles over flowers [143]. In Israel, two companies (Edete and Arugga) are developing robotic arrays that use optic sensing and artificial intelligence to deliver automated pollen collection and pollination (https://www.edetepta.com, https://www.arugga.com); these technologies are being marketed as eco-friendly alternatives to managed pollinators when wild pollination services are insufficient, although they still need to be benchmarked against them for efficiency and cost-effectiveness. There is always a low-tech fallback: hand-pollination by humans is still widespread [144], especially when labor costs are low [145]. Although this approach is largely unsustainable in the long term, it provides an alternative to ameliorate losses of wild or managed pollination services, or sudden changes in the cost-benefit balance of certain crops [144].

A last and probably best alternative is increasing local pollinator diversity (especially native wild pollinators) instead of (or, at least, in addition to) import of commercially reared bees. In an assemblage of pollinators, behavioral and physiological differences between species can provide a more complete pollination service, covering more hours of the day of flower visitation or providing the crop with a collection of functional traits that can not only increase pollination in some crops such as pumpkin or tomatoes, but also create crops more capable to endure future threats such as climate change and land-use transformations [146–148]. Besides actions and measures to protect overall landscape integrity and species conservation, visitation and pollination by wild species can be locally boosted through a variety of strategies [149]. Boosting pollinator diversity is a necessary move toward ecological intensification of agricultural production and will positively challenge not only the business-as-usual of food production stakeholders but also the current low diversity of crops [150].

Any alternative solutions to importing managed pollinators are meaningless if they never reach the stakeholders requesting them. In many cases, established invasive species are so abundant that deploying commercial colonies might lower rather than enhance crop yields. Since farming stakeholders are often prone to follow advice from their peers, or biased consulting from producers of commercial pollinators, expecting a bottom-up approach to change might result in a change too slow to avoid irreversible ecosystem damage. Therefore, we suggest encouraging a top-down approach in which national and regional regulators take action to stop importation and promote development of alternative approaches, while demanding that trade regulations are fair and reciprocal, unlike the current unbalanced scenario in which countries that would not authorize importation of invasive exotic species to their own territory have no problem exporting native species with invasive potential to other countries [7]. Moreover, it will be of great importance to keep sharing information with stakeholders on the real environmental costs of some apparently "green" eco-friendly solutions and setting an example by asking for (and heeding) expert advice.

Acknowledgments

The authors thank the editor, David Hemming, for his kind invitation to write this review. J.L. is supported by an Internal Doctoral Fellowship awarded by CONICET (National Council for Scientific and Technical Research); N.R.C. is supported by an Internal Postdoctoral Fellowship awarded by CONICET. All authors declare no competing interests. This work was partially funded by the International Cooperation Project SURPASS-RD 2019-1984 awarded by CONICET and the SURPASS2 project under the Newton Fund Latin America Biodiversity Programme: Biodiversity - Ecosystem services for sustainable development, awarded by the UKRI Natural Environment Research Council (NERC) NE/S011870/2.

Author contributions

J.L., N.R.C., N.A., and M.P.A. contributed to conceptualization, writing—original draft, writing—review & editing; E.E.Z contributed to conceptualization, supervision, writing—original draft, and writing—review & editing.

References

1. Goulson D. Bumblebees: behaviour, ecology, and conservation. Oxford: Oxford University Press; 2010. 330 p.
2. Crichton M. Jurassic park. London: Random House; 1991.
3. Goulson D. Effects of introduced bees on native ecosystems. Annual Review of Ecology, Evolution, and Systematics 2003;34:1–26.
4. Schneider SS, DeGrandi-Hoffman G, Smith DR. The African honey bee: factors contributing to a successful biological invasion. Annual Review of Entomology 2004;49:351–76.

5. Velthuis HHW, van Doorn A. A century of advances in bumblebee domestication and the economic and environmental aspects of its commercialization for pollination. Apidologie 2006;37(4):31.
6. Goulson D, Nicholls E, Botías C, Rotheray EL. Bee declines driven by combined stress from parasites, pesticides, and lack of flowers. Science 2015;347(6229):1255957.
7. Aizen MA, Smith-Ramírez C, Morales CL, Vieli L, Sáez A, Barahona-Segovia RM, et al. Coordinated species importation policies are needed to reduce serious invasions globally: the case of alien bumblebees in South America. Journal of Applied Ecology 2019;56:100–6.
8. Williams PH, Cameron SA, Hines HM, Cederberg B, Rasmont P. A simplified subgeneric classification of the bumblebees (genus Bombus). Apidologie 2008;39(1):46–74.
9. Cameron SA, Lim HC, Lozier JD, Duennes MA, Thorp R. Test of the invasive pathogen hypothesis of bumble bee decline in North America. PNAS 2016;113(16):4386–91.
10. Looney C, Strange JP, Freeman M, Jennings D. The expanding Pacific Northwest range of Bombus impatiens Cresson and its establishment in Washington State. Biological Invasions 2019;21(6):1879–85.
11. GBIF.org. Bombus Occurrences Structured Query [Internet]. 2021. DOI: 10.15468/dl.7dvysa. Available from: https://www.gbif.org/occurrence/download/0347458-200613084148143
12. Ratti CM, Colla SR. Discussion of the presence of an eastern bumble bee species (Bombus impatiens Cresson) in western Canada. The Pan-Pacific Entomologist. 2010;86(2):29–31.
13. Sachman-Ruiz B, Narváez-Padilla V, Reynaud E. Commercial Bombus impatiens as reservoirs of emerging infectious diseases in central México. Biological Invasions 2015;17(7):2043–53.
14. Fernández M, Hamilton H. Ecological niche transferability using invasive species as a case study. PLoS One 2015;10(3):e0119891.
15. Silva DP, Vilela B, Buzatto BA, Moczek AP, Hortal J. Contextualized niche shifts upon independent invasions by the dung beetle Onthophagus taurus. Biological Invasions 2016;5:1–12.
16. Liu C, Wolter C, Xian W, Jeschke JM. Most invasive species largely conserve their climatic niche. PNAS 2020;117(38):23643–51.
17. Kosior A, Celary W, Olejniczak P, Fijał J, Król W, Solarz W, et al. The decline of the bumble bees and cuckoo bees (Hymenoptera: Apidae: Bombini) of Western and Central Europe. Oryx 2007;41(1):79–88.
18. Hopkins I. History of the bumblebee in New Zealand: its introduction and results. New Zealand Department of Agriculture, Industry and Commerce). Report No.: 46; 1914. p. 29.
19. Lye GC, Lepais O, Goulson D. Reconstructing demographic events from population genetic data: the introduction of bumblebees to New Zealand. Molecular Ecology 2011;20(14):2888–900.
20. Gurr L. The distribution of bumblebees in the South Island of New Zealand. New Zealand Journal of Science 1964;7(4):625–42.
21. Macfarlane RP, Gurr L. Distribution of bumble bees in New Zealand. New Zealand Entomologist 1995;18(1):29–36.
22. Arretz PV, Macfarlane RP. The introduction of Bombus Ruderatus to Chile for red clover pollination. Bee World 1986;67(1):15–22.
23. Goulson D, Hanley ME. Distribution and forage use of exotic bumblebees in South Island, New Zealand. New Zealand Journal of Ecology 2004;28(2):225–32.
24. Morales CL, Arbetman MP, Cameron SA, Aizen MA. Rapid ecological replacement of a native bumble bee by invasive species. Frontiers in Ecology and the Environment 2013;11(10):529–34.
25. Heinrich B. Bumblebee Economics. Harvard University Press; 2004. 296 p.
26. Cooley H, Vallejo-Marín M. Buzz-pollinated crops: a global review and meta-analysis of the effects of supplemental bee pollination in Tomato. Journal of Economic Entomology 2021;114(2):505–19.
27. Dafni A. The threat of Bombus terrestris spread. Bee World 1998;79(3):113–4.
28. Banda HJ, Paxton RJ. Pollination of greenhouse tomatoes by bees. Acta Horticulturae 1991;(288):194–8.
29. Potts SG, Willmer P, Dafni A, Ne'eman G. The utility of fundamental ecological research of plant-pollinator interactions as the basis for landscape management practices. Acta Horticulturae 2001;(561):141–4.
30. Dafni A, Shmida A. International Bee Research Association, & Linnean Society of London. The possible ecological implications of the invasion of Bombus terrestris (L.) (Apidae) at Mr Carmel, Israel. In: Conservation of bees. Vol. 18. Cambridge: Academic Press; 1996. pp. 183–200. ISBN: 0124797407.
31. Dafni A, Kevan P, Gross CL, Goka K. *Bombus terrestris,* pollinator, invasive and pest: an assessment of problems associated with its widespread introductions for commercial purposes. Applied Entomology and Zoology 2010;45(1):101–13.

32. Hingston AB, Marsden-Smedley J, Driscoll DA, Corbett S, Fenton J, Anderson R, et al. Extent of invasion of Tasmanian native vegetation by the exotic bumblebee Bombus terrestris (Apoidea: Apidae). Austral Ecology 2002;27(2):162–72.
33. Goka K. Influences of invasive species on native species. Will the European Bumble Bee, Bombus terrestris, bring genetic pollution into the Japanese native species? Bulletin of Biogeographical Society of Japan 1998;53(2):91–101.
34. Matsumura C, Yokoyama J, Washitani I. Invasion status and potential ecological impacts of an invasive alien bumblebee, *Bombus terrestris* L. (Hymenoptera: Apidae) naturalized in southern Hokkaido, Japan. Global Environmental Research 2004;8(1):51–66.
35. Tsuchida K, Yamaguchi A, Kanbe Y, Goka K. Reproductive interference in an introduced Bumblebee: polyandry may mitigate negative reproductive impact. Insects 2019;10(2):59.
36. Inoue MN, Yokoyama J, Washitani I. Displacement of Japanese native bumblebees by the recently introduced *Bombus terrestris* (L.) (Hymenoptera: Apidae). Journal of Insect Conservation 2008;12(2):135–46.
37. Inoue MN, Yokoyama J. Competition for flower resources and nest sites between *Bombus terrestris* (L.) and Japanese native bumblebees. Applied Entomology and Zoology 2010;45(1):29–35.
38. Mizutani T, Goka K. Japan's *Invasive Alien Species Act*. Applied Entomology and Zoology 2010;45(1):65–9.
39. Goka K. Introduction to the special feature for ecological risk assessment of introduced bumblebees: status of the European bumblebee, *Bombus terrestris*, in Japan as a beneficial pollinator and an invasive alien species. Applied Entomology and Zoology 2010;45(1):1–6.
40. Winter K, Adams L, Thorp RW, Inouye D, Day L, Ascher J, et al. Importation of non-native bumble bees into North America: potential consequences of using *Bombus terrestris* and other non-native bumble bees for greenhouse crop pollination in Canada, Mexico, and the United States [Internet]. North American Pollinator Protection Campaign; 2006. p. 33. Available from: http://cues.cfans.umn.edu/old/pollinators/pdf-BBcolony/2006nonativebee.pdf
41. Smith-Ramírez C, Vieli L, Barahona-Segovia RM, Montalva J, Cianferoni F, Ruz L, et al. Las razones de por qué Chile debe detener la importación del abejorro comercial Bombus terrestris (Linnaeus) y comenzar a controlarlo. Gayana (Concepción) 2018;82(2):118–27.
42. Morales CL, Aizen MA. Does invasion of exotic plants promote invasion of exotic flower visitors? A case study from the temperate forests of the Southern Andes. Biological Invasions 2002;4(1):87–100.
43. Ings TC, Ward NL, Chittka L. Can commercially imported bumble bees out-compete their native conspecifics? Journal of Applied Ecology 2006;43(5):940–8.
44. Hingston AB, McQuillan PB. Does the recently introduced bumblebee *Bombus terrestris* (Apidae) threaten Australian ecosystems? Australian Journal of Ecology 1998;23(6):539–49.
45. Ono M. Ecological implications of introducing *Bombus terrestris*, and significance of domestication of Japanese native bumblebees (Bombus spp.). In: Proc Int Workshop on Biological Invasions of Ecosystem by Pests and Beneficial Organisms. Tsukuba: National Institute of Agro-Environmental Sciences; 1997. p. 244–52.
46. Prŷs-Jones OE, Corbet SA. Bumblebees. Exeter: Pelagic Publishing; 2011. 130 p.
47. Otterstatter MC, Thomson JD. Does pathogen spillover from commercially reared bumble bees threaten wild pollinators? PLoS One 2008;3(7):e2771.
48. Meeus I, Brown MJF, Graaf DCD, Smagghe G. Effects of invasive parasites on bumble bee declines. Conservation Biology 2011;25(4):662–71.
49. Graystock P, Blane EJ, McFrederick QS, Goulson D, Hughes WOH. Do managed bees drive parasite spread and emergence in wild bees? International Journal for Parasitology: Parasites and Wildlife 2016;5(1):64–75.
50. Arbetman MP, Meeus I, Morales CL, Aizen MA, Smagghe G. Alien parasite hitchhikes to Patagonia on invasive bumblebee. Biological Invasions 2012;15(3):489–94.
51. Maharramov J, Meeus I, Maebe K, Arbetman M, Morales C, Graystock P, et al. Genetic variability of the Neogregarine Apicystis bombi, an etiological agent of an emergent bumblebee disease. PLoS One 2013;8(12):e81475.
52. Schmid-Hempel R, Eckhardt M, Goulson D, Heinzmann D, Lange C, Plischuk S, et al. The invasion of southern South America by imported bumblebees and associated parasites. The Journal of Animal Ecology 2014;83(4):823–37.
53. Goka K, Okabe K, Yoneda M, Niwa S. Bumblebee commercialization will cause worldwide migration of parasitic mites. Molecular Ecology 2001;10(8):2095–9.

54. Arbetman MP, Gleiser G, Morales CL, Williams P, Aizen MA. Global decline of bumblebees is phylogenetically structured and inversely related to species range size and pathogen incidence. Proceedings of the Royal Society B 2017;284(1859):20170204.
55. Otti O, Schmid-Hempel P. A field experiment on the effect of Nosema bombi in colonies of the bumblebee *Bombus terrestris*. Ecological Entomology 2008;33(5):577–82.
56. Skyrm KM, Rao S, Stephen WP. A scientific note on a trend towards bivoltinism in Western North American bumblebees. Apidologie 2012;43(1):82–4.
57. Potapov G, Kolosova Y, Bolotov IN. Possible bivoltine development of several bumblebee species in Europe. Arctic Environmental Research 2018;18(1):45–51.
58. Lundberg H, Svensson BG. Studies on the behaviour of Bombus Latr. species (Hym., Apidae) parasitized by Sphaerularia bombi Dufour (Nematoda) in an alpine area. Norwegian Journal of Entomology 1975;22(2):129–34.
59. Colgan TJ, Carolan JC, Sumner S, Blaxter ML, Brown MJF. Infection by the castrating parasitic nematode *Sphaerularia bombi* changes gene expression in *Bombus terrestris* bumblebee queens. Insect Molecular Biology 2020;29(2):170–82.
60. Rutrecht ST, Brown MJF. The life-history impact and implications of multiple parasites for bumble bee queens. International Journal for Parasitology 2008;38(7):799–808.
61. Shykoff JA, Schmid-Hempel P. Incidence and effects of four parasites in natural populations of bumble bees in Switzerland. Apidologie 1991;22(2):117–25.
62. Brown MJF, Schmid-Hempel R, Schmid-Hempel P. Strong context-dependent virulence in a host–parasite system: reconciling genetic evidence with theory. Journal of Animal Ecology 2003;72(6):994–1002.
63. Yourth CP, Brown MJF, Schmid-Hempel P. Effects of natal and novel *Crithidia bombi* (Trypanosomatidae) infections on Bombus terrestris hosts. Insectes Sociaux 2008;55:86–90. DOI: 10.1007/s00040-007-0974-1
64. Rutrecht ST, Klee J, Brown MJF. Horizontal transmission success of *Nosema bombi* to its adult bumble bee hosts: effects of dosage, spore source and host age. Parasitology 2007;134(12):1719–26.
65. van der Steen JJM. Infection and transmission of *Nosema bombi* in *Bombus terrestris* colonies and its effect on hibernation, mating and colony founding. Apidologie 2008;39(2):273–82.
66. Gegear RJ, Otterstatter MC, Thomson JD. Bumble-bee foragers infected by a gut parasite have an impaired ability to utilize floral information. Proceedings of the Royal Society of London B: Biological Sciences 2006;273(1590):1073–8.
67. Folly AJ, Koch H, Stevenson PC, Brown MJF. Larvae act as a transient transmission hub for the prevalent bumblebee parasite *Crithidia bombi*. Journal of Invertebrate Pathology 2017;148:81–5.
68. Graystock P, Meeus I, Smagghe G, Goulson D, Hughes WOH. The effects of single and mixed infections of *Apicystis bombi* and deformed wing virus in *Bombus terrestris*. Parasitology 2016;143(3):358–65.
69. Meeus I, de Miranda JR, de Graaf DC, Wäckers F, Smagghe G. Effect of oral infection with Kashmir bee virus and Israeli acute paralysis virus on bumblebee (*Bombus terrestris*) reproductive success. Journal of Invertebrate Pathology 2014;121:64–9.
70. Singh R, Levitt AL, Rajotte EG, Holmes EC, Ostiguy N, van Engelsdorp D, et al. RNA viruses in hymenopteran pollinators: evidence of inter-taxa virus transmission via pollen and potential impact on non-apis hymenopteran species. PLOS ONE 2010;5(12):e14357.
71. Fürst MA, McMahon DP, Osborne JL, Paxton RJ, Brown MJF. Disease associations between honeybees and bumblebees as a threat to wild pollinators. Nature 2014;506(7488):364–6.
72. Otti O, Schmid-Hempel P. *Nosema bombi*: a pollinator parasite with detrimental fitness effects. Journal of Invertebrate Pathology 2007;96(2):118–24.
73. Rasmont P, Coppee A, Michez D, De Meulemeester T. An overview of the *Bombus terrestris* (L. 1758) subspecies (Hymenoptera: Apidae). Annales de la Société entomologique de France (NS). 2008;44(2):243–50.
74. Moreira AS, Horgan FG, Murray TE, Kakouli-Duarte T. Population genetic structure of *Bombus terrestris* in Europe: isolation and genetic differentiation of Irish and British populations. Molecular Ecology 2015;24(13):3257–68.
75. Lecocq T, Rasmont P, Harpke A, Schweiger O. Improving international trade regulation by considering intraspecific variation for invasion risk assessment of commercially traded species: the *Bombus terrestris* case. Conservation Letters 2015;9(4):281–9.

76. Todesco M, Pascual MA, Owens GL, Ostevik KL, Moyers BT, Hübner S, et al. Hybridization and extinction. Evolutionary Applications 2016;9(7):892–908.
77. Seabra SG, Silva SE, Nunes VL, Sousa VC, Martins J, Marabuto E, et al. Genomic signatures of introgression between commercial and native bumblebees, *Bombus terrestris*, in western Iberian Peninsula—Implications for conservation and trade regulation. Evolutionary Applications 2019;12(4):679–91.
78. Cejas D, López-López A, Muñoz I, Ornosa C, Rúa PD la. Unveiling introgression in bumblebee (*Bombus terrestris*) populations through mitogenome-based markers. Animal Genetics 2020;51(1):70–7.
79. Bartomeus I, Molina FP, Hidalgo-Galiana A, Ortego J. Safeguarding the genetic integrity of native pollinators requires stronger regulations on commercial lines. Ecological Solutions and Evidence. 2020;1(1):e12012.
80. Kraus FB, Szentgyörgyi H, Rożej E, Rhode M, Moroń D, Woyciechowski M, et al. Greenhouse bumblebees (*Bombus terrestris*) spread their genes into the wild. Conservation Genetics 2011;12(1):187–92.
81. Hart AF, Maebe K, Brown G, Smagghe G, Ings T. Winter activity unrelated to introgression in British bumblebee *Bombus terrestris audax*. Apidologie 2021;52(2):315–27.
82. Suni SS, Scott Z, Averill A, Whiteley A. Population genetics of wild and managed pollinators: implications for crop pollination and the genetic integrity of wild bees. Conservation Genetics 2017;18(3):667–77.
83. Kanbe Y, Okada I, Yoneda M, Goka K, Tsuchida K. Interspecific mating of the introduced bumblebee *Bombus terrestris* and the native Japanese bumblebee Bombus hypocrita sapporoensis results in inviable hybrids. Die Naturwissenschaften 2008;95(10):1003–8.
84. Kondo NI, Yamanaka D, Kanbe Y, Kunitake YK, Yoneda M, Tsuchida K, et al. Reproductive disturbance of Japanese bumblebees by the introduced European bumblebee *Bombus terrestris*. Die Naturwissenschaften 2009;96(4):467–75.
85. Mitsuhata M, Ono M. Hybridization between Japanese and European bumblebees (Bombus spp.). Summaries of the 7th International Pollination Symposium, Lethbridge, Albarta, Canada, 1996. 1996. Available from: https://ci.nii.ac.jp/naid/10012498064/
86. Yoon H-J, Kim S-Y, Lee K-Y, Lee S-B, Park I-G, Kim I-S. Interspecific hybridization of the Bumblebees *Bombus ignitus* and *B. terrestris*. International Journal of Industrial Entomology 2009;18(1):41–8.
87. Inoue MN, Saito F, Tsuchida K, Goka K. Potential increase in mating frequency of queens in feral colonies of *Bombus terrestris* introduced into Japan. Die Naturwissenschaften 2012;99(10):853–61.
88. Schmid-Hempel R, Schmid-Hempel P. Female mating frequencies in Bombus spp. from Central Europe. Insectes Sociaux 2000;47(1):36–41.
89. Payne CM, Laverty TM, Lachance MA. The frequency of multiple paternity in bumble bee (Bombus) colonies based on microsatellite DNA at the B10 locus. Insectes Sociaux 2003;50(4):375–8.
90. Takahashi J, Itoh M, Shimizu I, Ono M. Male parentage and queen mating frequency in the bumblebee *Bombus ignitus* (Hymenoptera: bombinae). Ecological Research 2008;23(6):937–42.
91. Morales CL, Sáez A, Garibaldi LA, Aizen MA. Chapter 13. Disruption of pollination services by invasive pollinator species. In: Vilà M, Hulme PE, editors. Impact of biological invasions on ecosystem services. Basel: Springer; 2017. p. 203–20. (Invading Nature - Springer Series in Invasion Ecology).
92. Aizen MA. Flower sex ratio, pollinator abundance, and the seasonal pollination dynamics of a protandrous plant. Ecology 2001;82(1):127–44.
93. Madjidian JA, Morales CL, Smith HG. Displacement of a native by an alien bumblebee: lower pollinator efficiency overcome by overwhelmingly higher visitation frequency. Oecologia 2008;156(4):835–45.
94. Sanguinetti A, Singer RB. Invasive bees promote high reproductive success in Andean orchids. Biological Conservation 2014;175:10–20.
95. Kenta T, Inari N, Nagamitsu T, Goka K, Hiura T. Commercialized European bumblebee can cause pollination disturbance: an experiment on seven native plant species in Japan. Biological Conservation 2007;134(3):298–309.
96. Dohzono I, Kunitake YK, Yokoyama J, Goka K. Alien bumble bee affects native plant reproduction through interactions with native bumble bees. Ecology 2008;89(11):3082–92.
97. Graham M. Efecto del robo de néctar por el abejorro invasor Bombus terrestris sobre la polinización y el éxito reproductivo de la arvejilla nativa Vicia nigricans (Fabaceae) [Tesis para optar al grado de licenciado en ciencias biológicas.]. Bariloche: Universidad Nacional del Comahue; 2015.
98. Dick CW, Etchelecu G, Austerlitz F. Pollen dispersal of tropical trees (Dinizia excelsa: Fabaceae) by native insects and African honeybees in pristine and fragmented Amazonian rainforest. Molecular Ecology 2003;12(3):753–64.

99. Dupont YL, Hansen DM, Valido A, Olesen JM. Impact of introduced honey bees on native pollination interactions of the endemic Echium wildpretii (Boraginaceae) on Tenerife, Canary Islands, Biological Conservation 2004;118(3):301–11.
100. Whelan RJ, Ayre DJ, Beynon FM. The birds and the bees: pollinator behaviour and variation in the mating system of the rare shrub Grevillea macleayana. Annals of Botany 2009;103(9):1395–401.
101. Aizen MA, Morales CL, Vázquez DP, Garibaldi LA, Sáez A, Harder LD. When mutualism goes bad: density-dependent impacts of introduced bees on plant reproduction. The New Phytologist 2014;204(2):322–8.
102. Sáez A, Morales CL, Ramos LY, Aizen MA. Extremely frequent bee visits increase pollen deposition but reduce drupelet set in raspberry. Journal of Applied Ecology 2014;51(6):1603–12.
103. May RM. Stability in multispecies community models. Mathematical Biosciences 1971;12(1):59–79.
104. Lande R. Risks of population extinction from demographic and environmental stochasticity and random catastrophes. The American Naturalist 1993;142(6):911–27.
105. Simberloff D. The role of propagule pressure in biological invasions. Annual Review of Ecology, Evolution, and Systematics 2009;40(1):81–102.
106. Fauvergue X, Vercken E, Malausa T, Hufbauer RA. The biology of small, introduced populations, with special reference to biological control. Evolutionary Applications 2012;5(5):424–43.
107. Herrera CM. Gradual replacement of wild bees by honeybees in flowers of the Mediterranean Basin over the last 50 years. Proceedings of the Royal Society B: Biological Sciences 2020;287(1921):20192657.
108. Walther G-R, Roques A, Hulme PE, Sykes MT, Pyšek P, Kühn I, et al. Alien species in a warmer world: risks and opportunities. Trends in Ecology & Evolution 2009;24(12):686–93.
109. Bellard C, Thuiller W, Leroy B, Genovesi P, Bakkenes M, Courchamp F. Will climate change promote future invasions? Global Change Biology 2013;19(12):3740–8.
110. Martins AC, Silva DP, Marco PD, Melo GAR. Species conservation under future climate change: the case of Bombus. Journal of Insect Conservation 2014;19(1):33–43.
111. Martinet B, Rasmont P, Cederberg B, Evrard D, Ødegaard F, Paukkunen J, et al. Forward to the north: two Euro-Mediterranean bumblebee species now cross the Arctic Circle. Annales de la Société entomologique de France (NS) 2015;51(4):303–9.
112. Goulson D, Williams P. *Bombus hypnorum* (L.) (Hymenoptera: Apidae), a new British bumblebee? British Journal of Entomology and Natural History. 2001;14(3):129–31.
113. Crowther LP, Hein P-L, Bourke AFG. Habitat and forage associations of a naturally colonising insect pollinator, the tree bumblebee *Bombus hypnorum*. PLoS One 2014;9(9):e107568.
114. Sultan SE, Moczek AP, Walsh D. Bridging the explanatory gaps: what can we learn from a biological agency perspective? BioEssays. 2022;n/a(n/a):2100185.
115. Maebe K, Hart AF, Marshall L, Vandamme P, Vereecken NJ, Michez D, et al. Bumblebee resilience to climate change, through plastic and adaptive responses. Global Change Biology 2021;27(18):4223–37.
116. Nooten SS, Rehan SM. Historical changes in bumble bee body size and range shift of declining species. Biodiversity and Conservation 2020;29(2):451–67.
117. Gérard M, Martinet B, Maebe K, Marshall L, Smagghe G, Vereecken NJ, et al. Shift in size of bumblebee queens over the last century. Global Change Biology 2020;26(3):1185–95.
118. Theodorou P, Baltz LM, Paxton RJ, Soro A. Urbanization is associated with shifts in bumblebee body size, with cascading effects on pollination. Evolutionary Applications 2021;14(1):53–68.
119. Hill MP, Gallardo B, Terblanche JS. A global assessment of climatic niche shifts and human influence in insect invasions. Global Ecology and Biogeography 2017;26(6):679–89.
120. Lanner J, Huchler K, Pachinger B, Sedivy C, Meimberg H. Dispersal patterns of an introduced wild bee, Megachile sculpturalis Smith, 1853 (Hymenoptera: Megachilidae) in European alpine countries. PLoS One 2020;15(7):e0236042.
121. Polidori C, Sánchez-Fernández D. Environmental niche and global potential distribution of the giant resin bee *Megachile sculpturalis*, a rapidly spreading invasive pollinator. Global Ecology and Conservation 2020;24:e01365.
122. Brown PMJ, Thomas CE, Lombaert E, Jeffries DL, Estoup A, Lawson Handley L-J. The global spread of *Harmonia axyridis* (Coleoptera: Coccinellidae): distribution, dispersal and routes of invasion. BioControl 2011;56(4):623–41.
123. Koppert. Quad [Internet]. [cited 2021 Nov 21]. Available from: https://www.koppertus.com/quad/
124. Biobest. Hive types | Biobest [Internet]. [cited 2021 Nov 21]. Available from: https://www.biobest-group.com/en/biobest/pollination/things-to-know-about-bumblebees-7052/hive-types-6683/

125. Cruz P, Escobar A, Almanza MT, Cure JR. Implementación de mejoras para la cría en cautiverio de colonias del abejorro nativo Bombus pauloensis (= B. atratus) (Hymenoptera: Apoidea). Revista Facultad de Ciencias Básicas 2008;4(1–2):70–83.
126. Salvarrey S, Arbulo N, Santos E, Invernizzi C. Cría artificial de abejorros nativos Bombus atratus y Bombus bellicosus (Hymenoptera, Apidae). Agrociencia (Uruguay). 2013;17(2):75–82.
127. Padilla SC, Cure JR, Riaño DA, Gutierrez AP, Rodriguez D, Romero E. Gyne and drone production in bombus atratus (Hymenoptera: Apidae). Journal of Apicultural Science 2017;61(1):55–72.
128. Cavigliasso P, Bello F, Rivadeneira MF, Monzon NO, Gennari GB, Basualdo M. Pollination efficiency of managed bee species (*Apis mellifera* and *Bombus pauloensis*) in highbush blueberry (*Vaccinium corymbosum*) productivity. Journal of Horticultural Research. 2020;28(1):57–64. Available from: http://yadda.icm.edu.pl/yadda/element/bwmeta1.element.agro-4c159450-3fa9-449d-b0a9-276493ae4e70
129. Estravis-Barcala MC, Palottini F, Macri I, Nery D, Farina WM. Managed honeybees and South American bumblebees exhibit complementary foraging patterns in highbush blueberry. Scientific Reports 2021;11(1):8187.
130. Salvarrey S, Antúnez K, Arredondo D, Plischuk S, Revainera P, Maggi M, et al. Parasites and RNA viruses in wild and laboratory reared bumble bees *Bombus pauloensis* (Hymenoptera: Apidae) from Uruguay. PLoS One 2021;16(4):e0249842.
131. Sharma HK, Thakur S, Rana K, Thakur M. Small scale rearing of Bombus haemorrhoidalis Smith. Pharma Innovation 2018;7(11):41–5.
132. Torres-Ruiz A, Jones RW, Barajas RA. Present and potential use of bees as managed pollinators in Mexico. Swen 2013;38(1):133–48.
133. Giannini TC, Alves DA, Alves R, Cordeiro GD, Campbell AJ, Awade M, et al. Unveiling the contribution of bee pollinators to Brazilian crops with implications for bee management. Apidologie 2020;51(3):406–21.
134. Hikawa M, Miyanaga R. Effects of pollination by *Melipona quadrifasciata* (Hymenoptera: Apidae) on tomatoes in protected culture. Applied Entomology and Zoology 2009;44(2):301–7.
135. Hogendoorn K, Steen Z, Schwarz MP. Native Australian carpenter bees as a potential alternative to introducing bumble bees for tomato pollination in greenhouses. Journal of Apicultural Research 2000;39(1–2):67–74.
136. Hogendoorn K, Gross CL, Sedgley M, Keller MA. Increased tomato yield through pollination by native Australian Amegilla chlorocyanea (Hymenoptera: Anthophoridae). Journal of Economic Entomology 2006;99(3):828–33.
137. Bell MC, Spooner-hart RN, Haigh AM. Pollination of greenhouse tomatoes by the Australian Bluebanded Bee Amegilla (Zonamegilla) holmesi (Hymenoptera: Apidae). Journal of Economic Entomology 2006;99(2):437–42.
138. Halcroft M. Investigations into the biology, behaviour and phylogeny of a potential crop pollinator: the Australian stingless bee, Austroplebeia australis [Internet]. Sydney: University of Western Sydney; 2012. Available from: https://researchdirect.westernsydney.edu.au/islandora/object/uws%3A14329/
139. Switzer CM, Hogendoorn K, Ravi S, Combes SA. Shakers and head bangers: differences in sonication behavior between Australian Amegilla murrayensis (blue-banded bees) and North American Bombus impatiens (bumblebees). Arthropod-Plant Interactions 2016; 10(1):1–8.
140. Howlett BG, Gee M. The potential management of the drone fly (*Eristalis tenax*) as a crop pollinator in New Zealand. New Zealand Plant Protection 2019;72:221–30.
141. Garcia Bulle Bueno F. Native bees as alternative crop pollinators: reproductive behaviour of *Tetragonula carbonaria* [Internet] [Thesis]. 2021. Available from: https://ses.library.usyd.edu.au/handle/2123/25582
142. Kimbrough L. Bubbles, lasers and robo-bees: the blossoming industry of artificial pollination [Internet]. Mongabay Environmental News. 2020. Available from: https://news.mongabay.com/2020/07/bubbles-lasers-and-robo-bees-is-artificial-pollination-here-to-stay/
143. Yang X, Miyako E. Soap bubble pollination. iScience. 2020;23(6):101188.
144. Wurz A, Grass I, Tscharntke T. Hand pollination of global crops—a systematic review. Basic and Applied Ecology. 2021. Available from: https://www.sciencedirect.com/science/article/pii/S1439179121001286
145. Partap U, Ya T. The human pollinators of fruit crops in Maoxian County, Sichuan, China. Mountain Research and Development 2012;32(2):176–86.
146. Albrecht M, Schmid B, Hautier Y, Müller CB. Diverse pollinator communities enhance plant reproductive success. Proceedings of the Royal Society B: Biological Sciences 2012;279(1748):4845–52.

147. Hoehn P, Tscharntke T, Tylianakis JM, Steffan-Dewenter I. Functional group diversity of bee pollinators increases crop yield. Proceedings of the Royal Society B: Biological Sciences 2008;275(1648):2283–91.
148. Brittain C, Kremen C, Klein A-M. Biodiversity buffers pollination from changes in environmental conditions. Global Change Biology 2013;19(2):540–7.
149. Garibaldi LA, Carvalheiro LG, Leonhardt SD, Aizen MA, Blaauw BR, Isaacs R, et al. From research to action: enhancing crop yield through wild pollinators. Frontiers in Ecology and the Environment 2014;12(8):439–47.
150. Aizen MA, Aguiar S, Biesmeijer JC, Garibaldi LA, Inouye DW, Jung C, et al. Global agricultural productivity is threatened by increasing pollinator dependence without a parallel increase in crop diversification. Global Change Biology 2019;25(10):3516–27.

5 Invasive alien species in changing marine Arctic economies and ecosystems

Brooks A. Kaiser[1*] and Melina Kourantidou[1,2,3]
[1]Department of Sociology, Environmental and Business Economics, University of Southern Denmark, 6705, Esbjerg Ø, Denmark; [2]Hellenic Centre for Marine Research, Institute of Marine Biological Resources and Inland Waters, 16452, Athens, Greece; [3]Woods Hole Oceanographic Institution, Marine Policy Center, Woods Hole, MA, 02543, USA

Abstract
The rate of change in Arctic marine environments in response to shifts driven by climate change threatens Arctic resilience. The growing recognition and visibility of these changes have scientific and social roots. Mitigating these consequences is therefore a social-scientific concern. Multiple scales, perspectives, and governance systems for Arctic marine environments, alongside receding climate and economic barriers to species movements and scientific research, create challenges and opportunities that differ in magnitude and breadth from marine invasions elsewhere. The receding barriers in the marine Arctic amplify the potential ecological and economic consequences from new species introductions and range expansions from adjacent biomes. While there is consensus that marine invasive species can cause severe damages to ecosystems and resource-dependent communities, which species pose what threats, and to whom, remain complex dynamic socioecological and biogeophysical economic questions. Decisions over prevention, detection, and monitoring along with institutional frameworks for cooperating and responding to threats also affect the expected severity of impacts. Technologies, and costs, for identifying and monitoring species compositions and risks are evolving, with novel research advances as well as increasingly sophisticated ecological-economic, environmental niche, and habitat suitability models. Despite advances in understanding drivers and dynamics of new species introductions, a dearth of baseline knowledge regarding Arctic marine invasions remains. Potential consequences extend beyond ecosystem changes and include legal, institutional, and social shifts. Studies on the red king and snow crab invasions in the Barents Sea from multiple disciplinary angles showcase complex social, economic, and ecological interconnections that are transforming communities and ecosystems.

Keywords: Arctic marine ecosystems, nonindigenous species (NIS), invasive alien species (IAS), prevention, detection, and monitoring of invasive alien species, ballast water and biofouling, vessel traffic and climate change

Review Methodology
The articles included in this review were selected primarily via bibliographic tracing of recent key articles on Arctic marine invasions and warming known to the authors.

*Corresponding author: baka@sdu.dk

Introduction

Arctic marine environments increasingly face disruptions from the establishment of invasive species as economic and ecological barriers fall. Globally expanding trade networks, transportation, climate change, and disruptive anthropogenic pressures on resources increase the likelihoods of introduction and establishment of marine invasive species, shift ecosystem dynamics in ways that are not always well understood, and pose large challenges to economies and human and social well-being [1].

An invasive alien species (IAS) is a species outside of its native range that has the potential to cause harm to human and ecosystem health and well-being. Comprehensive presentations of additional and alternative related terms, including aquatic or marine invasive species, can be found in, for example, Beck *et al.* [2], Falk-Petersen *et al.* [3], or Lovell *et al.* [4]. While this review focuses on marine invasive species, we use the broader term IAS both because many of the research concerns for marine species in the Arctic hold for other ecosystems as well and because marine and trade transport routes, already the most prevalent forms of transportation in most Arctic regions, are expected to become even more so as climate changes reduce land transport options and increase marine ones [5, 6]. The terms are, in all cases, spatially explicit and emphasize that the ecological and human, including economic, systems in the relevant region contextualize the potential consequences. While direct and indirect threats from IAS are neither new globally nor in the region [7–9], Arctic marine environments often have been perceived as low risk for bioinvasions due to limited anthropogenic activity, icy barriers, and unfavorable conditions for many species' survival, growth, reproduction, and dispersal [10, 11]. Today, the recognition and expectations of potential consequences are growing due to the rapid changes in the Arctic marine environment and to the increased visibility of these changes [7, 12–16]. Furthermore, potential consequences extend beyond ecosystem changes and include legal, institutional, and social shifts [17–19]. The growing recognition and visibility of changes have scientific and social roots. Mitigating these consequences is therefore a social-scientific concern.

Humanity has largely missed the planetary-scale prevention opportunity with climate change, which includes impacts to Arctic marine environments through ocean warming and acidification [20, 21]. Opportunities for successful precautionary approaches still exist and the sooner action is taken, the more options are preserved [22]. Governance for marine invasive species in the Arctic can precede rather than follow economic development [23, 24], but the window is small as human activities in the region continue to grow. High-level interdisciplinary cooperation is currently underway through Arctic Council scientific working groups such as the Conservation of Arctic Flora and Fauna and Protection of the Arctic Marine Environment (PAME) [25]. This cooperation recognizes this window of opportunity to act ahead of species movements and provides a strong platform for shaping policies, identifying research needs and priorities, and promoting coordinated responses. Comprehensive, resilient, and effective action will also require integrated management across stakeholders, rightsholders, and local users [26, 27], as well as vertically and horizontally across regions and borders [27].

Invasions in aquatic systems, and in the marine environment specifically, are generally more challenging to monitor than those in terrestrial systems. With limited capacity for monitoring and early detection, IAS are likely to first attract researchers' and/or resource managers' attention only at an advanced stage of the invasion [28]. At this point, control may be prohibitively costly and impacts hard to mitigate. For the Arctic, these concerns are amplified by long but unevenly documented historical use of marine environments that has been coupled with little opportunity for scientific identification of change in those environments [11]. Technological advances are expected to contribute to broader, more effective, and more comprehensive monitoring. For example, the rapid growth in understanding and use of environmental DNA and DNA barcoding for identifying and monitoring species and ecosystem changes holds promise to, inter alia, lower costs and increased coverage of ecosystem components and locations [12, 29–31], while biotech innovations can reduce introductions through common vectors like ballast water and hull fouling [32].

Overview of research, governance, and policy gaps and advances

Research on marine IAS in the Arctic is growing across many disciplines, reflecting the wide range of concerns and considerations that touch on understanding and managing these challenges. An overview of key knowledge gaps, pressures, and risks and their interconnections is presented in Fig. 5.1. Kourantidou *et al.* [14] have described the state of knowledge and identified and assessed patterns in the knowledge gaps regarding Arctic marine invasions that affect the ability to generate improved governance outcomes. Thus, this paper focuses on the period since 2015.

To date, much effort from multiple disciplines has been focused on identifying and/or characterizing newly introduced marine species and/or those that are proximate threats [11, 33–45], understanding key drivers of introductions (particularly related to maritime shipping) and interactions with warming conditions [1, 16, 18, 43, 46–63], and projecting habitat suitability. Advances in understanding invasion drivers and mechanisms are developing across multiple dimensions. Many researchers are integrating natural and social sciences at the species, ecosystem, and socio-ecological system levels [22, 34, 64–75]; they are also addressing the complex dynamics of ecological, social, economic, and legal implications from IAS [17, 19] and/or range expansions or shifts from climate change [65, 76]. Policies and initiatives targeting prevention and coordination among key actors and nations have also advanced along with the expanded knowledge base [24, 25, 77–84]. Despite the growing body of research and policy frameworks, key knowledge gaps remain that limit the ability to assess progress and delineate priorities for future research and governance directions [15, 52, 85]. We do not attempt to provide an exhaustive review of this literature but focus instead on exploring recent advances, persisting challenges, opportunities, and ways in which today's improved understanding may drive successful design of invasive species policies.

A growing rate of known species introductions

The number of known nonindigenous species (NIS), that is, species that are nonnative to an area, regardless of expected impacts from their presence, and the subset of IAS in the marine Arctic remains small [11]. Most scholars agree, however, that an increase in introductions should be expected with expanding anthropogenic activity and climate change, which may also facilitate establishment [11, 15]. The overall marine faunal biomass in the Arctic may also increase with increased global temperatures [86, 87], increasing the total pressure from climate change–induced invasion success [87]. At the same time, detection and monitoring of both new and existing IAS and their impacts are becoming markedly easier and less expensive [29, 64, 88, 89]. This is also the case for new prevention mechanisms [32] and sources of data globally [90] and in the Arctic [91]. Discussion on the role of changing human activity in the Arctic and its consequences for marine biodiversity is increasing among stakeholders and rightsholders, thereby improving the potential for increased and more comprehensive efforts in prevention, monitoring, and management [82, 92, 93].

The need for novel technologies and approaches is growing alongside anticipated growth in Arctic vessel traffic [94, 95], increasingly complex marine community structures and interests (e.g., increased globalization and trade effects, diversification into tourism, and wildlife conservation) [6, 96], and increasingly regulated Arctic transportation and environment [24, 97–99].

Vessel traffic as a primary past, present, and future driver

Though the public imagination may be captivated by stories of climate change–based growth from shorter transit routes between East and West, the underlying economic conditions mean that destination shipping to and from Arctic ports is expected to have much more significant growth [5, 6, 100–102]. This can be expected to further elevate the risk of invasive species

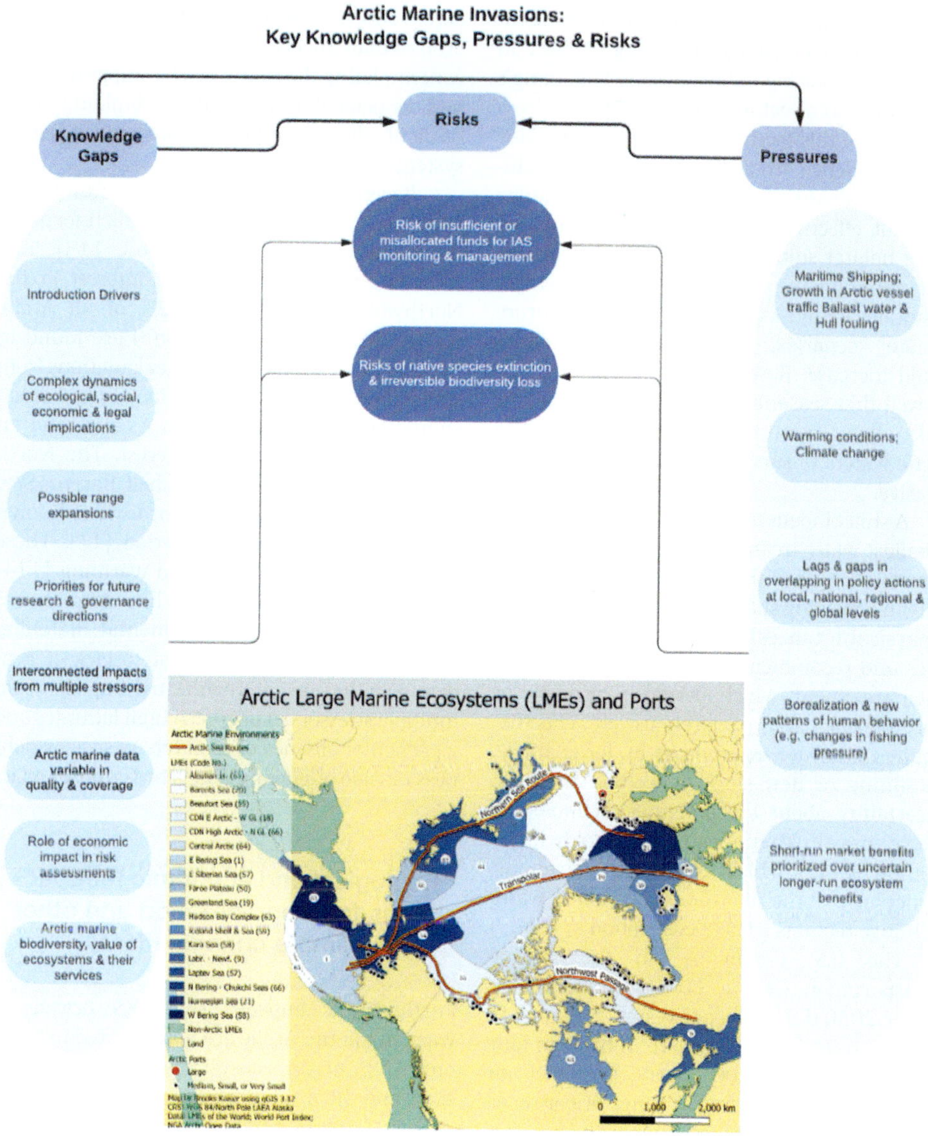

Fig. 5.1. Arctic large marine ecosystems and ports.

arrivals [5, 13]. Vessel traffic is perceived as a major potential source for NIS introduction with ballast water and hull fouling as leading vectors and has therefore received a lot of attention in research and regulation in recent years [14, 32, 96, 98, 103, 104]. After close to 27 years since its initial design and 13 years since its adoption (2004), in September of 2017, the Ballast Water Management (BWM) Convention [105] entered into force through negotiations under the International Maritime Organization (IMO), after a long and political process [98, 105]. Despite potential limitations of BWM practices, requirements, and standards, this milestone has long been high on the agenda of policy groups invested in IAS prevention, such as the PAME working group of the Arctic Council. In addition to implementation and monitoring concerns, uneven treatment of ballast water and biofouling remain a challenge, resulting in lags in policy action for biofouling, despite biofouling's importance as a vector [8, 106, 107].

Despite this focus, Ware *et al.* [43] provide recent evidence from the high Arctic archipelago, Svalbard, that ballast water exchange may not effectively prevent NIS introduction. Using habitat suitability matching, they found that although the coast of Svalbard was still unsuitable for most species, under certain future climate scenarios, colonization probabilities would increase. Rosenhaim *et al.* [58] further showed that seasonal accumulation via ballast tanks allows certain organisms transported in Arctic waters to survive, establish, and become invasive.

A shift of focus from ballast water exchange to ballast water treatment may help to alleviate some challenges in the Arctic, such as the difficulty in achieving sufficient environmental and/or physical distances for exchange under existing rules and recommendations [58, 108]. Ballast water treatment, which should in principle be more flexible and comprehensively effective, requires extensive investments in, for example, retrofitting or new ships altogether; however, uncertainty about the potential technologies' effectiveness in Arctic waters still holds many unknowns [32], which slow costly transitions, particularly if the alternative of ballast water exchange remains a lower-cost option.

The U.S. Committee on the Marine Transportation System [109] has estimated that by 2030 the number of vessels operating in the U.S. Arctic may be more than three to four times greater than 2008 levels. Pushback on these estimates stems from a range of concerns focused on longer run assessments and broader contemplation of the benefits and costs of increased Arctic shipping to both coastal communities and shippers [5, 6, 101]. Furthermore, high growth projections typically rely on a suite of scenarios focused on ice coverage and average economic growth in trade, without explicitly considering risk and resilience in terms of external shocks or disruptions to the economy, and by extension to vessel traffic [110]. The COVID-19 pandemic, for example, reduced global ocean traffic by approximately 44% in April, 2020, with significant consequences for all associated human activities at ports and across supply chains [111]. The impacts of this drop remain uncertain, but already scholars are contemplating the role of Arctic shipping routes and associated investments in building more resilient, disruption-proof trade and transport systems [112].

Recent scholarship includes a number of ballast water studies, several of which focus on the Canadian Arctic [12, 58, 63, 113]. This is in line with broad growing interest in the Northwest Passage. About 22% of all Arctic marine IAS recorded since 1960 are found in the Canadian Arctic (authors' calculations from supplementary information in [11]), which should be expected to be both a cause and an outcome of this research interest. The North Atlantic waters of the Iceland shelf, Barents Sea, and Norwegian Sea account for 66% of known introductions in this time period [11]. These waters show evidence of rapid warming [114] and borealization [47, 55] and have been shown likely to increase in environmental matching [70, 115]. Vilmi *et al.* [62] note a lack of academic studies on interconnected impacts from multiple stressors in northern high latitudes and a particular absence of research on species richness outside of boreal Canada or Fennoscandia.

Ecological and economic feedback loops from borealization and other climate change effects

Furthermore, impacts from IAS driven by warming in the Arctic are not expected to affect only Arctic marine ecosystems. Northern expansions and shifts of boreal systems may introduce new patterns of human behavior and related impacts as well, through, for example, changes in fishing pressures [76, 116]. Additionally, as shorter sailing routes are increasingly considered viable for connecting Asian ecosystems to North American and European ecosystems, species introductions and invasions can be expected to increase in these locations as well [71, 117].

To this end, a lot of recent scholarship has been aimed at identifying and assessing risks of new species introductions and establishment at both a regional and a pan-Arctic level. Chan *et al.* [11] examined a suite of primary introduction events across Arctic marine ecoregions to characterize temporal and spatial patterns, potential sources of NIS, and introduction pathways. According to their findings, throughout the last half of the century (1960–2015), the annual rate of NIS discovery has been low, ranging from 0 to 4 species, for a total of 54 known introductions of 34 species. The origins of about a third of the introductions were not known. While some introductions were attributed to a single path and others to multiple possible pathways, the major vectors were found to be vessels, followed by natural spread and aquaculture activities [11].

Goldsmit *et al.* [12] identified 100 planktonic and benthic marine IAS with the expected ability to arrive and survive in the Canadian Arctic as ballast water or hull fouling introductions and then narrowed this list to 23 high-risk species for which they built species distribution models (SDMs). The SDMs were used to identify suitable habitats for these high-risk species under the RCP 4.5 global warming scenario projected to two differing end dates (2050 and 2100), as well as potential invasion hotspots. At the pan-Arctic scale, Goldsmit *et al.* [12] predict that the net suitable habitat will increase for each of the species modeled, with zoobenthic species predicted to have the greatest expansions. At the same time, however, at the global scale, they find that in total net suitable habitat for the set of species studied, particularly zooplankton species, is expected to decline, as their original ranges change in response to warmer waters. The work builds on Goldsmit *et al.* [113], which focuses on projecting habitat suitability under current environmental conditions and future climate change scenarios for a smaller spatial scale in the Canadian Arctic and a smaller set of high-risk species, assessed based on dispersal pathways, biological attributes, and invasion history.

The results in Goldsmit *et al.* [12] supplement recent findings that boreal species are rapidly moving northward at a much faster pace compared to other regions. The recent findings pertain to a variety of species and communities: fish [53, 118], zooplankton [119], and more broad biogeochemical [120] and ecosystem [61, 121] conditions and indicate that the most rapid changes are coming through generalist species that may swiftly alter Arctic ecosystems [55, 118]. The hotspots for predicted invasions are Hudson Bay, Northern Grand Banks/Labrador, Chukchi/Eastern Bering Seas, and Barents/White Seas [12]; identifying vulnerable hotspots is valuable for targeting effective monitoring and preparedness.

Along those lines, research efforts have advanced the understanding of the potential for risks from Arctic marine IAS driven by climate change. Droghini *et al.* [122] evaluate habitat suitability for ship-mediated invasions in the Bering Sea. They focus on establishing how life cycles for the IAS matter for survival and reproduction, and the connections between life cycles and vessel activities. Their work provides a refined assessment in terms of both human activity and ecological characteristics for how low temperatures have thus far limited IAS establishment and damages, and which ecosystem changes from warming will matter most in increasing the risks of IAS arrivals and survival. Verna *et al.* [123] develop a framework for identifying at fine scale high-risk areas for ballast-borne invasions in Alaskan ports with high discharge activity, considering ballast water characteristics (discharge volume and age), environmental similarity between source and discharge regions, and marine IAS richness in source regions.

Reduced disparities in knowledge and resource allocations from broadening research perspectives

Despite the advances in understanding anthropogenic and climate-driven shifts in marine biodiversity and their habitats, the results can only be as strong as the underlying data, which for Arctic marine environments remain variable in quality and coverage [45]. In focusing on ballast water and vessel traffic, other important vectors, such as intentional introductions (e.g., from aquaculture), are missed and their relative threats are therefore harder to assess and prioritize. Consequences of intentional introductions may be more significant than accidental ones, due in part to vested interests in the establishment and survival of the species [124]. Furthermore, tools may come into use that promote these imbalances; the Canadian

Marine Invasive Screening Tool used as part of the process in Goldsmit *et al.* [12], for example, acknowledges that economic impacts are part of invasive species concerns but advises ignoring them in assessing risks [125].

The significant and relatively large gaps in knowledge over changing Arctic marine environments compared to other ecosystems [126, 127] can be expected to affect allocations of public funds for IAS monitoring and/or management in ways that may overlook the highest threats. Indeed, this may be the case in the Arctic, if the impacts of warming are not considered jointly through natural and social sciences perspectives. Based on solicitations of expert opinion, Essl *et al.* [52] identify trade and transport as the most relevant driver of invasions in marine ecosystems in general, but they also assess climate change as the most relevant driver for the polar regions in particular. Their findings highlight how different dimensions of invasive species challenges, such as ecological or economic, should be assessed comprehensively. Collaborative groups of IAS experts from multiple disciplines are well positioned to comprehensively assess key drivers and risks by considering both economic and climate factors. Such practices limit the potential for misidentifying risks and improperly prioritizing research and management efforts, which can be the case when transferring scientific information from one set of socioecological system to another and when designing policies [128, 129]. In this vein, horizon scanning, which systematically gathers and processes expert opinion, qualitative analysis, and other data and information to identify emerging scientific, technological, and sociopolitical issues likely to affect systems and processes, is a rapidly evolving tool gaining momentum among policymakers, researchers, and funders. Horizon scans can help reduce unintentional biases, identify risks and priorities for research, and build preparedness capacity. Recent relevant horizon scans have focused on drivers of Arctic invasions now and in the future [15], biological conservation [130], and economic developments like China's Belt and Road Initiative [131].

In the existence of many unknowns, the risk of IAS leading to native species extinction and/or irreversible biodiversity loss becomes a large challenge. Advances in scientific understanding of climate change in the Arctic are reducing knowledge gaps that have prevented informed decision-making and investments in facing threats of ecosystem disruption from IAS [43]. These advances affect our general understanding of Arctic marine ecosystems in ways that may profoundly shift our assessment of risks from invasions and include studies that have found greater than expected relative importance for ecosystem resilience in the polar night [132] and studies that elucidate critical unknowns about Arctic food web structures [133].

As researchers across the circumpolar Arctic continue to invest effort in detecting new species in marine waters and determining introduction drivers, potential for establishment and spread, and invasion stage (e.g., [39, 134, 135]), several scholars have discussed and explored the value of maintaining inventories of biota across all Arctic marine realms [11, 29, 136]. Despite the high costs and the many logistical challenges of such pan-Arctic inventories, experience from other aquatic realms show that such inventories may hold significant potential for identifying new introductions and preventing invasions.

Diversity in ecosystem knowledge, values, and uses

Barriers like ice that have historically prevented increased cross-latitudinal economic activities in Arctic regions have also historically limited scientific exploration. Our understanding of marine biodiversity in the Arctic is thus uneven across seasons, species, and locations and limited by scientific challenges and high costs of conducting research at high latitudes [137, 138]. High search costs and challenging conditions for monitoring and IAS containment contribute to the disparity in knowledge between marine invasions in the Arctic and those in more southerly latitudes, so that much of our knowledge is relatively recent and less developed compared to non-Arctic marine ecosystems [126, 139]. This is one reason for a variety of calls for citizen science and monitoring by local and Indigenous populations in the Arctic [82, 137].

The Arctic region consists of 18 of the world's 66 large marine ecosystems, or oceanic regions of greater than 200,000 km^2 and encompassing the waters from coastal grounds and estuaries to the outer continental shelves or major currents [140]. These account for approximately 19.5 million square kilometers of marine space, highlighting the vastness of ecosystem data needed and to the complexities of predicting and preventing invasive species movements within these systems. While long run observations on marine species and/or human activities in the Arctic are sometimes available, they must be filtered through stakeholder and community lenses, such as the whaling industry [141] or indigenous knowledge frameworks [26]. Specifying the science-community-industry interrelationship facilitates assessment of biases and uncertainties in existing research and proposed research agendas. Collaborations among these interrelated participants in Arctic ecosystems are a vital tool for detection and monitoring [79, 81, 82, 84], for building trust needed for information sharing and resource management [80], and for fleshing out models and increasing spatial and temporal observations [66, 78].

The Ocean Health Index (OHI: oceanhealthindex.org) is one such potential tool for improving monitoring and creating a "data-driven approach... [that] allows the quantification and clear communication of broad results to a range of stakeholders both inside and outside the Arctic," with a first application that illustrates the importance of filling transboundary management knowledge gaps [142], of which invasive species management is a significant example.

Local observations, particularly from indigenous communities, are increasingly helping to understand new introductions or range expansions. This is bringing indigenous knowledge to the forefront of invasive species research and policymaking for Arctic marine conservation and resource use [93, 134, 143]. Concerns specific to marine invasions regarding the validity of local observations and the ability to rely on those remain [14]. However, progress has been made in developing tools and mechanisms for effectively combining indigenous and western science knowledge, and ways of knowing, to support decision-making and design of policies [93, 144].

The moves to broaden and contextualize knowledge and policy frameworks are also important because management rules of thumb for, for example, ballast water exchange and other precautions that follow from western scientific knowledge insights may not hold as climate change impacts accelerate. For example, rules for ballast water exchange made by assuming sufficient environmental distances exist to protect ecosystems [145] may become increasingly invalid as well as difficult to implement in the Arctic, or as mentioned earlier, recommended ballast water exchange zones may not have been made with sufficient attention to Arctic conditions or appropriate scientific ecological assessment [108]. Spatially constrained Arctic coastal waters already limit opportunities for ballast water exchange at recommended distances, for example, and such actions may also become less effective or relevant throughout time as Arctic marine environments borealize and environmental distances are reduced [58, 108].

Complications of invasive species management from conflicting or uncertain economic incentives

While shipborne introductions are generally accidental, purposeful introductions occur when certain actors expect to benefit from the successful establishment of the new species. This can complicate management, particularly in the existence of knowledge gaps on impacts of the invasion, when there is disparity on how different stakeholders perceive these impacts or when the distribution of costs and benefits is uneven [67, 146]. Ex-ante policy options for intentional introductions differ from those of accidental introductions in terms of liability rules, regulations, and other precautionary measures [124]. Ex-post solutions may also differ as functions of how the benefits and costs of the invasion are perceived. Benefits of an introduction may be straightforward to monetize, for example, while costs, especially those felt indirectly through ecosystem changes and/or biodiversity loss, can be harder to quantify. In such cases, decisions rest not only on accurate estimates of net benefits but also on the political and economic incentives to act in the short and long terms [17]. Predictions that marine faunal biomass in the Arctic may increase with climate change may exacerbate these conflicts and increase human pressures, and distributional disparities, in pursuit of the increased biomass [87].

The controversies on the management of the red king crab (*Paralithodes camtschaticus*) and snow crab (*Chionoecetes opilio*) invasions in the Barents Sea provide concrete examples of existing trade-offs affecting both ecosystem quality and well-being [17, 67, 147]. Both crab fisheries are valuable commodities in a world of depleting fish stocks that, despite concerns over their ecological impacts, have previously created incentives to maintain healthy populations in the invasive range of their distribution. Their transboundary nature and distribution over international waters highlight the difficulties in cooperative management across Arctic states triggered by conflicting economic incentives. Furthermore, these difficulties appear to extend from management to research. The benefits of commercial exploitation may overshadow the environmental risks and limit the breadth and depth of research agendas, as has been the case for the Barents red king crab invasion/fishery [68]. This poses risks for other Arctic states as distribution ranges shift either naturally (e.g., through larvae dispersal [148, 149]) or through increased vessel traffic and as these pressures are intensified as a result of climate change.

The valuable red king crab, a human-mediated introduction in the southern Barents Sea, is a prominent example of cooperation failure with different management objectives in the two countries sharing the stock. Soviet scientists purposefully introduced the red king crab in the 1960s, and Russia today manages the crab exclusively as a profitable fishery. Norway on the other hand has developed mechanisms intended to reconcile the benefits of the fishery with the ecosystem risks and damages of the invasion, using commercial harvesting as the primary control mechanism [67]. The red king crab, after initially coming to Norwegian fishers' attention by damaging cod nets, has begun to enrich livelihoods in Arctic coastal communities, especially following onshore investments that are further entrenching the communities in accepting the ecological changes. The uncertainty over these ecosystem impacts, however, creates complex trade-offs between conservation and economic development putting pressure on resource managers and triggering polarized reactions among stakeholders affected by the invasion in different ways [17, 19, 128].

The more recent experience from the snow crab invasion in international waters of the Barents Sea, the Russian EEZ, the Norwegian EEZ, and the Svalbard Fisheries Protection Zone also creates an uneven matrix of risk and opportunity and has caused deep-seated conflicts among multiple countries beyond just Arctic states. In fact, disputes over the commercial potential of the fishery and property rights conflicts associated with continental shelves [17] have almost overshadowed the many ecosystem uncertainties and risks of the invasion.

Despite the attention that these crustaceans have brought to concerns for well-being and ecosystem health from IAS, coordinated responses remain absent. International IAS frameworks such as the Convention for Biological Diversity have failed to meaningfully influence decision-making for their management. At the same time, sector-specific, market frameworks like the Marine Stewardship Council (MSC) fishery certification program have created incentives that can distort decision-making and exacerbate conflicts through, among other things, incomplete definitions of sustainability [147]. Both the red king crab and snow crab fisheries in the Russian Barents have now been MSC certified, with limited concerns expressed over their invasion impacts or comprehensive assessment of their management as IAS. Most importantly, these two cases highlight the gap in institutional mechanisms for marine IAS management in the Arctic and the multidimensional weaknesses in existing governance structures that exacerbate risks as species ranges shift and new species are introduced. Opportunities to improve future outcomes depend on increased adoption of community-based scientific tools like OHI addressed above, which can create a common framework and baseline for negotiations over differences in values.

Conclusions

The ongoing shifts and pressures in the Arctic marine environment, driving introductions and establishments of new species, call for multifaceted and new approaches to planning and adaptive management.

Recent advances in marine invasive species research along with policy and governance developments reflect growing ecological and economic concerns. Increased vessel activity and climate change are understood to be the key drivers of new species introductions. As such, a great bulk of recent research has focused on improving our understanding of associated introduction mechanisms and sources; projecting future introductions, temporal and spatial patterns in establishment, and spread and habitat suitability; and advancing detection mechanisms and best practices for prevention.

The rapid pace of climate shifts in the Arctic indicates that effects may be more intense than those in more southern marine environments; climate change that reduces sea ice presence increases not only the potential for successful IAS establishments and invasions but also the economic interests that may augment human- and ship-mediated introductions. Thus, for Arctic seas, it is even more consequential that natural and social sciences work together with stakeholders and rightsholders involved at all levels of knowledge production and use and jointly form policies to minimize risks and impacts from IAS.

To this end, research agendas matter. The complexity of uses and governance of the Arctic marine environment at multiple scales calls for more and better cooperation to minimize risks from invasions and prevent irreversible losses. Marine invasions with a potential for commercial exploitation highlight the many limitations of existing institutional arrangements across Arctic states. A lack of coordinated responses to IAS management, stemming from either incomplete information or conflicting interests, may exacerbate the already-significant threats from IAS to ecosystem integrity and socioeconomic well-being.

References

1. Seebens H, Blackburn TM, Dyer EE, Genovesi P, Hulme PE, Jeschke JM, et al. Global rise in emerging alien species results from increased accessibility of new source pools. Proceedings of the National Academy of Sciences 2018;115(10):E2264–73.
2. Beck KG, Zimmerman K, Schardt JD, Stone J, Lukens RR, Reichard S, et al. Invasive species defined in a policy context: recommendations from the Federal Invasive Species Advisory Committee. Invasive Plant Science and Management 2008;1(4):414–21.
3. Falk Petersen J, Bøhn T, Sandlund OT. On the numerous concepts in invasion biology. Biological Invasions 2006;8(6):1409–24.
4. Lovell SJ, Stone SF, Fernandez L. The economic impacts of aquatic invasive species: a review of the literature. Agricultural and Resource Economics Review 2006;35(1):195–208.
5. Pahl J, Kaiser BA. Arctic port development. In: Vestergaard N, Kaiser BA, Fernandez LM, Larsen JN, editors. Arctic marine resource governance and development. Cham, Switzerland: Springer International Publishing; 2018. p. 139–84.
6. Kaiser BA, Pahl J, Horbel C. Arctic ports: local community development issues. In: Vestergaard N, Kaiser BA, Fernandez L, Larsen JN, editors. Arctic marine resource governance and development. Cham, Switzerland: Springer International Publishing; 2018. p. 185–217.
7. Miller AW, Ruiz GM. Arctic shipping and marine invaders. Nature Climate Change 2014;4(6):413–6.
8. Ojaveer H, Galil BS, Carlton JT, Alleway H, Goulletquer P, Lehtiniemi M, et al. Historical baselines in marine bioinvasions: Implications for policy and management. PLoS One 2018;13(8):e0202383.
9. Ruiz GM, Fofonoff PW, Carlton JT, Wonham MJ, Hines AH. Invasion of coastal marine communities in North America: apparent patterns, processes, and biases. Annual Review of Ecology and Systematics 2000;31(1):481–531.
10. Vermeij GJ, Roopnarine PD. The coming Arctic invasion. Science 2008;321(5890):780–1.
11. Chan FT, Stanislawczyk K, Sneekes AC, Dvoretsky A, Gollasch S, Minchin D, et al. Climate change opens new frontiers for marine species in the Arctic: current trends and future invasion risks. Global Change Biology 2019;25(1):25–38.

12. Goldsmit J, McKindsey CW, Schlegel RW, Stewart DB, Archambault P, Howland KL. What and where? Predicting invasion hotspots in the Arctic marine realm. Global Change Biology 2020;26(9):4752–71.
13. Kaiser BA, Fernandez LM, Vestergaard N. The future of the marine Arctic: environmental and resource economic development issues. The Polar Journal 2016;6(1):152–68.
14. Kourantidou M, Kaiser BA, Fernandez L. Towards Arctic resource governance of marine invasive species. Arctic Yearbook 2015;2015:175–94.
15. Ricciardi A, Blackburn TM, Carlton JT, Dick JT, Hulme PE, Iacarella JC, et al. Invasion science: a horizon scan of emerging challenges and opportunities. Trends in Ecology & Evolution 2017;32(6):464–74.
16. Stevenson TC, Davies J, Huntington HP, Sheard W. An examination of trans-Arctic vessel routing in the Central Arctic Ocean. Marine Policy 2019;100:83–9.
17. Kaiser BA, Kourantidou M, Fernandez L. A Case for the commons: The snow crab in the Barents. Journal of Environmental Management 2018;210:338–48.
18. Koenigstein S. Arctic marine ecosystems, climate change impacts, and governance responses: an integrated perspective from the Barents Sea. In: Arctic marine sustainability. Cham, Switzerland: Springer; 2020. p. 45–71.
19. Nyman E, Tiller R. 'Is there a court that rules them all'? Ocean disputes, forum shopping and the future of Svalbard. Marine Policy 2020;113:103742.
20. AMAP. AMAP assessment 2018: Arctic ocean acidification. Tromsø, Norway: Arctic Monitoring and Assessment Programme (AMAP); 2018.
21. Meredith M, Sommerkorn M, Cassotta S, Derksen C, Ekaykin A, Hollowed A, et al. Polar regions. In: Pörtner H-O, Roberts DC, Masson-Delmotte V, Zhai P, Tignor M, Poloczanska E, et al., editors. IPCC special report on the ocean and cryosphere in a changing climate. 2019.
22. Jones JS, Hartsig A, Gisclair BR. Advancing a network of safety measures in the Bering Strait Region: now is the time. Ocean & Coastal LJ 2020;25:64.
23. DiMento JF. Niels Vestergaard, Brooks A. Kaiser, Linda Fernandez, Joan Nymand Larsen (eds), Arctic Marine Resource Governance and Development (Springer, 2018) 235 pp. The Yearbook of Polar Law Online. 2019;10(1):466–9.
24. Hildebrand LP, Brigham LW, Johansson TM. Sustainable shipping in a changing Arctic. Cham, Switzerland: Springer; 2018.
25. CAFF, PAME. Arctic Invasive Alien Species: Strategy and Action Plan 2017, Conservation of Arctic Flora and Fauna and Protection of the Arctic Marine Environment. Akureyri, Iceland: Conservation of Arctic Flora and Fauna and Protection of the Arctic Marine Environment; 2017. Report No: 978-9935-431-65-3.
26. Moore SE, Hauser DD. Marine mammal ecology and health: finding common ground between conventional science and indigenous knowledge to track arctic ecosystem variability. Environmental Research Letters 2019;14(7):075001.
27. Carson M, Peterson G, editors. Arctic resilience report. Arctic Council, Stockholm Environment Institute and Stockholm Resilience Centre; 2016. Available from: URL: https://www.sei.org/publications/arctic-resilience-report/
28. Beric B, MacIsaac HJ. Determinants of rapid response success for alien invasive species in aquatic ecosystems. Biological Invasions 2015;17(11):3327–35.
29. Lacoursière-Roussel A, Howland K, Normandeau E, Grey EK, Archambault P, Deiner K, et al. eDNA metabarcoding as a new surveillance approach for coastal Arctic biodiversity. Ecology and Evolution 2018;8(16):7763–77.
30. Brown EA, Chain FJ, Zhan A, MacIsaac HJ, Cristescu ME. Early detection of aquatic invaders using metabarcoding reveals a high number of non-indigenous species in Canadian ports. Diversity and Distributions 2016;22(10):1045–59.
31. Chain FJ, Brown EA, MacIsaac HJ, Cristescu ME. Metabarcoding reveals strong spatial structure and temporal turnover of zooplankton communities among marine and freshwater ports. Diversity and Distributions 2016;22(5):493–504.
32. Holbech H, Pedersen KL. Ballast water and invasive species in the Arctic. Arctic Marine Resource Governance and Development/Springer; 2018. p. 115–37.
33. AMAP. Invasive alien species in terrestrial and marine environments. Adaptation Actions for a Changing Arctic (AACA) - Barents Area Overview report. Oslo, Norway: Arctic Monitoring and Assessment Programme (AMAP); 2017. p.

34. Bespalaya YV, Bolotov IN, Aksenova OV, Kondakov AV, Gofarov MY, Laenko TM, et al. Aliens are moving to the Arctic frontiers: an integrative approach reveals selective expansion of androgenic hybrid Corbicula lineages towards the North of Russia. Biological Invasions 2018;20(8):2227–43.
35. Dvoretsky AG, Dvoretsky VG. Commercial fish and shellfish in the Barents Sea: have introduced crab species affected the population trajectories of commercial fish? Reviews in Fish Biology and Fisheries 2015;25(2):297–322.
36. Ekimova IA, Antokhina TI, Schepetov DM. "Invasion" in the Russian Arctic: is global climate change a real driver? A remarkable case of two nudibranch species. Ruthenica 2019;29(2):103–13.
37. Fuhrmann MM. The role of the invasive red king crab in the food web of a high-latitude fjord. Studying macrobenthic communities and trophic control in Porsangerfjord, northern Norway. Tromsø: UiT The Arctic University of Norway; 2016.
38. Goldsmit J. Benthic non-indigenous species in ports of the Canadian Arctic: identification, biodiversity and relationships with global warming and shipping activity Rimouski, Québec: Université du Québec; 2016. 251 p. Available from: URL: http://semaphore.uqar.ca/id/eprint/1233/
39. Matul A, Kazarina GK. The North Pacific Diatom Species Neodenticula seminae in the Modern and Holocene Sediments of the North Atlantic and Arctic. Geosciences 2020;10(5):173.
40. Ramos-Esplá A, Micael J, Halldórsson H, Gíslason S. Iceland: a laboratory for non-indigenous ascidians. BioInvasions Records 2020;9:450–60.
41. Thomassen J, Dahle S, Hagen D, Hendrichsen D, Husa V, Miller A, et al. Invasive alien species in the Arctic – with focus on Svalbard and Jan Mayen. NINA Report 1413. Norwegian Institute for Nature Research; 2017.
42. Vinarski MV, Bolotov IN, Aksenova OV, Babushkin ES, Bespalaya YV, Makhrov AA, et al. Freshwater Mollusca of the Circumpolar Arctic: a review on their taxonomy, diversity and biogeography. Hydrobiologia 2020.
43. Ware C, Berge J, Jelmert A, Olsen SM, Pellissier L, Wisz M, et al. Biological introduction risks from shipping in a warming Arctic. Journal of Applied Ecology 2016;53(2):340–9.
44. Wisz M, Broennimann O, Grønkjær P, Møller PR, Olsen SM, Swingedouw D, et al. Arctic warming will promote Atlantic–Pacific fish interchange. Nature Climate Change 2015;5(3):261–5.
45. Christiansen JS, Bonsdorff E, Byrkjedal I, Fevolden S-E, Karamushko OV, Lynghammar A, et al. Novel biodiversity baselines outpace models of fish distribution in Arctic waters. The Science of Nature 2016;103(1–2):8.
46. Weiskopf SR, Rubenstein MA, Crozier LG, Gaichas S, Griffis R, Halofsky JE, et al. Climate change effects on biodiversity, ecosystems, ecosystem services, and natural resource management in the United States. Science of the Total Environment 2020;733:137782.
47. Andrews AJ, Christiansen JS, Bhat S, Lynghammar A, Westgaard J-I, Pampoulie C, et al. Boreal marine fauna from the Barents Sea disperse to Arctic Northeast Greenland. Scientific Reports 2019;9(1):1–8.
48. Carlson CJ, Albery GF, Merow C, Trisos CH, Zipfel CM, Eskew EA, et al. Climate change will drive novel cross-species viral transmission. bioRxiv. 2020.
49. Choi FMP, Murray CC, Therriault TW, Pakhomov EA. Intertidal invasion patterns in Canadian ports. Marine Biology 2016;163(9):1–12.
50. Christiansen JS, Sparboe M, Sæther BS, Siikavuopio SI. Thermal behaviour and the prospect spread of an invasive benthic top predator onto the Euro-Arctic shelves. Diversity and Distributions 2015;21(9):1004–13.
51. Denley D, Metaxas A, Simard N. Ocean temperature does not limit the establishment and rate of secondary spread of an ecologically significant invasive bryozoan in the northwest Atlantic. Aquatic Invasions 2019;14(4):594–614.
52. Essl F, Lenzner B, Bacher S, Bailey S, Capinha C, Daehler C, et al. Drivers of future alien species impacts: an expert-based assessment. Global Change Biology 2020;26(9):4880–93.
53. Fossheim M, Primicerio R, Johannesen E, Ingvaldsen RB, Aschan MM, Dolgov AV. Recent warming leads to a rapid borealization of fish communities in the Arctic. Nature Climate Change 2015;5(7):673–7.
54. Hoberg EP, Cook J, Agosta S, Boeger W, Galbreath K, Laaksonen S, et al. Arctic systems in the Quaternary: ecological collision, faunal mosaics and the consequences of a wobbling climate. Journal of Helminthology 2017;91:409–21.

55. Kortsch S, Primicerio R, Fossheim M, Dolgov AV, Aschan M. Climate change alters the structure of arctic marine food webs due to poleward shifts of boreal generalists. Proceedings of the Royal Society B: Biological Sciences 2015;282(1814):20151546.
56. Lipscombe MA. Assessing the current and future risk of ballast-sourced species invasions in Canada's eastern Arctic under a climate change scenario [graduate project]. Halifax, NS: Dalhousie University; 2016.
57. Pauchard A, Milbau A, Albihn A, Alexander J, Burgess T, Daehler C, et al. Non-native and native organisms moving into high elevation and high latitude ecosystems in an era of climate change: new challenges for ecology and conservation. Biological Invasions 2016;18(2):345–53.
58. Rosenhaim IL, Riemann-Campe K, Sumata H, Koeberle C, Brauner R, Herber A, et al. Simulated ballast water accumulation along Arctic shipping routes. Marine Policy 2019;103:9–18.
59. Schmidt NM, Hardwick B, Gilg O, Høye TT, Krogh PH, Meltofte H, et al. Interaction webs in arctic ecosystems: determinants of arctic change? Ambio 2017;46(1):12–25.
60. Seebens H, Schwartz N, Schupp PJ, Blasius B. Predicting the spread of marine species introduced by global shipping. Proceedings of the National Academy of Sciences 2016;113(20):5646–51.
61. Vihtakari M, Welcker J, Moe B, Chastel O, Tartu S, Hop H, et al. Black-legged kittiwakes as messengers of Atlantification in the Arctic. Scientific Reports 2018;8(1):1178.
62. Vilmi A, Alahuhta J, Hjort J, Kärnä O-M, Leinonen K, Rocha M, et al. Geography of global change and species richness in the North. Environmental Reviews 2017;25(2):184–92.
63. Goldsmit J, McKindsey C, Archambault P, Howland KL. Ecological risk assessment of predicted marine invasions in the Canadian Arctic. PLoS One 2019;14(2):e0211815.
64. Duarte S, Vieira PE, Lavrador AS, Costa FO. Status and prospects of marine NIS detection and monitoring through (e) DNA metabarcoding. bioRxiv. 2020.
65. Haug T, Bogstad B, Chierici M, Gjøsæter H, Hallfredsson EH, Høines ÅS, et al. Future harvest of living resources in the Arctic Ocean north of the Nordic and Barents Seas: a review of possibilities and constraints. Fisheries Research 2017;188:38–57.
66. Kaiser BA, Bakanev S, Bertelsen RG, Carson M, Eide A, Fernandez L, et al. Spatial issues in Arctic marine resource governance workshop summary and comment. Marine Policy 2015;58:1–5.
67. Kourantidou M. Stewardship of resources in rapidly evolving Arctic economies and ecosystems: the role of marine invasive species [PhD thesis]. Esbjerg: University of Southern Denmark; 2018.
68. Kourantidou M, Kaiser BA. Research agendas for profitable invasive species. Journal of Environmental Economics and Policy 2019;8(2):209–30.
69. Lawson D. Increased prevalence of human activity in the Arctic as a result of climate change, and the impacts on the Arctic ecosystem from resulting increases of introduced species. University of Washington Libraries/University of Washington; 2019.
70. Nong D, Countryman AM, Warziniack T, Grey EK. Arctic sea routes: potential new pathways for nonindigenous species spread. Arctic 2018;71(3):269–80.
71. Nong D, Warziniack T, Countryman AM, Grey EK. Melting Arctic sea ice: implications for nonindigenous species (NIS) spread in the United States. Environmental Science & Policy 2019;91:81–91.
72. Pongolini L, Pålsson J, Folkunger J, Larsson J, Bouyssou A, Hildebrand L, et al. Assessing and mitigating the environmental impacts of shipping in the Arctic. Malmo, SE: World Maritime University; 2017.
73. Reimer J, Droghini A, Fischbach A, Watson J, Bernard B, Poe A. Assessing the risk of non-native marine species in the Bering Sea. Anchorage, AK, US: North Pacific Research Board; 2017.
74. Ronalds B. Impacts of Arctic warming and sea ice loss on the Northern Hemisphere mid-latitude large-scale circulation [Doctoral dissertation]. Fort Collins, Colorado: Colorado State University; 2020.
75. Thiessen BM. Understanding Arctic shipping impacts and mitigation: impact assessment as a tool for knowledge brokerage. Saskatoon, Canada: University of Saskatchewan; 2019.
76. Ellefsen H, Grønbæk L, Ravn-Jonsen L. On international fisheries agreements, entry deterrence, and ecological uncertainty. Journal of Environmental Management 2017;193:118–25.
77. Andersen JH, Kallenbach E, Hesselsøe M, Knudsen SW, Møller PR, Bekkevold D, et al. Steps toward nation-wide monitoring of non-indigenous species in Danish marine waters under the Marine Strategy Framework Directive. Bergen, NO: Norsk Institutt for Vannforskning; 2016. Report No: 7022.
78. Bentley JW, Serpetti N, Fox C, Heymans JJ, Reid DG. Fishers' knowledge improves the accuracy of food web model predictions. ICES Journal of Marine Science 2019;76(4):897–912.

79. Flagg R, Otokiak M, Hoeberechts M, Marshall LM, editors. Integrated Monitoring Systems for Coastal Communities. OCEANS 2019 MTS/IEEE SEATTLE; Seattle, WA: IEEE; 2019.
80. Gordon HSJ. Building relationships in the Arctic: indigenous communities and scientists. Northern sustainabilities: understanding and addressing change in the circumpolar world. Cham, Switzerland: Springer; 2017. p. 237–52.
81. Johnson N, Behe C, Danielsen F, Krümmel E, Nickels S, Pulsifer PL. Community-based monitoring and indigenous knowledge in a changing arctic: a review for the sustaining arctic observing networks. Sustain Arctic Observing Network Task 2016;9:74.
82. Kaiser BA, Hoeberechts M, Maxwell K, Eerkes-Medrano L, Hilmi N, Safa A, et al. The importance of connected ocean monitoring knowledge systems and communities. Frontiers in Marine Science 2019;6:309.
83. Österblom H, Cvitanovic C, van Putten I, Addison P, Blasiak R, Jouffray J-B, et al. Science-industry collaboration: sideways or highways to ocean sustainability? One Earth 2020;3(1):79–88.
84. Thompson K-L, Lantz T, Ban N. A review of Indigenous knowledge and participation in environmental monitoring. Ecology and Society 2020;25(2).
85. Pyšek P, Hulme PE, Simberloff D, Bacher S, Blackburn TM, Carlton JT, et al. Scientists' warning on invasive alien species. Biological Reviews 2020;95(6):1511–34.
86. Bryndum-Buchholz A, Tittensor DP, Blanchard JL, Cheung WW, Coll M, Galbraith ED, et al. Twenty-first-century climate change impacts on marine animal biomass and ecosystem structure across ocean basins. Global Change Biology 2019;25(2):459–72.
87. Boyce DG, Lotze HK, Tittensor DP, Carozza DA, Worm B. Future ocean biomass losses may widen socioeconomic equity gaps. Nature Communications 2020;11(1):1–11.
88. Leduc N, Lacoursière-Roussel A, Howland KL, Archambault P, Sevellec M, Normandeau E, et al. Comparing eDNA metabarcoding and species collection for documenting Arctic metazoan biodiversity. Environmental DNA 2019;1(4):342–58.
89. Sepulveda AJ, Nelson NM, Jerde CL, Luikart G. Are environmental DNA methods ready for aquatic invasive species management? Trends in Ecology & Evolution 2020;35(8):668–78.
90. Yang D, Wu L, Wang S, Jia H, Li KX. How big data enriches maritime research–a critical review of automatic identification system (AIS) data applications. Transport Reviews 2019;39(6):755–73.
91. Zhang Z, Huisingh D, Song M. Exploitation of trans-Arctic maritime transportation. Journal of Cleaner Production 2019;212:960–73.
92. Dawson J, Carter N, van Luijk N, Parker C, Weber M, Cook A, et al. Infusing Inuit and local knowledge into the Low Impact Shipping Corridors: an adaptation to increased shipping activity and climate change in Arctic Canada. Environmental Science & Policy 2020;105:19–36.
93. Aporta C, Bishop B, Choi O, Wang W. Knowledge and data: an exploration of the use of inuit knowledge in decision support systems in marine management. In: Chircop A, Goerlandt F, Aporta C, Pelot R, editors. Governance of arctic shipping: rethinking risk, human impacts and regulation. Cham: Springer International Publishing; 2020. p. 151–69.
94. Eguíluz VM, Fernández-Gracia J, Irigoien X, Duarte CM. A quantitative assessment of Arctic shipping in 2010–2014. Scientific Reports 2016;6(1):1–6.
95. Melia N, Haines K, Hawkins E. Sea ice decline and 21st century trans-Arctic shipping routes. Geophysical Research Letters 2016;43(18):9720–8.
96. Monios J. Environmental governance in shipping and ports: sustainability and scale challenges. Maritime transport and regional sustainability. Cambridge, MA: Elsevier; 2020. p. 13–29.
97. Brigham LW. Arctic policy developments and marine transportation. In: Coates KS, Holroyd C, editors. The Palgrave handbook of Arctic policy and politics. Cham: Springer International Publishing; 2020. p. 393–408.
98. Chircop A, Goerlandt F, Aporta C, Pelot R. Governance of Arctic shipping: rethinking risk, human impacts and regulation. Cham, Switzerland; 2020.
99. Fedi L. Arctic shipping law: from atomised legislations to integrated regulatory framework: the Polar Code (r) evolution? In: Arctic shipping. Oxon, UK/New York, US: Routledge; 2019. p. 117–36.
100. Lasserre F. Arctic shipping: climate change, commercial traffic and port development. Canada: Friesens Corporation; 2018. p. 509–21.
101. Lasserre F. Arctic shipping: a contrasted expansion of a largely destinational market. In: Finger M, Heininen L, editors. The global Arctic handbook. Cham: Springer International Publishing; 2019. p. 83–100.

102. Lasserre F, Faury O. Arctic shipping: climate change, commercial traffic and port development. Oxon, UK/New York, US: Routledge; 2019.
103. Fernandez L, Kaiser BA, Vestergaard N, editors. Marine invasive species in the Arctic. Copenhagen, Denmark: Nordic Council of Ministers; 2014.
104. Saebi M, Xu J, Curasi SR, Grey EK, Chawla NV, Lodge DM. Network analysis of ballast-mediated species transfer reveals important introduction and dispersal patterns in the Arctic. Scientific Reports 2020;10(1):1–15.
105. International Convention for the Control and Management of Ships' Ballast Water and Sediments (BWM), [Internet] 2020 [cited 2020 August 13]. Available from: URL: http://www.imo.org/en/About/Conventions/ListOfConventions/Pages/International-Convention-for-the-Control-and-Management-of-Ships'-Ballast-Water-and-Sediments-(BWM).aspx
106. Galil BS, McKenzie C, Bailey S, Campbell M, Davidson IC, Drake L, et al. ICES viewpoint background document: evaluating and mitigating introduction of marine non-native species via vessel biofouling. Copenhagen, DK: ICES: International Council for the Exploration of the Sea; 2019.
107. Chan FT, MacIsaac HJ, Bailey SA. Survival of ship biofouling assemblages during and after voyages to the Canadian Arctic. Marine Biology 2016;163(12):250.
108. Goldsmit J, Nudds SH, Stewart DB, Higdon JW, Hannah CG, Howland KL. Where else? Assessing zones of alternate ballast water exchange in the Canadian eastern Arctic. Marine Pollution Bulletin 2019;139:74–90.
109. (CMTS) UCotMTS. A ten-year projection of maritime activity in the U.S. Arctic region, 2020–2030. Washington, DC: U.S. Committee on the Marine Transportation System; 2019. 118 p
110. Golan MS, Jernegan LH, Linkov I. Trends and applications of resilience analytics in supply chain modeling: systematic literature review in the context of the COVID-19 pandemic. Environment Systems & Decisions 2020;1:222–43.
111. March D, Metcalfe K, Tintoré J, Godley B. Tracking the global reduction of marine traffic during the COVID-19 pandemic. Research Square; 2020. [preprint]
112. Jones DA. Opening China's 'Belt and Road Initiative' to competition with international investment: expanding freight cargo delivery between Asia and Europe to accelerate free trade and national security in the aftermath of the Covid-19 pandemic. Journal of Critical Reviews 2020;7(8):2020.
113. Goldsmit J, Archambault P, Chust G, Villarino E, Liu G, Lukovich JV, et al. Projecting present and future habitat suitability of ship-mediated aquatic invasive species in the Canadian Arctic. Biological Invasions 2018;20(2):501–17.
114. Lind S, Ingvaldsen RB, Furevik T. Arctic warming hotspot in the northern Barents Sea linked to declining sea-ice import. Nature Climate Change 2018;8(7):634–9.
115. Ware C, Berge J, Sundet JH, Kirkpatrick JB, Coutts AD, Jelmert A, et al. Climate change, non-indigenous species and shipping: assessing the risk of species introduction to a high-A rctic archipelago. Diversity and Distributions 2014;20(1):9–10.
116. Hänsel MC, Schmidt JO, Stiasny MH, Stöven MT, Voss R, Quaas MF. Ocean warming and acidification may drag down the commercial Arctic cod fishery by 2100. PLoS One 2020;15(4):e0231589.
117. Robinson TB, Martin N, Loureiro TG, Matikinca P, Robertson MP. Double trouble: the implications of climate change for biological invasions. NeoBiota 2020;62:463.
118. Frainer A, Primicerio R, Kortsch S, Aune M, Dolgov AV, Fossheim M, et al. Climate-driven changes in functional biogeography of Arctic marine fish communities. Proceedings of the National Academy of Sciences 2017;114(46):12202–7.
119. Møller EF, Nielsen TG. Borealization of Arctic zooplankton—smaller and less fat zooplankton species in Disko Bay, Western Greenland. Limnology and Oceanography 2020;65(6):1175–88.
120. Polyakov IV, Alkire MB, Bluhm BA, Brown KA, Carmack EC, Chierici M, et al. Borealization of the Arctic Ocean in response to anomalous advection from sub-Arctic seas. Frontiers in Marine Science 2020;7:491.
121. Pecuchet L, Blanchet MA, Frainer A, Husson B, Jørgensen LL, Kortsch S, et al. Novel feeding interactions amplify the impact of species redistribution on an Arctic food web. Global Change Biology 2020;26(9):4894–906.
122. Droghini A, Fischbach AS, Watson J, Reimer J. Regional ocean models indicate changing limits to biological invasions in the Bering Sea. ICES Journal of Marine Science 2020;77(3):964–74.
123. Verna DE, Harris BP, Holzer KK, Minton MS. Ballast-borne marine invasive species: exploring the risk to coastal Alaska, USA. Management of Biological Invasions 2016;7(2):199–211.

124. Kaiser BA. Marine Invasive Species Management: adapting in the Arctic. In: Fernandez L, Kaiser B, Vestergaard N, editors. Marine invasive species in the Arctic. Copenhagen, Denmark: Nordic Council of Ministers; 2014.
125. Fisheries and Oceans Canada. Canadian marine invasive screening tool manual. Government of Canada; 2017. Available from: URL: https://www.bio.gc.ca/science/monitoring-monitorage/cmist/usage-en.php
126. Wassmann P, Duarte CM, Agusti S, Sejr MK. Footprints of climate change in the Arctic marine ecosystem. Global Change Biology 2011;17(2):1235–49.
127. Falardeau M, Bennett EM. Towards integrated knowledge of climate change in Arctic marine systems: a systematic literature review of multidisciplinary research. Arctic Science 2019;6(1):1–23.
128. Kaiser BA, Kourantidou M. Invasive crab species in the Barents Sea: stakeholder perceptions, incentives, and path dependencies. In: Briand F, editor. Engaging marine scientists and fishers to share knowledge and perceptions – early lessons; CIESM Workshop Monographs no. 50. Monaco: CIESM; 2018. p. 133–48.
129. Richardson DM, Pyšek P. Fifty years of invasion ecology–the legacy of Charles Elton. Diversity and Distributions 2008;14(2):161–8.
130. Sutherland W, Fleishman E, Clout M, Gibbons D, Lickorish F, Peck L, et al. Ten years on: a review of the first global conservation horizon scan. Trends in Ecology & Evolution 2019;34(2):139–53.
131. Hughes A, Lechner A, Chitov A, Horstmann A, Hinsley A, Tritto A, et al. Horizon scan of the belt and road initiative. Trends in Ecology & Evolution 2020;35(7):583–93.
132. Berge J, Daase M, Renaud PE, Ambrose Jr WG, Darnis G, Last KS, et al. Unexpected levels of biological activity during the polar night offer new perspectives on a warming Arctic. Current Biology 2015;25(19):2555–61.
133. Murphy EJ, Cavanagh RD, Drinkwater KF, Grant SM, Heymans J, Hofmann EE, et al. Understanding the structure and functioning of polar pelagic ecosystems to predict the impacts of change. Proceedings of the Royal Society B: Biological Sciences 2016;283(1844):20161646.
134. Falardeau M, Bouchard C, Robert D, Fortier L. First records of Pacific sand lance (Ammodytes hexapterus) in the Canadian Arctic Archipelago. Polar Biology 2017;40(11):2291–6.
135. Fefilova EB, Sukhikh NM, Rasova EE, Velegzhaninov IO, Abramova EN. New data on the expansion of Eurytemora Giesbrecht (Copepoda: Calanoida) in the Russian Arctic Region. Doklady Biological Sciences 2020;492:86–8.
136. Piepenburg D, Archambault P, Ambrose WG, Blanchard AL, Bluhm BA, Carroll ML, et al. Towards a pan-Arctic inventory of the species diversity of the macro-and megabenthic fauna of the Arctic shelf seas. Marine Biodiversity 2011;41(1):51–70.
137. Bergmann M, Lutz B, Tekman MB, Gutow L. Citizen scientists reveal: marine litter pollutes Arctic beaches and affects wild life. Marine Pollution Bulletin 2017;125(1–2):535–40.
138. Mahanes SA, Sorte CJ. Impacts of climate change on marine species invasions in northern hemisphere high-latitude ecosystems. Frontiers of Biogeography 2019;11(1):1–13.
139. Mallory ML, Gilchrist HG, Janssen M, Major HL, Merkel F, Provencher JF, et al. Financial costs of conducting science in the Arctic: examples from seabird research. Arctic Science 2018;4(4):624–33.
140. Sherman K, Peterson B. Large marine ecosystems of the world. University of Rhode Island; 2020. Available from: URL: http://lmehub.net/
141. Smith TD, Reeves RR, Josephson EA, Lund JN. Spatial and seasonal distribution of American whaling and whales in the age of sail. PLoS One 2012;7(4):e34905.
142. Burgass MJ, Milner-Gulland E, Lowndes JSS, O'Hara C, Afflerbach JC, Halpern BS. A pan-Arctic assessment of the status of marine social-ecological systems. Regional Environmental Change 2019;19(1):293–308.
143. Ksenofontov S, Backhaus N, Schaepman-Strub G. 'There are new species': indigenous knowledge of biodiversity change in Arctic Yakutia. Polar Geography 2019;42(1):34–57.
144. Kourantidou M, Hoover C, Bailey M. Conceptualizing indicators as boundary objects in integrating Inuit knowledge and western science for marine resource management. Arctic Science 2020;6(3):279–306.
145. Carlton JT, Reid DM, Van Leeuwen H. Shipping study: The role of shipping in the introduction of non-indigenous aquatic organisms to the coastal waters of the United States (other than the Great Lakes) and an analysis of control options. Mystic, CT: Williams College Mystic Ct Maritime Studies Program; 1995.

146. McNeely JA. The great reshuffling: how alien species help feed the global economy. Invasive Species And Biodiversity Management 1999:11–31.
147. Kourantidou M, Kaiser BA. Sustainable seafood certifications are inadequate to challenges of ecosystem change. ICES Journal of Marine Science 2019;76(4):794–802.
148. Michelsen HK, Nilssen EM, Pedersen T, Svensen C. Temporal and spatial dynamics of the invasive red king crab and native brachyuran and anomuran larvae in Norwegian waters. Aquatic Biology 2020;29:1–16.
149. Pedersen O, Nilssen E, Jørgensen L, Slagstad D. Advection of the red king crab larvae on the coast of North Norway—a Lagrangian model study. Fisheries Research 2006;79(3):325–36.

6 Biological control of *Drosophila suzukii*

Xingeng Wang[1]*, Jana C. Lee[2], Kent M. Daane[3], Matthew L. Buffington[4] and Kim A. Hoelmer[1]

[1]USDA-ARS Beneficial Insects Introduction Research Unit, Newark, Delaware, USA; [2]USDA-ARS Horticultural Crops Research Unit, Corvallis, Oregon, USA; [3]University of California Berkeley, Berkeley, California, USA; [4]USDA-ARS Systematic Entomology Laboratory, Washington D.C., USA

Abstract

Drosophila suzukii (Matsumura) is native to East Asia but has widely established in the Americas and Europe, where it is a devastating pest of soft-skinned fruits. It has a wide host range and these noncrop habitats harbor the fly which then repeatedly reinvades crop fields. Biological control in noncrop habitats could be the cornerstone for sustainable management at the landscape level. Toward this goal, researchers have developed or investigated biological control tactics. We review over 100 studies, conducted in the Americas, Asia and Europe on natural enemies of *D. suzukii*. Two previous reviews provided an overview of potential natural enemies and detailed accounts on foreign explorations. Here, we provide an up-to-date list of known or evaluated parasitoids, predators and entomopathogens (pathogenic fungi, bacteria, nematodes, and viruses) and summarize research progress to date. We emphasize a systematic approach toward the development of biological control strategies that can stand alone or be combined with more conventional control tools. Finally, we propose a framework for the integrated use of biological control tools, from classical biological control with host-specific Asian parasitoids, to augmentative and conservation biological control with indigenous natural enemies, to the use of entomopathogens. This review provides a roadmap to foster the use of biological control tools in more sustainable *D. suzukii* control programs.

Keywords: spotted-wing drosophila, parasitoid, predator, entomopathogen, invasive species

Review Methodology

We searched and reviewed scientific literature on natural enemies and biological controls of *Drosophila suzukii* from major online databases such as CAB Abstracts, Google Scholar, and ISI Web of Science. The literature has been synthesized by authors working on *D. suzukii* biological control over the past decade.

Introduction

Native to East Asia, the spotted-wing drosophila (SWD), *Drosophila suzukii* (Matsumura) (Diptera: Drosophilidae) has established widely in the Americas and Europe since it was first detected on both continents in 2008 [1, 2]. Unlike the majority of frugivorous *Drosophila* species breeding on damaged or rotting fruits, SWD can oviposit into intact and ripening fruits with its unique saw-like ovipositor [3], resulting in secondary infections and rapid degradation of

*Corresponding author: xingeng.wang@usda.gov

the fruits [4–6]. SWD is highly polyphagous, utilizing numerous cultivated and wild fruits with soft and thin skins [7–12]. Coupled with its rapid development and high reproductive capacity [13–15], SWD can cause devastating economic losses [16, 17].

Insecticides are still most commonly used for SWD control [18–22], but they present commonly recognized negative effects on natural enemies, secondary pest outbreaks, pre-harvest intervals restriction, insecticide resistance, and increased management costs [23–26]. Therefore, researchers have been developing various alternative tactics, including cultural controls [27–30], exclusion barriers [31] and behavioral controls [32–34]. However, these methods typically target only adult flies at local scale (crop field) [31, 35, 36]. Other novel biotech approaches are being tested, such as sterile male releases [37–40] and gene editing (e.g., RNAi and CRISPR/Cas9) [40], but are not yet commercially available and must still pass regulatory scrutiny and public acceptance.

Biological control is a viable option for several reasons. First, many Asian paraitoids attack SWD larvae inside the fruit, where pesticides are less effective [35, 41]. Second, natural enemies may suppress fly population in non-crop habitats where insecticide use is often restricted, thereby reducing the population pressure in the cash crops [42–63]. Natural enemies may play a unique role in these non-crop reservoirs by establishing self-perpetuating (e.g., no cost) populations that form the cornerstone for sustainable management of this polyphagous pest. Two previous reviews provide an overview of potential applications of various agents [41] and detailed accounts on exploration and evaluation of Asian parasitoids [64]. Our aim here is to provide an up-to-date list of known parasitoids, predators and entomopathogens (pathogenic fungi, bacteria, nematodes, and viruses), and summarize the current progress and challenges of various biological control approaches (i.e., classical, augmentative and conservation biological control). Most importantly, we emphasize a systematic approach and provide the framework to integrate biological control tools with conventional management strategies.

Biological control with parasitoids

Parasitoids play an important role in regulating some *Drosophila* populations [65, 66]. The parasitoid fauna attacking *Drosophila* species is primarily comprised of larval parasitoids in the genera *Asobara* (Braconidae), *Leptopilina* and *Ganaspis* (Figitidae), and pupal parasitoids in *Trichopria* (Diapriidae) and *Pachycrepoideus* and *Spalangia* (Pteromalidae) [67]. Before SWD emerged as a serious pest most research focused on three cosmopolitan larval parasitoids, *Asobara tabida* (Nees), *Leptopilina boulardi* Barbotin et al. and *L. heterotoma* (Thompson) that attack common drosophilids, such as *Drosophila melanogaster* Meigen, and used them as model species for studies of host-parasitoid interactions [68]. Until recently the parasitoid fauna associated with SWD was poorly understood. Since then, widespread efforts have collected, identified, and studied promising parasitoids both in the fly's invaded and native ranges for their potential use in classical (Asian parasitoids) and/or augmentative (resident parasitoids) biological control programs.

Resident parasitoids and augmentative biological control

Surveys of parasitoids attacking SWD and related frugivorous *Drosophila* species were made in Europe [69–75], North America [76–79], and South America [80] (Table 6.1). The three larval parasitoids commonly recovered (*A. tabida*, *L. boulardi*, and *L. heterotoma*) were exclusively reared from drosophilids other than SWD. Laboratory tests confirmed that SWD has a strong host cellular immune defense [93–96] and there was no larval parasitism in the invaded region with the exception of *L. heterotoma* attacking a low percentage of SWD in Italy [91, 97] and France [96]. However, egg laying by these larval parasitoids could reduce the survival of parasitized SWD [41, 64, 98]. Pupal parasitoids were recovered from SWD and include the common *Pachycrepoideus vindemiae* (Rondani) and *Trichopria drosophilae* Perkins (Diapriidae) in North America and Europe and *Trichopria anastrephae* Lima in Brazil. Other

Table 6.1. Parasitoid complexes and their observed parasitism on SWD in different regions of its native and introduced ranges.

Parasitoid species	Countries	Host species[1]	Parasitism (%)	References
Larval parasitoids				
Braconidae				
Asobara tabida	Japan	SWD/Others	–	[81]
A. brevicauda	South Korea	SWD	<1	[82]
A. japonica	Japan, South Korea	SWD/Others	0–16.7	[81, 82]
A. leveri	China, South Korea	SWD/DP/Others	<1	[82, 83]
A. mesocauda	China, South Korea	SWD/DP/DSP	–	[83, 84]
A. pleuralis	China	SWD/DP	–	[84]
A. sp.	Japan	SWD	<7	[85]
A. sp. TS1	Japan	SWD	1.2	[86]
A. triangulata	South Korea	SWD	<1	[83, 87]
A. unicolorata	China	SWD/DP	<1	[83]
Areotetes striatiferus	China	SWD/DP	0.6–6.9	[84]
Tanycarpa chors	China	SWD/DP	2.1	[84]
Figitidae				
Ganaspis brasiliensis (G1 and G3)[2]	China, Japan, South Korea	SWD/DP/DSP	0–75.6	[82–84]
G. brasiliensis (G1)[3]	Canada	SWD	–	[134]
G. brasiliensis (G4 and G5)[2]	Japan, Thailand, the Philippines, Hawaii, Uganda, Indonesia, Malaysia, Benin, the Caribbean Sea, Brazil, Mexico	Others	–	[92, 94, 98 141–144]
G. xanthopoda[4]	Japan	SWD/Others	–	[81]
L. heterotoma	Italy	SWD/Others	<1	[91]
L. j. formosana	South Korea, Japan	SWD/Others	<1	[82, 139]
L. j. japonica[3]	China, Japan, South Korea	SWD/DP	0–34.5	[82–85, 139]
L. j. japonica	Canada, Italy[3]	SWD/Others	–	[90, 134]
Leptopilina sp.	China	SWD/DP	7.2–35.9	[84]
Pupal parasitoids				
Diapriidae				
Trichopria cf *drosophilae*	Italy, Spain	SWD/Others	0–10.7	[70, 71]
T. drosophilae	China, France, Italy, Mexico, Slovenia, South Korea, Switzerland, USA	SWD/Others	0–11.1	[72–74, 76, 77, 82, 83, 91]
T. anastrephae	Brazil	SWD	–	[80]
Pteromalidae				

Continued

Table 6.1. Continued

Parasitoid species	Countries	Host species[1]	Parasitism (%)	References
Pachycrepoideus vindemiae	China, France, Italy, Mexico, the Netherlands, South Korea, Spain, Switzerland, Turkey, USA	SWD/Others	0–31.0	[69–73, 75–77, 82]
Spalangia erythromera	Italy	SWD/Others	–	[71]
S. simplex	Mexico	SWD	–	[76]

[1]SWD = *Drosophila suzukii*, DP = *D. pulchrella*, DSP = *D. subpulchrella*, and Others = other drosophilids when the species was either not known, not provided or not one of the above species.
[2]This species is likely a complex consisting of different strains (named as G1-G5, see [83, 143]).
[3]Accidentally introduced and established in British Columbia, Canada and Italy [90, 134].
[4]This species was later re-assigned as *Ganaspis brasiliensis* [143].

tested pupal parasitoids attacking SWD include the pteromalids *Vrestovia fidenas* (Walker) and *Spalangia erythromera* Förster in Europe [72, 99], *Spalangia simplex* Perkins in Mexico [76], and *Muscidifurax raptorellus* Girault & Sanders in Canada [100] (Table 6.2). A lack of pupal immunity against parasitoids may explain why these pupal parasitoids have broad host ranges [94].

The most promising resident parasitoids are *P. vindemiae* and *T. drosophilae* in North America and Europe, and *T. anastrephae* in South America. They have been systematically evaluated for their temperature range [110–112] and cold tolerance [113, 114], host stage preference [91, 115], fecundity [91, 109, 115–117], host species preference and suitability [115, 118, 119, 125], interspecific competition [120], functional response [121], as well as host feeding and other factors (food and water) affecting the fecundity of *P. vindemiae* [126, 127]. The key biological traits of *T. anastrephae* were similar to that of *T. drosophilae* [122–124]. Both *P. vindemiae* and *T. drosophilae* successfully attacked SWD pupae in fruit or soil [118, 128]. Under similar laboratory conditions, the former species was less efficient than the latter species [109, 120, 121]. Field surveys showed that *P. vindemiae* was more widely distributed than *T. drosophilae* in Europe [72] and North America [77], although laboratory tests indicated that *T. drosophilae* could adapt to the colder climates [113] or higher elevations [114]. Both species are sympatric in many regions and interspecific competition may affect their impact on fly populations, with *T. drosophilae* appearing to have an advantage over *P. vindemiae* under laboratory conditions [120]. Wolf et al. [118] found that *P. vindemiae* mainly parasitized hosts in the foliage, whereas most *T. drosophilae* attacked pupae on the ground in the field, which may relax interspecific competition.

In general, both *P. vindemiae* and *T. drosophilae* preferred larger (e.g., SWD) over smaller (e.g., *D. melanogaster*) hosts [100, 115, 117, 118, 129]. Wang et al. (unpubl. data) demonstrated both parasitoids could successfully develop from 25 different *Drosophila* species tested. Although the parasitoids' efficiency and offspring fitness varied among host species, effects on life-history characteristics or ecological traits appeared to be unrelated to the phylogenetic position of tested host species. Both parasitoids appeared to benefit from attacking larger hosts, as the body size of emerging progeny was positively correlated to host size. These results show remarkable levels of phenotypic plasticity in body growth and development. This study also shows that larger hosts are likely to be more suitable for rearing these parasitoids for augmentative releases.

Although these pupal parasitoids are effective against SWD in the laboratory tests, to date the reported field parasitism of SWD has been too low to provide economic suppression [41, 64]. This suggests that the natural pupal parasitoid populations are not sufficiently large enough to suppress SWD populations [52, 63] and augmentative release could help enhance

Table 6.2. Frugivorous *Drosophila* parasitoids species evaluated for their efficacy on SWD in laboratory or field trials.

Parasitoid	Parasitoid origin	Rate of parasitism[1]	Emergence[1]	References
Larval parasitoids				
Braconidae				
Aphaereta sp.	USA	Low	None	[94]
Asobara citri	Ivory Coast	Low	None	[94]
A. japonica	Japan, South Korea	High	High	[82, 89, 93, 94, 101–108]
A. pleuralis	Indonesia	Low	None	[94]
A. tabida	France, Switzerland	Low	None	[72, 93, 94]
Figitidae				
Ganaspis brasiliensis	China, South Korea	High	High	[83, 102–105, 107]
Ganaspis cf. *brasiliensis*	China, Japan	High	High	[101, 102]
Ganaspis sp.	USA	Low	Low	[94]
G. xanthopoda[2]	Japan, Uganda, USA	Low	Low	[94]
Leptopilina boulardi	Congo, France, Italy, Kenya, Mexico[3], Switzerland, USA	Low	None	[71, 72, 93, 94, 98]
L. clavipes	The Netherlands	Low	None	[94]
L. guineaensis	Cameroon, South Africa	Low	None	[94]
L. heterotoma	France, Italy, Switzerland, USA	Low	Low	[71, 72, 77, 91, 93–97, 101, 102]
L. j. japonica	China, South Korea	High	High	[101–105, 107]
L. victoriae	Philippines, USA	Low	None	[94]
Pupal parasitoids				
Diapriidae				
Trichopria cf. *drosophilae*	France, Italy, Spain	High	High	[70, 93]
T. drosophilae	China, Italy[3], Mexico[3], South Korea, Switzerland, USA	High	High	[69, 71, 72, 91, 97, 109–121, 128–132]
T. sp.	France, USA	High	High	[94]
T. anastrephae	Brazil	High	High	[122–124]
Pteromalidae				
Muscidifurax raptorellus	Canada	High	High	[100]
Pachycrepoideus sp.	USA	High	High	[94]
P. vindemiae	Canada, China, France, Italy, South Korea, Spain, Switzerland, USA	High	High	[70–72, 77, 91, 93, 94, 109, 112, 116, 118, 120, 121, 125–127]
Spalangia erythromera	Switzerland	High	High	[72]
Vrestovia fidenas	Switzerland	High	High	[72, 99]

[1]Studies were conducted with SWD larvae or pupae presented either in artificial diet or within fruit.
[2]This species was later re-assigned as *Ganaspis brasiliensis* [92].
[3]Open field release.

their impacts. Indeed, trials in Italy showed that *T. drosophilae* reduced SWD emergence within a radius of 10 m of the release within an enclosed crop system [130] and release of *T. drosophilae* in Italy at a rate of 0.33 specimens/m^2 per week in unmanaged vegetation reduced cherry fruit damage by 34% [131]. Similarly, work in Mexico using a release rate of 4.5 *T. drosophilae*/m^2 per week in commercial berries (*Rubus fruticosus* L.) reported a 50% reduction of SWD in the field [132]. These results suggest that augmentative release of *T. drosophilae* can lower SWD outbreaks. While no augmentative trials have been reported for *P. vindemiae*, this parasitoid and other tested pupal parasitoids such as *M. raptorellus* are commonly sold for release in livestock operations, making releases in crops potentially feasible. A population model and field observations suggest the optimal timing for *T. drosophilae* release corresponds with the early SWD population increases [133] and early season release may reduce potential risk on non-target species which typically breed on rotting fruits in later seasons [128].

Native Asian parasitoids and classical biological control

Surveys were conducted to discover native SWD parasitoids in South Korea [82], China [83, 84] and Japan [81, 84]. At least 19 larval parasitoids, including 12 *Asobara* and 7 figitids, as well as two pupal parasitoids (*P. vindemiae* and *T. drosophilae*), were found (Table 6.1). Within the SWD's native range, *L. heterotoma* and *L. boulardi* have never been recorded from SWD [82, 83], while the record of *A. tabida*, collected from SWD in Japan, is a misidentification [88, 134]. *Asobara japonica* Belokobylskij, *Ganaspis brasiliensis* Ihering and *Leptopilina japonica* Novković & Kimura are the dominant larval *Drosophila* parasitoids in East Asia [64]. In South Korea, these three species accounted for 85.7% of all larval parasitoids collected with percentage parasitism that ranged from 0–17.1% and varied by geography, season, and collection methods; the two figitids were the major parasitoids found in wild fresh *Rubus* fruits infested by SWD in natural vegetation, whereas the braconid was the major parasitoid collected from fruit bait traps infested predominantly by other drosophilids [82]. In China, collected wild fruits (see 82, 83 for information on wild fruit plants and their habitats) were often co-infested by *D. pulchrella* Tan, Hsu & Sheng or by *D. subpulchrella* Takamori (which are also characterized by a serrated ovipositor, like SWD, and able to attack fresh fruits) in natural vegetation [83, 84]. Similarly, 97.1% of emerged parasitoids collected from fresh fruits were *G. brasiliensis* and *L. japonica*, with the highest parasitism in Yunnan by the former and latter species being 47.8% and 42.0%, respectively [83]. In Japan, *G. brasiliensis* (referred to as *G.* cf. *brasiliensis* in Girod et al. [84, 101, 102]) was the most abundant parasitoid collected from SWD in wild cherries with the highest parasitism of 75.6% reported in Nara [84]. Similarly, *A. japonica* was commonly collected in banana traps throughout Japan [81]. Populations of *A. japonica* in Japanese main islands and South Korea seem to be parthenogenetic, whereas those in the south-western Japanese islands reproduce sexually [82, 136]. One unidentified *Asobara* sp. was recorded from SWD in wild fruits [87–89]. These surveys were conducted typically during the peaks of occurrence of parasitoids. Therefore, information is still lacking on seasonal population dynamics of these native parasitoids in their native ranges.

Asobara japonica, *G. brasiliensis*, and *L. japonica* have been systematically evaluated in quarantine laboratories for their relative efficiency, host specificity, climatic adaptability as well as potential interactions [82, 83, 101–107, 135] (Table 6.2). All three prefer to attack young host larvae [106, 107]. Under similar conditions, *L. japonica* developed faster and consequently had a competitive advantage over other two species in multi-parasitized hosts [104]. However, *G. brasiliensis* discriminated strongly against hosts parasitized by *L. japonica* and *A. japonica* discriminated against hosts parasitized by *L. japonica*. The combined impacts on host suppression by *L. japonica* and *G. brasiliensis* appeared to be additive, likely due to the interspecific discrimination by *G. brasiliensis* [104]. Indeed, both parasitoids co-exist in most collection sites in China or South Korea [82, 83]. *Asobara japonica* was more efficient than *L. japonica* or *G. brasiliensis* at attacking flies reared in artificial diet while the two figitids performed similarly [103,105]. The intrinsic rate of increase for *A. japonica* on SWD and *D. melanogaster* [106] was higher or similar to that of SWD on artificial diet [13] and fruit [137], and higher than that *G. brasiliensis* and *L. japonica* [107], *T. drosophila* [115] and *P. vindemiae* [91]. Although these laboratory results suggest that *A. japonica* has a higher potential to suppress SWD populations than the other two larval parasitoid species, this parasitoid may not play an important role in suppressing SWD populations in fresh fruits and will likely attack SWD infesting rotting fruit in late fruit seasons.

For classical biological control, imported species must have few or no significant non-target impacts. In Switzerland, Girod et al. [101, 102] tested six different European nontarget fly species with these three larval parasitoids. They found that *A. japonica* developed from all tested drosophilids, *L. japonica* successfully parasitized *D. melanogaster* and *D. subobscura* Collin but a Japanese population of *G.* cf. *brasiliensis* was strictly specific to SWD, as was also reported by Kasuya et al. [85], whereas another population from China parasitized SWD and *D. melanogaster* and sporadically parasitized *D. subobscura*. Giorgini et al. [83] found similar host specificity of a Chinese *G. brasiliensis* population from Yunnan. Later, Seehausen et al. [86] further confirmed that a Chinese population of *G. brasiliensis* seems to be strictly specific to *Drosophila* spp. in fresh fruits whereas other Chinese and Japanese populations also attack SWD in diet and, sporadically, *D. subobscura*. The wide host range of *A. japonica* has also been reported from field surveys in Japan [81, 89, 108, 138]. Considering possible non-target impacts, this parasitoid is not recommended for introduction in North America or Europe [106]. *Leptopilina japonica* may be considered as an oligophagous species; it was also observed to parasitize SWD and several other *Drosophila* species from the *melanogaster* species group in Japan [138–140]. Based its specificity to SWD and closely related species, *G. brasiliensis* appears to be the most suitable candidate for classical biological control in North America and Europe.

However, the existence of differentiated populations of *G. brasiliensis* with varying host specificity appears to be the case. This species was redescribed by Buffington and Forshage [141] who established the new combination of *G. brasiliensis*. Before the redescription of *G. brasiliensis*, research on SWD parasitoids referred to a "SWD-specificstrain" of *Ganaspis xanthopoda* Ashmead in Japan [81, 85, 88, 138], or used the names *G. xanthopoda* or *Ganaspis* sp. when referring to specimens reared in Thailand and the Philippines [139], Hawaii and Uganda [94], Indonesia [92, 142], Malaysia [143], Benin and the Caribbean Sea [67], Brazil [141], and Mexico [98, 144] (Table 6.1). Using molecular analyses of the mitochondrial cytochrome oxidase I gene fragment, Nomano et al. [143] grouped *Ganaspis* individuals from different geographical regions into five lineages (named as G1-G5) (Table 6.1). The G2 included individuals only from the subtropical Japanese island of Iriomote-Jima reared from *Drosophila ficusphila* Kikkawa & Peng, the G4 lineage included individuals only from Indonesia that parasitize *Drosophila eugracilis* Bock and Wheeler, and the G5 lineage included individuals from Japan, and all outside of Asian regions from unknown host(s) (Table 6.1). The populations collected in South Korea [82], China [83, 84], and Japan [84] consisted of G1 and G3. G1 populations appeared to be relatively host-specific to SWD [84, 145] while G3 can also attack closely related *D. melanogaster* and *D. simulans* Sturtevant [83]. A petition for release of the most host specific G1 strain has been approved by the North American Plant Protection Organization and a final regulatory decision by USDA-APHIS is pending as of this writing.

The genetic distance based on the COI gene nucleotide sequence between the G1 and G3 individuals was large enough (5–7%) to suggest reproductive isolation [84, 143]. The G1 and G3 are seemingly differentiated by acid-soluble insect protein spectra [147], some host-searching behaviors, and do not cross successfully [86]. Based on these observations, Seehausen et al. [86] suggest that G1 and G3 are two cryptic species. However, phylogenomic reconstruction on a representative group of the G1-G5 lineages using a large (ca. 1300 loci) ultraconserved element (UCE) data set showed differences among *G. brasiliensis* populations, but COI-assigned G1 and G3 specimens were intermingled and not monophyletic (Buffington et al. unpubl. data). When compared to inter-species genetic difference in *Leptopilina*, the *G. brasiliensis* would be considered all one species with some populationlevel signal. Further detailed molecular analyses (e.g., whole genome sequences) among different populations may help determine whether these two lineages should be considered as cryptic species or biotypes. The G3 and *L. japonica* did not show a significant preference between SWD and *D. melanogaster* in artificial diet [103]. Biondi et al. [103] argue that by specializing on SWD in fresh fruit, the G1 population could avoid competition from other common parasitoids found attacking *D. melanogaster* in decaying fruits [67, 82, 83, 93, 102]. Hence, plasticity in host searching would be selected for, and two genotypes (G1 and G3) could be in sympatry. From a biological control point of view, the flexibility of G3 would allow it to use alternative hosts such as *D. melanogaster* and persist when SWD is rare in late season fruit. If a species is too restricted in its host searching, it could be easily extirpated under the right conditions. A CLIMEX model predicts that *G. brasiliensis* (G1 and G3) would likely establish in the western, south-eastern, and east coastal states in North America and most southern European countries [64]. *Ganaspis brasiliensis* entered a facultative diapause below 17.2°C and the South Korean population was slightly more cold-tolerant than the Chinese population [146]. We suggest the introduction of both G1 and G3 or different geographical populations of both lineages, which could result in better SWD control than G1 or G3 alone. It is worth mentioning that a recent survey in British Columbia, Canada found that both the G1 population of *G. brasiliensis* and *L. japonica* have established in the Vancouver area, possibly through accidental introduction [134]. Also, *L. japonica* was found established in multiple locations in Trento, Italy [90], although it is unknown how the parasitoid was introduced. The establishment of these two parasitoids will likely play an important role for the control of SWD in Europe and North America.

Biological control with predators

Predators of SWD have been tested at various scales: (1) basic consumption of exposed (naked) SWD life stages, (2) predation inside fruit by confining infested fruit with and without predators in arenas, (3) direct observation or PCR detection of field predation, and (4) predation levels in the field using predator inclusion and exclusion treatments (Figure 6.1). Predators are discussed below by group and listed in Table 6.3.

Among arachnids, web building and hunting spiders in the field have fed on SWD, with up to 21% testing positive in PCR tests [58, 149]. The commercially available *Stratiolaelaps scimitus* (Womersley) mite did not feed on exposed SWD larvae or pupae [148]. Among coleopterans, the rove beetle *Dalotia coriaria* Kraatz varied in predation success depending on the ease of finding SWD larvae inside fruits. Infested raspberry exposed to rove beetles had up to a 50% reduction in SWD [151]. Perhaps because raspberries are softer than blueberries, there was no predation impact on intact infested blueberries, but a 35% reduction of SWD in

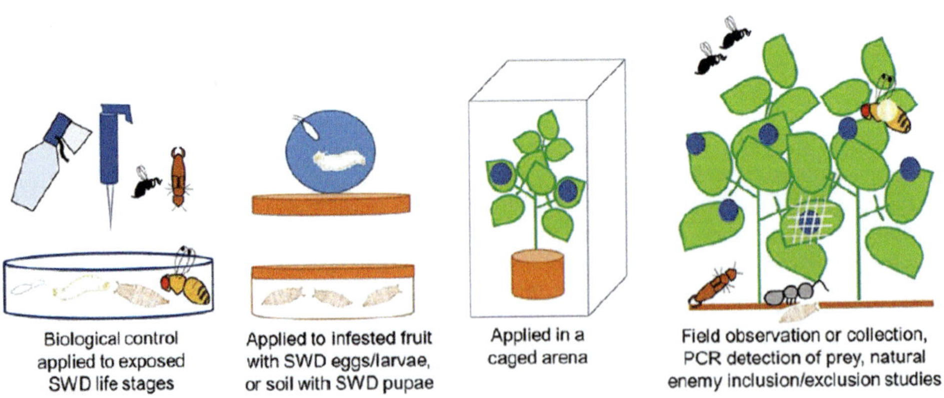

Fig. 6.1. Various scales of experimental testing for biological control of spotted-wing drosophila (SWD) *Drosophila suzukii*.

Table 6.3. Predators tested for predation on SWD and general outcomes (NS = non-significant).

Predator	Testing description	Outcomes	References
Acari, Laelapidae, *Strateolaelaps scimitus*	Lab: exposed (naked) SWD larvae or pupae	No predation	[148]
Aranae web-builders, Agelendiae, Araneidae, Linyphiidae, Theridiidae	Field-collected specimens	Tested positive for SWD feeding, up to 21%	[58, 149]
Aranae hunters, Clubionidae, Lycosidae, Oxypodidae, Philodromidae, Pisauridae, Salticidae, Thomisidae	Field-collected specimens	Tested positive for SWD feeding, up to 8%	[58, 149]
Coleoptera, Carabidae, *Bembidion quadrimaculatum*, *Pterostichus mutus*	Lab: exposed SWD pupae	Confirmed consumption	[150]
Coleoptera, Staphylinidae, *Dalotia coriaria*	Lab: infested blueberry, strawberry, or SWD on agar	NS to 50% reduction in SWD	[151–153]
Dermpatera, Forcifulidae, *Forficula auricularia*	Lab: exposed SWD life stages, infested cherry; field-collected specimens	NS to 45% reduction in SWD; 43% tested positive in field	[149, 154, 155]
Dermpatera, Labiduridae, *Labidura riparia*	Lab: exposed SWD larvae or pupae	Confirmed consumption	[70]
Heimptera Anthocoridae, *Anthocoris nemoralis*	Lab: exposed SWD adults	14% reduction	[149]
Hemiptera, Anthocoridae, *Orius insidiosus*	Lab: exposed SWD, infested blueberry, raspberry, strawberry, caged blueberry plants	NS to 67% reduction	[100, 152, 153]
Hemiptera, Anthocoridae, *Orius laevigatus*	Lab: exposed SWD larvae or adults, infested blueberry, strawberry	NS to 10% reduction, egg predation confirmed	[70, 149, 156]
Hemiptera, Anthocoridae, *Orius majusculus*	Lab: exposed life stages, infested blueberry, cherry	Confirmed consumption of eggs and larvae, but not pupae	[149, 154, 156]
Hemiptera, Miridae, *Dicyphus hesperus*	Lab: exposed life stages, infested raspberry	Consumption of exposed eggs, NS with infested fruit	[100]
Hemiptera, Nabidae, *Himacerus mirmicoides*	Field-collected specimens	16% tested positive for SWD	[149]
Hemiptera, Pentatomidae, *Podisus maculiventris*	Lab: exposed life stages infested raspberry	Consumption of exposed larvae, NS with infested fruit	[100]
Hymenoptera, Formicidae	Field observations	Removed sentinel pupae	[157]
Mantodea, Mantidae	Field-collected specimens	Tested positive for SWD	[58]
Neuroptera, Chrysopidae, *Chrysoperla carnea*	Lab: exposed SWD, infested raspberry or cherry	NS to 32% reduction of SWD	[100, 154]
Orthoptera, Gryllidae, *Gryllus pennsylvanicus*	Lab: exposed pupae	Confirmed consumption	[150]

damaged infested blueberries with rove beetles present [152]. Two carabid beetle species have thus far been confirmed to consume SWD pupae in the laboratory [150].

Among hemipterans, *Anthocoris nemoralis* (F.), *Orius insidious* (Say), *O. laevigatus* (Fieber), and *O. majusculus* (Reuter) have shown promise, with a 66%–67% reduction of SWD developing in strawberries or blueberries [153]. Moreover, *Orius* can feed on the egg stage of SWD [70, 100]. About 16% of field-collected nabids, mostly *Himacerus mirmicoides* (Costa), tested positive for SWD feeding [149]. The commercially available mirid, *Dicyphus hesperus* (Knight), and pentatomid, *Podisus maculiventris* (Say), can consume exposed SWD but showed little impact when confined with infested raspberries [100]. Among neuropterans, green lacewing larvae, *Chrysoperla carnea* (Stephens), can feed on SWD eggs, larvae, and pupae. When SWD infest cherry or raspberry, the presence of this larva resulted in a 32% reduction in SWD infestation [100, 154]. This species is commercially available. Among other groups, about 43% of field-collected earwigs, mostly *Forficula auricularia* L., have tested positive for SWD DNA [149], and their presence among infested cherries in box arenas led to a 40–45% reduction of SWD [155]. Ants have been directly observed to dig up sentinel pupae [157]. Mantids and crickets have been consumed in the laboratory and field [58, 150].

More field trials are needed to provide growers with applications. Predators that are commercially available and promising include *D. coriaria*, *Orius* spp., and *C. carnea*, but they have not been tested in the field. Currently, we have observed that predators reduced by 19%–49% SWD emerging from fruit in predator inclusion/exclusion studies [157] and removed as little as 1%–4% and as much as 61%–100% of sentinel pupae placed in crop fields [79, 150, 157].

Biological control with entomopathogens

Pathogens have largely been tested in the laboratory, and methods have varied depending on whether the pathogen targets SWD on foliage or in the soil. Seven viruses have caused mortality when injected into the thorax of adult SWD (Table 6.4).

Bacterial pathogens are often tested by feeding larvae with bacteria-infused diet, feeding adults with infected sugar solutions, spraying adults or spraying/dipping infested fruit. Twenty-two variants of *Bacillus thuringiensis* Berliner have been tested. *Bacillus thuringiensis israelensis* had little effect with increased mortality in just one of 15 trials (Table 6.4). *Bacillus thuringiensis kurstaki* increased mortality in 5 out of 12, but counterproductively increased oviposition among exposed flies [159]. *Bacillus thuringiensis thuringiensis* increased larval, pupal or adult mortality in all 4 of 4 trials [160]. Recently, bacteria isolated from moribund larvae were tested, and *Leuconostoc pseudomesenteroides* Farrow et al. increased mortality of adults and reduced food uptake in larvae [158].

One fungal pathogen, *Entomophthora muscae* (Cohn) Fresen., has been observed in an outbreak among field SWD [161], but the pathogen is difficult to culture and unlikely to be commercialized. Fungal pathogens are often tested by applying to vials or small surfaces and exposing to adult flies, directly spraying adults, or by dipping infested fruit. *Beauveria bassiana* (Bals.) Vuill., *Isaria fumosorosea* Wise, *Metarhizium anisopliae* (Metch.) Sorok, and *M. brunneum* Petch have had varying results depending on application and stage of SWD tested (Table 6.4). Besides causing direct mortality, some of these pathogens can reduce egg maturation [163], reduce reproductive output [164], and deter oviposition [97]. However, some fungal pathogens can increase oviposition [159], have limited residual impacts [152], and have limited spore viability in the field [163].

Nematodes are tested by applying directly to SWD larvae or pupae, or by applying to infested fruit, diet or soil since these stages would be concealed if field applications were made [175,176]. All tested species except one have been pathogenic (Table 6.4). *Heterorhabditis bacteriophora* (Poinar) was pathogenic in 11 of 18 trials, *Steinernema carpocapsae* (Weiser) in 20 of 23, *S. feltiae* (Filipjev) in 18 of 21, and *S. kraussei* (Steiner) in 2 of 3 (Table 6.4). Interestingly, *S. carpocapsae* can infect newly emerging adults; 89% of emerging adults were infected when they pupated in soil treated with nematodes [176].

Table 6.4. Pathogens tested on SWD and general outcomes (NS = non-significant).

Pathogen	Testing description	Outcomes	References
Bacteria			
Bacillus altitudinis	Fed adult SWD female	100% mortality	[158]
B. safensis	Fed adult female	Increased survival	[158]
B. simplex	Fed adult female	100% mortality	[158]
B. thuringiensis var. *aizawai* (multiple serotypes, strains)	Fed larva or adult, spray fruit, vial or fly, test oviposition	NS on mortality, increased oviposition	[159, 160]
B. thuringiensis var. *bolivia*	Fed larva or adult	Increased larval mortality, NS for adults	[160]
B. thuringiensis var. *dendrolimus*	Fed larva	NS on mortality	[160]
B. thuringiensis var. *entomocidus*	Fed larva or adult	NS on mortality	[160]
B. thuringiensis var. *galleriae*	Fed larva or adult	NS on mortality	[160]
B. thuringiensis var. *higo*	Fed larva or adult	Larval mortality with high dose, NS for adults	[160]
B. thuringiensis var. *israelensis* (multiple serotypes, sources)	Fed larva or adult, spray fly or dip fruit, test oviposition	Mostly NS, higher adult mortality from direct spray, increased oviposition	[159, 160, 162]
B. thuringiensis var. *jegathesan*	Fed larva or adult	NS mortality	[160]
B. thuringiensis var. *kenyae*	Fed larva or adult	NS on mortality	[160]
B. thuringiensis var. kurstaki (multiple serotypes, strains)	Fed larva or adult, spray fruit, vial or fly, test oviposition	NS to 100% larval mortality, higher adult mortality from direct spray, increased oviposition	[159, 160]
B. thuringiensis var. *kyushenunsis*	Fed larva or adult	NS on mortality	[160]
B. thuringiensis var. *leesis*	Fed larva or adult	NS on mortality	[160]
B. thuringiensis var. *malaysiensis*	Fed larva or adult	NS on mortality	[160]
B. thuringiensis var. *medillin*	Fed larva or adult	NS on mortality	[160]
B. thuringiensis var. *navarrensis*	Fed larva or adult	NS on mortality	[160]
B. thuringiensis var. packistani	Fed larva or adult	Larval mortality with high dose, NS for adults	[160]
B. thuringiensis var. *sotto*	Fed larva or adult	NS on mortality	[160]
B. thuringiensis var. *sumiyoshiensis*	Fed larva or adult	NS on mortality	[160]
B. thuringiensis var. *thompsoni*	Fed larva or adult	Larval mortality with high dose, NS for adults	[160]

Continued

Table 6.4. Continued

Pathogen	Testing description	Outcomes	References
B. thuringiensis var. *thuringiensis*	Fed larva or adult	Increased larval, pupal and adult mortality	[160]
B. thuringiensis var. *vazensis*	Fed larva or adult	NS on mortality	[160]
B. thuringiensis var. *wuhaensis*	Fed larva or adult	NS on mortality	[160]
Brevibacterium frigoritolerans	Fed larva and adult	Quicker adult mortality, no change in larval consumption	[158]
Curtobacterium plantarum	Fed adult	NS on mortality	[158]
Kocuria rhizophila	Fed adult	Increased survival	[158]
Leuconostoc pseudomesenteroides	Fed adult	Quicker adult mortality, reduce larval consumption	[158]
Micrococcus yunnanensis	Fed adult	Increased survival	[158]
Paenibacillus dongdonensis	Fed adult	Quicker adult mortality	[158]
P. odorifer	Fed adult	Quicker adult mortality	[158]
Tatumella ptyseos	Fed adult	NS on mortality	[158]
T. terrea	Fed adult	Quicker adult mortality	[158]
Fungi			
Beauveria bassiana (multiple strains)	Spray fly, vial, or leaf; dip fruit and expose to adult; dip infested fruit;	NS to 96% mortality of various stages, some oviposition deterrence on treated fruit, higher oviposition among affected females	[97, 152, 159, 160, 163–168]
B. bassiana (multiple strains)	Spray foliage in field cage	NS in infestation or adults trapped	[163]
Entomophthora muscae	Natural occurrence, expose to adult	Observed outbreaks, ~25% mortality in lab	[161, 169]
Isaria fumosorosea (multiple strains)	Spray/apply to adult, vial or fruit, dip fruit and expose to adult, dip infested fruit	NS to 93% mortality, higher oviposition	[152, 159, 160, 163, 164, 170–173]
I. javanica	Spray adult	100% mortality	[173]
Lecanicillium lecanii	Spray adult or paper	NS to 72% mortality	[160,167]
L. muscarium	Dip infested berry, spray fly	NS from dip, ~18% adult mortality	[165]
Metarhizium anisopliae (multiple strains)	Spray/apply to adult, pupae or fruit, dip fruit and expose to adult, dip infested fruit	NS to 82% mortality, oviposition deterrence	[97, 167, 168, 170, 171]
M. brunneum (multiple strains)	Spray/apply to adult, paper, leaf or fruit; expose fly to autoinnoculator	NS to 100% mortality; transmission between adults	[152, 160, 163, 174]
M. robertsii	Apply to adult	~16% mortality	[163]

Continued

Table 6.4. Continued

Pathogen	Testing description	Outcomes	References
Verticillium lecanii	Spray adult or fruit; apply to vial	NS on mortality, increased oviposition	[159]
Nematodes			
Heterorhabditis bacteriophora	Direct application or in sand with exposed (naked) SWD, or infested fruit, all life stages	NS to 94% mortality	[97, 152, 154, 170, 175–179]
H. georgiana	Apply to exposed larvae	NS larval mortality	[178]
H. indica	Apply to exposed pupa	33% pupal mortality, 47% by adult stage	[179]
Oscheius oniric	Apply to exposed larva or infested fruit	90% larval mortality, 82% with infested fruit	[178]
Steinernema carpocapsae	Apply to exposed or SWD in sand or diet, dip or apply to infested diet, dip infested fruit; all life stages	NS to 90% mortality; infection reduced adult flight; no encapsulation of *S. carpocapsae*	[152,165,170, 175-177,179]
S. feltiae	Apply to exposed or SWD in sand or diet, dip or apply to infested diet, dip infested fruit; all life stages	NS to 94% mortality	[97, 152, 165, 168, 170, 175–177, 179]
S. kraussei	Apply to larva or pupa on sand, dip infested fruit	~38–50% mortality assessed at pupal stage, NS with fruit dip	[165, 170]
Virus			
Drosophila C virus (DCV)	Intrathoracic injection	Total mortality	[180]
Cricket paralysis virus (CrPV)	Intrathoracic injection	Total mortality	[180]
Flock house virus (FHV)	Intrathoracic injection	Total mortality	[180]
Drosophila A virus (DAV)	Intrathoracic injection	50% mortality	[181]
La Jolla virus (LJV)	Intrathoracic injection	50% mortality	[181]
Drosophila C virus (DCV)	Intrathoracic injection	Survival enhanced by *Wolbachia*	[182]
Flock house virus (FHV)	Intrathoracic injection	Survival enhanced by *Wolbachia*	[182]

Additional field studies are needed since SWD life stages may be exposed to lower doses under field treatment conditions. Given the cost of pathogens, a "lure-and-infect" strategy is being examined where adults are attracted to a baited auto-inoculator and exit with inoculum (Figure 6.2). This would reduce the amount of inoculum used and protect the pathogen from UV, cold weather or water. This novel technology could also infect flies to serve as "entomovectors" for spreading other entomopathogens [183]. Notably, *M. brunneum* can infect adult SWD that walk through a container with a spore carpet, and inoculated males can cross-infect females in small cages or cups, and vice-a-versa [149,174]. Recently, 21% of adult SWD infected with *S. carpocapsae* as they emerged from the soil were able to fly in the laboratory [176]. The authors

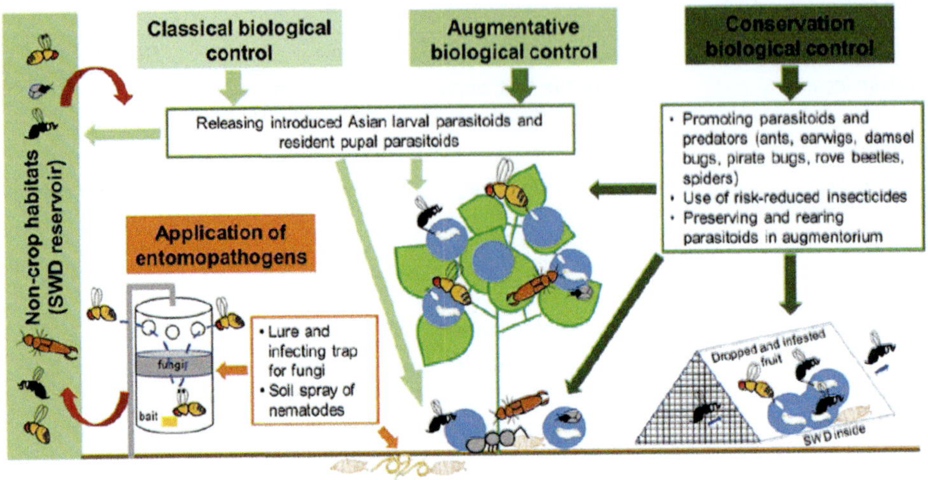

Fig. 6.2. Conceptual diagram of biocontrol strategies and principal biocontrol agents targeting different stages of spottedwing drosophila (SWD) *Drosophila suzukii*.

suggested that dispersion of infected adults could help disperse nematodes through the field and non-crop areas that are a source of infestation.

Future research is needed to assess compatibility of pathogens with predators and parasitoids, especially if predator/parasitoid releases are planned. Thus far, some trials show that multiple biological control tactics may be compatible. When SWD pupae were exposed to treatments of *B. bassiana*, *M. anisopliae*, *H. bacteriophora* or *S. feltiae*, emergence by the parasitoid *T. drosophilae* was unaffected in the laboratory [97]. Likewise, adults of *T. drosophilae* and rove beetle, *D. coriaria*, were unaffected by applications of *H. bacteriophora*, *S. feltiae*, and *S. carpocapsae* in Petri dish assays [176]. However, the predator *O. laevigatus* died from exposure to *S. carpocapsae* in Petri dish assays but not when nematodes were applied to a plant. This suggests that *O. laevigatus* would escape harmful effects in a field situation.

Conservation biological control

The aim of conservation biological control is to create environmentally friendly habitats for natural enemies [23–25, 184], which requires an understanding of the agroecosystem. SWD occurs earlier in crops adjacent to forests [45, 47] and its density in crop fields was positively related to the surrounding forest cover [49, 52, 54, 61]. Non-crop habitats provide SWD overwintering sites and shelter and so could either positively or negatively impact SWD and natural enemy densities [58], depending on the availability of associated host plants and drosophilid communities [61, 63]. If there are no effective natural enemies in the habitats for the target pest or the natural habitat is insufficient to provide large enough enemy populations needed for the pest, the habitats may fail to support conservation biological control [185]. This subject is still poorly understood yet is fundamentally important for the success of biological control of this polyphagous pest.

Judicious pesticide use is also a component of conservation biological control. Schlesener et al. [186] showed that organophosphates, pyrethroids, and neonicotinoids caused high parasitoid mortality. Similarly, spinosad was highly toxic to *P. vindemiae* immatures and adults [187]. Several bio-insecticides appear to have lower toxicity. De Souza et al. [188] reported that the essential oils of several piper plants repelled SWD oviposition and negatively affected egg viability, but the lethal concentration on SWD had a low toxicity to the parasitoid *T. anastrephae*. Similarly, a formulation based on *Annona mucosa* seedextract caused >85% mortality of SWD, yet the same dosage caused little mortality of *T. anastrephae* [189]. Gowton et al. [190] showed that peppermint essential oil used as VOC fumigation could repel SWD oviposition but also reduced the survival and parasitism of adult *P. vindemiae*. However, it did not affect developing *P. vindemiae*.

A variety of organic insecticides were tested on two generalist predators of SWD; the green lacewing *Chrysoperla rufilabris* (Burmeister) was susceptible to spinosad, and the minute pirate bug *Orius insidiosus* was susceptible to fresh and aged residues of spinosad and sabadilla alkaloids [191]. Moreover, sublethal effects of insecticide exposure resulted in reduced egg hatch of *O. insidiosus*. The lethality of insecticides to predators may explain why secondary outbreaks of aphids and scales have been observed more frequently by growers when they treat fields for SWD [41].

Mulching and floor management have been examined as cultural practices to control SWD [29, 30], specifically targeting SWD as they often wander from fruit to pupate in the soil [157]. Such ground practices to make the soil less hospitable to SWD may however be incompatible with soil drench treatments with nematodes where a moist soil environment is necessary for infective juveniles to survive and find hosts.

To preserve and increase parasitism, Rossi-Stacconi et al. [192] suggested to use a tent-like augmentorium design constructed of a fine mesh (optimal 1–0.8 mm holes) (Figure 6.2). The mesh is enough to prevent the escape of adult SWD yet allows in and out of smaller parasitoids. The method has been used for the sanitation of pest tephritid fruit flies [193]. Dropped or damaged fruits can be collected and deposited inside the augmentorium container where parasitoids can reproduce (or overwinter) on SWD in a protected environment and disperse to the field. This method may become important especially when effective Asian larval parasitoids such as *G. brasiliensis* are introduced.

Concluding remarks and guides for future research

Based on current research progress, we propose a framework for optimal use of various available biological control approaches as either stand-alone or supplemental methods to other IPM tactics as illustrated in Figure 6.2:

(1) Classicalbiologicalcontrolbyintroducingthe hostspecific larval parasitoid *G. brasiliensis* (G1 and G3 strains of different geographical populations that are adapted to different climate zones within invaded regions) and if necessary, additional host-specific Asian larval parasitoids.
(2) Augmentative biological control using locally occurring pupal parasitoids (e.g., *P. vindemiae*, *T. drosophilae*, and *T. anastrephae*).
(3) Conservation biological control by using reduced-risk organically certified (e.g., OMRI-listed) insecticides to reduce non-target impacts on parasitoids and predators, and other compatible cultural practices (e.g., augmentorium technique) to preserve parasitoids and increase SWD parasitism while conducting field sanitation.
(4) Use of entomopathogens (e.g., fungal pathogens could be applied using the autoinoculation lure-and-infect trap and nematodes can be sprayed on soil to infect SWD pupae or teneral SWD).

There is still a dearth of information pertaining to the field ecology of these natural enemies and potential effect of non-crop habitat on biological control. Also, most studies thus far were done in the laboratory and the efficacy of many natural enemies needs to be further demonstrated in the field. Almost all known SWD parasitoid species can attack *D. melanogaster* (which is a promising competitor of SWD [194, 195]). It is important to understand how different entomophagous insects and entomopathogens might interact antagonistically to reduce overall impacts on SWD, or synergistically to enhance SWD suppression, and how these interactions might be modulated via shared hosts or prey and by the landscape context [63, 73, 103, 104, 118, 120, 196, 197]. Thus, future studies may include:

(1) improving knowledge of field ecology of key natural enemies including multi-trophic interactions, apparent competition, field dispersal patterns as well as the effect of landscape content on the efficacy of the natural enemies;

(2) studying the ecology, parasitism, and host specificity of these accidentally introduced larval parasitoids in Canada and Italy;
(3) developing optimal rearing techniques for promising natural enemies, including possible genetic improvement of natural enemies by selecting biological traits among different populations for selection or breeding [198];
(4) developing optimal release strategies for promising natural enemies;
(5) developing conservation strategies to reduce non-target impacts of other control measures on natural enemies;
(6) conducting field trials to develop recommendations for the use of entomopathogens;
(7) integrating various biological control tactics into ecologically based IPM for SWD.

Acknowledgements

We thank Emily Rutkowski (USDA-ARS Corvallis, Oregon, USA) for illustrating SWD life stages for the figures. Funding was provided by the National Institute of Food and Agriculture, USDA Specialty Crops Research Initiative award no. 2015-51181-24252, USDA Organic Research and Extension Initiative award no. 2018-51300-28434, the California Cherry Board, a University of California Agricultural and Natural Research Grant, and USDA base funding (projects 8010-22000-028-00D and 2072-22000- 040-00D). The USDA is an equal opportunity provider and employer. The USDA does not endorse products mentioned in this publication.

References

1. Asplen MK, Anfora G, Biondi A, Choi DS, Chu D, Daane KM, et al. Invasion biology of spotted wing drosophila (*Drosophila suzukii*): a global perspective and future priorities. Journal of Pest Science 2015;88:469–94.
2. Andreazza F, Bernardi D, dos Santos RSS, Garcia FRM, Oliveira EE, Botton M, et al. *Drosophila suzukii* in Southern neotropical region: current status and future perspectives. Neotropical Entomology 2017;46:591–605.
3. Atallah J, Teixeira L, Salazar R, Zaragoza G, Kopp A. The making of a pest: the evolution of a fruit-penetrating ovipositor in *Drosophila suzukii* and related species. Proceedings of the Royal Society B 2014;281:20132840.
4. Ioriatti C, Walton V, Dalton D, Anfora G, Grassi A, Maistri S, et al. *Drosophila suzukii* (Diptera: Drosophilidae) and its potential impact to wine grapes during harvest in two cool climate wine grape production regions. Journal of Economic Entomology 2015;108:1148–55.
5. Rombaut A, Guilhot R, Xuereb A, Benoit L, Chapuis MP, Gibert P, et al. Invasive *Drosophila suzukii* facilitates *Drosophila melanogaster* infestation and sour rot outbreaks in the vineyards. Royal Society Open Science 2017;4:e170117.
6. Lewis MT, Koivunen EE, Swett CL, Hamby KA. Associations between *Drosophila suzukii* (Diptera: Drosophilidae) and fungi in raspberries. Environmental Entomology 2019;48:68–79.
7. Lee JC, Bruck DJ, Curry H, Edwards D, Haviland DR, Van Steenwyk RA, et al. The susceptibility of small fruits and cherries to the spotted-wing drosophila, *Drosophila suzukii*. Pest Management Science 2011;67:1358–67.
8. Stewart TE, Wang XG, Molinar A, Daane KM. Factors limiting peach as a potential host for spotted wing drosophila. Journal of Economic Entomology 2014;107:1771–1779.
9. Lee JC, Dreves AJ, Cave AM, Kawai S, Isaacs R, Miller JC, et al. Infestation of wild and ornamental noncrop fruits by *Drosophila suzukii* (Diptera: Drosophilidae). Showing results for Annals of the Entomological Society of America 2015;108:117–29.

10. Poyet M, Le Roux V, Gibert P, Meirland A, Prevost G, Eslin P, et al. The wide potential trophic niche of the Asiatic fruit fly *Drosophila suzukii*: The key of its invasion success in temperate Europe? PLoS One 2015;10:e0142785.
11. Kenis M, Tonina L, Eschen R, van der Sluis B, Sancassani M, Mori N, et al. Non-crop plants used as hosts by *Drosophila suzukii* in Europe. Journal of Pest Science 2016;89:735–48.
12. Little CM, Chapman TW, Hillier NK. Plasticity is key to success of *Drosophila suzukii* (Diptera: Drosophilidae) invasion. Journal of Insect Science 2020;20:5:1–8.
13. Emiljanowicz LM, Ryan GD, Langille A, Newman J. Development, reproductive output and population growth of the fruit fly pest *Drosophila suzukii* (Diptera: Drosophilidae) on artificial diet. Journal of Economic Entomology 2014;107:1392–98.
14. Wiman NW, Walton VM, Dalton DT, Anfora G, Burrack HJ, Chiu J, et al. Integrating temperature-dependent life table data into a matrix projection model for *Drosophila suzukii* population estimation. PLoS One 2014;9:e106909.
15. Wiman NG, Anfora G, Biondi A, Chiu JC, Daane KM, Dalton DT, et al. Modelling *Drosophila suzukii* populations in response to the environment and management strategies. Journal of Pest Science 2016;89:653–65.
16. Del Fava E, Ioriatti C, Melegaro A. Cost-benefit analysis of controlling the spotted wing drosophila (*Drosophila suzukii* (Matsumura)) spread and infestation of soft fruits in Trentino, Northern Italy. Pest Management Science 2017;73:2318–27.
17. Farnsworth D, Hamby KA, Bolda M, Goodhue RE, Williams JC, Zalom FG. Economic analysis of revenue losses and control costs associated with the spotted wing drosophila, *Drosophila suzukii* (Matsumura), in the California raspberry industry. Pest Management Science 2017;73:1083–90.
18. Haviland DR, Beers EH. Chemical control programs for *Drosophila suzukii* that comply with international limitations on pesticide residues for exported sweet cherries. Journal of Integrated Pest Management 2012;3:1–6.
19. Van Timmeren S, Isaacs R. Control of spotted wing drosophila, *Drosophila suzukii*, by specific insecticides and by conventional and organic crop protection programs. Crop Protection 2013;54:126–33.
20. Diepenbrock LM, Rosensteel OD, Hardin JA, Sial AA, Burrack HJ. Season-long programs for control of *Drosophila suzukii* in southeastern blueberries. Crop Protection 2016;84:76–84.
21. Shawer R, Tonina L, Tirello P, Duso C, Mori N. Laboratory and field trials to identify effective chemical control strategies for integrated management of *Drosophila suzukii* in European cherry orchards. Crop Protection 2018;103:73–80.
22. Van Timmeren S, Mota-Sanchez D, Wise JC, Isaacs R. Baseline susceptibility of spotted wing drosophila (*Drosophila suzukii*) to four key insecticide classes. Pest Management Science 2018;74:78–87.
23. Desneux N, Decourtye A, Delpuech J M. The sublethal effects of pesticides on beneficial arthropods. Annual Review of Entomology 2007;52:81–106.
24. Biondi A, Mommaerts V, Smagghe G, Vinuela E, Zappala L, Desneux N. The non-target impact of spinosyns on beneficial arthropods. Pest Management Science 2012;68:1523–36.
25. Roubos CR, Rodriguez-Saona C, Holdcraft R, Mason KS, Isaacs R. Relative toxicity and residual activity of insecticides used in blueberry pest management: mortality of natural enemies. Journal of Economic Entomology 2014;107:277–85.
26. Gress BE, Zalom FG. Identification and risk assessment of spinosad resistance in a California population of *Drosophila suzukii*. Pest Management Science 2019;75:1270–76.
27. Hampton E, Koski C, Barsoian O, Faubert H, Cowles RS, Alm SR. Use of early ripening cultivars to avoid infestation and mass trapping to manage *Drosophila suzukii* (Diptera: Drosophilidae) in *Vaccinium corymbosum* (Ericales: Ericaceae). Journal of Economic Entomology 2014;107:1849–57.
28. Leach H, Moses J, Hanson E, Fanning P, Isaacs R. Rapid harvest schedules and fruit removal as non-chemical approaches for managing spotted wing drosophila. Journal of Pest Science 2018;91:219–226.
29. Rendon D, Walton VM. Drip and overhead sprinkler irrigation in blueberry as cultural control for *Drosophila suzukii* (Diptera: Drosophilidae) in Northwestern United States. Journal of Economic Entomology 2019;112:745–52.
30. Rendon D, Hamby KA, Arsenault-Benoit AL, Taylor CM, Evans RK, Roubos CR, et al. Mulching as a cultural control strategy for *Drosophila suzukii* in blueberry. Pest Management Science 2020;76:55–66.

31. Ebbenga DN, Burkness EC, Hutchison WD. Evaluation of exclusion netting for spotted-wing drosophila (Diptera: drosophilidae) management in Minnesota wine grapes. Journal of Economic Entomology 2019;112:2287–94.
32. Rice KB, Short BD, Leskey TC. Development of an attract and kill strategy for *Drosophila suzukii* (Diptera: Drosophilidae): evaluation of attracticidal spheres under laboratory and field conditions. Environmental Entomology 2017;110:535–42.
33. Cloonan KR, Abraham J, Angeli S, Syed Z, Rodriguez-Saona C. Advances in the chemical ecology of the spotted wing drosophila (*Drosophila suzukii*) and its applications. Journal of Chemical Ecology 2018;44:922–39.
34. Wallingford AK, Cha DH, Loeb GM. Evaluating a push–pull strategy for management of *Drosophila suzukii* Matsumura in red raspberry. Journal of Applied Entomology 2018;74:120–25.
35. Haye T, Girod P, Cuthbertson AGS, Wang XG, Daane KM, Hoelmer KA, et al. Current SWD IPM tactics and their practical implementation in fruit crops across different regions around the world. Journal of Pest Science 2016;89:643–51.
36. Schetelig MF, Lee KZ, Otto S, Talmann L, Stokl J, Degenkolb T, et al. Environmentally sustainable pest control options for *Drosophila suzukii*. Journal of Applied Entomology 2018;142:3–17.
37. Lanouette G, Brodeur J, Fournier F, Martel V, Vreysen M, Cáceres C, et al. The sterile insect technique for the management of the spotted wing drosophila, *Drosophila suzukii*: establishing the optimum irradiation dose. PLoS One 2017;12:e0180821.
38. Sassù F, Nikolouli K, Pereira R, Vreysen MJB, Staufer C, Cáceres C. Irradiation dose response under hypoxia for the application of the sterile insect technique in *Drosophila suzukii*. PLoS One 2019;14:e0226582.
39. Krüger AP, Schlesener DCH, Martins LN, Wollmann J, Deprá M, Garcia FRM. Effects of irradiation dose on sterility induction and quality parameters of *Drosophila suzukii* (Diptera: Drosophilidae). Journal of Economic Entomology 2018;112:741–46.
40. Nikolouli K, Sassù F, Mouton L, Staufer C, Bourtzis K. Combining sterile and incompatible insect techniques for the population suppression of *Drosophila suzukii*. Journal of Pest Science 2020;93:647–61.
41. Lee JC, Wang XG, Daane KM, Hoelmer KA, Isaacs R, Sial AA, et al. Biological control of spotted-wing drosophila (Diptera: Drosophilidae): Current and pending tactics. Journal of Integrated Pest Management 2019;10:13.
42. Arnó J, Solà M, Riudavets J, Gabarra R. Population dynamics, non-crop hosts and fruit susceptibility of *Drosophila suzukii* in Northeast Spain. Journal of Pest Science 2016;89:713–23.
43. Diepenbrock LM, Swoboda-Bhattarai KA, Burrack HJ. Ovipositional preference, fidelity, and fitness of *Drosophila suzukii* in a co-occurring crop and non-crop system. Journal of Pest Science 2016;89:761–69.
44. Klick J, Yang WQ, Walton VM, Dalton DT, Hagler JR, Dreves AJ, et al. Distribution and activity of *Drosophila suzukii* in cultivated raspberry and surrounding vegetation. Journal of Applied Entomology 2016;140:37–46.
45. Pelton E, Gratton C, Isaacs R, Van Timmeren S, Blanton A, Guédot C. Earlier activity of *Drosophila suzukii* in high woodland landscapes but relative abundance is unaffected. Journal of Pest Science 2016;89:725–33.
46. Wang XG, Stewart TJ, Biondi A, Chavez BA, Ingels C, Caprile J, et al. Population dynamics and ecology of *Drosophila suzukii* in Central California. Journal of Pest Science 2016;89:701–12.
47. Cahenzli F, Bühlmann I, Daniel C, Fahrentrapp J. The distance between forests and crops affects the abundance of *Drosophila suzukii* during fruit ripening, but not during harvest. Environmental Entomology 2018;47:1274–79.
48. Grassi A, Gottardello A, Dalton DT, Tait G, Rendon D, Ioriatti C, et al. Seasonal reproductive biology of *Drosophila suzukii* (Diptera: Drosophilidae) in temperate climates. Environmental Entomology 2018;47:166–74.
49. Haro-Barchin E, Scheper J, Ganuza C, De Groot GA, Colombari F, van Kats R, et al. Landscape scale forest cover increases the abundance of *Drosophila suzukii* and parasitoid wasps. Basic and Applied Ecology 2018;31:33–43.
50. Hennig EI, Mazzi D. Spotted wing drosophila in sweet cherry orchards in relation to forest characteristics, bycatch, and resource availability. Insects 2018;9:118.

51. Panel ADC, Zeeman L, van der Sluis BJ, van Elk P, Pannebakker BA, Wertheim B, et al. Overwintered *Drosophila suzukii* are the main source for infestations of the first fruit crops of the season. Insects 2018;9:145.
52. Santoiemma G, Mori N, Tonina L, Marini L. Semi-natural habitats boost *Drosophila suzukii* populations and crop damage in sweet cherry. Agriculture Ecosystems & Environment 2018;257:152–58.
53. Tait G, Grassi A, Pfab F, Crava CM, Dalton DT, Magarey R, et al. Large-scale spatial dynamics of *Drosophila suzukii* in Trentino, Italy. Journal of Pest Science 2018;91:1213–24.
54. Tonina L, Mori N, Sancassani M, Dall'Ara P, Marini L. Spillover of *Drosophila suzukii* between non-crop and crop areas: implications for pest management. Agricultural and Forest Entomology 2018;20:575–81.
55. Whitehouse TS, Sial AA, Schmidt JM. Natural enemy abundance in southeastern blueberry agroecosystems: distance to edge and impact of management practices. Environmental Entomology 2018;47:32–8.
56. Leach H, Hagler JR, Machtley SA, Isaacs R. Spotted wing drosophila (*Drosophila suzukii*) utilization and dispersal from the wild host Asian bush honeysuckle (Lonicera spp.). Agricultural and Forest Entomology 2019;21:149–58.
57. Santoiemma G, Trivellato F, Caloi V, Mori N, Marini L. Habitat preference of *Drosophila suzukii* across heterogeneous landscapes. Journal of Pest Science 2019;92:485–94.
58. Schmidt JM, Whitehouse TS, Green K, Krehenwinkel H, Schmidt-Jeffris R, Sial AA. Local and landscape-scale heterogeneity shape spotted wing drosophila (*Drosophila suzukii*) activity and natural enemy abundance: implications for trophic interactions. Agriculture Ecosystems & Environment 2019;272:86–94.
59. Thistlewood HMA, Rozema B, Acheampong S. Infestation and timing of use of non-crop plants by *Drosophila suzukii* (Matsumura) (Diptera: Drosophilidae) in the Okanagan Basin, Canada. Canadian Entomologist 2019;151:34–48.
60. Wang XG, Kacar G, Daane KM. Temporal dynamics of host use by *Drosophila suzukii* in Joaquin Valley of California: Implications for area-wide pest management. Insects, 2019;10:206.
61. Delbac L, Rusch A, Binet D, Thiery D. Seasonal variation of Drosophilidae communities in viticultural landscapes. Basic and Applied Ecology 2020. doi: 10.1016/j.baae.2020.08.002.
62. Tait G, Cabianca A, Grassi A, Ferdinand P, Tiziana O, Simone P, et al. *Drosophila suzukii* daily dispersal between distinctly different habitats. Entomologia Generalis 2020;40:25–37.
63. Trivellone V, Meier M, Cara C, Paltrinieri LP, Gugerli F, Moretti M, Wolf S, Jana Collatz J. Multiscale determinants drive parasitization of Drosophilidae by Hymenopteran parasitoids in agricultural landscapes. Insects 2020;11:334.
64. Wang XG, Daane KM, Hoelmer KA, Lee JC-T. Biological control of spotted-wing Drosophila - an update on promising agents. In: Garcia FRM, editor. *Drosophila suzukii* management. Springer Nature. In press 2020.
65. Janssen A, Driessen G, Dehaan M, Roodbol N. The Impact of parasitoids on natural populations of temperate woodland *Drosophila*. Netherlands Journal of Zoology 1988;38:61–73.
66. Fleury F, Gibert P, Ris N, Allemand R. Ecology and life history evolution of frugivorous *Drosophila* parasitoids. In: Prévost G, editor. Advances in parasitology: Parasitoids of *Drosophila*, vol 70. London: Academic Press; 2009. p. 3–44.
67. Carton Y, Boulétreau B, van Alphen JJM, van Lenteren JC. The *Drosophila* parasitic wasps. In: Ashburner M, Carson HL, Thompson JN, editors. The genetics and biology of *Drosophila*, vol 3e. London: Academic Press; 1986. p. 347–94.
68. Prévost G. Advances in parasitology. Volume 70: Parasitoids of *Drosophila*. London, UK: Academic Press; 2009.
69. Rossi Stacconi MV, Grassi A, Dalton DT, Miller B, Ouantar M, Loni A, et al. First field records of *Pachycrepoideus vindemiae* as a parasitoid of *Drosophila suzukii* in European and Oregon small fruit production areas. Entomologia 2013;1:1–15.
70. Gabarra R, Riudavets J, Rodríguez GA, Pujade-Villar J, Arnó J. Prospects for the biological control of *Drosophila suzukii*. BioControl 2015;60:331–39.
71. Mazzetto F, Marchetti E, Amiresmaeili N, Sacco D, Francati S, Jucker C, et al. Drosophila parasitoids in northern Italy and their potential to attack the exotic pest *Drosophila suzukii*. Journal of Pest Science 2016;89:837–50.

72. Knoll V, Ellenbroek T, Romeis J, Collatz J. Seasonal and regional presence of hymenopteran parasitoids of *Drosophila* in Switzerland and their ability to parasitize the invasive *Drosophila suzukii*. Scientific Reports 2017;7:e40697.
73. Kremmer L, Thaon M, Borowiec N, David J, Poirie M, Gatti JL, et al. Field monitoring of *Drosophila suzukii* and associated communities in south eastern France as a pre-requisite for classical biological control. Insects 2017;8:124.
74. Modic S, Žigon P, Razinger J. *Trichopria drosophilae* (Diapriidae) and *Leptopilina heterotoma* (Figitidae), native parasitoids of *Drosophila suzukii*, confirmed in Slovenia. Acta agriculturae Slovenica 2019;113:181–85.
75. Zengin E, Karaca I. Dynamics of trapped adult populations of *Drosophila suzukii* Matsumura (Diptera: Drosophilidae) and its parasitoids in Uak Province, Turkey. Egyptian Journal of Biological Pest Control 2019;29:43.
76. Cancino MDG, Hernandez AG, Cabrera JG, Carrillo GM, Gonzalez JAS, Bernal HCA. Parasitoids of *Drosophila suzukii* (Matsumura) (Diptera: Drosophilidae) in Colima, Mexico. Southwestern Entomologist 2015;40:855–58.
77. Miller B, Anfora G, Buffington M, Daane KM, Dalton DT, Hoelmer KM, et al. Seasonal occurrence of resident parasitoids associated with *Drosophila suzukii* in two small fruit production regions of Italy and the USA. Bulletin of Insectology 2015;68:255–63.
78. Lue CH, Borowy D, Buffington ML, Leips J. Geographic and seasonal variation in species diversity and community composition of frugivorous *Drosophila* (Diptera: Drosophilidae) and their *Leptopilina* (Hymenoptera: Figitidae) parasitoids. Environmental Entomology 2018;47:1096–106.
79. Kamiyama MT, Schreiner Z, Guedot C. Diversity and abundance of natural enemies of *Drosophila suzukii* in Wisconsin, USA fruit farms. BioControl 2019;64:665–76.
80. Wollmann J, Schlesener DCH, Ferreira MS, Garcia FRM. Parasitoids of Drosophilidae with potential for parasitism on *Drosophila suzukii* in Brazil. Drosophila Information Service 2016;99:38–42.
81. Mitsui H, Van Achterberg K, Nordlander G, Kimura MT. Geographical distributions and host associations of larval parasitoids of frugivorous Drosophilidae in Japan. Journal of Natural History 2007;41:1731–38.
82. Daane KM, Wang XG, Biondi A, Miller B, Miller JC, Riedl H, et al. First exploration of parasitoids of *Drosophila suzukii* in South Korea as potential classical biological agents. Journal of Pest Science 2016;89:823–35.
83. Giorgini M, Wang XG, Wang Y, Chen FS, Hougardy E, Zhang HM, et al. Exploration for native parasitoids of *Drosophila suzukii* in China reveals a diversity of parasitoid species and narrow host range of the dominant parasitoid. Journal of Pest Science 2019;92:509–22.
84. Girod P, Borowiec N, Buffington M, Chen GH, Fang F, Kimura MT, et al. The parasitoid complex of *D. suzukii* and other fruit feeding *Drosophila* species in Asia. *Scientific Reports* 2018;8:11839.
85. Kasuya N, Mitsui H, Ideo S, Watada M, Kimura MT. Ecological, morphological and molecular studies on *Ganaspis* individuals (Hymenoptera: Figitidae) attacking *Drosophila suzukii* (Diptera: Drosophilidae). Applied Entomology and Zoology 2013;48:87–92.
86. Seehausen ML, Ris N, Driss L, Racca A, Girod P, Warot S, et al. Evidence for a cryptic parasitoid species reveals its suitability as a biological control agent. Scientific Reports. In press 2020.
87. Guerrieri E, Giorgini M, Cascone P, Carpenito S, van Achterberg C. Species diversity in the parasitoid genus *Asobara* (Hymenoptera: Braconidae) from the native area of the fruit fly pest *Drosophila suzukii* (Diptera: Drosophilidae). PLoS One 2016;11:e0147382.
88. Kimura MT, Mitsui H. *Drosophila* parasitoids (Hymenoptera) of Japan. Entomological Science 2020;23:359–68.
89. Nomano FY, Mitsui H, Kimura MT. Capacity of Japanese *Asobara* species (Hymenoptera; Braconidae) to parasitize a fruit pest *Drosophila suzukii* (Diptera; Drosophilidae). Journal of Applied Entomology 2015;139:105–13.
90. Puppato S, Grassi A, Pedrazzoli F, De Cristofaro A, Ioriatti C. First report of *Leptopilina japonica* in Europe. Insects 2020;11:611.
91. Rossi Stacconi MV, Buffington M, Daane KM, Dalton DT, Grassi A, Kaçar G, et al. Host stage preference, efficacy and fecundity of parasitoids attacking *Drosophila suzukii* in newly invaded areas. Biological Control 2015;84:28–35.
92. Kimura MT, Suwito A. Altitudinal patterns of abundances and parasitism in frugivorous drosophilids in west Java, Indonesia. Journal of Natural History 2015;49:1627–39.

93. Chabert S, Allemand R, Poyet M, Eslin P, Gibert P. Ability of European parasitoids (Hymenoptera) to control a new invasive Asiatic pest, *Drosophila suzukii*. Biological Control 2012;63:40–7.
94. Kacsoh BZ, Schlenke TA. High hemocyte load is associated with increased resistance against parasitoids in *Drosophila suzukii*, a relative of *D. melanogaster*. PLoS One 2012;7:e0034721.
95. Poyet M, Havard S, Prevost G, Chabrerie O, Doury G, Gibert P, et al. Resistance of *Drosophila suzukii* to the larval parasitoids *Leptopilina heterotoma* and *Asobara japonica* is related to haemocyte load. *Physiological Entomology* 2013;38:45–53.
96. Iacovone A, Ris N, Poirie M, Gatti JL. Time-course analysis of *Drosophila suzukii* interaction with endoparasitoid wasps evidences a delayed encapsulation response compared to *D. melanogaster*. PLoS One 2018;13:e0201573.
97. Ibouh K, Oreste M, Bubici G, Tarasco E, Rossi Stacconi MV, Ioriatti C, et al. Biological control of *Drosophila suzukii*: efficacy of parasitoids, entomopathogenic fungi, nematodes and deterrents of oviposition in laboratory assays. Crop Protection 2019;125:104897.
98. Sánchez-González JA, Lomeli-Flores JR, Rodríguez-Leyva E, Arredondo-Bernal HC, González-Cabrera J. *Drosophila suzukii* response to *Leptopilina boulardi* and *Ganaspis brasiliensis* parasitism. Bulletin of Insectology 2020;73:209–15.
99. Wolf S, Baur H, Collatz J. Life history of *Vrestovia fidenas*, a potential control agent of *Drosophila suzukii*. BioControl 2019;64:263–75.
100. Bonneau P, Renkema J, Fournier V, Firlej A. Ability of *Muscidifurax raptorellus* and other parasitoids and predators to control *Drosophila suzukii* populations in raspberries in the laboratory. Insects 2019;10:68.
101. Girod P, Lierhmann O, Urvois T, Turlings TCJ, Kenis M, Haye T. Host specificity of Asian parasitoids for potential classical biological control of *Drosophila suzukii*. Journal of Pest Science 2018;91:1241–50.
102. Girod P, Rossignaud L, Haye T, Turlings TCJ, Kenis M. Development of Asian parasitoids in larvae of *Drosophila suzukii* feeding on blueberry and artificial diet. Journal of Applied Entomology 2018;142:483–94.
103. Biondi A, Wang XG, Daane KM. Host preference of three Asian larval parasitoids to closely related *Drosophila* species: implications for biological control of *Drosophila suzukii*. Journal of Pest Science. In press 2020.
104. Wang XG, Hogg BN, Hougardy E, Nance AH, Daane KM. Potential competitive outcomes among three solitary larval endoparasitoids as candidate agents for classical biological control of *Drosophila suzukii*. Biological Control 2019;130:18–26.
105. Wang XG, Biondi A, Daane KM. Functional responses of three candidate Asian larval parasitoids evaluated for classical biological control of *Drosophila suzukii*. Journal of Economic Entomology 2020;113:73–80.
106. Wang XG, Biondi A, Nance AH, Zappalà L, Hoelmer KA, Daane KM. Assessment of *Asobara japonica* as a potential biological control agent for the spotted wing 1 drosophila, *Drosophila suzukii*. Entomologia Generalis. In press 2020
107. Wang XG, Nance AH, Jones JML, Hoelmer KA, Daane KM. Aspects of the biology and reproductive strategy of two Asian larval parasitoids evaluated for classical biological control of *Drosophila suzukii*. Biological Control 2018;121:58–65.
108. Ideo S, Watada M, Mitsui H, Kimura MT. Host range of *Asobara japonica* (Hymenoptera : Braconidae), a larval parasitoid of drosophilid flies. Entomological Science 2008;11:1–6.
109. Garcia-Cancino MD, Gonzalez-Cabrera J, Sanchez-Gonzalez JA, Arredondo-Bernal HC. Biological and population parameters, as well as oviposition preference, of two pupal parasitoids of *Drosophila suzukii* (Diptera: Drosophilidae) in Mexico. Journal of Entomological Science 2020;55:87–97.
110. Rossi Stacconi MV, Panel A, Baser N, Ioriatti C, Pantezzi T, Anfora G. Comparative life history traits of indigenous Italian parasitoids of *Drosophila suzukii* and their effectiveness at different temperatures. Biological Control 2017;112:20–7.
111. Amiresmaeili N, Jucker C, Savoldelli S, Lupi D. Understanding *Trichopria drosophilae* performance in laboratory conditions. Bulletin of Insectology 2018;71:251–56.
112. Wang XG, Serrato MA, Son Y, Walton VM, Hogg BN, Daane KM. Thermal performance of two indigenous pupal parasitoids attacking the invasive *Drosophila suzukii* (Diptera: Drosophilidae). Environmental Entomology 2018;47:764–72.
113. Amiresmaeili N, Romeis J, Collatz J. Cold tolerance of the drosophila pupal parasitoid *Trichopria drosophilae*. Journal of Insect Physiology 2020;125:104087.

114. Colombari F, Tonina L, Battisti A, Mori N. Performance of *Trichopria drosophilae* (Hymenoptera: Diapriidae), a generalist parasitoid of *Drosophila suzukii* (Diptera: Drosophilidae), at low temperature. Journal of Insect Science 2020;20:1–5.
115. Wang XG, Kaçar G, Biondi A, Daane KM. Life-history and host preference of *Trichopria drosophilae*, a pupal parasitoid of spotted wing drosophila. BioControl 2016;61:387–97.
116. Zhu CJ, Li J, Wang H, Zhang M, Hu HY. Demographic potential of the pupal parasitoid *Trichopria drosophilae* (Hymenoptera: Diapriidae) reared on *Drosophila suzukii* (Diptera: Drosophilidae). Journal of Asia-Pacific Entomology 2017;20:747–51.
117. Yi CD, Cao P, Lin J, Liu X, Ao G, Zhang Q, et al. Life history and host preference of *Trichopria drosophilae* from Southern China, one of the effective pupal parasitoids on the *Drosophila* species. Insects 2020;11:103.
118. Wolf S, Barmettler E, Eisenring M, Romeis J, Collatz J. Host searching and host preference of resident pupal parasitoids of *Drosophila suzukii* in the invaded regions. Pest Management Science. In press 2020. doi: 10.1002/ ps.6013.
119. Woltering SB, Romeis J, Collatz J. Influence of the rearing host on biological parameters of *Trichopria drosophilae*, a potential biological control agent of *Drosophila suzukii*. Insects 2019;10:183.
120. Wang XG, Kacar G, Biondi A, Daane KM. Foraging efficiency and outcomes of interactions of two pupal parasitoids attacking the invasive spotted wing drosophila. Biological Control 2016;96:64–71.
121. Kaçar G, Wang XG, Biondi A, Daane KM. Linear functional response by two pupal *Drosophila* parasitoids foraging within single or multiple patch environments. PLoS One 2017;12:e0183525.
122. Kruger AP, Scheunemann T, Vieira JGA, Morais MC, Bernardi D, Nava DE, et al. Effects of extrinsic, intraspecific competition and host deprivation on the biology of *Trichopria anastrephae* (Hymenoptera: Diapriidae) reared on *Drosophila suzukii* (Diptera: Drosophilidae). Neotropical Entomology 2019;48:957–65.
123. Vieira JGA, Krüger AP, Scheuneumann T, Morais MC, Speriogin HS, Garcia FRM, et al. Some aspects of the biology of *Trichopria anastrephae* (Hymenoptera: Diapriidae), a resident parasitoid attacking *Drosophila suzukii* (Diptera: Drosophilidae) in Brazil. Journal of Economic Entomology 2019;113:81–7.
124. Vieira JGA, Kruger AP, Scheuneumann T, Garcez AM, Morais MC, Garciaet FRM, et al. Effect of temperature on the development time and life-time fecundity of *Trichopria anastrephae* parasitizing *Drosophila suzukii*. Journal of Applied Entomology. In press 2020.
125. Wang XG, Messing RH. The ectoparasitic pupal parasitoid, *Pachycrepoideus vindemmiae* (Hymenoptera: Pteromalidae), attacks other primary tephritid fruit fly parasitoids: host expansion and potential non-target impact. Biological Control 2004;31:227–36.
126. Bezerra Da Silva CS, Price BE, Soohoo-Hui A, Walton VM. Factors affecting the biology of *Pachycrepoideus vindemmiae* (Hymenoptera: Pteromalidae), a parasitoid of spotted-wing drosophila (*Drosophila suzukii*). PLoS One 2019;14:19326203.
127. Bezerra Da Silva CH, Price BE, Walton VM. Water-deprived parasitic wasps (*Pachycrepoideus vindemmiae*) kill more pupae of a pest (*Drosophila suzukii*) as a water-intake strategy. Scientific Reports 2019;9:3592.
128. Wolf S, Boycheva-Woltering S, Romeis J, Collatz J. *Trichopria drosophilae* parasitizes *Drosophila suzukii* in seven common non-crop fruits. Journal of Pest Science 2020;93:627–38.
129. Chen JN, Zhou SC, Wang Y, Shi M, Chen XX, Huang JH. Biocontrol characteristics of the fruit fly pupal parasitoid *Trichopria drosophilae* (Hymenoptera: Diapriidae) emerging from different hosts. Scientific Reports 2018;8:e13323.
130. Rossi Stacconi MV, Amiresmaeili N, Biondi A, Carli C, Caruso S, Dindo ML, et al. Host location and dispersal ability of the cosmopolitan parasitoid *Trichopria drosophilae* released to control the invasive spotted wing drosophila. Biological Control 2018;117:188–96.
131. Rossi Stacconi MV, Grassi A, Ioriatti C, Anfora G. Augmentative releases of *Trichopria drosophilae* for the suppression of early season *Drosophila suzukii* populations. BioControl 2019;64:9–19.
132. Gonzalez-Cabrera J, Moreno-Carrillo G, Sanchez-Gonzalez JA, Mendoza-Ceballos MY, Arredondo-Bernal HC. Single and combined release of *Trichopria drosophilae* (Hymenoptera: Diapriidae) to control *Drosophila suzukii* (Diptera: Drosophilidae). Neotropical Entomology 2019;48:949–56.
133. Pfab F, Rossi Stacconi MV, Anfora G, Grassi A, Walton V, Pugliese A. Optimized timing of parasitoid release: a mathematical model for biological control of *Drosophila suzukii*. Theoretical Ecology 2018;11:489–501.

134. Abram PK, McPherson AE, Kula R, Hueppelsheuser T, Thiessen J, Perlman SJ, et al. New records of *Leptopilina, Ganaspis,* and *Asobara* species associated with *Drosophila suzukii* in North America, including detections of *L. japonica* and *G. brasiliensis*. Journal of Hymenoptera Research 2020;78:1–17.
135. Biondi A, Wang XG, Miller JC, Miller B, Shearer PW, Zappala, et al. Innate olfactory responses of *Asobara japonica* toward fruits infested by the invasive spotted wing drosophila. Journal of Insect Behavior 2017;30:495–506.
136. Murata Y, Ideo S, Watada M, Mitsui H, Kimura MT. Genetic and physiological variation among sexual and parthenogenetic populations of *Asobara japonica* (Hymenoptera: Braconidae), a larval parasitoid of drosophilid flies. European Journal of Entomology 2009;106:171–78.
137. Tochen S, Dalton DT, Wiman N, Hamm C, Shearer P, Walton VM. Temperature-related development and population parameters for *Drosophila suzukii* (Diptera: Drosophilidae) on cherry and blueberry. Environmental Entomology 2014;43:501–10.
138. Mitsui H, Kimura MT. Distribution, abundance and host association of two parasitoid species attacking frugivorous drosophilid larvae in central Japan. European Journal of Entomology 2010;107:535–40.
139. Novkovic B, Mitsui H, Suwito A, Kimura MT. Taxonomy and phylogeny of *Leptopilina* species (Hymenoptera: Cynipoidea: Figitidae) attacking frugivorous drosophilid flies in Japan, with description of three new species. Entomological Science 2011;14:333–46
140. Kimura MT, Novkovic B. Local adaptation and ecological fitting in host use of the Drosophila parasitoid *Leptopilina japonica*. Ecological Research 2015;30:499–505.
141. Buffington ML, Forshage M. Redescription of *Ganaspis brasiliensis* (Ihering, 1905), new combination (Hymenoptera: Figitidae), a natural enemy of the invasive *Drosophila suzukii* (Matsumura, 1931) (Diptera: Drosophilidae). Proceedings of the Entomological Society of Washington 2016;118:1–13.
142. Kimura MT, Suwito A. Diversity and abundance of frugivorous drosophilids and their parasitoids in Bogor, Indonesia. Journal of Natural History 2012;46:1947–57.
143. Nomano FY, Kasuya N, Matsuura A, Suwito A, Mitsui H, Buffington ML, et al. Genetic differentiation of *Ganaspis brasiliensis* (Hymenoptera: Figitidae) from East and Southeast Asia. Applied Entomology and Zoology 2017;52:429–37.
144. Gonzalez-Cabrera J, Cordoba-Urtiz EG, Moreno-Carrillo G, Sanchez-Gonzalez JA, Arredondo-Bernal HC. First report of the parasitoid *Ganaspis brasiliensis* Ihering (Hymenoptera: Figitidae) in Mexico. Entomological News 2020;129:67–70.
145. Matsuura A, Mitsui H, Kimura MT. A preliminary study on distributions and oviposition sites of *Drosophila suzukii* (Diptera: Drosophilidae) and its parasitoids on wild cherry tree in Tokyo, central Japan. Applied Entomology and Zoology 2018;53:47–53.
146. Hougardy E, Hogg BN, Wang XG, Daane KM. Comparison of thermal performances of two Asian larval parasitoids of *Drosophila suzukii*. Biological Control 2019;136:104000.
147. Reeve MA, Seehausen ML. Discrimination between Asian populations of the parasitoid wasp *Ganaspis* cf. *brasiliensis* using a simple MALDI-TOF MS-based method for use with insects. Biology Methods and Protocols 2019;4:1–8.
148. Cuthbertson AG, Blackburn LF, Audsley N. Efficacy of commercially available invertebrate predators against *Drosophila suzukii*. Insects 2014;5:952–60.
149. Wolf S, Zeisler C, Sint D, Romeis J, Traugott M, Collatz J. A simple and cost-effective molecular method to track predation on *Drosophila suzukii* in the field. Journal of Pest Science 2018;91:927–35.
150. Ballman ES, Collins JA, Drummond FA. Pupation behavior and predation on *Drosophila suzukii* (Diptera: Drosophilidae) pupae in Maine wild blueberry fields. Journal of Economic Entomology 2017;110:2308–17.
151. Renkema JM, Telfer Z, Gariepy T, Hallett RH. *Dalotia coriaria* as a predator of *Drosophila suzukii*: Functional responses, reduced fruit infestation and molecular diagnostics. Biological Control 2015;89:1–10.
152. Woltz JM, Donahue KM, Bruck DJ, Lee JC. Efficacy of commercially available predators, nematodes and fungal entomopathogens for augmentative control of *Drosophila suzukii*. Journal of Applied Entomology 2015;139:759–70.
153. Renkema JM, Cuthbertson AGS. Impact of multiple natural enemies on immature *Drosophila suzukii* in strawberries and blueberries. BioControl 2018;63:719–28.
154. Englert C, Herz A. Acceptability of *Drosophila suzukii* as prey for common predators occurring in cherries and berries. Journal of Applied Entomology 2019;143:387–96.

155. Bourne A, Fountain MT, Wijnen H, Shaw B. Potential of the European earwig (*Forficula auricularia*) as a biocontrol agent of the soft and stone fruit pest *Drosophila suzukii*. Pest Management Science 2019;75:3340–45.
156. Malagnini V, Zanotelli L, Tolotti G, Profaizer D, Angeli G. Evaluation of predatory activity of *Orius laevigatus* and *O. maisculus* towards *Drosophila suzukii* under laboratory conditions. IOCB Working group "Integrated Protection of fruit crops", 26–28 May 2014. Fondazione Edmund Mach, Vigalzano di Pregine Valsugana.
157. Woltz JM, Lee JC. Pupation behavior and larval and pupal biocontrol of *Drosophila suzukii* in the field. Biological Control 2017;110:62–9.
158. Hiebert N, Carrau T, Bartling M, Vilcinskas A, Lee KZ. Identification of entomopathogenic bacteria associated with the invasive pest *Drosophila suzukii* in infested areas of Germany. Journal of Invertebrate Pathology 2020;151:107389.
159. Cahenzli F, Strack T, Daniel C. Screening of 25 different natural crop protection products against *Drosophila suzukii*. Journal of Applied Entomology 2018;142:563–77.
160. Cossentine J, Robertson M, Buitenhuis R. Impact of acquired entomopathogenic fungi on adult *Drosophila suzukii* survival and fecundity. Biological Control 2016;103:129–37.
161. Dara SK, Mann T, Li DW, Layton B. Epizootics of an entomophthoralean fungus in spotted-wing drosophila populations on fig. E-Journal of Entomology and Biologicals 2017.
162. Biganski S, Jehle JA, Kleespies RG. *Bacillus thuringiensis* serovar *israelensis* has no effect on *Drosophila suzukii* Matsumura. Journal of Applied Entomology 2018;142:33–6.
163. Alnajjar G, Drummond FA, Groden E. Laboratory and field susceptibility of *Drosophila suzukii* Matsumura (Diptera: Drosophilidae) to entomopathogenic fungal mycoses. Journal of Agricultural and Urban Entomology 2017;33:111–32.
164. Rhodes EM, Avery PB, Liburd OE. Efficacy of entomopathogenic fungal products for biological control of spotted wing drosophila (Diptera: Drosophilidae) under laboratory conditions. Florida Entomologist 2018;101:526–28.
165. Cuthbertson AGS, Collins DA, Blackburn LF, Audsley N, Bell HA. Preliminary screening of potential control products against *Drosophila suzukii*. Insects 2014;5:488–98.
166. Gargani E, Tarchi F, Frosinini R, Mazza G, Simoni S. Notes on *Drosophila suzukii* Matsumura (Diptera Drosophilidae): Field survey in Tuscany and laboratory evaluation of organic products. Redia Giornale di Zoologia 2013;96:85–90.
167. Naranjo-Lazaro JM, Mellin-Rosas MA, Gonzalez-Padilla VD, Arredondo-Bernal HC. Seleccion de cepas de hongos entomopatogenos como agentes de control biologica de adultos de *Drosophila suzukii* Matsumura (Diptera: Drosophilidae). In: XXXVI Congreso Nacional de Control Biologico. Oaxaca, Mexico; 2013
168. Oreste M, Baser N, Ibouh K, Verrastro V, Tarasco E. Potential of entomopathogenic fungi and nematodes against *Drosophila suzukii* in laboratory assays. IOBC-WPRS Bulletin 2017;129:77–8.
169. Becher PG, Jensen RE, Natsopoulou ME, Verschut V, De Fine Licht HH. Infection of *Drosophila suzukii* with the obligate insect-pathogenic fungus *Entomophthora muscae*. Journal of Pest Science 2017;91:781–87.
170. Cuthbertson AG, Audsley N. Further screening of entomopathogenic fungi and nematodes as control agents for *Drosophila suzukii*. Insects 2016;7:24.
171. Naranjo-Lazaro JM, Mellin-Rosas MA, Gonzalez-Padilla VD, Sanchez-Gonzalez JA, Moreno-Carrillo G, Arredondo-Bernal HC. Susceptibility of *Drosophila suzukii* Matsumura (Diptera: Drosophilidae) to entomopathogenic fungi. Southwestern Entomologist 2014;39: 201–04.
172. Naranjo-Lazaro JM, Mellin-Rosas MA, Gonzalez-Padilla VD, Arredondo-Bernal HC. Seleccion de aislados monosporicos de hongos entomopatogenos ye su patogenicidad sobre *Drosophila suzukii* (Matsumrua) (Diptera: Drosophilae). In: Memoria del XXXVII Congreso Nacional de Control Biologico, 6–7 de Noviember 2014. Merida Yucatan: Impartido en el Hotel Holiday Inn; 2014
173. Naranjo-Lazaro JM, Mellin-Rosas MA, Arrendondo-Bernal HC. Determinacion de concentracion letal CL50 y CL90 y tiempo letal TL50 y Tl90 de hongos entompatogenos en adultos de *Drosophila suzukii* Matsumrua (Diptera: Drosophildae). In: Memoria del XXXVIII Congreso Nacional de Control Biologica, 5 y 6 de Noviember 2015. Leon, Gto; 2015
174. Yousef M, Aranda-Valera E, Quesada-Moraga E. Lure-and-infect and lure-and-kill devices based on *Metarhizium brunneum* for spotted wing Drosophila control. Journal of Pest Science 2018;91:227–35.
175. Garriga A, Morton A, Ribes A, Garcia-del-Pino F. Soil emergence of *Drosophila suzukii* adults: a susceptible period for entomopathogenic nematodes infection. Journal of Pest Science 2019;93:639–46.

176. Garriga A, Mastore M, Morton A, Pino FGD, Brivio MF. Immune response of *Drosophila suzukii* larvae to infection with the nematobacterial complex *Steinernema carpocapsae-Xenorhabdus nematophila*. Insects 2020;11:210.
177. Hübner A, Englert C, Herz A. Effect of entomopathogenic nematodes on different developmental stages of *Drosophila suzukii* in and outside fruits. BioControl 2017;62:669–80.
178. Foye S, Steffan SA. A rare, recently discovered nematode, *Oscheius onirici* (Rhabditida: Rhabditidae), kills *Drosophila suzukii* (Diptera: Drosophilidae) within fruit. Journal of Economic Entomology 2020;113:1047–51.
179. De Brida AL, Wilcken SRS, Garrigós Leite L, Mello Garcia FR. Virulence of entomopathogenic nematode to pupae and adults of *Drosophila suzukii* in laboratory. Revista Cientifica Rural 2019;21:126–38.
180. Lee KZ, Vilcinskas A. Analysis of virus susceptibility in the invasive insect pest *Drosophila suzukii*. Journal of Invertebrate Pathology 2017;148:138–41.
181. Carrau T, Hiebert N, Vilcinskas A, Lee KZ. Identification and characterization of natural viruses associated with the invasive insect pest *Drosophila suzukii*. Journal of Invertebrate Pathology 2018;154:74–8.
182. Cattel J, Martinez J, Jiggins F, Mouton L, Gibert P. *Wolbachia*-mediated protection against viruses in the invasive pest *Drosophila suzukii*. Insect Molecular Biology 2016;25:595–603.
183. Taning CNT, Smagghe G. Threat of *Drosophila suzukii* as an invasive species and the potential of entomovectoring. In: Smagghe G, Boecking O, Maccagnani B, Mand M, Kevan P, editors. Entomovectoring for precision biocontrol and enhanced pollination of crops. Cham: Springer; 2020.
184. Roubos CR, Rodriguez-Saona C, Isaacs R. Mitigating the effects of insecticides on arthropod biological control at field and landscape scales. Biological Control 2014;75:28–38.
185. Tscharntke T, Karp DS, Chaplin-Kramer R, Batáry P, DeClerck F, Gratton C, et al. When natural habitat fails to enhance biological pest control – Five hypotheses, Biological Conservation 2016;204:449–58.
186. Schlesener DCH, Wollmann J, Pazini JB, Padilha AC, Grutzmacher AD, Garcia FRM. Insecticide toxicity to *Drosophila suzukii* (Diptera: Drosophilidae) parasitoids: *Trichopria anastrephae* (Hymenoptera: Diapriidae) and *Pachycrepoideus vindemmiae* (Hymenoptera: Pteromalidae). Journal of Economic Entomology 2019;112:1197–206.
187. Cossentine JE, Ayyanath MM. Limited protection of the parasitoid *Pachycrepoideus vindemiae* from *Drosophila suzukii* host-directed spinosad suppression. Entomologia Experimentalis et Applicata 2017;164:78–86.
188. De Souza MT, de Souza MT, Bernardi D, Krinski D, de Melo DJ, Oliveira DDC, et al. Chemical composition of essential oils of selected species of Piper and their insecticidal activity against *Drosophila suzukii* and *Trichopria anastrephae*. Environmental Science and Pollution Research. In press 2020
189. Bernardi D, Ribeiro L, Andreazza F, Neitzke C, Oliveira EE, Botton M, et al. Potential use of *Annona* by products to control *Drosophila suzukii* and toxicity to its parasitoid *Trichopria anastrephae*. Industrial Crops and Products 2017;110:30–5.
190. Gowton CM, Michał Rout M, Carrillo J. Peppermint essential oil inhibits *Drosophila suzukii* emergence but reduces *Pachycrepoideus vindemmiae* parasitism rates. Scientific Reports 2020;10:9090.
191. Sarkar N, Rhodes EM, Spies J, Roubos CR, Sial AA, Philip D, et al. Evaluation of non-target effects of OMRI-listed insecticides for management of *Drosophila suzukii* Matsumura in berry crops. Journal of Applied Entomology 2019;144:12–25.
192. Rossi-Stacconi M, Brewer LJ, Miller B, Dalton D, Lee J, Park K, et al. Biocontrol of Spotted-wing Drosophila. OSU Extension Catalog; 2019. Available from: URL: https://catalog.extension.oregonstate.edu/em9269/html.
193. Klungness LM, Jang EB, Mau RL, Vargas RI, Sugano JS, Fujitang E. New sanitation techniques for controlling tephritid fruit flies (Diptera: Tephritidae) in Hawaii. Journal of Applied Sciences and Environmental Management 2005;9:5–14.
194. Shaw B, Brain P, Wijnen H, Fountain MT. Reducing *Drosophila suzukii* emergence through interspecies competition. Pest Management Science 2018;74:1466–71.
195. Dancau T, Tanya LM, Stemberger TLM, Clarke P, Gillespie DR. Can competition be superior to parasitism for biological control? The case of spotted wing Drosophila (*Drosophila suzukii*), *Drosophila melanogaster* and *Pachycrepoideus vindemiae*. Biocontrol Science and Technology 2017;22:3–16.
196. Bernardi D, Andreazza F, Botton M, Baronio CA, Nava DE. Susceptibility and interactions of *Drosophila suzukii* and *Zaprionus indianus* (Diptera: Drosophilidae) in damaging strawberry. Neotropical Entomology 2017;46:1–7.

197. Shrader M, Burrack HJ, Pfeiffer D. Effects of interspecific larval competition on developmental parameters in nutrient sources between *Drosophila suzukii* (Diptera: Drosophilidae) and *Zaprionus indianus*. Journal of Economic Entomology 2020;113:230–38.
198. Kruitwagen A, Beukeboom LW, Wertheim B. Optimization of native biocontrol agents, with parasitoids of the invasive pest *Drosophila suzukii* as an example. Evolutionary Applications 2018;11:1473–97.

7 *Mimosa diplotricha*: a review and synthesis of its problem and control options

Osariyekemwen Uyi[1,2,3*]

[1]*Department of Animal and Environmental Biology, University of Benin, P.M.B. 1154, Benin City, Nigeria;* [2]*Department of Entomology, University of Fort Hare, Private Bag X1314, 5700 Alice, South Africa;* [3]*Department of Entomology, Pennsylvania State University, University Park, Pennsylvania, 16802, USA*

Abstract

Mimosa diplotricha C. Wright ex Sauvalle (=*Mimosa invisa* Mart.) (Mimosaceae) is a perennial weedy shrub of neotropical origin and a serious biotic threat in its invasive range. Despite its invasiveness and associated problems, there are surprisingly few reviews on this weed. This paper, therefore, reviews the existing but scattered literature on the invasion history and negative impacts of *M. diplotricha* in different ecological systems in its introduced ranges. Following the introduction of *M. diplotricha* into Indonesia and Australia in the early nineteenth century, the weed has since rapidly spread into many other countries in Asia, Africa and Oceania. It is known to be present and/or invasive in more than 14 Asian (e.g. India, Thailand, the Philippines) and 17 African countries (e.g. Nigeria, Ethiopia, Kenya) and more than 16 countries in Oceania including Australia and Papua New Guinea, with some serious negative effects on agriculture, biodiversity conservation and livelihoods. Landowners, locals and peasants in invaded areas employ physical/mechanical, cultural and chemical control methods to manage *M. diplotricha*, but these methods are unsustainable, costly and largely ineffective. The first biological control of *M. diplotricha* worldwide began in Australia in the 1980s with the release and establishment of *Heteropsylla spinulosa*, a sap-sucking bug that significantly reduced densities of the weed. This bug was subsequently introduced to many Islands in Oceania where it established and reduced the densities of the weed. This paper discusses the problems of *M. diplotricha* in different ecosystems in invaded areas, control options and gives recommendations for the sustainable management of the weed in Asia and Africa.

Keywords: *Mimosa diplotricha*, invasive alien plant, introduction and impact, management strategies, weed biocontrol, *Heteropsylla spinulosa*

Review Methodology

We reviewed the literature by searching both academic and grey literature in Google Scholar and ISI Web of Science. Published papers, conference proceedings, CABI Invasive Species Compendium and dissertations, some of which are not available online, were also assessed. Prior to a systematic review of the reference lists in the acquired papers, we sorted all data sources and included only articles relevant to the invasion history, impact and control of *M. diplotricha* in West and East Africa, Southeast and South Asia and Oceania. This was supported by discussions with members of the International Organisation of Biological Control (IOBC), who are experts on certain invasive plant species.

---

*Corresponding author: osariyekemwen.uyi@uniben.edu

Introduction

Mimosa diplotricha C. Wright ex Sauvalle (=*Mimosa invisa* Mart.) (Mimosaceae) is a fast-growing leguminous perennial shrub of neotropical origin that is currently invasive in parts of East and West Africa, South and Southeast Asia and Oceania [1–4]. *M. diplotricha* is commonly known as the giant sensitive plant, creeping sensitive plant and various local names also exist wherever it has been introduced [5]. This weed is still called *M. invisa* Mart. in some parts of Africa and Asia [6, 7].

M. diplotricha is and remains a huge threat to natural and semi-natural ecosystems in its introduce ranges, where it is known to negatively impact a variety of agricultural crops, livestock production, biodiversity conservation and livelihoods [1–4]. However, the weed is also known to provide some benefits in its invasive ranges especially in Asia (India, Sri Lanka, Indonesia, etc.) where it is used as a cover crop and green manure in plantations to improve soil biological, chemical and physical properties [8–11]. Despite the positive attributes of the *M. diplotricha*, it is still seen as a menace that needs to be controlled and managed in all countries where it is invasive. The rapid expansion and spread of *M. diplotricha* in its invasive ranges [1, 2, 12, 13] warrant a general review on the threats posed by this thorny leguminous shrub in different ecosystems.

Understanding the effects of *M. diplotricha* on agriculture, biodiversity, livelihoods and human well-being is essential for developing sustainable management strategies and guiding policy formulation. This paper, therefore, reviews the existing but scattered literature on the invasion history and negative impacts of *M. diplotricha* in different ecological systems in invaded areas (West and East Africa, South and Southeast Asia and Oceania). This paper also discusses the management and control initiatives or options undertaken in these areas and gives recommendations for the sustainable management of the weed in Asia and Africa. This paper is expected to raise awareness on the implications of the uncontrolled spread of the weed in the tropics and stress the need for an effective national or regional control programmes.

Description, biology and ecology of *Mimosa diplotricha*

Mimosa is one of the largest genera of Mimosoid legumes with over 500 species [14, 15]. *M. diplotricha* is an annual shrubby and scrambling climber that often forms and spreads impenetrable, tangled, dense thickets and can sometimes behave as a biennial or perennial leguminous vine [1, 2, 16, 17].

The main stalk has up to eight pairs of sub-leaf stalks and each sub-leaf stalk may bear as many as 20–30 opposite leaflets, which are small, bright green, alternate, bipinnate and sessile and are about 6–12 mm long and 1.5 mm wide [1, 2]. The stem is four angled, woody decumbent base with re-curved thorns (3–6 mm long), up to 3 m in height. The pinkish-violet flowers occur in globose heads about 12 mm in diameter and may occur singly, in pairs or threes on individual stalks originating in the axils of young leaves. Although flowering occurs throughout the year [1, 2, 18], most occurs in late wet season [16]. The smooth light brown seed is flat, hard, ovate and about 2.5 mm long [1]. It is a prolific seed producer that can produce up to 20,000 seeds/m^2 per year and a single plant can produce up to 10,000 seeds per annum [1, 19, 20]. Seeds are retained in spiny pod segments and are adapted for dispersal by floating on water as well as by spiny segments adhering to animal fur and clothing. The movement of vehicles and machineries and the transportation of contaminated plants or soil materials can also assist the spread of the seeds. Some of these seeds may germinate immediately while others may remain in the soil for several years before germination. The seeds can remain dormant for up to 50 years [21].

M. diplotricha grows best where fertility, soil and air humidity are all high and dies away in prolonged dry seasons. The plant commonly grows in crops, plantations and pastures, as well as on disturbed moist wastelands, drains and watercourses in tropical and subtropical regions and seems to prefer open areas and disturbed ecosystems such as forest fringes and roadsides with lots of sunlight. The weed is heliophytic in adaptation and cannot grow under a closed canopy but can grow under a wide range of altitude (0–2000 m asl). In Nigeria, most germination occurs at the beginning of the wet season (between March and May) but seeds may germinate at any time of the year when the suitable conditions of moisture and temperature are met. The ability of the plant to tolerate a wide range of soil *p*H, soil types, vegetation types and altitudinal regimes has been documented [4, 7, 22].

Introduction and spread of *Mimosa diplotricha*

Although the nativity of the *M. diplotricha* has been traced to Brazil [1, 2], its natural habitat range in the Americas stretches from Mexico to Argentina including the Caribbean Islands [1, 2, 14]. The weed is now known to be invasive or present in many tropical countries in Africa, Asia and Oceania [1, 2, 5]. The invasion history of *M. diplotricha* is shown in Fig. 7.1. Outside of its native range, *M. diplotricha* was recorded in Indonesia on the Island of Java in 1900 from where it spread into other countries in Asia [23, 24]. As of the 1960s and 1970s, *M. diplotricha* was already seen as a menace in Thailand, Cambodia, Sri Lanka, Malaysia, the Philippines, Taiwan, Vietnam and India [1, 2, 25, 26]. *M. diplotricha* was introduced to Taiwan in 1965 as an ornamental [26]. The weed is thought to have spread into Laos, Myanmar, East Timor, Singapore and China between the 1980s and 1990s where it is now invasive or present in large parts in these countries [1, 2, 5]. In Oceania, the first record of the weed dates back to the 1920s when *M. diplotricha* was known to have been present in a location near Tully in Queensland, Australia in 1929 [2]. The weed was first recorded in Fiji in 1936 and first reported in Western Samoa in 1972 [2]. As of the 1970s and 1980s, *M. diplotricha* had spread to Papua New Guinea (PNG), French Polynesia, Guam, New Caledonia, Samoa, Palau, Vanuatu, Cook Islands and Solomon Islands forming dense tangled, thorny clumps that smother other vegetation [1, 2, 5]. In the late 1980s and early 1990s, the weed was reported to be invasive and present in Niue, Northern Mariana Islands and Wallis and Futuna Islands [1, 2, 5].

The timing and source of introduction of *M. diplotricha* into Africa is unknown but the weed is thought to have been on the continent for over three decades [1, 2, 4]. The weed was reported as a noxious pest in Nigeria, Cameroon, Cote d' Ivoire and Mauritius in the 1970s and 1980s [1, 2]. Between the 1990s and 2000s, *M. diplotricha* was reported to be present or invasive in Ethiopia, Guinea, Ghana, Reunion, Burundi,

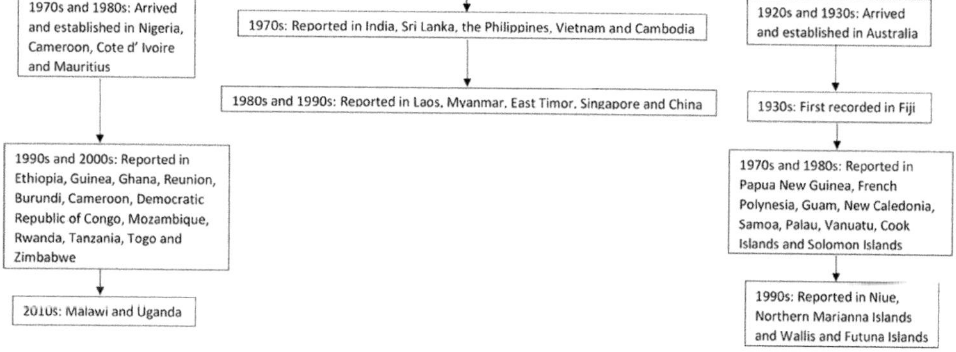

Fig. 7.1. The invasion history of *Mimosa diplotricha*.

Democratic Republic of Congo, Mozambique, Rwanda, Tanzania, Togo and Zimbabwe [1, 2, 5, 13, 27]. A recent report published in 2017 documented the invasiveness of *M. diplotricha* in Uganda and Malawi [13].

In Australia, *M. diplotricha* is confined to the north Queensland coastal region between Ingham and Cooktown, around Mackay, and at Brisbane [16, 21], and it is thought to have the potential to spread to the Northern Territory and Western Australia [28]. In Western Samoa, it is estimated that 85% of the villages on the island of Upolu are infested with the weed [29]. It commonly forms clumps up to 20 m in diameter in the Markham and Ramu Valleys in Papua New Guinea [19]. On Peninsular Malaysia, it occurs in the States of Perlis, Kedah, Seberang Perai, northern Perak, Selangor, Malacca, Negeri Sembilan and Johore [30]. In Nigeria, *M. diplotricha* is invasive in the entire southern states in Nigeria and maintains some presence in the north-central region of the country [4]. Seed dispersal is through running water and carried in animal fur, clothing, vehicles, agricultural implements and machinery. Anthropogenic influences, such as the movement of humans and machinery, trade, infrastructure development—such as road construction—and tourism across the 'open borders' between countries in Asia, Africa and Oceania may have been fundamental drivers that have facilitated the widespread invasion of *M. diplotricha* in the tropics. In addition, the high reproductive capacity of the weed, readily dispersed propagules, its ability to outcompete other plant species for both light and space and adaptation for growth in a range of soil types and climatic conditions may be associated with its overall invasion success in new areas. Given that *M. diplotricha* can spread through contaminated crop seeds or through deliberate introduction as a forage or ground cover, it has the potential to be introduced, to spread and become a major weed in many tropical countries that lack the capacity to control or manage the weed.

Impact of *M. diplotricha*

M. diplotricha is a serious weed that has been implicated in the transformation of the integrity of many natural and semi-natural ecosystems, leading to a decline in biodiversity, agricultural production and biodiversity conservation efforts in the Pacific, South and Southeast Asia, Australia and in several countries in West and East Africa [1–4, 6, 31]. In countries where *M. diplotricha* is invasive, the weed is commonly seen in arable croplands, plantation crop farms, fallow lands, roadsides, abandoned lands and deforested areas. The weed rapidly invades new areas and smothers pasture and forage crops in tropical and subtropical countries, reducing crop yields. For example, in Nigeria, Alabi *et al.* [6] showed that interference of the weed negatively affected the growth parameters of cassava, one of the most widely grown staple crops in Nigeria. In the same study, high populations of *M. diplotricha* was shown to reduce storage root yield in cassava and accounted for 85% reduction in crop yield. Locals in Benin City, Nigeria and Ecologists and Crop Scientists consider *M. diplotricha* as the most noxious weed in that part of Nigeria where it has invaded farms, plantations, abandoned buildings and roadsides [4, 32]. In Papua New Guinea, *M. diplotricha* is known to negatively impact on the growth and yield of sugarcane and infestation by the weed resulted in increased harvesting time of the crop [3]. The estimated total cost of lost time due to *M. diplotricha* interference was US$320,000 per year [3].

M. diplotricha is a very serious pest of oil palm and coconut (especially at the nursery stage) in Nigeria [4] and a serious weed problem in rubber and coconut plantations in Papua New Guinea. It is known to pose a serious weed problem to the production of many crops such as rice in the Philippines, Indonesia, Thailand, Vietnam and Laos [18, 33]. It is considered a pest of sugarcane in Papua New Guinea, Taiwan, Philippines, Australia and India; tomato in the Philippines; rubber in Indonesia and Malaysia; soya beans, citrus, maize, apple, cassava and tea in Indonesia; banana in India and pineapple in the Philippines [1, 34–39]. It is seen as potentially the worst weed in plantations and arable lands in the Philippines and Fiji [1]. The weed poses a serious threat to tropical pastures in Australia and the Philippines and the Pacific Islands [1, 29, 40, 41]. On cattle ranches in the Ramu-Markham Valleys, in Papua New Guinea, up to US$130,000 was spent on the chemical control and slashing of the weed [42]. The prickly thorns of *M. diplotricha* make control of infested farms and harvesting of crops very difficult, hence invasion always results in increased production and management cost and decreased productivity in many agroecosystems. Losses incurred by farmers due to the suppression of staple crops by *M. diplotricha* have been described as overwhelming and this has forced numerous farmers to abandon their farms or plantations in Nigeria (personal observation).

M. diplotricha thickets interfere with the communities of natural and semi-natural ecosystems and may constrain the movement of wildlife and humans due to their ability to smother other plants and numerous sharp recurved prickles. Infestations by *M. diplotricha* hinder the regeneration, reproduction and growth of native species in infested areas and consequently result in the gradual loss of biodiversity [22, 43]. In southern Nigeria, *M. diplotricha* compete with indigenous plant species and replace nonnative species such as *Chromolaena odorata* (L.) (Asteraceae) [44]. In India, Sankaran [45] reported that *M. diplotricha* smothers other invasive plants such *C. odorata* and *Mikania micrantha* H.B.K. (Asteraceae) and establishes itself over them. The movement of rhinoceros (*Rhinoceros unicornis*), Asian elephant (*Elephas maximus*), tiger (*Panthera tigris*) and swamp deer (*Cervus duvauceli*) has been reported to be hampered by dense stands of *M. diplotricha* in Kaziranga National Park in northeast India [22]. The avoidance of *M. diplotricha* by livestock during grazing is thought to be due to its sharp and curved thorns and/or the presence of mimosine (a non-protein amino acid), which is toxic to herbivores if ingested as it can cause vascular endothelial damage, necrosis of heart and liver and anaemia in cattle [22]. Overall, *M. diplotricha* is known to pose a major threat to agricultural productivity (crop and livestock production) and biodiversity conservation, thereby negatively affecting the quality of lives of locals and the integrity of ecosystems in invasive ranges.

However, *M. diplotricha* is also known to provide some benefits in its invasive ranges. It has been reported by local people and scientists in Asia to have several beneficial attributes. *M. diplotricha* is a nitrogen (N) fixer [41] and in India, Sri Lanka, Indonesia and other Asian countries, the weed has been frequently used as cover crop, green manure and soil renovator to add soil nitrogen and organic matter and to reduce soil erosion [1, 8–11, 40, 46]. In Benin City, Nigeria, and elsewhere the weed prevents cattle from invading arable farms (personal observation). In southern Nigeria, *M. diplotricha* displaces more troublesome weeds like *Imperata cylindrical* (L.) (Poaceae) and other invasive plants such as *C. odorata* and *Tithonia diversifolia* (Hemsl.) (Asteraceae). The insecticidal activities of the root and leaf extracts and powders are increasingly being recognised [47, 48].

Control options

Because of the problems of *M. diplotricha*, it has remained a subject of several eradication and control programmes in some countries (e.g. Nigeria, Australia and some Pacific Islands) [3, 31, 49, 50, 51]. In many countries where *M. diplotricha* is invasive, several conventional methods such as cultural, mechanical and chemical control methods were adopted and are still being used in managing the weed (Fig. 7.2). This section discusses the conventional control options and biological control efforts undertaken in the world.

Cultural control

Parsons and Cuthbertson [16] suggested that sugarcane fields with heavy infestations of *M. diplotricha* may need to be quarantined for a number of years and further advised that destroying the crop is key to preventing further spread of this weed. In upland rice farms in the Philippines, treating *M. diplotricha* infestation with 5 t/ha of fresh *Gliricidia sepium* green manure plus 5 t/ha of fresh *Senna spectabilis* mulch significantly reduced the weed biomass [52]. Chadhokar [53] reported that discouraging overgrazing can limit the spread and infestations and thus control may be more realistic in areas where animals are restricted from grazing.

Mechanical/physical control

The use of physical means, such as weeding with crude implements—shovels, hoes and cutlasses—has been widely adopted in managing *M. diplotricha* in Nigeria and elsewhere. Hand weeding is possible when the plants are young or in small infestations, but the prickles or thorns on the vines can cause serious injuries to humans and animals. In Nigeria, the use of hand-held hoe to remove the weed is a popular and remains an effective method of control among peasants. However, most farmers complain that the thorns on the stems and pods reduce the efficiency of manually controlling the weed. The vigorous regrowth from the root crown and the rapid and prolific seedling development make slashing or burning ineffective methods of control [2, 3, 16]. Despite the efficiency of mechanical control methods such as using hand-held hoes to remove the weed [54], this method of control is laborious, time wasting and impractical in invasive conditions especially in developing countries where most farmers are predominantly peasants.

Chemical control

Farmers and landowners in invaded countries use a host of herbicides such as isoxaflutole, atrazine, diuron, paraquat, etc. (applied singly or as a mixture) have been used in the control of *M. diplotricha* [3, 31, 49, 50, 51]. In some Pacific Islands, paraquat + diuron applied postemergence have provided good control of *M. diplotricha* and 2, 4-D + atrazine has been applied as an overall spray especially in pasture situations to control young to semi-mature stands [51]. In Nigeria, an empirical research showed that atrazine + metolachor, bentazon + propanil and acetochlor + atrazine controlled *M. diplotricha* satisfactorily but reduced the yield of cassava compared to hand-weeded control [31]. By applying primextra, a formulated mixture of atrazine and metolachlor followed by hand weeding, a considerable amount of control of the weed was achieved in cassava fields in south-eastern Nigeria [50]. Despite the efficiency of chemical control methods in the management of *M. diplotricha* especially in large-scale farming systems, it is expensive and not environmentally friendly. Therefore, its use is not sustainable. A combination of other weed control methods such as chemical, mechanical and biological control may be more effective [4]. It should be noted that some of these herbicides (e.g. 2, 4-D and Paraquat) used to control M. diplotricha are not allowed to be used in some countries.

Biological control

The first biological control programme for the management of *M. diplotricha* commenced in Australia in the early 1980s with the identification of over 70 insect species and 2 fungal species (as potential biocontrol agents) in Brazil [4, 24]. Of the over 70 species, only 3 proved to be promising biocontrol candidates. Following

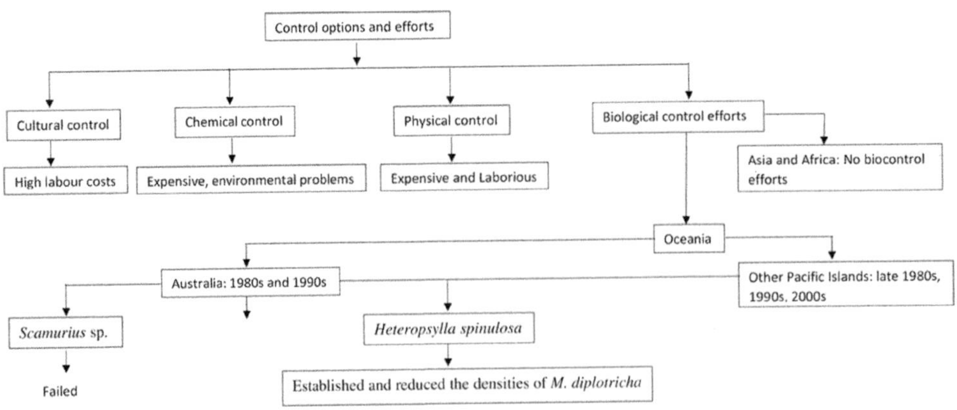

Fig. 7.2. The control of *Mimosa diplotricha* in Africa, Asia and Oceania.

the failure of the Coreid bug, *Scamurius* sp. (Hemiptera: Coreidae) to establish after it was released in Australia in 1988 [55], another Hemipteran species, *Heteropsylla spinulosa* Muddiman, Hodkinson and Hollis (Hemiptera: Psylloideae) was released in 1988–1989 and immediately established [29]. Within 2 years of commencing field releases, the tiny sap-sucking bug widely dispersed and caused severe damage to *M. diplotricha*. Feeding (sap-sucking) activities by both immatures (nymphs) and adults on leaflets, rachises and growing tips result in stunted growth and deformed plants. Females lay about 50 eggs in their lifetime and the nymphs pass through five instars and become adults in 18–20 days. The insect is able to complete eight generations per year [56]. Feeding damage caused by *H. spinulosa* resulted in dense clumps of the weed reduced to small masses of bare stems with stunted growth tips that result in death in some cases and this allowed other plant species to re-establish in Australia [57]. This Psyllid bug was shown to reduce seed production in *M. diplotricha* by up to 85% and 100% [57].

H. spinulosa was introduced from Australia into Western Samoa in 1988–1989 and Papua New Guinea in 1991 [29, 42] where it is known to have established and reduced the biomass of *M. diplotricha* [3]. This insect is thought to have exerted a spectacular control over *M. diplotricha* in Papua New Guinea [3]. Following its introduction into several other Pacific Islands such as Cook Islands, Fiji, Palau, Pohnpei, Samoa, Solomon Islands and Yap, *H. spinulosa* has not only established but provided a sustained suppression requiring little or no additional efforts to manage this weed [3]. Most areas cleared of *M. diplotricha* by *H. spinulosa* in these island nations have been taken over by the native plants. Due to the ability of *H. spinulosa* to cause severe stunting and distortion of leaves and growing tips and its overall impact on *M. diplotricha*, this biocontrol agent should be considered for the control and management of *M. diplotricha* in Nigeria, India, Thailand and other countries where the weed is problematic. Such control attempts would benefit from international collaborations between institutions in these invasive ranges and a host of others countries (e.g. Australia, Papua New Guinea) that have successfully initiated biocontrol programmes against the weed.

Attempts to release a third biological control candidate, *Psigida walkeri* (Grote) (Lepidoptera: Citheroniidae), were unsuccessful because this leaf-feeding moth failed host specificity screening [56]. In the Philippines, a fungus isolated from *M. invisa* (=*M. diplotricha*), *Fusarium pallidoroseum*, which caused high damage to *M. invisa*, was rejected as a potential biocontrol agent because the fungus culture filtrates of *F. pallidoroseum* caused disease symptoms on a broad range of plant species [58]. In Queensland, Australia, an indigenous stem-spot fungus, *Corynespora cassiicola*, is known to have exercised a degree of control on *M. diplotricha*. The fungus is thought to be widespread and specific to *M. diplotricha* and causes defoliation and dieback [21].

Conclusion

Since the introduction of *M. diplotricha* into Indonesia and Australia in the early nineteenth century, the weed has spread extensively into many countries in Asia, Africa and Oceania, especially in the last eights decades. Apart from the invasive characteristics of *M. diplotricha*, the continuous spread of, and invasion by, this weed in countries where it is invasive has been due to: (i) the increased human disturbances associated with the recent economic growth and infrastructural development; (ii) lack of an integrated control and management programme; (iii) lack of a sustained biological control programme especially in Asian and African countries; and (iv) its use as a cover crop and green manure in plantations to improve soil biological, chemical and physical properties in some Asian countries. Given that *M. diplotricha* has attained a problematic status in agricultural lands and commercial plantations as well as being a perceived threat to biodiversity conservation and human livelihoods in its invasive range, it would be justified to reduce its competitiveness in order to limit its menace in both natural and semi-natural ecosystems. Because conventional methods of controlling *M. diplotricha* are unsustainable and impractical, I recommend that Asian and African countries where the *M. diplotricha* is invasive should prioritise the release of *H. spinulosa*, which can effectively inflict extensive but selective damage on this invasive alien weed. If the spread and impact of *M. diplotricha* in Asian and African countries and Pacific Islands are to be curtailed and eventually reversed, several initiatives for control and management are required at both national and regional levels. These include: (i) initiation and sustenance of an effective biological control programme in affected countries and regions; (ii) creation of sustained public awareness about the weed problem especially in areas that are sparsely infested or in un-infested areas that have the potential to become infested; and (iii) the development of a coordinated and integrated control and management plan for *M. diplotricha* on a national and regional scale.

Summary points

- From its native range in the Americas, *M. diplotricha* was introduced into Indonesia and Australia in the early nineteenth century. The weed is now present in over 14 countries in Asia, 17 countries in Africa and many islands in the Oceania.
- *M. diplotricha* is a threat to the diversity of extant vegetation and impacts negatively on agriculture, food security and livelihoods. It is considered one of the worst invasive species in many countries where it is invasive.
- Control strategies have historically relied on conventional means such as the use of hand hoe to uproot entire plants, slash and burn, and the seldom use of chemical herbicides. These methods are, however, laborious and costly in the long term and are unsustainable.
- Biological control of *M. diplotricha* worldwide first began in Australia in the 1980s with the release and establishment of *H. spinulosa*, a sap-sucking bug that significantly reduced densities of the weed. This bug was transferred to many islands in Oceania where it established and reduced the densities of the weed.
- Because the ecological and economic impacts of *M. diplotricha* in many countries in Asia and Africa are too significant to be ignored, it is suggested that a coordinated biological control programme should be initiated and sustained against the weed.

Acknowledgments

The author would like to thank two anonymous reviewers for their comments and suggestions on the previous version of this manuscript. The author also wishes to thank some locals and farmers in Benin City, Nigeria, for providing useful comments on the negative impacts and control of *Mimosa diplotricha*.

References

1. Holm LG, Plucknett DL, Pancho JV, Herberger JP. The world's worst weeds: distribution and biology. Honolulu, USA: University Press; 1977.
2. Waterhouse DF, Norris KR. Biological control: pacific prospects. Melbourne, Australia: Inkata Press; 1987.
3. Kuniata LS. *Mimosa diplotricha* C. Wright ex Sauvalle (Mimosaceae). In: Muniappan R, Reddy GVP, Raman A, editors. Biological control of tropical weeds using arthropods, Cambridge, UK: Cambridge University Press; 2009. p. 247–55.
4. Ekhator F, Uyi OO, Ikuenobe CE, Okeke CE. The distribution and problems of the invasive alien plant, *Mimosa diplotricha* C. Wright ex Sauvalle (Mimosaceae) in Nigeria. American Journal of Plant Sciences 2013; 4:866–77.
5. CABI Invasive Species Compendium. *Mimosa diplotricha* (Creeping sensitive plant) 2019. Available from: URL: https://www.cabi.org/isc/datasheet/34196

6. Alabi BS, Ayeni AO, Agboola AA, Majek BA. Giant sensitive plant interference in cassava. Weed Science 2001; 49:171–6.
7. Chauhan BS, Johnson DE. Seed germination and seedling emergence of giant sensitive plant (*Mimosa invisa*). Weed Science 2008; 56:244–8.
8. Jayasinghe CK. The role of leguminous cover crops in soil improvement with special reference to the nitrogen economy of tropical rubber soils. Bulletin of the Rubber Research Institute of Sri Lanka 1991; 28:23–6.
9. Thomas GV, Shantaram MV. In situ cultivation and incorporation of green manure legumes in coconut basins, an approach to improve soil fertility and microbial activity. Plant and Soil 1984; 80:373–80.
10. Wiersum KF. Early experiments in agroforestry: colonial tobacco cultivation with tree fallows in Deli, Sumatra. The International Tree Crops Journal 1983; 2:313–21.
11. Thomas GV, George M. Basin management using green manure legumes for root (wilt) affected coconut palms. Indian Coconut Journal (Cochin) 1990; 20(11):7–8.
12. Wei C, Zhao Z, Ding L, Pan Y. Newly recorded invasive plants in Guangxi. Guangxi Zhiwu/Guihaia 2013; 33(2):275–8.
13. Witt A, Luke Q, editors. Guide to naturalized and invasive plants of Eastern Africa. Wallingford, UK: CABI, 2017.
14. Barneby RC. Sensitivae censitae: a description of the genus *Mimosa* Linnaeus (Mimosaceae) in the new world. Memoirs of the New York Botanical Garden 1991; 65:835.
15. Simon MF, Grether R, de Queiroz LP, Särkinen TE, Dutra VF, Hughes CE. The evolutionary history of Mimosa (Leguminosae): toward a phylogeny of the sensitive plants. American Journal of Botany 2011; 98(7)1201–21.
16. Parsons WT and Cuthbertson EG. Noxious weeds of Australia. Melbourne, Australia: Inkata Press; 1992.
17. Noda K, Teerawatsakul M, Prakongvongs C, Chaiwiratnukul L. Major weeds in Thailand. Bangkok, Thailand: Department of Agriculture; 1985.
18. Kostermans, AJGH, Wirjahardja S, Dekker RJ. The weeds: description, ecology and control. In: Soerjani M, Kostermans, AJGH, Tjitrosoepomo G, editors. Weeds of rice in Indonesia. Jakarta, Indonesia: Balai Pustaka; 1987. p. 524–65.
19. Kuniata LS, Dori L, Dori F. A potential biological control agent for Mimosa invisa weed in Papua New Guinea. Harvest (Port Moresby) 1993; 15(1):54–5.
20. Kuniata LS and Korowi KT. Bugs offer sustainable control of *Mimosa invisa* and *Sida* spp. in the Markham Valley, Papua New Guinea. In: Cullen JM, Briese DT, Kriticos DJ, Lonsdale MW, Morin L, Scott JK, editors. Proceedings of the XI International Symposium on Biological Control of Weeds; 2003 April 27–2003 May 2; Canberra, Australia: CSIRO Entomology; 2004. p. 567–73.
21. DAF. Giant sensitive plant – *Mimosa diplotricha* (*Mimosa invisa*). The state of Queensland Department of Agriculture and Fisheries. Available from: URL: https://www.daf.qld.gov.au/__data/assets/pdf_file/0017/67121/IPA-Giant-Sensitive-Plant-PP27.pdf
22. APFISN (Asia - Pacific Forest Invasive Species Network).Invasive pest fact sheet: giant sensitive plant (*Mimosa diplotricha*); 2007. Available from: URL: http://www.fao.org/forestry/13377-1-0.pdf.
23. Soerjani M, Kostermans AJGH, Tjitrosoepomo G. Weeds of rice in Indonesia., Jakarta, Indonesia: Balai, Pustaka; 1987.
24. Waterhouse DF. Biological control of weeds: Southeast Asia prospects, Canberra: Australian Center for International Agricultural Research; 1994.
25. Napompeth B. Biological control of weeds in Thailand. BIOTROP Special Publication 1990;38:233–6.
26. Wu SH, ShuMiaw C, Rejmanek M. Naturalized Fabaceae (Leguminosae) species in Taiwan: the first approximation. Botanical Bulletin of Academia Sinica 2003; 44(1):59–60.
27. Wakjira M. An invasive alien weed giant sensitive plant (*Mimosa diplotricha* Sauvalle) invading Southwestern Ethiopia. African Journal of Agricultural Research 2011; 6(1):127–31.
28. Groves RH, Hosking JR, Batianoff GN, Cooke DA, Cowie ID, Johnson RW et al. Weed categories for natural and agricultural ecosystem management. Canberra, Australia: Bureau of Rural Sciences; 2003. Available from: URL: http://www.affa.gov.au/corporate_docs/publications/pdf/rural_science/lms/weeds/brs_weeds.pdf
29. Willson BW, Garcia CA. Host specificity and biology of *Heteropsylla spinulosa* (Homoptera: Psyllidae) introduced into Australia and Western Samoa for the biological control of *Mimosa invisa.* Entomophaga 1992; 37:293–9.

30. Baki BB, Prakash N. Studies on the reproductive biology of weeds in Malaysia anther sterility in *Mimosa invisa*. Wallaceana 1994; 73:13–6.
31. Alabi BS, Ayeni AO, Agboola AA. Effect of selected herbicides on the control of thorny Mimosa in cassava. Nigerian Journal of Weed Science 1999; 12:51–7.
32. Ogbe FMD, Bamidele JF. Incidence and spread of an invasive weed, Mimosa invisa Mart. in Benin City metropolis, Nigeria. International Journal of Botany 2006; 2(3):336–9.
33. Moody K. Weeds reported in Rice in South and Southeast Asia. Manila, Philippines: International Rice Research Institute, 1989.
34. Wong PW. Commercial development of Roundup postemergence weedicide as a one-shot treatment for control of lalang (*Imperata cylindrica*) in Malaysian rubber cultivation. Proceedings of the Third Indonesian Weed Science Conference, Bandung; Indonesia: Weed Science Society; 1975. p.164–79.
35. Aliudin L, Kusumo S. Testing the effectiveness of several herbicides in an apple orchard. Bulletin Penelitian Hortikultura 1978; 6:35–42.
36. Tppongsorut R, Weed inventory in the Horticultural Station at Cipanas, West Java. BIOTROP Newsletter 1978; 23:8–9.
37. Mendoza SP Jr, Weed management in pineapple. Symposium, Weed Control in Tropical Crops. Manila, the Philippines; 1978. p. 147–8.
38. Suwanarak K. Effectiveness of certain herbicides in litchi (*Litchi chinensis* Sonn.). Weed watcher 1988; 6–7:6.
39. Muniappan R, Viraktamath CA. Invasive alien weeds in the Western Ghats. Current Science 64(8):555–8
40. Henty EE, Pritchard GH. Weeds of New Guinea and their control. Lp, Papua New Guinea: Division of Botany, Department of Forests; 1975. Botany Bulletin No. 7.
41. Swarbrick JT. Major weeds of the tropical South Pacific. Proceedings of the 12th Asian-Pacific Weed Science Society Conference, Seoul, Korea Republic, No. 1. Taipei, Taiwan: Asian-Pacific Weed Science Society; 1989. p. 21–30.
42. Kuniata LS. Importation and establishment of *Heteropsylla spinulosa* (Homoptera: Psyllidae) for the biological control of *Mimosa invisa* in Papua New Guinea. International Journal of Pest Management 1994; 40(1):64–5.
43. Werren G. Environmental weeds of the wet tropics bioregion: risk assessment and priority ranking. Report prepared for the Wet Tropics Management Authority, Cairns, Australia; 2001. Available from: URL: http://www.rainforest-crc.jcu.edu.au/publications/research%20reports/ReportPDFs/Weeds.pdf
44. Uyi OO, Ekhator F, Ikuenobe CE, Borokini TI, Aigbokhan EI, Egbon IN et al. *Chromolaena odorata* invasion in Nigeria: a case for coordinated biological control. Management of Biological Invasions 2014; 5:377–93.
45. Sankaran KV. *Mimosa invisa*; a Growing menace in South India. Biocontrol News and Information 2001; 22:4–5.
46. Allen ON, Allen EK. The Leguminosae: a source book of characteristics, uses, and nodulation. Madison, USA: University of Wisconsin Press; 1981.
47. Uyi OO, Adetimehin AD, Uyamasi EE, Ejomah AJ. Insecticidal activities of the leaf powders of *Chromolaena odorata* and *Mimosa diplotricha* against the common bed bug, *Cimex lectularius* (L.). International Journal of Zoological Research 2018; 14(1):37–42.
48. Uyi O, Samugana J. Repellent and insecticidal efficacy of leaf and stem Powders of *Mimosa diplotricha* C. Wright ex Sauvalle (Mimosaceae) against maize weevil, *Sitophilus zeamais* Motschulsky (Coleoptera: Curculionidae). Nigeria Journal of Entomology 2018; 34: 99–107.
49. Melifonwu AA. Control of *Mimosa inivisa* in Maize (*Zea mays*) with atrazine and its mixtures with alachlor and metolachlor. Nigerian Journal of Weed Science 1994; 7:9–14.
50. Melifonwu AA, Korieocha DS, Ogbonna MC, Balogun RB. Effect of selected herbicides and integrated weed management on the control of *Mimosa invisa* Mart in cassava (*Manihot esculenta* Crantz). International Journal of Applied Research and Technology 2012; 1(2):187–96.
51. Swarbrick JT. Weeds of the Pacific Islands. South Pacific Commission, Noumea, New Caledonia; 1997.
52. MacLean RH, Litsinger JA, Moody K, Watson AK, Libetario EM. Impact of *Gliricidia sepium* and *Cassia spectabilis* hedgerows on weeds and insect pests of upland rice. Agriculture, Ecosystems and Environment 2003; 94(3):275–88.

53. Chadhokar PA. Weed problems of grazing lands and control of some problem weeds in the Markham Valley of Papua New Guinea. PANS 1978; 24(1):63–6.
54. Alabi BS, Ayeni AO, Agboola AA, Majek BA. Manual control of thorny Mimosa (*Mimosa invisa*) in cassava (*Manihot esculenta*). Weed Technology 2004; 18:77–82.
55. Julien MH, Griffiths MW. Biological control of weeds: a world catalogue of agents and their target weeds, 4th ed. Wallingford, UK: CAB International Publishing; 1998.
56. Ablin MP. Growth and reproduction of the weed, *Mimosa invisa,* in Northeast Queensland after the introduction of the biocontrol agent *Heteropsylla spinulosa* [MS thesis]. Townsville, Australia: James Cook University; 1995.
57. Vitelli MP, Garcia C, Lockett CJ, West GM, Willson BW. Host specificity and biology of the Moth *Psigida walkeri* (Lepidoptera: Citheroniidae), a potential biological control agent for *Mimosa diplotricha* in Australia and the South Pacific. Biological Control 2001; 22:1–8.
58. Mabbayad M, Watson A. Rejection of *Fusarium pallidoroseum* as a Biological Control Agent of *Mimosa invisa* in Upland Rice. Biocontrol Science and Technology 2000; 10:255–65.

8 Fall armyworm (*Spodoptera frugiperda*) management by smallholders

Allan J. Hruska
Food and Agriculture Organization of the United Nations, Viale delle Terme di Caracalla, Rome 00153, Italy

Abstract

The fall armyworm (FAW) (*Spodoptera frugiperda*) is a crop pest species that has become global, having spread from its native American distribution to Africa and Asia since 2016. Its rapid spread, plus concerns about potential yield losses, have led to the search for sustainable management options. While most farmers affected by FAW in America have large-scale farm operations, the overwhelming majority of farmers in Africa and Asia are smallholders. This dramatically different context means that different management approaches must be sought. Large-scale producers with high-productivity, access to international market prices, risk-transfer mechanisms and the benefits of government subsidies are able to use technologies unavailable to smallholder farmers without access to those conditions. This review examines these differences and surveys the literature for accessible management options for smallholders, largely based on locally available solutions using ecological knowledge. Innovative digital solutions may also play a role in helping farmers learn about these solutions, and share them locally.

Keywords: Africa, Agriculture, Americas, Asia, Fall armyworm (FAW), Infestations, Maize, Pests, Smallholders

Review Methodology

We reviewed all the literature relevant to FAW management by smallholder maize producers.

The Rapid Spread of Fall Armyworm

The fall armyworm (FAW) (*Spodoptera frugiperda*), has been a consistently important insect pest for a number of crop species, especially maize, in the Americas for centuries [1]. In the last few years, FAW has become an invasive species in Africa, the Near East and Asia. FAWwas first reported in early 2016 in mainland West Africa (Nigeria, Togo, Benin) and on the island of São Tomé (São Tomé and Príncipe) [2]. It rapidly spread across sub-Saharan Africa during 2016 and 2017 and by late 2018, had been confirmed by virtually every country in sub-Saharan Africa [3]. It was confirmed in Yemen and India by July 2018 [4] and by early 2019, had been confirmed in an additional five Asian countries, including China [3].

FAW prefers maize, but it is also common on sorghum, rice and millets, and is sporadically important on a vast array of additional crops and plants, including cotton and vegetables. It is

*Corresponding author: allan.hruska@fao.org

reported to infest 186 host plant species in North and Central America [5], while Montezano *et al.* [6] have reported 353 host plant species based on a literature review and additional surveys in Brazil, from 76 plant families, principally Poaceae, Asteraceae and Fabaceae.

FAW is native to tropical and subtropical regions of the Americas and migrates north and south annually, following maize plantings in temperate zones [7]. FAWis not able to enter diapause and is killed by low temperatures [1, 8]. In its native range, it is established year-round where winter temperatures rarely fall below 10 °C [8–10]. Adult females are relatively short-lived (13–19 days at 26.8 °C) but highly fecund, with around 1000 eggs being laid per female, in clusters of 100–300, usually on leaf surfaces [11]. The first and second instars scrape leaves, leaving a 'windowing' damage, while later instars leave ragged holes in young leaves emerging from the plant whorl. In a trial examining larval dispersal of *S. frugiperda*, Pannuti *et al.* [12] reported finding over 90% of recovered larvae within a 1.1 m radius of a maize plant 14 days after being infested with an egg mass. Larvae are cannibalistic at high larval densities [13], resulting in typically one later instar larva per maize plant.

FAW Management in America

In the USA, Brazil and Argentina, FAW is routinely controlled via the use of genetically modified maize, which incorporate genes that express for the production of toxins lethal to FAW. The adoption of this technology has now surpassed 85% of the maize area planted in those three countries [14]. The use of this technology and of effective pesticides is supported by the access that the farmers have to relatively stable international markets for their maize, which is primarily used for animal feed, ethanol production, or processed to extract high fructose syrup, coupled with subsidies and risk transfer mechanisms (e.g. insurance).

Globally, the vast majority of maize farmers are smallholders, whose context is very different from that of large-scale commercial farmers. These differences include economic, ecological and cultural factors. The vast majority of smallholder maize farmers do not have access to high and stable prices for their maize, subsidies, or risk-transfer mechanisms, severely limiting their access to expensive control technologies. They also produce in much more diverse landscapes and cropping systems, and have different production goals.

FAW in Africa and Asia

FAWquickly and naturally spread across sub-Saharan Africa and Asia upon its arrival, and now routinely infests millions of hectares of maize across Africa and Asia. Although FAW is capable of feeding on many crop species, the crop most affected in Africa and Asia has been maize. However, in drier areas, such as in the Sahel, FAW has been consistently reported from crops other than maize; e.g. sorghum, millets, wheat and teff. The infestation levels vary dramatically, but the average infestation level across the continent is about 30% of plants infested with FAW[3]. Most maize farmers across the continent are now aware of the pest: 99% of maize farmers in Ethiopia were aware of FAW as were 100% of maize farmers in Kenya [15].

Maize is grown on approximately 37 million hectares annually in sub-Saharan Africa [16], the vast majority (approximately 95% of the area) in smallholdings, usually <2 ha. These small plots form a heterogeneous patchwork of often polyculture plots, mixed in a landscape of trees and shrubs across much of Africa. Most maize grown in Africa is white maize, produced for family consumption, unlike the yellow maize grown by commercial producers globally.

Asia produces about 32% of the maize globally [3]. China is the second largest maize producer in the world, growing maize on over 40 million ha. Like Africa, most maize farmers are smallholders. Unlike Africa, in Asia, most maize is yellow maize, produced as animal feed.

Smallholder Maize Producers

Smallholder maize farmers in sub-Saharan Africa typically grow their small plots of maize using almost exclusively family labour; they plant saved seed, use few purchased inputs, and their production is mostly for household consumption [17–19] If they have excess maize, they typically receive low prices from local intermediaries for their production; e.g. US$0.02/kg [FAO, in press]. Low produce price was perceived to be a risk among 100% of farmers surveyed in Ghana [18].

The low prices severely limit their ability or interest in using higher rates of purchased inputs. The concept of economic injury level and economic threshold was developed to determine the break-even point between the direct costs of pest damage and the value of the production protected or 'saved' by taking the management actions [20]. Typically, the calculation seeks a pest population level at which action should be taken to avoid economic losses. Pedigo *et al.* [21] defined Economic Injury level (EIL) as EIL = C/VIDK, where the EIL is expressed in number of injury equivalents per production unit (e.g. insects/ha), C= cost of the management activity per unit of production (e.g. US$/ha), V =market value per unit of the produce (e.g. US$/kg), I = injury units per insect per production unit [e.g. proportion defoliated (insect/ha) and D= damage per unit injury [e.g. (kg reduction/ha)/ proportion defoliated] and K = proportionate reduction of the insect population from the action.

Economic equilibrium (optimal expenditures) is reached when C= VIDK. For smallholder maize producers in Africa facing FAW, typical values for VIDK are: (US$0.05/kg) (1400 kg/ ha) (15% yield reduction) (75% effectiveness) = US$7.88/ha. Thus, to be economically rational, a farmer should not spend more than US$7.88/ ha to manage FAW under these conditions. The low prices received by the farmers and the low productivity of production result in very limited options for smallholder maize farmers to manage FAW in their maize. A recent paper by Baudron *et al.* [22] estimated 12% yield reduction from FAW infestation in maize in Eastern Zimbabwe.

Given this economic context, it is not surprising that most smallholder maize farmers in Africa, before the arrival of FAW, had used few purchased inputs. Most had used no pesticides. Muhammad *et al.* [23] conducted 350 household surveys of maize-producing families in Makueni and Machakos, Kenya and found that 0.5% had used herbicides and 2% had used insecticides. Household surveys carried out from 2010 to 2012 by the World Bank demonstrated that across six countries (Ethiopia, Malawi, Niger, Nigeria, Tanzania and Uganda) 16% of farmers used agrochemicals across all crops. This group included countries with significant agricultural input subsidy programmes. In countries without input subsidy programmes, the use of pesticides was very low (Malawi, 3%, Niger, 8% and Uganda, 11%). The use of these inputs in maize is probably even lower, given the use of production for self-consumption and the low prices received for any market sales [24].

FAO [25] carried out household surveys in Namibia in August 2017. Interviews of 592 households in 301 villages demonstrated that only 1.7% of households used pesticides. This low proportion is in line with the 2013/2014 Namibia Census of Agriculture which found that only roughly 13 000 out of approximately 150 000 agricultural households in the communal sub-sector use pesticides of any sort, including herbicides, fungicides, insecticides and traditional products [26].

During 2016 and 2017, many governments, some with donor support, provided pesticides free of charge to farmers for FAWcontrol or applied them with government employees. For example, in Zambia in 2017, nearly 102 000 litres of pesticides valued at 18 million Zambian Kwacha (US$1.97 million) were distributed, and a further 3 million Kwacha (US$330 000) was spent on sprayers and personal protective equipment such as gumboots and respirators [27]. In Ethiopia, the government distributed 271 779 litres of pesticides for FAW in 2017 and 167 896 litres in 2018 [28].

Not surprisingly, the use of pesticide among smallholder maize farmers increased dramatically in those countries with such policies and programmes. This was documented in a report for CABI by Rwomushana *et al.* [27], which said: '53% of farmers in Ghana and 43% of farmers in Zambia used pesticides (in 2018). However, when compared with 2017 data, farmers were generally using fewer synthetic chemicals (72% in 2017 for Ghana, and 62% in 2017 in Zambia)...The marked reduction in pesticide use in Zambia for the control of FAWcompared to the previous year may be due to reduced purchase and distribution of pesticides by the government...The current survey (2018) also established that in Zambia, twice as many farmers were using traditional methods such as applying ash, urea or sand on larvae, while only 27% of farmers handpicked egg masses compared to 36% in 2017. The increase in farmers deploying a wider range of control methods may also be attributed in part to the lack of pesticide distribution as well as Zambian smallholder farmers' traditional non-chemical approaches to maize cultivation. A third of farmers interviewed did not apply any control measure, compared to 23% in the 2017 survey' [27]. Many donors and governments began to scale back the distribution of pesticides from 2017 to 2018, as exemplified by the 38% reduction in pesticide distribution by the Ethiopian government during that period [28].

Most smallholder maize farmers in Africa traditionally use cultural controls, including the destruction of crop residues, manipulation of planting dates, use of locally available substances and tillage methods [29]. The majority of smallholder farmers in Africa practice intercropping [30]. As free pesticide distribution diminishes, due to lack of government funds and donor fatigue, smallholders will be faced with managing FAWin their fields within the context of their production systems. This review examines the agroecological basis of sustainable integrated pest management for smallholder maize farmers and suggests a research agenda to fill knowledge gaps and validate promising practices over a wide range of ecosystems.

Since the arrival of FAW in Africa, some research has begun on management practices accessible by smallholders. Although some practices have been tried, there are still many more potential effective, accessible practices based on the agroecology of FAW in Africa and Asia that need to be explored. This review presents a survey of the agroecology of FAW that leads to management of landscapes and cropping systems to take advantage of locally available practices that are immediately accessible to smallholders, but which require further research and validation over the wide range of cropping systems and environments across Africa.

Impact of FAW Infestation on Maize Yield

The direct foliar damage from FAW feeding on maize is alarming to many farmers who have never seen this type of damage before. This alarm is often shared by politicians, leading to the urgent response of pesticide procurement and use. But the foliar damage caused by FAW in many cases does not result in dramatic yield reduction. The maize plant is quite capable, under good moisture and nutrition conditions, to compensate for a level of foliar damage that may appear alarming to farmers first seeing the damage. The response of maize plants to foliar damage at different stages of plant growth has been well studied in other parts of the world. Dozens of careful, controlled studies on the yield loss caused by FAW have been carried out in the Americas (notably, the USA, Brazil and Mexico).

In the USA, many maize farmers have subsidized crop insurance for their maize crops; specifically against hail, which can have a devastating effect on maize yields at certain moments. Crop insurance in the USA started in the 1880s when private insurance companies first sold policies to protect farmers against the effects of hail storms. In 2016, farmers spent US$981 million on crop-hail insurance to protect US$36 billion worth of crops [31]. Due to the large area covered and the importance of this insurance programme, the National Crop Insurance Service (NCIS) has reviewed and prepared 'corn loss instructions' for the private insurance companies that provide coverage to farmers.

The US Department of Agriculture published the results of NCIS research in the Corn Loss Adjustment Standards Handbook (FCIC 25080) in 2013. The results from the research show that even 70% defoliation at the 12-leaf stage is estimated to cause a 15% yield reduction. A 25% defoliation never causes more than 9% yield reduction and may cause less than a 5% yield reduction when the damage occurs before the 18-leaf stage. These results are very significant for FAW damage, as rarely does the defoliation from FAW exceed 50%, and more typically is closer to 25%. At these levels of infestation, farmers with little experience with FAWand unaccustomed to observing significant foliar damage to their maize are likely to overestimate the impact of FAW damage on their maize yield. In some cases, however, FAW does directly infest cobs. This direct damage has a much greater impact of grain yield and grain quality than does the indirect foliar damage.

The results demonstrate the capacity of maize plants to compensate for low-level artificial defoliation across growth stages and the capacity of maize plants to compensate for higher levels of defoliation when it occurs at certain growth stages. These results were subject to further testing recently by Thomison *et al.* [32], who carried out field studies in three different locations in the midwestern United States, conducting artificial defoliations at two levels (50% and 100%) and at three stages of maize growth, then measuring the yield reduction from each treatment and comparing their results with the NCIS chart. In all cases, the field trials had lower yield losses than the NCIS chart and in most cases, significantly lower. In none of the cases did artificial defoliation at the 50% level, even when conducted up to three times, result in more than 24% yield loss.

These studies of the impact of artificial defoliation on yield clearly demonstrate maize's ability to compensate for foliar damage. While the defoliation caused by FAW looks dramatic, especially to farmers unaccustomed to seeing such foliar damage in their fields, its impact on maize yield is likely to be significantly lower than what farmers would estimate.

In addition to artificial defoliation studies, the yield loss due to FAWinfestation has also been extensively studied in field trials conducted in the Americas. The maize yield response to FAW infestation in the American state of Indiana was studied by Cruz and Turpin [33]. They artificially infested maize plants with FAW larvae at 8–10 leaf stage, measured FAWfeeding damage and yield in trials over 2 years. In the first year, they recorded a 15% yield reduction from 98% of plants infested versus 0% infestation. The same study showed no significant yield reduction of 31% of plants infested. In the second year, an 18% yield reduction was recorded with 100% of plants infested. Cruz *et al.* [34] examined the impact of 100% of plants infested (manually infested) with FAWat mid-whorl stage on maize yield in Brazil on three different maize varieties. Yield reductions due to FAW were measured at 58%, 29% and 21% compared to uninfected controls. The yield impact of FAW and southwestern corn borer [*Diatraea grandiosela*] was examined over 2 years by Williams and Davis [35], used artificial infestation of larvae, resulting in almost 100% of plants infested. FAW infestation resulted in 13% yield reduction, while *D. grandiosela* infestation resulted in a 57% yield reduction.

Hruska and Gladstone [36] studied the response of maize yield to different levels of plant infestation during three periods of vegetative maize growth in irrigated maize under hydric stress in Nicaragua. Chemical insecticides were applied and protected the plants from FAW and the neotropical cornstalk borer (*Diatraea lineolata*). Yields from treatments which had infestation during just one period (approximately 15 days) did not differ from complete protection. The treatment that had maximum infestation during three periods yielded 34% less than the treatment that had complete protection. Hruska and Gould [37] reported on four field trials in Nicaragua of FAW and *Diatraea lineolata* infestation and their impact on maize yield. In the four studies, the yield reduction from maximum FAW and *D. lineolata* infestations resulted in yield reductions of 25%, 17%, 25% and 43%. The differences were explained based on different infestation pressures and environmental (rainfall) conditions.

FAW Economic Thresholds and the Need for Validation in Africa and Asia

Based on the results of field trials on FAWimpact on yield and the economics of maize production and protection, recommendations have been made regarding management decisions for FAW in maize in the USA. For example, Purdue University in Indiana recommends: 'The application of an insecticide is usually not economical for control of the fall armyworm. However, it may be necessary if the infestation is extremely severe and/or the plants are under stress. In such cases, if 75% of the plants exhibit whorl feeding damage and larvae are less than 1–1/4 inches (31 mm) long, and the plants are under stress, treatment may be advisable' [38].

Growing conditions of maize in Africa and Asia are quite different from conditions in the Americas. The ability of the maize plant to compensate for foliar damage depends on the genetics, nutrition and water availability of the plant. Maize varieties, plant density and other agronomic practices will also influence the relationship between FAW infestation and yield. Experimental field studies are urgently needed to validate the results from the Americas. In addition, African and Asian smallholder farmers may have different decision-making processes than American maize farmers. To date, there is only one published paper that measured maize yield loss due to FAWin Africa, reporting a 12% reduction [22].

During the first years of infestation, the only estimates of infestation levels and yield losses were available from farmers' estimates. Such estimates are fraught, due to the inexperience of farmers with FAW. Without the experience of observing damage to their maize and then harvesting the maize, farmers in Africa do not have the practical experience to guide their estimates of yield loss. In addition, farmers may sense that overestimates of damage or yield loss may result in benefits from government programmes or foreign donations, creating an incentive to overestimate damage and yield loss.

Smallholder Maize Integrated Pest Management

Maize integrated pest management programmes in a number of countries, including the USA, Brazil and Argentina, incorporate new technologies, such as the latest generation of chemical insecticides and genetically modified hybrid varieties. The access to these technologies is made possible by the economics of producing for international markets that seek maize for use as animal feed, ethanol production and as a source of sweeteners. These same technologies are typically not accessible by smallholder maize farmers, due to the low and unstable prices and lack of access to risk transfer mechanisms. Smallholder maize farmers are very constrained in their access to integrated pest management options and must often depend on locally available, low-cost options.

Before the arrival of FAW and the free provision of pesticides to smallholder maize farmers in Africa and Asia, the overwhelming majority did not use pesticides in their maize but have instead used cultural control methods to deter or kill insect pests. The practices of intercropping maize with common beans, handpicking and killing insects, and the application of tobacco extracts, wood ash and soils to plants is common among these farmers [30]. The economic constraints on smallholder maize producers mean that their sustainable integrated pest management will be largely dependent on agro-ecological approaches. Those will rely heavily on knowledge and experience of FAWin the local cropping systems and the conditions that lead to the prevention of high levels of FAW damage. They will also rely on adequate manipulation of the ecosystem to maximize the effectiveness of naturally occurring enemies, coupled with the use of locally available, low-cost tactics, such as mechanical control and the use of local substances to deter or kill FAW.

Due to the typically small plot size of maize production in Africa and Asia, the direct, mechanical control (squashing egg masses and handpicking small larvae) is feasible. Good experiences have been reported with this technique in Kenya (http://www.fao.org/farmer-field-schools/newsevents/detail-events/en/c/1113777/) and Ethiopia (2018), where 337 000 ha of infested FAW fields were mechanically controlled in 2017 and 402 000 ha in 2018 with satisfaction expressed.

Good management of FAW begins with preventing infestation and the impact of larval feeding. Late-planted maize is often infested with high levels of FAWas the moths seek vegetative maize. Thus, staggered planting of maize in a landscape should be avoided. The ability of maize to compensate for foliar damage is largely influenced by adequate moisture and nutrition. Many smallholders grow maize under unfavourable conditions for both and as a result, suffer greater yield loss than if the maize had adequate nutrition and moisture. Increased plant diversity, especially intercropping, has been shown to suppress herbivore populations, increase natural enemy populations and reduce crop damage [39]. Garcia Gonzalez *et al.* [40] found that fields intercropping pumpkin with maize had lower FAW infestations than fields of just maize. Baudron *et al.* [22] however, found significantly greater FAWdamage in maize fields intercropped with pumpkins. Van Huis [41] showed in Nicaragua that intercropping maize with beans reduced the percentage of maize plants infested by *S. frugiperda* by 20% to 30%. Hailu *et al.* [42] demonstrated that intercropping common beans or groundnut with maize has reduced FAW oviposition by 30% in maize in Uganda.

One use of plant diversity is the 'push-pull' system that uses two types of plants: one which emits volatile chemicals that repel pest insects, such as desmodium, *Desmodium uncinatum Jacq.* (*Leguminaceae*) (push); and planting an attractive trap plant, such as Napier grass, *Pennisetum purpureum* Schumach (*Poaceae*) (pull), as a border crop around this intercropped field. Moths are repelled from the main crop by the repellent plant and are simultaneously attracted to the trap plant [43]. The use of desmodium and Napier grass in maize systems has recently been shown to be effective in reducing FAWinfestations and plant damage caused by FAW[44]. Reductions are reported to be 82.7% in the average number of larvae per plant and 86.7% in plant damage per plot in climate-adapted push-pull compared to maize monocrop plots.

Role of Natural Enemies

Many studies have shown that FAW is attracted by a large array of natural enemies and their combined level of mortality can be quite high. Luginbill [1] stated in 1928: 'As has been previously mentioned, the abundance or scarcity of natural enemies to a large extent determines whether or not fall army worms become abundant enough during a season to cause destruction to a crop. Local outbreaks are often controlled by natural enemies alone'.

In Nicaragua, 15 parasitoid species, one species of a nematode parasite, three entomopathogen species and a large number of predator species have been found on *S. frugiperda* in maize. Important parasites are: the braconids (up to 30% parasitism of the collected larvae, mainly *Aleiodes* (= *Rogas) laphygmae* and *Chelonus insularis*); the tachinids (up to 60% parasitism in maize and up to 100% in weeds, mainly *Lespesia archippivora);* and mermithids (up to 30% parasitism, *Hexamermis* sp) [41]. Pair and Gross [45] found 73% FAW pupal mortality mainly due to predators. In Honduras, Wheeler *et al.* [46] reported that 42% of FAWs were killed by a complex of parasitoids, *Chelonus insularis* (Cresson) (Hymenoptera; Braconidae) being the most common.

Castro *et al.* [47] found up to 71% natural parasitism of FAW larvae by an endoparasitic nematode (Mermithidae). In Brazil, Varella *et al.* [48] conducted careful life table analyses of FAW mortality, at both egg and early larval stages over 3 years. They found that total egg mortality ranged from 73% to 81% and the greatest egg mortality was due to unviability, dislodgement and predation. At the early larval stage, >95% of early larvae died from predation, drowning and dislodgment by rainfall. They note the special importance of rainfall and predation in early larval mortality. Hoballah *et al.* [49] studied natural mortality of FAWin Veracruz, Mexico over 2 years and found that early larvae suffered 34% and 80% natural mortality during the 2 years.

In Ethiopia, Sisay *et al.* [50] examined parasitism of FAW and found *Cotesia icipe* was the dominant larval parasitoid with parasitism ranging from 33.8% to 45.3%, while in Kenya, the tachinid fly, *Palexorista zonata,* was the primary parasitoid with 12.5% parasitism. Many of these parasoitoids are very sensitive to chemical insecticides used in commercial maize fields, as demonstrated by Meagher *et al.* [51] in Florida, where they found that natural parasitism levels of FAWlarvae ranged from 1% in a commercial sweet corn field to 92% in an unsprayed field on an agricultural experiment station in Palm Beach County.

Perfecto [52] found that experimental reduction in ant (*Pheidole radowszkoskii* and *Solenopsis geminata*) populations in maize fields in Nicaragua resulted in significantly greater FAWlarval populations per plant and greater foliar damage to maize caused by FAW. Perfecto [53] also showed that application to the soil at planting of carbofuran resulted in significantly reduced ant foraging and increased population levels of FAW. Ants are often common in maize cropping systems in Africa. *Myrmicaria natalensis* is a common ant in Africa agriculture, and has been observed predating on FAW in Malawi (personal observation). This species has previously been observed attacking *Heliothis armigera* [54].

For smallholder farmers in Mesoamerica, the combination of the use of plant-diverse cropping systems and the levels of control by natural enemies maintains FAW populations at levels low enough levels for smallholders in Central America [55].

Although many natural control organisms of FAW appear to be already present, field surveys need to be conducted to determine their distribution across the new environments. Important gaps between the natural enemy complexes in the natural FAW distribution and the complexes present in the new range could be studied for possible classical biological control programmes.

Enhancing populations of natural enemies through landscape management

Plant diversity has also been shown to increase FAW natural enemy populations harbouring and increasing the activity of natural enemies [56, 57]. In Nicaragua, van Huis [41] demonstrated that weeds increased parasitism of FAW: 35 and 42 days after plant emergence, parasitism by the tachinid *Lespesia archippivora* on larvae from weeds was 10% to 20% higher than on larvae from maize; at 42 days, 54% of the maize larvae and 75% of the weed larvae were parasitized by *L. archippivora*. Perfecto and Sediles [58] showed that maize intercropped with beans led to a 28% decrease in FAW infestation. They attributed this to the lower populations of ants in maize monocultures.

Degri *et al.* [59] studied the effect of intercropping pearl millet with groundnut on stem borer infestation of millet in Nigeria and found up to 48% reduction in the per cent of plants infested with stem borer and a 58% reduction in the number of borers per plant. This reduction was attributed to the increased levels of natural enemies found in the intercropped fields. Midega *et al.* [57] found that in the push-pull system there is an increased abundance, diversity and activity of predatory arthropods, contributing to reducing pest populations. Cañas and O'Neil [60] tested a traditional practice in Honduras of spraying sugar water in maize fields and found that the sugar-treated maize had higher numbers of natural enemies, 35% less FAW leaf damage and 18% lower plant infestation rates than maize treated with water alone.

The observations that ants can be important predators of FAW in maize fields, coupled with the observation that these predatory ants are attracted to certain protein-based substances, has led farmers to experiment with the use of fish soup and kitchen grease to increase ant populations and decrease FAW larval populations [61].

Importance and Use of Entomopathogens

Gardner and Fuxa [62] reviewed the pathogens naturally found on FAW in the USA and noted that the *S. frugiperda* nucleopolyhedrovirus (SfMNPV) is commonly found naturally and frequently causes epizootics in FAW. They found up to 38% mortality in FAWattacking sorghum in the state of Georgia. Other important pathogens of FAW include fungi, bacteria and nematodes. Some of these pathogens have already been observed killing FAWlarvae in Africa, sometimes at high levels (personal observation). FAWwas first reported in Yemen in August 2018 and in the first reports of FAWin maize, cadavers of FAWwere found infected with fungi. Similarly, fungi have been reported killing FAW soon after its arrival in India [63].

Due to their effectiveness and low mammalian toxicity, interest in using entomopathogens as bio-pesticides against FAW has increased. A number of commercial bio-pesticides are registered and available for FAWin many countries, including fungi and bacteria, and more recently SfNPV (Andermatt & AgBiTech). Some of these pathogens (*Bacillus thuringiensis*) have also been produced at low cost, in local production (Cuba, Brazil). Local production of Trichogramma has been successful in a number of countries (Brazil, Egypt, etc.) for FAW and other lepidopteran crop pest species.

In addition, at a very local level, some farmers have multiplied pathogens by infecting larvae with the pathogens and then harvesting them, processing them artisanally and applying them. Some smallholders in Central America simply collect dead larvae from their fields, grind them, strain them and apply a solution of the extract into maize plants infested with FAW (personal observation).

Vega [64] reviewed 85 published papers reporting fungal entomopathogens as endophytes and their potential for biological control. Of the 38 plant species studied, maize was the most common. In a number of cases, the endophytic fungus *Beauveria bassiana* demonstrated an effect on Lepidoptera species tested. Entomopathogenic fungi are widespread in agricultural fields and help suppress crop pests. These natural enemies may be hindered by certain agronomic practices associated with conventional agriculture including the use of pesticides.

The entomopathogenic communities have been shown to be affected by the cropping systems.

A significantly higher occurrence of insect pathogenic fungi in soils from arable fields of organically managed farms versus conventionally managed farms in Norway was found by Klingen *et al.* [65]. Clifton *et al.* [66] reported similar results in Iowa, in the USA, where in 1 year of the survey, soil from organic fields and accompanying margins had significantly more entomopathogenic fungi than conventional fields and accompanying margins. Regression analysis revealed that the percentage of silt and the application of organic fertilizer were positively correlated with entomopathogenic fungi abundance; but nitrogen concentration, tillage, conventional fields and margins of conventional fields were negatively correlated with entomopathogenic fungi abundance.

Use of Locally Available Substances Toxic to FAW

Many smallholder farmers around the world use locally available substances to try to control FAW. These include local botanical extracts, soil, sand, wood ash, lime, oils and soaps.

In addition to botanical extracts, some of which have been demonstrated to be effective, other substances are reported by farmers to be useful in killing FAW larvae. With the exception of botanical extracts, the effectiveness of the other substances has been little tested and virtually no published results are in the scientific literature. These substances (wood ash, soil, sand, detergents, oils, salt, lime, urine and others) beg scientific testing, elucidation of the mechanism of action and an exploration of the robustness of results over space and time.

Among these substances, the effectiveness of botanical extracts on FAW have been studied. Neem seed powder was shown to be effective in killing FAW larvae, always causing mortality of over 70% in laboratory studies [67]. Silva *et al.* [68] confirmed this result, finding neem toxicity to FAW larvae and calculating the LC [48] values for neem seed cake and leaves at 0.13% and 0.25% respectively. Souza *et al.* [69] demonstrated that plant oils from *Corymbia citriodora, Eucalyptus urograndis* and *Eucalyptus urograndis* had positive significant effects of protecting maize from FAW larvae. An aqueous seed extract from Carica papaya was found to produce significant mortality of FAW larvae, similar to that of the chemical insecticide malathion [70].

Salinas-Sanchez *et al.* [71] found that hexane, acetone and ethanol extracts of *Tagetes erecta* caused 48%, 60% and 72% mortality respectively of FAW larvae. Lizarazo *et al.* [72] tested three local plants in Colombia against FAW larvae and found that one of them (Polygonum hydropiperoides) produced larval mortality as high as a commercial insecticide (chlorpyrifos). Delgado Cáceres and Gaona Mena [73] achieved 82% mortality of FAW larvae with Polygonum hydropiperoides extracts at 50 g/100 ml water. Franco *et al.* [74] demonstrated 100% FAWlarval mortality with water extracts of seed of Carica papaya at 10% concentration. Plant oils from turmeric, clove, palmarosa and neem were shown to have significant positive effects of protecting maize from first and second instar FAW larvae [75].

Stevenson *et al.* [76] reviewed the current status of research and use of botanically active substances in Africa and found great potential for their use in pest management in Africa. Anjarwalla *et al.* [77] collected a good deal of information on 18 pesticidal plants, much of it of use to smallholder farmers in southern and eastern Africa.

Use of Soil for FAW Control

Smallholder maize farmers around the world may place soil into infested whorls of maize plants out of desperation, but a look into soil and FAW control analysis reveals a fascinating agroecology, basis of functionality and potential for use. Soil may kill FAW larvae directly, via abrasiveness or absorption of wax from insect cuticles, causing larval desiccation.

Diatomaceous earth has long been used for insect control, especially for stored grain pests [78] and many commercial products are available. It is considered one of the safest and effective naturally occurring insecticides [79]. Diatomaceous earth adheres to the insect body and damages the protective waxy layer of the insect cuticle by absorption, and to a lesser degree by abrasion. Loss of water from insect body results in death [79]. Constanski *et al.* [80] studied the effects of several inert powders, including diatomaceous earth and bentonite on FAW. Benotite caused 93% FAW mortality, diatomaceous earth caused 47% mortality.

In addition to the direct physical properties of soil particles, soil often contains a rich ecosystem of microorganisms, some of which may kill FAW larvae. A number of studies have looked at these entomopathogenic communities. Valicente and Barreto [81] found that 75% of 1448 soil samples in ten Brazilian states, 75% contained *Bacillus thuringiensis*. Ramirez-Rodriguez and Sánchez-Peña [82] isolated the fungal entomopathogen *Beauveria bassiana* from the soil in Mexico and found that it caused 98% mortality of FAW larvae. Williams *et al.* [83] sampled soil from maize fields in southern Mexico, Guatemala and Belize and found that 29% of the samples contained SfMNPV occlusion bodies/polyhedra in high levels. Williams *et al.* [83] sampled soil from maize fields in southern Mexico, Guatemala and Belize and found that 29% of the samples contained SfMNPV occlusion bodies/polyhedra in levels high enough to produce mortality in early instar FAW larvae.

Knowledge Gaps

Research on host plant resistance, including genetically modified plants, and chemical control is advancing in the private sector and international research institutes. Most of these new technologies are not accessible by smallholder farmers. Large knowledge gaps exist for the use of biological control, plant chemical ecology and the use of locally available substances to deter or kill FAW. Very little research is being carried out on the use of these methods by smallholders.

The research and experiences from the Americas demonstrate that FAW has a complex agroecology that can be used to sustainably manage FAW. Smallholders in Central America and Mexico have been sustainably managing FAW for thousands of years. Many smallholders still use the traditional mixed cropping systems (milpa), where maize, common beans and a cucurbit are grown together in polycultures. The experiences and the research from this context are especially informative for smallholders in Africa and Asia. Experiences and research from large-scale commercial maize farms in the Americas are probably less useful for smallholders.

Experienced farmers do not panic if FAW infests their maize. They understand that the population levels are not great and that they do not cause much damage or yield loss. They also go to the fields often to observe and mechanically control FAW. They have a basic understanding that many of the FAWeggs and larvae are naturally killed in the fields by rain, due to desiccation, or the direct action of natural enemies. Farmers sometimes try to attract natural enemies (e.g. by spraying sugar water) and when they need to, they use locally available substances that have shown to be effective in the past (use of local botanical extracts, soil, ash, lime, salt, oils, soaps and urine).

These elements of smallholder maize integrated pest management need to be carefully studied across the ecosystems of Africa and Asia, to better understand under what conditions they work best, understand their mechanisms and to begin to develop recommendations that can be up-scaled. Many farmers are already trying some of these innovations in their fields. For example, the use of fish soup to attract ants to fields has been reported from Malawi. Researchers should work with farmer innovators in their fields to test and measure the results of these practices under different conditions. Special attention should be paid to the costs, both economic and labour time of different approaches, to ensure that the tactics are accessible by smallholders. This type of co-creation is part of Farmer Field Schools, which are a natural crucible for doing farmer-led and researcher-aided research.

In addition to co-creation with farmers, researchers should pay special attention to conducting controlled yield-loss studies as a response to different levels of FAW infestation at different periods of maize growth. These data should be gathered from a number of ecosystems across the continent so that better decisions can be made by farmers and resource allocators. Farmer Field Schools have been shown to be an excellent medium for smallholder farmer innovation and co-creation [84].

Digital Technologies

A promising innovation that is being developed by FAO and partners is the use of digital technologies that have the potential of delivering FAW advice, based on local conditions and creating communities of farmers to share experiences locally. FAO has developed the Fall Armyworm Monitoring and Early Warning System [3]. These tools are able to help farmers correctly identify and monitor populations of FAW while providing them with off-line, free advice based on local conditions delivered from satellite data. This scalable technology shows huge potential in delivering up-to-date information and advice to farmers in their pockets.

Conclusion

FAW continues to spread into new territory, moving further east and north in Asia, as well as north from sub-Saharan Africa. The vast majority of farmers in the new territories are smallholders. These farmers have limitations and special contexts which define their options for management of FAW. Fortunately, there is a vast literature and knowledge about the ecology of FAW systems, including chemical ecology and natural biological control, which can be harnessed and tested in the new environments and landscapes of FAW in Africa and Asia. This review has examined this extensive literature and reviews the FAW management options for smallholders, largely based on locally available solutions using agroecological knowledge.

Acknowledgements

The author would like to thank the following reviewers of this text, for their comments and suggestions: Mona Chaya, Senior Coordinator for Food Chain Crises, FAO; Joyce Mulila-Mitti, Plant Production and Protection Officer for Southern Africa FAO; Keith Cressman, Senior Locust Forecasting Officer FAO; Makiko Taguchi, Agricultural Officer, FAO; Hans Dreyer, Director, Plant Production and Protection Division of the Agriculture and Consumer Protection Department, FAO; and three anonymous reviewers. The views expressed in this publication are those of the author(s) and do not necessarily reflect the views or policies of the Food and Agriculture Organization of the United Nations.

References

1. Luginbill P. The fall armyworm. USDA Technical Bulletin 1928;34:1–91.
2. Goergen G, Kumar PL, Sankung SB, Togola A, Tamo M. First report of outbreaks of the fall armyworm *Spodoptera frugiperda* (J E Smith) (Lepidoptera, Noctuidae), a new alien invasive pest in West and Central Africa. PLOS ONE 2016;11:1–9. Available from: URL: https://doi.org/10.1371/journal.pone.0165632.

3. FAO. Fall Armyworm Monitoring and Early Warning System (FAMEWS) Platform; 2018. Available from: URL: http://www.fao.org/fall-armyworm/en/. (last accessed 21 November 2018).
4. Ganiger PC, Yeshwanth HM, Muralimohan K, Vinay N, Kumar ARV, Chandrashekara K. Occurrence of the new invasive pest, fall armyworm, *Spodoptera frugiperda* (J. E. Smith) (Lepidoptera, Noctuidae), in the maize fields of Karnataka, India. Current Science 2018;115(4):621–23.
5. Casmuz A, Juarez ML, Socias MG, Murua MG, Prieto S, Medina S, *et al.* Revision de los hospederos del gusano cogollero del maíz, *Spodoptera frugiperda* (Lepidoptera: Noctuidae). Revista de la Sociedad Entomológica Argentina 2010;69:209–31.
6. Montezano DG, Specht A, Sosa-Gómez DR, Roque-Specht VF, Sousa-Silva JC, Paula-Moraes SV, *et al.* Host plants of *Spodoptera frugiperda* (Lepidoptera: Noctuidae) in the Americas. African Entomology 2018;26(2):286–300.
7. Mitchell ER, McNeil JN, Westbrook JK, Silvain JF, Lalanne-Cassou RB, Chalfant SD, *et al.* Seasonal periodicity of fall armyworm, (Lepidoptera: Noctuidae) in the Caribbean basin and northward to Canada. Journal of Entomological Science 1991;26:39–50.
8. Sparks AN. Review of the biology of the fall armyworm. The Florida Entomologist 1979;62:82–87. Available from: URL: https://doi.org/10.2307/3494083.
9. Ashley TR, Wiseman BR, Davis FM, Andrews KL. The fall armyworm: A bibliography. The Florida Entomologist 1989;72(1):52. doi: 10.2307/3494982.
10. Nagoshi R, Meagher R. Review of fall armyworm (Lepidoptera: Noctuidae) genetic complexity and migration. The Florida Entomologist 2008;91:546–54. doi: 10.1653/0015-4040-914546.
11. Johnson SJ. Migration and the life history strategy of the fall armyworm, *Spodoptera frugiperda* in the Western Hemisphere. Insect Science Application 1987;8(4/5/6):543–49.
12. Pannuti LER, Paula-Moraes SV, Hunt TE, Baldin ELL, Dana L, Malaquias JV. Plant-to-plant movement of *Striacosta albicosta* (Lepidoptera: Noctuidae) and *Spodoptera frugiperda* (Lepidoptera: Noctuidae) in maize (Zea mays). Journal of Economic Entomology 2016;109:1125–31. Available from: URL: https://doi.org/10.1093/jee/tow042.
13. Andow DA, Farias JR, Horikoshi RJ, Bernardi D, Nascimento ARB, Omoto C. Dynamics of cannibalism in equal-aged cohorts of Spodoptera frugiperda. Ecological Entomology 2015;40(3):229–36. Available from: URL: https://doi.org/10.1111/een.12178.
14. ISAAA. Global Status of Commercialized Biotech/GM Crops in 2017: Biotech Crop Adoption Surges as Economic Benefits Accumulate in 22 Years. ISAAA Brief No. 53; 2017. ISAAA. Ithaca, NY.
15. Kumela T, Simiyu J, Sisay B, Likhayo P, Mendesil E, Gohole L. Farmers' knowledge, perceptions, and management practices of the new invasive pest, fall armyworm (*Spodoptera frugiperda*) in Ethiopia and Kenya. International Journal of Pest Management 2018;65(1):1–9. Available from: URL: https://doi.org/10.1080/09670874.2017.1423129.
16. FAOSTAT. Maize production. 2018. Available from: URL: http://www.fao.org/faostat/ (last accessed 12 November 2018).
17. Erenstein O, Kassie GT, Langyintuo A, Mwangi W. Characterization of Maize Producing Households in Drought Prone Regions of Eastern Africa. CIMMYT Socio-Economics Working Paper 1; 2011. Mexico, D.F. CIMMYT.
18. Wiredu NA, Gyasi KO, Abdoulaye T, Sanogo D, Langyintuo A. Characterization of maize producing households in the Northern Region of Ghana. Country Report Ghana; 2010; CSRI/SARI – IITA, Ibadan, Nigeria.
19. Kassie GT, Erenstein O, Mwangi W, La Rovere R, Setimela P, Langyintuo A. Characterization of Maize Production in Southern Africa: Synthesis of CIMMYT/ DTMA Household Level Farming System Surveys in Angola, Malawi, Mozambique, Zambia and Zimbabwe. Socio-Economics Program Working Paper 4; 2012; Mexico, D.F.: CIMMYT.
20. Stern VM, Smith RF, van den Bosch R, Hagen KS. The integrated control concept. Hilgardia 1959;29:81–101.
21. Pedigo LP, Hutchins SH, Higley LG. Economic injury levels in theory and practice. Annual Review of Entomology 1986;31(1):341–68.
22. Baudron F, Zaman-Allah MA, Chaipa I, Chari N, Chinwada P. Understanding the factors influencing fall armyworm (*Spodoptera frugiperda* J.E. Smith) damage in African smallholder maize fields and quantifying its impact on yield. A case study in Eastern Zimbabwe. Crop Protection 2019;120:141–50. Available from: URL: https://doi.org/10.1016/j.cropro.2019.01.028.

23. Muhammad L, Mwabu D, Mulwa R,MwangiW, Langyintuo A, La Rovere R. Characterization of maize producing households in Machakos and Makueni districts in Kenya. Country Report–Kenya, Nairobi; 2010; KARI-CIMMYT.
24. Christiaensen L, Demery L, editors. Agriculture in Africa: Telling Myths from Facts. Directions in Development. World Bank Washington, DC; 2018; doi:10.1596/978-1-4648-1134-0. License: Creative Commons Attribution CC BY 3.0 IGO.
25. FAO. The Republic of Namibia: Fall armyworm impact and needs assessment. Rome, Italy; 2018; Licence: CC BY-NC-SA 3.0 IGO. Available from: URL: http://www.fao.org/3/i9556en/I9556EN.pdf (last accessed 15 November 2018).
26. Namibia Statistics Agency. Namibia Census of Agriculture 2013/2014. Communal Sector Report November 2015. Windhoek; 2015.
27. Rwomushana I, Bateman M, Beale T, Beseh P, Cameron K, Chiluba M, *et al.* Fall armyworm: impacts and implications for Africa. Evidence Note Update, October 2018. CABI; 2018.
28. Salato Z. Fall Armyworm Status in Ethiopia. Ministry of Agriculture presentation at the Fall Armyworm Monitoring and Early Warning Midterm Evaluation meeting, Kigali, Rwanda November 2018; 2018.
29. Van den Berg J, Nur AF, Polaszek A (editor). Cultural control. African Cereal Stemborers: Economic Importance, Taxonomy, Natural Enemies and Control. CABI Publishing,Wallingford, UK; 1998. p. 333–47.
30. Abate T, van Huis A, Ampofo JKO. Pest management strategies in traditional agriculture: an African perspective. Annual Review of Entomology 2000;45:631–59.
31. NCIS. What is crop insurance?; 2017; Available from: URL: https://cropinsuranceinamerica.org/what-is-crop-insurance/ (last accessed 04 December 2017).
32. Thomison PR, Nafziger ED, Coulter JA, Zarnstorff ME, Geyer AB, Lindsey AJ. Response of Corn to Multiple Defoliation Events. Information poster presented at the Annual Meeting of the American Society of Agronomy November 2016; 2016. Available from: URL: https://scisoc.confex.com/scisoc/2016am/webprogram/Paper99910.html.
33. Cruz I, Turpin FT. Yield impact of larval infestations of the fall armyworm (Lepidoptera: Noctuidae) to midwhorl growth stage of corn. Journal of Economic Entomology 1983;76(5):1052–54.
34. Cruz I, Figueiredo MC, Oliveira AC, Vasconcelos CA. Damage of *Spodoptera frugiperda* (Smith) in different maize genotypes cultivated in soil under three levels of aluminium saturation. International Journal of Pest Management 1999;45(4):293–96.
35. Williams WP, Davis FM. Response of corn to artificial infestation with fall armyworm and southwestern corn borer larvae. Southwestern Entomologist 1990;15:163–66.
36. Hruska AJ, Gladstone SM. Effect of period and level of infestation of the fall armyworm, *Spodoptera frugiperda*, on irrigated maize yield. The Florida Entomologist 1988;71(3):249–54.
37. Hruska AJ, Gould F. Fall armyworm (Lepidoptera: Noctuidae) and *Diatraea lineolata* (Lepidoptera: Pyralidae): impact of larval population level and temporal occurrence on maize yield in Nicaragua. Journal of Economic Entomology 1997;90(2):611–22.
38. Purdue University. Corn Crop Management: Fall Armyworm 2019; Available from: URL: https://extension.entm.purdue.edu/fieldcropsipm/insects/fall-armyworm.php (last accessed 10 February 2019).
39. Letourneau DK, Armbrecht I, Salguero Rivera B, Montoya Lerma J, Jiménez Carmona E, Constanza Daza M. Does plant diversity benefit agroecosystems? A synthetic review. Ecology Applied 2011;21:9–21. doi:10.1890/09-2026.1.
40. Garcia González MT, Rojas JA, Castellanos Gonzalez L, Enjamio Jiménez D. Policultivo (maíz-calabaza) en el control de *Spodoptera frugiperda* (Smith) en Fomento, Sancti Spiritus. Centro Agrícola 2010;37(1):57–64.
41. Van Huis A. Integrated pest management in the small farmer's maize crop in Nicaragua. Mededelingen Landbouwhogeschool Wageningen 1981;81(6):221.
42. Hailu G, Niassy S, Khan ZR, Ochatum N, Subramanian S. Maize-legume intercropping and Push-pull for management of fall armyworm, stemborers and striga in Uganda. Agronomy Journal 2018;110:1–10. doi: 10.2134/agronj2018.02.0110.
43. Khan Z, Midega CAO, Bruce TJA, Hooper AM, Pickett JA. Exploiting phytochemicals for developing a push–pull crop protection strategy for cereal farmers in Africa. Journal of Experimental Botony 2010;61:4185–96.

44. Midega CAO, Pittchar JO, Pickett JA, Hailu GW, Khan ZR. A climate-adapted push-pull system effectively controls Fall Armyworm, *Spodoptera frugiperda* (J E Smith), in maize in East Africa. Crop Protection 2018;105:10–15. Available from: URL: https://doi.org/10.1016/j.cropro.2017.11.003.
45. Pair SD, Gross HR Jr. Field mortality of pupae of the fall armyworm, *Spodoptera frugiperda* (J.E. Smith), by predators and a newly discovered parasitoid, *Diapetimorpha introita*. Journal of the Georgia Entomological Society 1984;19:22–26.
46. Wheeler GS, Ashley TR, Andrews KL. Larval parasitoids and pathogens of the fall armyworm in Honduran maize. Entomophaga 1989;34:331–40. Available from: URL: https://doi.org/10.1007/BF02372472.
47. Castro MT, Pitre HN, Meckenstock DH. Populations of fall armyworm, *Spodoptera frugiperda* (J.E. Smith) larvae and associated natural enemies in sorghum and maize cropping systems in southern Honduras. Tropical Agriculture (Trinidad) 1989;66(3):259–64.
48. Varella AC, Menezes-Netto AC, Alonso JDdS, Caixeta DF, Peterson RKD, Fernandes OA. Mortality dynamics of *Spodoptera frugiperda* (Lepidoptera: Noctuidae) Immatures in Maize. PLOS ONE 2015;10(6):e0130437. doi:10.1371/journal.pone.0130437.
49. Hoballah ME, Degen T, Bergvinson D, Savidan A, Tamo C, Turlings TCJ. Occurrence and direct control potential of parasitoids and predators of the fall armyworm (Lepidoptera: Noctuidae) on maize in the subtropical lowlands of Mexico. Agricultural and Forest Entomology 2004;6:83–88.
50. Sisay B, Simiyu J, Malusi P, Likhayo P, Mendesil E, Elibariki N, *et al.* First report of the fall armyworm, *Spodoptera frugiperda* (Lepidoptera: Noctuidae), natural enemies from Africa. Journal of Applied Entomology 2018;142(8):800–4. Available from: URL: https://doi.org/10.1111/jen.12534.
51. Meagher RL Jr, Nuessly GS, Nagoshi RN, Hay-Roe MM. Parasitoids attacking fall armyworm (Lepidoptera: Noctuidae) in sweet corn habitats. Biological Control 2016;95:66–72. Available from: URL: https://doi.org/10.1016/j.biocontrol.2016.01.006.
52. Perfecto I. Ants (Hymenoptera: Formicidae) as natural control agents of pests in irrigated maize in Nicaragua. Journal of Economic Entomology 1991;84(1):65–70.
53. Perfecto I. Indirect and direct effects in a tropical agroecosystem: the maize-pest-ant system in Nicaragua. Ecology 1990;71(6):2125–34.
54. Samways MJ. Soil dumping by *Myrmicaria natalensis* (Smith) (Hymenoptera: Formicidae) as a competitive advantage over other ant species. Phytophylactica 1982;14:3–5.
55. Wyckhuys KAG, O'Neil RJ. Population dynamics of *Spodoptera frugiperda* Smith (Lepidoptera: Noctuidae) and associated arthropod natural enemies in Honduran subsistence maize. Crop Protection 2006;25:1180–90.
56. Landis DA, Wratten SD, Gurr GM. Habitat management to conserve natural enemies of arthropod pests in agriculture. Annual Review of Entomology 2000;45:175–201. Available from: URL: https://doi.org/10.1146/annurev.ento.45.1.175.
57. Midega CAO, Khan ZR, Van den Berg J, Ogol CKPO, Pickett JA, Wadhams LJ. Maize stemborer predator activity under 'push–pull' system and Bt-maize: a potential component in managing Bt resistance. International Journal of Pest Management 2006;52:1–10.
58. Perfecto I, Sediles A. Vegetational diversity, ants (Hymenoptera: Formicidae), and herbivorous pests in a neotropical agroecosystem. Environmental Entomology 1992;21:61–67.
59. Degri MM, Mailafiya DM, Mshelia JS. Effect of intercropping pattern on stem borer infestation in pearl millet (Pennisetum glaucum L.) grown in the Nigerian Sudan Savannah. Advances In Entomology 2014;2:81–86. Available from: URL: https://file.scirp.org/pdf/AE_2014042816484099.pdf.
60. Cañas LA, O'Neil RJ. Applications of sugar solutions to maize, and the impact of natural enemies on fall armyworm. International Journal of Pest Management 1998;44(2):59–64. doi:10.1080/096708798228329.
61. Demeter Seeds. Ants control Fall Army Worm in maize in Malawi. 2018. Available from: URL: https://www.youtube.com/watch?v=ccPhXNMUv3w (last accessed 05 December 2018).
62. Gardner WA, Fuxa JR. Pathogens for the suppression of the fall armyworm. The Florida Entomology 1980;63(4):439–47.
63. Shylesha AN, Jalali SK, Gupta A, Varshney R, Venkatesan T, Shetty P, *et al.* Studies on new invasive pest *Spodoptera frugiperda* (J. E. Smith) (Lepidoptera: Noctuidae) and its natural enemies. Journal of Biological Control 2018;32(3):1–7.

64. Vega F. The use of fungal entomopathogens as endophytes in biological control: a review. Mycologia 2018;110(1):4–30. Available from: URL: https://www.tandfonline.com/doi/full/10.1080/00275514.2017.1418578; https://doi.org/10.1080/00275514.2017.1418578.
65. Klingen I, Eilenberg J, Meadow R. Effects of farming system, field margins and bait insect on the occurrence of insect pathogenic fungi in soils. Agriculture Ecosystem Environment 2002;91:191–98.
66. Clifton EH, Jaronski ST, Hodgson EW, Gassmann AJ. Abundance of soil-borne entomopathogenic fungi in organic and conventional fields in the Midwestern USA with an emphasis on the effect of herbicides and fungicides on fungal persistence. PLOS ONE 2015;10(7):e0133613. doi:10.1371/journal.pone.0133613.
67. Maredia KM, Segura OL, Mihm JA. Effects of neem, *Azadirachta indica*, on six species of insect pests. Tropical Pest Management 1992;38(2):190–95.
68. Silva MS, Broglio SMF, Trindade RCP, Santos Ferrreira E, Gomes IB, Micheletii LB. Toxicity and application of neem in fall armyworm. Comunicata Scientiae 2015;6(3):359–64. doi:10.14295/CS.v6i3.808.
69. Souza TF, Favero S, Conte C.. Bioatividade de óleos essenciais de espécies de eucalipto para o controle de *Spodoptera frugiperda* (J. E. Smith, 1797) (Lepidoptera: Noctuidae). Bioactivity of essential oils of eucalyptus species for control of *Spodoptera frugiperda* (J.E. Smith, 1797) (Lepidoptera: Noctuidae). Revista Brasileira de Agroecologia Rev. Bras. de Agroecologia 2010;5(2):157–64.
70. Figueroa-Brito R, Villa-Ayala P, López-Olguín JF, Huerta-de la Peña A, Pacheco-Aguilar JR, Ramos-López MA. Nitrogen fertilization sources and insecticidal activity of aqueous seeds extract of Carica papaya against *Spodoptera frugiperda* in maize. Ciencia e investigación agraria 2013;40(3):567–77.
71. Salinas-Sánchez DO, Aldana-Llanos L, Valdés-Estrada ME, Gutiérrez-Ochoa M, Valladares-Cisneros G, Rodríguez-Flores E. Insecticidal activity of Tagetes erecta extracts on *Spodoptera frugiperda* (Lepidoptera: Noctuidae). The Florida Entomologist 2012;95(2):428–32. Available from: URL: https://doi.org/10.1653/024.095.0225.
72. Lizarazo HKC, Mendoza FR, Carrero S. Efecto de extractos vegetales de Polygonum hydropiperoides, Solanum nigrum y Calliandra pittieri sobre el gusano cogollero (*Spodoptera frugiperda*). Agronomia colombiana 2008;26(3):427–34, 2008. ISSN electrónico 2357-3732. ISSN impreso 0120-9965.
73. Delgado Cáceres LR, Gaona Mena EF. Control de *Spodoptera frugiperda* Smith (Lepidoptera: Noctuidae) con extractos de *Polygonum hydropiperoides* Michx (Ka'atái) en condiciones de laboratorio. Ciencia e investigacion agraria 2012;14(1):5–9.
74. Franco ASL, Jiménez PA, Luna LC, Figueroa-Brito R. Efecto tóxico de semillas decuatro variedades de Carica papaya (Caricaceae) en *Spodoptera frugiperda* (Lepidoptera: Noctuidae). Folia Entomológica Mexicana 2006;45:171–77.
75. Barbosa MS, Dias BB, Guerra MS, Haralampidou da Costa Vieira G. Applying plant oils to control fall armyworm (*Spodoptera frugiperda*) in corn. Australian Journal Crop Science 2018;12(04):557–62. doi:10.21475/ajcs.18.12.04. pne822
76. Stevenson PC, Isman MB, Belmain SR. Pesticidal plants in Africa: A global vision of new biological control products from local uses. Industrial Crops and Products 2017;110:2–9. https://www.sciencedirect.com/science/article/pii/S0926669017305459.
77. Anjarwalla P, Belmain S, Sola P, Jamnadass R, Stevenson PC. Handbook on Pesticidal Plants. World Agroforestry Centre (ICRAF), Nairobi, Kenya; 2016.
78. Korunić Z. Diatomaceous earths–natural insecticides. Pesticides & Phytomedicine Belgrade 2013;28(2):77–95. doi: http://dx.doi.org/10.2298/PIF1302077K.
79. Ebeling W. Sorptive dusts for pest control. Annual Review of Entomology 1971;16(1):123–58. doi:10.1146/annurev. en.16.010171.001011.
80. Constanski KC, Zorzetti J, Santoro PH, Hoshino AT, Janeiro Neves PMO. Inert powders alone or in combination with neem oil for controlling *Spodoptera eridania* and *Spodoptera frugiperda* (Lepidoptera: Noctuidae) larvae. Semina: Ciencias Agrarias 2016;37(4):1801–10. doi:10.5433/1679-0359.2016v37n4p1801.
81. Valicente FH, Barreto MR. Bacillus thuringiensis survey in Brazil: geographical distribution and insecticidal activity against *Spodoptera frugiperda* (J.E. Smith) (Lepidoptera: Noctuidae). Neotropical Entomology 2003;32(4):639–44.
82. Ramierez-Rodriguez D. Sanchez-Peña SR. Endophytic *Beauveria* bassiana in Zea mays: pathogenicity against larvae of fall armyworm, *Spodoptera frugiperda*. Southwestern Entomologist 2016;41(3):875–78. Available from: URL: https://doi.org/10.3958/059.041.0330.

83. Williams T, Richards A, Christian P. Distribution and abundance of baculovirus populations in soil; 2008. Available from: URL: http://trevorwilliams.info/Baculovirus_in_soil.htm (last accessed 22 November 2018).
84. FAO. Monitoring, Evaluation and Learning in Farmer Field School Programmes: A Framework and Toolkit. Food and Agriculture Organization, Rome; 2019. http://www.fao.org/3/i2561e/i2561e01.pdf (accessed 23 May 2019).

9 *Aedes albopictus* in America: current perspectives and future challenges

Isik Unlu[1,2]

[1]Center for Vector Biology, Department of Entomology, Rutgers University, New Brunswick, NJ 08901, USA; [2]Broward County Mosquito Control, 1201 WAirport Road, Pembroke Pines, FL 33023, USA

Abstract

The Zika virus outbreak in Miami-Dade, Florida, in 2016 heightened our awareness of the presence, expansion and vectorial capacity of invasive *Aedes* mosquitoes in the USA. The Asian tiger mosquito, *Aedes albopictus*, has not been implicated as a vector during an arbovirus outbreak in the USA, however this species is established in at least 40 states, including Hawaii, and has a greater geographic distribution in North America than *Ae. aegypti*. Current mosquito control strategies may be insufficient to control *Ae. albopictus* in the USA, given the affinity of this species for cryptic habitats and diurnal activity. Accordingly, control efforts have had to be creative. For example, more mosquito abatement districts are modifying truck-mounted sprayers to apply larvicides to cryptic habitats in, inspecting and treating corrugated extension gutters as part of source reduction campaigns, and improving ultra-low volume (ULV) adulticiding methods to achieve higher reduction in adult *Ae. albopictus*. This review discusses *Ae. albopictus* surveillance, action thresholds for intervention, rising concerns on resistance, public involvement, challenges facing mosquito control professionals and how to address gaps for a sustainable surveillance and management programme.

Keywords: Invasive *Aedes* species, Expansion, Arbovirus, Public health, Insecticide resistance, Integrated pest management

Review Methodology

CAB Abstracts and Google Scholar were searched with the terms *Aedes albopictus*, mutations, arboviruses, expansion, *Aedes albopictus* reviews, *Aedes albopictus* surveillance, and resistance. I also spoke to colleagues to get an update on their current projects.

Introduction

One of the discussions at the 2014 Northeastern Mosquito Control Association meeting in Cambridge, Massachusetts, USA, turned to which mosquito species is the most difficult to manage. We did not understand at the time how to control *Aedes albopictus* (Skuse), the Asian tiger mosquito, in corrugated gutter extensions, a vertical pipe that carries rainwater from the guttering of a building to the ground or a storm water runoff system (Figure 9.1) [1], let alone in numerous other containers in which they lay their eggs [2]. *Ae. albopictus* is difficult to control

*Corresponding author: iunlu@broward.org

because it is found in our urban and suburban yards. The eggs of saltmarsh mosquitoes hatch when flooded by predictable rainfall or tides [3]. Once a coastal area is flooded, mosquito control districts will schedule large-area aerial larviciding. Unlike mosquitoes that lay their eggs in saltmarshes and inland floodplains, *Ae. albopictus* exploits natural and artificial containers as oviposition sites, posing unique challenges to mosquito control [4, 5]. In its native range of southeastern and eastern Asia, this species lays its eggs in tree holes, bamboo nodes and artificial containers [6]. In addition to its use of containers as oviposition sites, the affinity of this species for urban environments where containers such as buckets and tyres are scattered and abundant, complicates larval control measures [2, 7, 8]. *Ae. albopictus* exhibits skip-oviposition, a behaviour wherein a species may oviposit eggs from a single gonotrophic cycle in multiple containers [9], therefore the techniques we use to control saltmarsh and floodwater mosquitoes cannot be translated to this species [10]. The most common method to control *Ae. albopictus* larvae is source reduction by 'tip and toss'. However, this technique is labour-intensive, time-consuming and costly [5]. Most mosquito control districts lack the resources to perform season-long, door-to-door source reduction in their jurisdictions [5]. Containers including but not limited to buckets, plant pot saucers, tyres and recycle bins are located within private residential parcels that are often inaccessible to mosquito control personnel (Figure 9.2). Residents are usually at work during the day, and some residents have guard dogs and gates that may preclude access. Once mosquito inspectors gain access to residents' properties, they have to remove or treat each container. Despite the time spent educating residents about source reduction, new containers frequently appear in their backyards [11, 12]. Mosquito control operators may hesitate entering private property or asking permission to perform inspections for safety concerns [13]. Source reduction often includes trash collection, but inspectors are not responsible for eliminating these kinds of containers (e.g. potato chip bags) (Figure 9.3).

We need to think outside-the-box to find time- and cost-effective methods for *Ae. albopictus* management. New or improved surveillance techniques need to be implemented to help determine accurate action thresholds for larval and adult mosquito control. Routine resistance monitoring has to be incorporated into mosquito control. Finally, public education has to be reconfigured using active education tools. Addressing these gaps will lead to more sustainable surveillance and management programmes.

Fig. 9.1. An example of a CES within a residential backyard Mercer County, New Jersey, USA. This CES is holding water and immature *Ae. albopictus*.

Public Health Impact

The global spread of *Ae. albopictus* has mostly occurred over the last three decades, following the earlier spread of *Aedes aegypti* and *Culex pipiens*. It is the third example of the spread of a mosquito vector caused by human activities [10, 14, 15]. *Ae. albopictus* has been incriminated as a primary or secondary vector of several viruses, particularly those that cause dengue fever (DENV), dengue haemorrhagic fever and chikungunya (CHIKV) [16–19]. *Ae. albopictus* is implicated less of a disease vector and more of a nuisance [20, 21]. However, with the emergence of *Ae. albopictus*-adapted CHIKV in 2006–2007 in the Southwestern Indian Ocean, India, Central Africa and Italy, the vector status of *Ae. albopictus* has irreparably changed [22].

Ae. albopictus are far more opportunistic blood feeders than *Ae. aegypti* (Linnaeus) [23, 24]. *Ae. aegypti*, on the other hand, tends to feed preferentially on humans and multiple times in a single gonotrophic cycle [25]. Because of its opportunistic feeding behaviour, *Ae. albopictus* transmit emerging viruses from animals (wildlife and domestic) to humans, particularly in newly occupied areas with people, such as Central Africa, a known hot spot of viral diversity [26]. *Ae. albopictus* is a competent vector of 26 viruses in the families Flaviviridae (*Flavivirus*), Togaviridae (*Alphavirus*), Bunyaviridae (*Bunyavirus* and *Phlebovirus*), Reoviridae (*Orbivirus*) and Nodaviridae (*Picornavirus*) [27, 28]. Laboratory demonstration of vector competence is one factor that indicates a role in virus transmission. In addition, geographic and seasonal overlap, host blood meals and isolation of virus from field-caught mosquitoes may incriminate a species as a vector. *Ae. albopictus* is recognized as a primary vector in the transmission of DENV and CHIKV [17, 29]. The role of *Ae. albopictus* in the transmission of DENV was first recognized in tropical Asia, where an illness clinically compatible with DENV was reported prior to the introduction of *Ae. aegypti* in the mid-nineteenth century [22]. At this time, *Ae. albopictus* was likely the main vector of DENV, even if epidemics were rare and geographically limited. During the Second World War, *Ae. albopictus* was suspected in the transmission of DENV during major outbreaks that in Hawaii and Japan [30]. *Ae. albopictus* was also suspected to have a significant role in the transmission of DENV in China and to be responsible for all of the transmission in some Chinese provinces [31]. *Ae. albopictus* was the sole vector during the 1977–1978 La Réunion epidemic that affected approximately 30% of the population. *Ae. aegypti* was rare on this island after the 1950s, playing no role during this outbreak [32]. In Central Africa, where the incidence of DENV recently increased [33], the virus was first isolated from *Ae. albopictus* females that were collected during an outbreak in Libreville, Gabon, while *Ae. aegypti* specimens tested negative [34].

Until recently *Ae. aegypti* has considered the primary vector of CHIKV in Asia, and *Ae. albopictus* was considered a secondary vector [22]. However, *Ae. albopictus* was the main and the only vector of CHIKV during the 2005–2007 outbreak in the Indian Ocean islands (Madagascar, Mauritius, the Seychelles, La Réunion and Maldives); Sri Lanka, India, Indonesia and Malaysia [22, 35]. *Ae. albopictus* was also confirmed as a vector during the first outbreak in Europe, reported in Italy in August 2007 [33]. In Central Africa, a major urban outbreak of CHIKV was reported in the Democratic Republic of Congo in 2004 and in Cameroon and Gabon in 2007 [33]. Entomological surveys revealed that *Ae. albopictus* was the primary vector in the Gabonese outbreak [36]. A strain of CHIKV with a single point mutation was isolated during the 2007 outbreak. In this mutation, E1-A226V, a valine residue replaces alanine at position 226 of the E1, the envelope gene has presumably resulted from independent viral exposure to *Ae. albopictus* and selection by this species [37–39]. This mutation provides a selective advantage for the replication and transmission of this CHIKV strain by *Ae. albopictus* [37]. The emergence of this strain has demonstrated that *Ae. albopictus* is able to interfere with the native virus–vector system and enhance viral emergence [40]. Introduction of new mosquito species increases the probability of new virus-mosquito interaction, therefore increases probability of disease transmission.

Zika virus (ZIKV) was first isolated in Uganda in 1947 [41] and diverged into distinct West African, East African and Asian genotypes that are responsible for cases globally [42]. Both ZIKV and CHIKV appear to have spread from

Fig. 9.2. Common containers for *Ae. albopictus* use as larval habitats in the urban areas.

Fig. 9.3. An example of a residential backyard and challenges for mosquito control professionals to conduct door-to-door source reduction campaigns.

East Africa to Asia at approximately the same point within the last century and diverged into distinct Asian genotypes [43]. This genotype was identified from a Malaysian isolate in 1966 and is thought to be a progenitor of subsequent Asian cases [44]. The epidemic capacity of ZIKV was revealed by the 2007 breakout in Micronesia [45]. The Asian strain of ZIKV has since expanded to Pacific islands and the Americas. Calvet *et al.* [46] reported that the ZIKV responsible for locally acquired cases in Brazil in 2016 shared over 97% of its genome with the ZIKV strain responsible for the French Polynesia outbreak in 2013 [46]. Historically, *Ae. aegypti* has been the primary vector of endemic ZIKV, but *Ae. albopictus* has been shown experimentally to be an equally competent vector [47, 48]. The spread of ZIKV to the Americas is a concern because of the relationship *Ae. albopictus* has with humans and its diapausing eggs which tolerate colder winter temperatures. *Ae. albopictus* has the ability to expand its range to northern regions of the hemisphere [49]. *Ae. albopictus* was the sole vector of the CHIKV epidemics in La Réunion and Mauritius [50] and remains a concern for transmission of ZIKV in North and South America [51].

While infection and transmission rates were not determined, investigators in the Netherlands demonstrated that *Ae. albopictus* can experimentally transmit yellow fever virus (YFV) [52]. This was corroborated by more recent studies, for example, Johnson *et al.* [53] infected *Ae. albopictus* from Brazil with YFV, and they reported a lower percentage of disseminated infection for *Ae. albopictus* compared with *Ae. aegypti*. However, the difference was not significantly different. A smaller proportion of salivary gland tissue was infected in *Ae. albopictus* compared with *Ae. aegypti*. *Ae. albopictus* is a competent vector, however it likely plays a less important role in urban transmission because of its broad host range [53].

Ae. albopictus is a competent vector of West Nile virus (WNV), although its role in epidemic transmission is unknown [54–56]. When North American strains of *Ae. albopictus* were fed on 2- to 3-day-old chickens, infected with the New York Crow-397–99 strain of WNV, between 36 and 92% of mosquitoes expressed WNV in salivary expectorate [57].

Integrated pest management and the importance of surveillance

Chemical insecticides have been heavily used for mosquito control since the 1940s [58] and led to resistance problem in insects that serve as vectors of arthropod-borne diseases [59]. The ban on dichlorodiphenyltrichloroethane (DDT) from the USA market in 1972 catalysed the need for integrated pest management (IPM) and additional insect control methods and products [60]. In 1972, the United States Congress initiated a programme for environmental protection and proposed IPM measures intended to support responsible chemical pesticide use [58]. The American Cooperative Extension Service initiated organized IPM programmes [58]. These events led to IPM-driven surveillance programmes in every mosquito control operation in the USA. Mosquito surveillance involves collecting mosquitoes in areas of interest, thereby providing a better understanding of mosquito biology, ecology and behaviour. The primary purpose of mosquito surveillance is to determine species composition, abundance, infection rates and distribution. Collecting and counting eggs, larvae and adult mosquitoes enables a mosquito abatement district to decide if mosquito control is necessary. Because collecting mosquito eggs and larvae is labour-intensive and expensive, most abatement districts rely on adult mosquito surveillance by deploying traps several times a week [61, 62].

Traps are important tools to monitor and collect female mosquitoes of various physiological stages. Host-seeking females use visual, olfactory and physical stimuli to find a host. Exhaled carbon dioxide and skin secretions play a major role in finding host location [63, 64]. Carbon dioxide, alone [63] or as a supplement to light [65], enhances mosquito trapping and is an important stimulus for many haematophagous (blood feeding) Diptera [66].

Adult surveillance became more standardized and efficient following the development of the New Jersey light trap (NJLT) in 1932 [67]. The NJLT soon became the gold standard for monitoring mosquito populations and has since been used consistently by all counties in New Jersey [68]. In the 1960s the Centers for Disease Control and Prevention (CDC) miniature light trap was developed to collect host-seeking mosquitoes [69]. This trap utilizes light and CO_2 to attract mosquitoes. The standard NJLT and the CDC miniature light trap have been widely used for arbovirus (arthropod-borne virus) surveillance in the USA [70, 71]. However, since these traps target host-seeking nulliparous mosquitoes [72, 73], they are unlikely to collect mosquitoes that have been exposed to arboviruses [74]. Unfortunately, diurnal species such as *Ae. albopictus* respond poorly to these traps [75–77], possibly because their apertures are 1.5 m above the ground and are operated at night using a light source as an attractant. Because *Ae. albopictus* bites during the day and seeks hosts near ground level [78], this species is not attracted to light sources. Reiter [74] introduced a portable battery powered gravid trap (GT) and demonstrated that at least 90% of the *Culex* mosquitoes collected with this trap were gravid (contained developing eggs) [74]. Lampman and Novak [79] tested sod-infused GTs to collect *Ae. albopictus* and demonstrated only some attraction of the species to the trap [79]. Although GTs seem more promising than light traps, they are still ineffective tools for *Ae. albopictus* surveillance.

Surveillance of immature Ae. albopictus

Oviposition cups are inexpensive and can be any small container, generally dark-coloured with water and a partially submerged substrate for oviposition [80]. Dechlorinated tap water is commonly used to collect eggs, and water infused with hay, grass or leaves is more effective than tap water alone [80, 81]. For example, oak leaf infusion is a highly effective attractant for gravid *Ae. albopictus* [82]. Oviposition cups can be deployed easily, but should not be used for more than a week at a time if they are non-lethal because adult mosquitoes will emerge by that time [83]. Therefore, adequate sampling requires routine trapping at sites representative of the habitats in the target sites. [84].

Oviposition cups are a traditionally useful tool for collecting eggs, often in terms of the presence or absence across seasons [5]. The skip-oviposition behaviour of container-inhabiting *Aedes* and the presence of other containers means that they alone cannot reliably project adult population numbers [85], therefore collections should be made and evaluated tandem with adult collections [5]. However, these methods are labour-intensive, and they are often not the primary surveillance system for mosquito populations [85, 86]. Oviposition cups, while useful, should be used after consideration of their limitations in the field [85]. Deploying artificial containers to sample eggs creates competition between these and existing natural or artificial containers in the selected environment, and this should be taken into consideration when oviposition cups are used as a surveillance tool [1, 87]. Once collected, eggs must be counted under a microscope, usually on dirty or worn collection substrates, and larvae should be reared to at least the third instar for identification [85]. Mosquito larvae and pupae can also be collected by direct sampling with dippers, nets or water traps. Such samples are commonly measured by counting larvae per dip [88]. No standardization for dipping technique is in common usage; as such, dipping is not a particularly precise sampling method, since differences in the habitat, mosquito larvae behaviour, stage differences, and other factors can influence sample numbers [86]. Dippers are still effective for deep standing water, such as recycling bins and barrels, whereas pipettes or turkey basters are more suited for slender containers like hollow fence posts and narrow tyres. Monitoring the number of container-inhabiting *Aedes* mosquitoes has historically relied on sampling immatures [89], but this presents unique challenges [90].

Each index used to quantify *Aedes* populations has a different objective. The house index (HI) represents the percentage of houses with larvae present. The container index (CI) represents the percentage of water-holding containers that contain larvae. The Breteau index (BI) is the number of containers containing mosquito larvae per 100 houses surveyed. Based on several studies immature container indices have failed to correlate with adult catches in BGS traps, and they have not correlated with nuisance action thresholds [2, 91–94]. In four north Queensland, Australia towns relationships between immature and adult female densities were evaluated using BI, HI, CI, larval density and adult productivity indices. None of these indices of larvae predicted adult abundance [91].

Surveillance of adult Ae. albopictus

Almost 20 years after the introduction of *Ae. albopictus* to the United States, surveillance, distribution and population assessments were still challenging [77, 95]. Therefore, *Ae. albopictus* populations have most likely been underestimated. In 2008, Mercer and Monmouth Counties, New Jersey, USA, purchased the BioGents Sentinel® (BGS) trap (Biogents AG, Regensburg, Germany), which was originally designed to attract *Ae. aegypti* [96], to determine the abundance of *Ae. albopictus* in their jurisdiction [75, 97]. Mercer County detected *Ae. albopictus* larvae in 2004 but collected limited numbers of adults from light traps, resulting in no adult surveillance data [97]. After 2008, the number of BGS traps used for *Ae. albopictus* increased (five times) in the county. Based on initial findings [95, 77], subsequent studies evaluated the efficacy of BGS traps for collecting *Ae. albopictus* [75, 98–101]. The BGS trap uses a combination of visual cues, chemical attractants and convection currents to capture host-seeking mosquitoes. Studies have shown that a BGS trap without chemical attractants may still capture invasive *Aedes* mosquitoes, which indicates that visual cues play a key role in attracting these mosquitoes [77, 102]. BG-Sentinel traps use mainly a BG Lure (BioGents GmbH, Regensburg, Germany supplied by AgriSense) containing several molecules found in human sweat including lactic acid, fatty acids and ammonia.

The BGS trap is 36 cm wide and 40 cm tall. It collapses to 10 cm and contains a 12 V battery. When turned on, the fan helps propel the chemical attractant up through the white cloth cover [96, 98, 99]. Host-seeking mosquitoes approach the scent and are aspirated through the trap aperture and fan, collected in a catch bag below.

The development and commercial availability of the BGS trap provided a much-needed surveillance tool to monitor expanding populations of invasive *Ae. albopictus* in North America. Because of their high capture rates and specificity for these species, BGS traps have quickly become the gold standard for *Aedes* surveillance [77, 95]. However, continuing loss-of-function and inconsistencies during field use limited the utility and reliability of this trap for researchers and mosquito control personnel [103]. The need to reduce construction defects and increase the field durability of the trap led to the creation of the next generation of BGS traps. Newly designed PVC-coated BGS Prototype (BGS2P) trap, with and without a variety of lures, compared against the original BGS1 for efficacy in four different geographical areas in the USA [98]. The BGS1 trap collected two to three times more *Ae. albopictus* than the prototype in Florida, Louisiana, New Jersey and Virginia, regardless of the type of lure or CO_2 used for the investigations [98]. In a study conducted in Jacksonville, Florida, USA, the BGS1 collected twice as many *Ae. albopictus* than the BGS2P (referred to as BGS2) [100]. Additionally, these authors have also shown that the BGS1, without the addition of any attractants, is capable of collecting 2.3 times more *Ae. albopictus* than the BGS2P [100]. Studies in Louisiana and New Jersey also showed the BGS1 collected about twice as many *Ae. albopictus* than the BGS2P without attractants.

In 2016, we compared the field efficacy of the BGS2P, BGS1 and BGS2 (a new design based on data collected from Florida, Louisiana, New Jersey and Virginia) (Figure 9.4) as a surveillance tool for *Ae. albopictus* [104]. Previous studies suggested that the textile of the trap repelled mosquitoes, which could be addressed with a period of ventilation [99]. We found that the BGS2P was the least efficient for sampling *Ae. albopictus* of the three options, supporting the majority of the previous findings [98, 100]. Arimoto and colleagues [100] conjectured that the coating of the PVC-coated textile body was repellent, and the colour did not contribute to the poor performance of the BGS2P. A follow-up study that allowed the fabric to ventilate for 1 week after trap usage found that the BGS2 caught the highest number of *Ae. albopictus* adults, although these numbers were not significantly higher than the BGS1. On the other hand, Akaratovic and colleagues [99] found no significant difference in total female mosquito collections among the three BGS models using the 3-compound BG cartridge lure [98, 105, 106]. We still do not know why BGS2P is not performing as well as BGS1 and BGS2. However, Biogents AG reverted to the original material they use for BGS1 and kept the blue colour for BGS2. These versions are available for purchase (https//:bg-sentinel.com/en/news.html#new_BG-Sentinel_2). For research and operational purposes both are adequate [104]. The BGS2 is more user-friendly and dependable due to the structural improvements. The combination of retaining the blue colour, the PVC body composition, elimination of mounting poles to a self-supporting structure and the addition of an air-actuated flap in the event of battery failure improved the BGS2. Having a more durable trap is important for effective surveillance and mosquito management programmes. Fewer trap failures allow researchers to work with more robust data sets, saving time and money on repairs.

Rochlin and colleagues [107] tested Mosquito Magnets (Woodstream Corp, Lititz, PA, USA, Liberty and Executive models) when they did not have the success they expected from BGS traps. Researchers found that seasonal trends in mosquito collections were relatively consistent for the Mosquito Magnet in their jurisdiction. They reported a gradual increase in *Ae. albopictus* numbers from mid-May and peaking in late July. A gradual decline from August onwards was characterized by several brief, isolated peaks that were driven by unstable weather patterns (lower temperature and higher winds), especially in 2013. Similar trends were observed for BGS traps albeit with much lower numbers in 2012. The Mosquito Magnet also provided information on *Ae. japonicus* and *Ae. triseriatus* populations, which reached their peaks earlier in the season and then declined in August with a consistent trend between the two seasons. BGS traps collected too few of these two species to determine temporal trends [107].

Another trap worth mentioning for invasive *Aedes* species is the Gravid *Aedes* Trap, (GAT), which relies on its exploitation of the 'fly to the light' strategy of capturing insects [108]. This strategy has proven to be more efficient at capturing gravid *Ae. aegypti* than standard sticky ovitraps in North Queensland, Australia [109]. Because the GAT kills gravid females, it also acts as a lethal ovitrap allowing mosquito control professionals to conduct surveillance while reducing mosquito populations. Most lethal ovitraps rely on the use of pesticide-treated ovistrips (oviposition media) and they do not capture adult females [110]. Although GAT is practical and costs less than most traps, field studies assessing its effectiveness at capturing gravid *Ae. albopictus* mosquitoes have yet to be conducted since there are only a few studies available. Johnson and colleagues [89] evaluated

Fig. 9.4. The three variations of BGS traps: (a). original BGS1, (b) BGS2 prototype and (c). BGS2.

GATs in a wide variety of geographic areas. The BGS trap outperformed the GAT in total collections during the study period, but this was expected given the BGS trap is designed to attract host-seeking females, and the GAT is designed to attract gravid females. No significant differences in the number of female *Ae. albopictus* captured were observed between GAT (weekly counts) and BGS traps (daily counts) with and without rain cover addition. The BGS trap captured significantly more male *Ae. albopictus* than the GAT in Hammond Island, Australia. In contrast, the BGS trap captured significantly more female *Ae. albopictus* than the GAT in Trenton, New Jersey, USA [89].

CDC autocidal gravid ovitraps (AGO) are based on the same principle as the GAT; they contain a sticky surface to collect mosquitoes with no lethal agent. A study was conducted in 2017 in New Jersey, USA, to evaluate if AGO traps could reduce *Ae. albopictus* populations. We presented our preliminary results at the 2017 national meeting of the American Mosquito Control Association (AMCA). We collected 9939 mosquitoes from June to October from 24 AGO traps. These collections included 9129 *Ae. albopictus* (92.7%), 236 *Culex* species (2.4%) and 53 *Ae. japonicus* (0.5%) (Unlu *et al.* unpublished data). Because the CDC AGO contains a thick layer of glue, we were concerned about our ability to identify the specimens. We noted each specimen we could not identify, and 395/9939 (3.9%) were unidentifiable because they were embedded in the glue board. Later this paper will discuss this trap in the control section, and specifically for this reason, there is a great deal of use if research shows them to be effective tools for the reduction of *Ae. albopictus* populations. Conducting simultaneous surveillance and control would reduce the operational stress of conducting each separately.

Action threshold

Controlling the biting mosquito population is the highest priority for any mosquito abatement programme. The goal of IPM is to keep mosquito populations below a certain threshold that is influenced by an array of scientific, political and social factors [67]. This threshold is sometimes defined as the biting frequency that residents in an area are willing to tolerate during a specific period of time. Previous studies have found this frequency to vary greatly among different communities, generally increasing from urban to suburban areas [111, 112]. Practical values for treatment thresholds are usually gathered from the experience of mosquito abatement organizations that incorporate these conditions. Generally, action thresholds are more commonly used for research projects [113–115]. However, the need for this value is important in response to high numbers of *Ae. albopictus*, and to understand the efficacy control strategies.

Ae. albopictus management tools and strategies

Ae. Albopictus persists in cryptic habitats. Unlu *et al.* [1, 114] reported this species the most abundant species in corrugated extension spouts (CESs) [1]. Container-inhabiting *Aedes* species prefer cryptic larval habitats, which are difficult to treat with traditional mosquito control techniques [14]. Invasive *Aedes* species also exhibit skip oviposition and oviposit in a wide number and variety of artificial containers within peridomestic habitats [9, 64], pose a great deal of difficulty to treat containers housing immature mosquitoes.

Since container-inhabiting *Aedes* mosquitoes require unconventional tools, this section will start with geneticbased population control methods. These methods are strategies that capitalize on the transmission of heritable, deleterious elements throughout a population. Of these methods, the sterile insect technique (SIT) has seen success with several pests of agriculture and livestock over its 50-year history [116]. Sterile male technique involves the controlled release of adult males that have been sterilized by radiation or by chemical methods. Sterilized males produce nonviable sperm, and thus, females that they inseminate will lay nonviable eggs. In practice, radiation techniques have been observed to induce somatic damage in male mosquitoes, which can negatively affect their reproductive performance [117]. A genus of gram-negative bacteria called *Wolbachia* produce a similar effect by inducing reproductive incompatibility between infected and uninfected adults [118]. *Wolbachia* are obligatory intracellular bacteria used for a variant of SIT known as incompatible insect technique (IIT). *Wolbachia* are maternally inherited and infect natural arthropod populations by inducing male-killing, feminization, parthenogenesis and, most commonly, unidirectional and bidirectional cytoplasmic incompatibility (CI) [119]. Males should be released for the successful application of IIT against target species based on a unidirectional CI approach. Releasing females would contribute to natural populations and lead to fertile mating between released males and females, resulting in local populations infected with the *Wolbachia* population [119]. The resulting public relations would be disastrous. Good example of the risk of releasing females is reported by Dobson and colleagues [120], in the populations of *Ae. albopictus* collected originally from Thailand and Houston (Koh; Thailand, pre-1970 and superinfected Houston Hou; Texas, 1986) resulting in increased fecundity associated with *Wolbachia* infections in the laboratory trials [120]. Horizontal transmission is a concern when releasing infected males lest *Wolbachia* spread and reduce incompatibility, but no horizontal transmission by males was observed in that study. Dutra *et al.* [121] tested the efficacy of using*Wolbachia* infection to make *Ae. aegypti* populations less capable of circulating ZIKV [121]. They reported that mosquitoes harbouring *Wolbachia* were less likely to possess Zika-positive saliva. Infected mosquitoes still exhibited lower viral titres, reducing the capacity for further transmission. Blagrove *et al.* [122] found that the introduction of high frequency *Wolbachia* wMel (*Wolbachia pipientis*) transinfection in wild populations of *Ae. albopictus* may prevent this species from transmitting diseases including CHIKV and DEN [122]. Successful transfers such as *Ae. aegypti* infected with a *Wolbachia* type from *Ae. albopictus* [123], helps to promote the development of control programmes using the Wolbachia-based strategies [118]. Most of the work with *Wolbachia* was undertaken in the laboratory [120, 123, 124] but further work is needed to show the efficacy of the method in the field. Following proof-of-concept studies, public acceptance as in Mediterranean fruit fly programme is required for transgenic mosquito releases. One advantage of *Wolbachia*related studies for public perception is that when males are released, they are cytoplasmically incompatible with the targeted population and the released transgene has a reduced probability of becoming dominant within the wild-type population [125].

The release of insects carrying a dominant lethal (RIDL) is derived from SIT and could provide a new solution to *Ae. albopictus* management [126]. For example, female-specific RIDL strains developed for *Ae. aegypti* and *Ae. albopictus* render females flightless [127]. Ideal objectives for genetic control strategies can be fit into two categories: (1) population suppression (e.g. SIT) by reducing the abundance of competent mosquito species and (2) population replacement, reducing the overall vectorial capacity [128]. Both of these are preventative and costly, and therefore most effective if implemented before an outbreak.

Reports on *Ae. albopictus* often claim that 'it thrives in a bottle cap'. While true, this claim ignores the diversity of containers this species uses as oviposition sites. *Ae. albopictus* appear to exhibit restless movement from one oviposition site to another, placing a few eggs at a time [129]. In laboratory experiments *Ae. albopictus* females start laying eggs in favourable containers without obstruction between the entrance and the water, and they continue to lay eggs until most or all mature eggs are deposited [87]. The oviposition behaviour of *Ae. albopictus* females is also affected by changes under environmental conditions because they are sensitive to short photoperiods at pupal and adult stages [130]. When winter temperatures are too low for larval development or the day length shortens, some *Ae. albopictus* will deposit overwintering eggs that will not hatch until spring [87]. Because diapausing eggs will not hatch immediately, females will change their ovipositing behaviour to protect eggs from long and harsh winter conditions. A shift in *Ae. triseriatus* (Say) to favouring containers correlating with a seasonal change, was reported with conspecifics [131]. Unlu *et al.* [1, 114] study agrees with this finding and suggested that *Ae. albopictus* egg abundance switched in late summer from corrugated extension gutters to open containers such as buckets [1]. The first generation of *Ae. albopictus* in temperate areas spends 20–26 weeks as eggs and about 7 weeks as larvae [130]. This long development time provides a great opportunity to control this species by targeting the eggs and larvae of a unique generation. The mechanical removal of artificial larval habitats, where females oviposit during the fall, would be an efficient technique to suppress the first generation of mosquitoes in the spring. Because mosquito control personnel cannot contact every person in urban and suburban areas where biting occurs, communicating with residents to modify habits is difficult [130]. Controlling larvae of *Ae. albopictus* is further complicated by the diversity of containers and the preference of this species for cryptic habitats [1, 132]. We have also shown in previous studies that immature *Ae. albopictus* numbers were higher in cryptic CESs (containers positioned on the ground horizontally) than in open containers (containers positioned on the ground vertically) during the peak season of August. Because *Ae. albopictus* has an affinity for cryptic habitats, control using an area-wide application of larvicides has been found to be ineffective [105, 133, 134].

Despite these challenges, a number of traditional and novel techniques are used to control *Ae. albopictus* larvae. Many containers, such as bird baths, buckets, tyres and recycle bins, are within private residential parcels not easily accessible to mosquito control personnel. Direct application of larvicides and pupicides is the best way to eliminate immature *Ae. albopictus*. However, these efforts during sanitation campaigns or door-to-door source reductions are labour- and cost-intensive. When source reduction is the only option, *Bacillus thuringiensis israelensis* de Barjac (*Bti*), spinosad and methoprene are commonly used active ingredients [133, 135–138]. Farajollahi *et al.* [137] showed water-dispersible granular formulations of *Bti* (VectoBac® Valent Biosciences, Libertyville, IL, USA) were effective against *Ae. albopictus* during their open-field trial in Mercer County, New Jersey, USA. A liquid formulation of *Bti* (Acrobe™, American Cyanamid, Princeton, NJ, USA) reduced larval numbers for 47 days in Florida [139]. Boyce *et al.* [140] reviewed 14 studies and found *Bti* can be effective in controlling the larvae of *Ae. albopictus* and *Ae. aegypti* in a variety of containers [140]. *Bti* typically kills larvae within 24–48 h of treatment [141], and several studies have reported a residual effect for 2–4 weeks [137, 141, 142]. These results suggest that *Bti* is effective only for containers individually treated. The number of oviposition sites may be too great and their locations too cryptic to justify large-scale reliance on *Bti* to treat these sites.

(*S*)-methoprene (Altosid) is a mimic of juvenile hormone III (naturally found in mosquitoes) [143] and have been used to kill immature mosquitoes since the mid-1970s. It targets fourth instar larvae by interfering with diploid cell division and apoptosis in polytene cells by preventing normal development. Usually (*S*)-methoprene may take up to 7 days to kill larvae, depending on the age structure (most effective against late instars) [144]. Nelder *et al.* [136] reduced *Ae. albopictus* numbers by 95% in the field for at least 32 days by the tandem use of Agnique®MMF (surface film, Cognis Corporation, Cincinnati, OH, USA) and Altosid® (*S*-methoprene, Central Life Sciences, Schaumburg, IL, USA) [136]. Application of one Altosid pellet (0.006 g) per tyre provided 116 days of *Ae. albopictus* control in Florida, and 2 g of Altosid pellets provided 150 days of control against *Ae. albopictus* in small bowls in Louisiana [139, 145]. Noncirculating hydroponic tanks of lettuce treated with *Bti* and methoprene formulations demonstrated effective control against *Ae. albopictus* larvae and pupae for the duration of the lettuce crop (4–5 weeks) in Hawaii [146].

Explaining how insect growth regulators (IGRs) work for the public is difficult because IGRs do not kill immediately. IGRs, also known as insect developmental inhibitors, are effective products against developing larvae, as many studies in laboratory and field settings have demonstrated [136, 141]. The efficacy of IGRs depends on the target species and is influenced by the time of susceptibility during development and effectiveness [147]. IGRs are therefore less effective against asynchronous larval populations [147]. In contrast, although it is an IGR, chitin synthesis inhibitors, which act during ecdysal changes are just as active against both synchronous and asynchronous broods [147]. Juvenoid compounds are formulated, diverse time-release compositions that continue releasing active ingredient over time after a single application. Time-release formulas are more expensive, but this is often outweighed by the time and effort saved from repeated applications.

Chitin is a polymer of *N*-acetyl glucosamine that forms the protective exoskeleton of all arthropods and is replaced periodically during growth and development. Chitin biosynthesis starts with the disaccharide trehalose, and culminates in the polymerization of the *N*-acetyl glucosamine subunits by chitin synthase to produce chitin microfibrils [148]. Chitin in the old exoskeleton is degraded by chitinases, deacetylases and hexosaminidases and recycled. Chitin synthesis has been used as a target for developing biorational insecticides such as benzoylphenyl ureas. Diflubenzuron was the original compound. Several benzoylphenyl ureas with diverse activity spectra have since been synthesized and widely used for pest control. Newer pesticides target both chitinase and chitin synthase and other novel sites as well [148]. Mulla *et al.* [149] showed novaluron (Rimon Tehcnical, Makhteshim-Agan, Raleigh, NC, USA) to possess exceptional capacity for emergence inhibition against *Ae. aegypti* in Thailand [149]. At a high concentration (1 mg/l), up to 96% inhibition was maintained for 190 days. A study in Mexico found that a similar dosage of novaluron inhibited 97% emergence of *Ae. albopictus* [150]. Novaluron has been used successfully to control *Ae. aegypti* in Brazil. [151]. The lowest lethal concentration of novaluron was as effective as higher concentrations during semi-natural trials against *Ae. albopictus* and other mosquitoes of public health importance. [150]. Novaluron is not federally registered for use against mosquitoes in the USA.

Pyriproxyfen is a juvenile hormone mimic that suppresses embryogenesis, metamorphosis and adult emergence in insects [152]. The possibility of using pyriproxyfen is gaining popularity with researchers [102, 105, 153, 154]. Pyriproxyfen is virtually non-toxic to mammals and birds, non-carcinogenic [155] and is recommended by theWorld Health Organization (WHO) for use as an insecticide in drinking water sources and containers [156]. It also exhibits poor adsorption into soil and movement into groundwater [157]. While pyriproxyfen has a short half-life in soil [157], it adsorbs readily to suspended organic matter and can remain active for up to 2 months [157]. Despite this persistence, pyriproxyfen has a low risk of selecting for resistance because toxicity is maintained [158]. It exhibits excellent efficacy against a variety of mosquito species, especially container-inhabiting species such as *Ae. aegypti* and *Ae. albopictus* [141, 159, 160]. The extremely low LC_{50} values (LC_{50} of 0.012 ppb against *Ae. albopictus* and 0.023 ppb against *Ae. aegypti*) [161] of pyriproxyfen against these species [162] suggests that a few ultra-low volume (ULV) drops of pyriproxyfen within a container may deliver a lethal concentration. Because of its favourable environmental profile, low risk to vertebrates, efficacy and persistence, pyriproxyfen is an ideal candidate for truck-mounted area-wide larviciding efforts in urban and suburban settings [153, 154].

Autodissemination is a time- and cost-effective technology designed based on attraction to and dispersal from an autodissemination unit (usually treated with pyriproxyfen). This pest management method uses contaminated insects to transfer a lethal application horizontally or vertically by mating, oviposition, aggregation or other association [163–165]. The effectiveness of autodissemination is contingent on three conditions: (1) mosquitoes are attracted to the autodissemination station; (2) sufficient insecticidal product is transferred to the mosquitoes and (3) contaminated mosquitoes disseminate to target habitats [163, 166]. Gaugler *et al.* [163] met these three criteria for *Ae. albopictus* during semi-field trials, in which containers were detected to be contaminated by mosquito dissemination. Snetselaar [167] accomplished >90% larval mortality with *Ae. aegypti* by using synthesis of an entomopathogenic fungus, *Beauveria bassiana*, as an attractant and pyriproxyfen, the most commonly used chemical for autodissemination [167]. IGRs such as pyriproxyfen do not cause immediate mortality or incapacitate adults [168], but they remain effective against immature mosquitoes at low concentrations [163, 166]. In particular, the skip-oviposition behaviour of both *Ae. albopictus* and *Ae. aegypti* facilitates autodissemination by distributing eggs from a single gonotrophic cycle in multiple sites [82]. Emergence of *Ae. aegypti* in Peru was inhibited by 42–98% *via* pyriproxyfen autodissemination. Abad-Franch *et al.* [169] recently tested emulsifiable pyriproxyfen against *Ae. aegypti*, *Ae. albopictus* and *Culex* spp. in a tropical climate, and found a ten-fold reduction in treatment sites compared with control in adult mosquito emergence. Large-scale field trials of autodissemination strategies still require further evaluation before implementation. A novel approach called autodissemination augmented by males has shown reduced adult populations of *Ae. albopictus* as a direct result of pyriproxyfen-contaminated males delivering lethal applications to oviposition sites [165]. In most studies, pupal mortality has accounted for the majority of immature and adult population reduction *via* autodissemination. In a study of an urban site in Trenton, New Jersey, USA, the effects of deployed autodissemination stations in *Ae. albopictus* hot spots (small sites of high adult counts) was evaluated through monitoring of eggs, larvae, pupae and adult populations [105]. Targeting these hot spots reduced counts of eggs, larvae and pupae in the treatment sites compared with control sites.

Pyriproxyfen may also be applied in area-wide applications by truck-mounted sprayers. We combined the larvicides VectoBac WDG (*Bti*) and NyGuard IGR® (pyriproxyfen, MGK Chemical Co., Minneapolis, MN, USA) with the adulticide Duet™(sumithrin and prallethrin, Clarke Mosquito Control, Roselle, IL, USA) and performed area-wide applications to achieve extended suppression of *Ae. albopictus* in New Jersey, USA [154]. We deployed bioassay cups to assess the spatial penetration and efficacy of the applications. Inhibition of adult emergence was significantly higher in the treatment cups than in laboratory controls or field control cups [154]. We observed a lower trend in adult numbers following the season-long combined application of pyriproxyfen and adulticide. Numbers of adult *Ae. albopictus* in the treatment site were up to five times lower than in the control site [154]. We cannot specifically attribute the season-long reduction in adult numbers to pyriproxyfen or the early season *Bti* alone, however we found that larvicide and adulticide applications performed in tandem when adult numbers above a set threshold can be effective. Our results agree with previous studies on combined control methods. Abramides *et al.* [138] used source reduction, larviciding (*Bti* and diflubenzuron) and adulticiding (alpha-cypermethrin) to control *Ae. albopictus* populations in Catalonia, Spain, and showed a significant reduction in the number of eggs recorded in the treated areas [138]. Lam *et al.* [170] treated vegetation with VectoBac WDG (*Bti*) using backpack mist blowers and truck-mounted ULV equipment in Singapore, and found that human landing rates were significantly lower in treatment areas than in control sites [170]. Following applications of pyriproxyfen using a truck-mounted ULV sprayer, mortality recorder from bioassay cups ranged between 50 and 82%, also 50% reduction in adult numbers recorded in the treatment sites in St. Augustine, Florida, USA [153].

The methodology for attractive toxic sugar baits (ATSBs) consisted of a sugar bait, usually sucrose, paired with a toxin use a toxin, such as boric acid [171]. Lea [172] began laboratory trials of this toxic sugar bait on *Ae. aegypti* with 20% sucrose in aqueous solution and low concentrations of malathion on filter paper [172]. Over 85% of adults offered the bait died within 24 h. Xue *et al.* [173] evaluated the efficacy of ATSB on *Ae. albopictus* when applied to plants. Boric acid in 1% solution mixed with 5% sugar water was sprayed on foliage in outdoor cages. About 80% of *Ae. albopictus* died after 48 h, which reduced landing rates on human subjects [173]. These studies validated the efficacy of the active ingredient, but field conditions necessitate an attractant to draw adults to the product. Beier *et al.* [174] conducted a 47-day trial comparing the impact of a single application of ATSB in sugar-rich and sugar-poor environments against *Anopheles spp.* to determine how well an attractant works in the presence of naturally abundant alternatives. They found that abundant natural sugar delayed the impact of the ATSB, which used 75% prickly pear cactus juice to attract mosquitoes. However, both sugar-rich and sugar-poor environments trapped significantly fewer females than control groups 2 weeks after treatment [174]. Application of 1% boric acid in an ATSB killed 92% of *Ae. albopictus* within 14 days of a laboratory trial [175]. Direct applications of ATSB and eugenol to non-flowering vegetation around tyres in Florida significantly decreased adult *Ae. albopictus* populations up to 21 days post-treatment [176]. Impact on non-target insects was considered low while eugenol remained effective against *Ae. albopictus*.

Barrier treatment is a precise application of residual insecticides to perimeters of vegetation or an artificial substrate to provide protection from diseases vectors [177]. Barrier treatment has been used in the private sector for *Ae. albopictus* control, despite the limited number of peerreviewed studies on its impact on *Ae. albopictus* in the field [178]. Lambda-cyhalothrin and bifenthrin barrier treatments were effective against *Ae. albopictus* between 5 and 9 weeks post-application [179–181]. Pyrethroids such as permethrin, delta-methrin, cyfluthrin and tau-fluvalinate were also effective under semi-field conditions [182]. Urban field applications of barrier sprays applied during the peak *Ae. albopictus* season (July to September, when the species is most difficult to control) reduced the adult numbers for 4 weeks post-treatment [106]. We built on our 2016 trials to quantify the efficacy and persistence of early-season barrier treatments, when *Ae. albopictus* populations are low, to determine if these applications improve population suppression later in the season in 2017 [102]. For these treatments (both in 2016 and 2017), we used lambda-cyhalothrin (Damand® CS, Syngenta Crop Protection Inc., Greensboro, NC, USA) and pyriproxyfen (Archer®, Syngenta Crop Protection Inc., Greensboro, NC, USA). We found no advantage of early-season treatments and in fact conclude that the early applications might be counter-productive since the pesticides would have likely degraded before exponential population growth [102]. Control of *Ae. albop ictus* remains demanding, however, barrier treatments are a successful tool with potential for integration into an integrated mosquito management strategy [181]. Some private pest control operators claim barrier treatments reduce the spread of mosquito-borne diseases, but this requires scientific validation.

Applying adulticides from aircraft and truck-mounted equipment is common practice in response to a mosquitoborne disease outbreak. Aerial adulticiding reduces the number of host-seeking females [183, 184]. In order to keep the insecticide from dispersing upwards, ULV cold aerosol spray adulticide applications are generally applied during night-time campaigns when a thermal inversion occurs [185]. Light winds help spread the insecticide droplets. Based on data collected from the 1990s, aerial ULV treatments did not disrupt dengue transmission [186]. One probable reason for failure of control may be the nocturnal resting behaviour of day-biting mosquitoes in natural and artificial sites sheltered from the insecticide [187]. Because *Ae. albopictus* is a diurnal species, the ineffectiveness of night-time ULV applications has become conventional wisdom in the USA, and many mosquito abatement programmes do not spray adulticides for *Ae. albopictus* control [115]. Although previous results not promising [188, 189] it is important to test the technology (night-time adulticiding) with the settings where it will be used. For example, ULV adulticiding reduced 54–90% of *Ae. albopictus* numbers in New Jersey, USA; when the study was repeated with a few modifications Unlu *et al.* [113, 154] achieved 52% reduction in *Ae. albopictus* in an urban area. The major difference between the two studies was that the former covered surface roads and alleyways, whereas the latter covered only surface roads [113]. Both studies found uniform droplet analysis results from the Teflon-acrylic rods (slides) rotating at 5.6 m/s placed in the front and the back of the properties. In addition, Unlu *et al.* [113] placed caged mosquitoes in the front and the back of the properties and found that both products Duet™ (sumithrin and prallethrin, Clarke Mosquito Control, Roselle, IL, USA) and Anvil® (sumithrin and piperonyl butoxide [PBO], Clarke Mosquito Control, Roselle, IL, USA) were more effective in front yards compared with back yards. This indicated that driving alleyways while adulticiding can increase the efficacy of adulticiding [113]. Both studies concluded that ULV adulticiding can be an effective tool to reduce the adult numbers of *Ae. albopictus* in urban and suburban areas.

Using traps for both surveillance and control was first introduced at the International Symposium on Hematophagous Insect Attractants [190]. The inspiration came from tsetse fly control, achieved with attractant-baited, insecticide-impregnated traps in Africa [191, 192]. This approach is often referred to as 'lure and kill' (trapping out) [193]. The first studies to use attractive bait traps under field conditions were performed in 1993. The USDA's Center for Medical, Agricultural and Veterinary Entomology (CMAVE, Gainesville, FL) and the Collier Mosquito Control District (CMCD), Naples, FL, USA collaborated to assess the effect of these traps on mosquito populations in tourist areas of northern Key Island, Florida. When abundance of mosquito species was determined, *Aedes taeniorhynchus* (Wied.) (84.7%), *Culex nigripalpus* Theobald (13.8%) and *Anopheles atropos* Dyar and Knab (1.4%) were the most common species on Key Island [190]. CDC AGO attracts and capture *Ae. aegypti* mosquitoes; studies conducted in southern Puerto Rico reported 53–79% reductions in *Ae. aegypti* numbers [84, 194].

Johnson *et al.* [195] used GAT to reduce *Ae. albopictus* numbers in Maryland, USA. Traps were deployed as part of a project called Citizen Action through Science (Citizen AcTS). Researchers reported declining number of biting in neighbourhoods with higher coverage of GATs [195].

The BG-SNAP (SNAP) is a proposed redesign of the BG-Sentinel trap and is developed for the homeowner market. The SNAP is similar to the BGS2, and is distinguished by an abbreviated conical body and a drawstring opening at the bottom, because it is meant to be hung rather than placed on the ground. Instead of the BGS2's linear core, the SNAP fan and intake funnel are attached to its top [99]. Mosquitoes pass through the fan before collecting in the catch bag or, in its absence, in the trap body or a 1-piece combined funnel [99]. A cartridge lure can be placed in the SNAP's cover opening or a sweet scent lure can be placed in an inner pocket. We tested SNAP traps in suburban areas in New Jersey. Preliminary analyses show no significant difference between treatment and controls because we did not have enough trapping prior to trap deployments. However, we observed a trend of fewer *Ae. albopictus* in the group of eight backyards in 2015 and 2016 where we deployed SNAP traps from July to September.

The efficacy of canola oil as a killing agent in the GAT was assessed against the standard long-lasting insecticideimpregnated net (LLIN) in a large-scale field study based on a recent report highlighting its efficacy against gravid *Ae. aegypti* [196]. The study incorporated 48 GATs with standard LLIN and 48 GATs treated with canola oil. The traps were distributed across 96 individual residences spread over the 17 wards of Parramatta Park and Cairns North, in Cairns Queensland, Australia with each trap separated from other traps by at least two residences [196]. Canola oil was applied to the inside of the upper compartment of the GAT, and killed mosquitoes by suffocation or adherence. Researchers baited all GAT with standard infusion (10 g hay/3 l water), and they identified the mosquitoes collected every week. The study lasted over 9 weeks from June to August 2015, and the canola oil had the lowest mean percentage of traps positive for *Ae. aegypti* (66.7%). GATs treated with sticky card (Trappit) collected more *Ae. aegypti* than the LLIN and GATs treated with canola oil [196]. Although canola oil-treated GATs had lowest percentage of for *Ae. aegypti*, in areas with insecticide-resistant populations, the possibility of the successful incorporation of canola oil as an insecticide-free killing agent may be a valuable asset [196].

This section will be concluded on control and management by discussing the relative paucity of conventional biological control methods against *Ae. albopictus*. Only biological control studies published for invasive *Aedes* species is the use of cyclopid copepods and predator mosquito species *Toxorhynchites* spp. In 1989, in a collaboration between The Centers for Disease Control and Prevention and the New Orleans Mosquito Control Board, cyclopid copepods were assessed for their control of *Aedes* larvae in tyres [197]. Once *Macrocyclops* were established in tyres, they killed all first instar larvae of *Ae. albopictus*, *Ae. aegypti* (Linn.) and *Ae. triseriatus* (Say) [198]. *Macrocyclops* were able to eliminate *Ae. albopictus* larvae from treated tyre piles in 2 months, resulting in a reduction in adult mosquitoes the following month. It took an additional month for the adult mosquitoes to disappear. Tyres were left in the field and checked for *Ae. albopictus* larvae, which did not reestablish in the tyres the following summer [198]. Another study concluded genera of cyclops copepods preyed upon *Aedes* larvae aggressively [199]. When copepods eat all the larvae, the copepods themselves die. Copepods are effective predators of *Ae. albopictus* larvae because they are opportunistic feeders. In other words, they also consume phytoplankton, protozoa and rotifers, and survive in the absence of mosquito larvae. Most studies were conducted in New Orleans in the 1980s and 1990s; evaluating copepods in other regions would be helpful, especially regarding control to the following season.

In 1984, 1986 and 1987 laboratory-reared adult *Toxorhynchites splendens* (Palawan strain) between the ages of 7 and 10 days were released in the small island of Minnajima, northern Okinawa, Japan to control containerinhabiting mosquitoes such as *Ae. albopictus* and *Culex quinquefasciatus* (Say). Results of this study showed that releasing 250 female and 200 male *Tx. splendens* each month effectively controlled container-inhabiting mosquitoes on the island [200]. Another study from New Orleans, Louisiana, USA used *Tx. rutilus* (Dyar & Knab) to control *Ae. aegypti* and *Cx. quinquefasciatus* [201]. On a regular basis researchers released 1–2 *Tx. rutilus* larvae in each container (*n*: 76) positive for mosquito larvae, achieving 74% control. Although results were promising from these studies, due to the cannibalistic behaviour of *Toxorhynchites*, the release of this predator should be constant to maintain pressure on container-inhabiting mosquitoes, making this method very labour- and time-intensive.

Resistance

When an active ingredient from a single class of chemistry is used repeatedly and often without rotation, sub-chronic persistence selects for insecticide resistance [59]. Resistance can impede the operational effectiveness of mosquito control. Two mechanisms of resistance have been characterized in mosquitoes. Metabolic resistance occurs when enzymes degrade or inactivate an active ingredient, and target-site resistance describes a mutation expressed at the level of a target site, such as a voltage-gated sodium channel, that renders the target site inaccessible to active ingredient binding [202]. If a mosquito is already resistant against one chemical class, cross-resistance can occur if the mechanism of action of a new AI overlaps with that of the former. For example, *Ae. aegypti* resistant to the organophosphate Temephos are less susceptible to pyriproxyfen [203]. Most of our understanding of resistance mechanisms is based on *Ae. aegypti*, but in recent years we have been learning more about *Ae. albopictus* [204, 205]. Pyrethroid resistance in *Ae. albopictus* in Florida may have acquired this trait from exposure to DDT, with which DDT shares a similar mechanism of action [204]. Eight strains of *Ae. albopictus* from USA were evaluated for susceptibility to five adulticides and six larvicides. There was significant resistance to DDT in two samples from Florida and one from New Jersey [205]. These samples also exhibited some resistance to malathion and reduced susceptibility to pyriproxyfen and methoprene. All the specimens tested from Florida and New Jersey were susceptible to pyrethroids, despite the risk of cross-resistance DDT would have conferred.

Community Outreach

In urban areas, artificial containers such as bird baths, rain barrels and plant saucers are located in private residential properties, access to which requires permission from homeowners. Because of this, effective *Ae. albopictus* management certainly depends on community involvement [206, 207]. Although some studies have shown public education helps to decrease mosquito numbers [207, 208], other studies have shown that educational campaigns limited to the distribution of brochures on source reduction were ineffective [11, 209]. In New Jersey, USA, Healy *et al.* [210] tested the idea using active community peer education in a source reduction programme with the help of AmeriCorps volunteers. The volunteers were educated during the peak months of *Ae. albopictus* activity in urban and suburban areas with approximately 1000 residences. The volunteers were primarily involved with training community members on container-inhabiting mosquitoes such as drilling holes in recycle bins to reduce the amount of container habitats for mosquito larvae. Researchers found that using volunteers encouraged residents to play a more active role in source reduction [210]. Dowling *et al.* [207] reported that mosquito-related education helps promote community-based container management as part of integrated mosquito management, but hidden containers under porches, in shrubbery or behind sheds, were overlooked [207].We have shown that numbers of *Ae. albopictus* larvae were higher in cryptic CESs (containers positioned on the ground horizontally) than in open containers (containers positioned on the ground vertically) during the peak season of August [1] Involving the public in reducing *Aedes* larval habitats is difficult. At the same time, most mosquito control districts do not have the funding or the personnel to conduct door-to-door source reduction l [13].

Present and future challenges

Mosquito abatement agencies often lack accurate action threshold (i.e. minimum number of mosquitoes to initiate mosquito control). Thresholds are sometimes used in research but not usually for surveillance and intervention for mosquito control districts [211]. Should an action threshold for *Ae. albopictus* be based on biting or on pathogen transmission? Is it even possible to set a threshold to prevent disease transmission? Findings on the correlation between dengue transmission and entomological indices are conflicting, some studies did not find a significant association [85, 212], while others reported that larval indices are predictive of the risk for DENV transmission [213, 214]. Sanchez *et al.* [213] reported that a BI of four or more most accurately predicted dengue transmission, and therefore was appropriate to initiate control in areas with low infestation levels. Chadeem reported statistically significant positive relationships between entomological indices (pupae per person) and dengue incidence [215]. We found that larval-based indices did not predict local adult abundance. However, we found a correlation when we focused on a small group of container types to calculate indices [2]. This indicates that a few container types produce the most adults, and eliminating or treating these will be the most cost-effective strategy. However, targeted source reduction efforts are sometimes ineffective [216, 217]. After covering water tanks that held 70% of *Ae. aegypti* pupae, researchers in Brazil observed a significant decrease in weekly adult *Ae. aegypti* numbers. On the other hand, targeted container control in Thailand reduced the *Ae. aegypti* pupae per Person index by only 15% [217]. We need to better understand if a targeted approach to larval containers will be effective in reducing adult populations of *Ae. albopictus*, especially considering the flexibility this species exhibits with container choice [9, 218]. During monthly door-to-door source reduction campaigns we found that abatement personnel could ignore small trash items, because *Ae. albopictus* seldom use them as oviposition sites, with the exemption of potato chip bags during their peak in August [2]. We only removed these items in August because based on our findings removing these items throughout the year would be a waste of time and money. Unlu *et al.* [1, 114] also found corrugated extension gutters did not contain *Ae. albopictus* larvae until mid-July. Therefore, mosquito control operators did not check and treat these habitats before July [1]. For the same reason, we developed a 'hot spot treatment' approach to focus abatement efforts to tyres and areas with a high focal number of *Ae. albopictus* [13].

Focks [85] discussed the challenge of collecting and counting eggs using oviposition cups [85]. *Ae. albopictus* practice skip-oviposition, and oviposition cups set out as part of an experiment will be far fewer and less cryptic than already existing containers. Oviposition cups can help determine the presence or absence of dengue vectors (*Ae. aegypti* and *Ae. albopictus*), but they are not reliable to estimate adult population numbers. Therefore, egg surveillance should be conducted in tandem with adult mosquito collection [5]. Based on our experience in urban sites in New Jersey, the number of eggs in oviposition cups does not correlate with the number of females, especially during dry summers [5]. In contrast, Suter *et al.* [219] found egg density 2.26 times higher in the control site compared with the treatment site, which is in agreement with studies conducted in Italy that found a positive correlation between treatment and lower egg numbers [220, 221]. In other studies, researchers further discussed the lack of correlation between egg and adult numbers when using analysis of variances (ANOVA) for count data (number of eggs and adults) instead of a negative binomial model. Unlu *et al.* [113, 154] used the negative binomial model but did not find a significant correlation between egg and adult numbers for *Ae. albopictus*. Researchers should remain cautious while considering either or both methods because of conflicting results when using eggs and adult numbers for surveillance of *Aedes* mosquitoes [154].

Whether immature stages or adults are collected, we still cannot predict disease outbreaks [222]. Human cases should not be the trigger for intervention; this is already too late. Therefore I will address Microsoft's Project Premonition, a study that aims to predict infectious disease outbreaks to mitigate life changing diseases such as ZIKV. The project's vision is a worldwide system to collect mosquitoes and detect pathogens before they cause disease (Microsoft Blog, 2015). A global system requires scalable, robust and safe autonomous systems and tremendous computational power. This is a great example of a paradigm shift in vector control, from reactive to proactive to combat disease outbreaks. The conventional tools we use for surveillance and control for invasive *Aedes* species are time-and labour-intensive. Waiting for people to become infected, essentially using humans as sentinels, is a poor option, so we need to think outside the box and bring new methods into mosquito control practices.

Entomopathogenic and mermithid nematodes are also biological alternatives to chemical agents against *Ae. albopictus*. Entomopathogenic nematodes are represented by the families *Steinernematidae* and *Heterorhabditidae*. Characterized by their symbiotic relationship with bacteria [223], entomopathogenic species possess many of the desirable attributes of biological control agents. They can be easily cultured *in vitro* [224]. They are available commercially, have an extensive pest host range and are highly virulent [225]. Entomopathogenic nematodes kill mosquito larvae. For example, laboratory exposure of 170–200 *Heterorhabditis bacteriophora* infective nematodes per host killed 100% of *Cx. pipiens* larvae [226]. However, laboratory results have not translated into acceptable field outcomes. For example, *Heterorhabditis indica* showed low efficacy against container- and bromeliadinhabiting *Ae. aegypti* despite excellent mortality achieved in the lab [227]. Entomopathogenic nematode candidacy as mosquito bioagents is flawed because they evolved in soil and are therefore adapted to this environment [228]. They cannot swim and in aquatic habitats descend through the larval feeding zone and are lost in the bottom silt. Their brief presentation to larval hosts coupled with a low probability of recycling in larval habitats and strong host immune response [229] suggest limited biological control potential. Further studies are necessary to determine how infective stages could be suspended where larvae feed. Bring in biological control techniques against invasive *Aedes* species.

Mosquito-parasitic mermithids received considerable attention from biocontrol researchers in the 1970s and 1980s, but enthusiasm faded when *Bti* became widely available as the first commercial biological control agent against mosquitoes. Currently mermithid nematodes are used routinely in Mexico [230] and West Africa [231] against anopheline species in malaria-endemic zones with excellent results. These mermithids are produced *in vivo* (produced in living insects) labour-intensive and costly methods [232]. The development of efficient mass production of mosquitoes to support *Wolbachia* and RIDLrelease strategies could open a fresh opportunity to exploit parasitic nematodes against mosquitoes. Any technology to decrease the effort in rearing mosquitoes will benefit trials with *Wolbachia*, RIDL mosquitoes and mermithids. Citizen science may be a powerful, cost-effective tool to engage the public in invasive *Aedes* mosquito surveillance and control. Citizen science dates back to the 1970s and with current technology, participation is faster and easier at converting citizen observations into usable data [233]. Websites make it relatively easy to aggregate data, incorporate software to minimize anomalies and collate into raw data sets [233]. A good example is 'The Invasive Mosquito Project' which aims to (1) clarify one's role in protecting individuals, communities and pets from illness; (2) raise awareness of mosquito pathogens and (3) participate in a national programme to gather mosquito distribution data [234]. Teachers are responsible for presenting information to their K-12 students and describing how oviposition cups are used for mosquito surveillance.

Jordan *et al.* [235] initiated a citizen science programme called 'Mosquito Stoppers' in Maryland, USA engaging citizen scientists in finding mosquito habitats in backyards and asking them to rate their outdoor experience [235]. Citizen scientists rated their mosquito encounter as 'just fine', 'irritating' or 'intolerable'. Although citizen scientists received minimal training, they collected data accurately on *Ae. albopictus* [235]. NASA initiated the Mosquito Habitat Mapper in 2017 application called GLOBE Observer. The application provides simple training on identifying and eliminating mosquito habitats. Citizen scientists can also identify the mosquito to species and collect data on whether it is a competent vector for ZIKV, DF, YF and CHIKV. This Citizen Science project goal is to help NASA scientists support satellite-based research of environmental conditions that are indicators of disease outbreak (https://www.nasa.gov/feature/nasa-citizen-science-app-tackles-mosquito-borne-disease).

The project led by Rutgers University provides communities with scientific support to develop projects that directly benefit communities by modifying citizen science idea and call Citizen Action through Science (Citizen AcTS) [195]. The idea of Citizen AcTS is built around a 'buy-in' model, in which residents purchase, deploy and maintain lethal GATs. Residents also removed mosquito habitat from their yards. Johnson *et al.* [195] observed effective mosquito control through the cooperation of homeowners, community leaders and scientific advisors. Reduction in *Ae. albopictus* biting pressure was higher in the areas with higher GAT coverage. When communities accept an active role in monitoring mosquito numbers, the probability of reducing invasive *Aedes* is enhanced [236].

The education and training of mosquito professionals is still a deficit in *Ae. albopictus* management. Education usually focuses on the public [11, 188]. But the 2016 Miami-Dade ZIKV outbreak required responding personnel to be trained in a matter of days, demonstrating a greater need for professionally trained mosquito control personnel. In late 2016, the CDC awarded a 1 year, $1.6 million (grant), to the AMCA to design, develop and deploy a targeted mosquito surveillance and control training programme. The award and development of the programme was in response to the 2016 Zika outbreaks in the Western Hemisphere and is a mechanism to standardize Best Management Practices for mosquito control, specifically regarding invasive *Aedes* species. If there is a new outbreak, this training programme will help combat the disease spread by expediting the training of additional new staff. The curriculum of the programme was derived from the AMCA Best Practices Manual (https://www.mosquito.org/), which was updated by an advisory council selected from organizations involved in invasive *Aedes* surveillance and control. The mosquito control industry has never had a nationwide formal training programme in the USA because training has always had a local focus. Programme goals include constant improvement and expansion of online outreach.

With funding provided by CDC, five regional Centers of Excellence (COEs) were established to address the problem of emerging and exotic diseases in the USA. The first goal of these centres is to conduct operational research to develop and validate effective vector-borne disease prevention and control methods to respond to disease outbreaks. The second goal is to train a cohort of public health entomologists with the skill-set necessary to quickly respond to future disease outbreaks in the USA. The third and most important goal is to build effective collaborations between academic communities and public health organizations at federal, state and local levels for invasive *Aedes* species surveillance and control.

Operational research will play a key role in how we respond to future outbreaks transmitted by invasive *Aedes* mosquitoes. Numerous studies have emphasized our need for IPM for these species. I hope new and innovative tools can be tested against *Ae. albopictus* with the help of regional COEs. Our current strategy to control *Ae. albopictus* and *Ae. aegypti* tends to be singular, such as source reduction. However, the focus should be on a combination of strategies tailored to different geographic regions [5, 13, 218]. For example, the 2008 'Area-wide control of the Asian tiger mosquito' project, conceived by Rutgers University, Mercer, and Monmouth mosquito control counties with USDA Agricultural Research Service staff developed strategies for effective area-wide control of *Ae. albopictus* which can be recommended to other health agencies. The project's aim was to develop a standard methodology in Mercer and Monmouth Counties in New Jersey. I would like to conclude the review by stating the importance of political commitment, which was repeatedly stressed as the most important component of a control programme, regardless of the methods used. Participation of those affected is crucial, and no programme has succeeded sustainable mosquito control programmes without enlightened, adequately funded and well-organized leadership.

Conclusion

As environments change, species must adapt or suffer unfavourable consequences. The vector control environment is currently poised for massive change and must adapt as well. Biology and ecology of vector species are dynamic; therefore we need to adjust how we respond to these changes. Changes in the regulatory environment coupled with rising opposition to area-wide adulticide use will continue to drive change in mosquito control. Rather than borrowing heavily from agricultural-based insect management as we have since the inception of mosquito control, we should focus on developing novel ideas designed for mosquitoes. Mosquito abatement districts tend to be conservative, depending on older technologies such as adulticiding. We cannot respond to new threats with old tools. We are now challenged to adapt to new technologies ranging from drones, to sterile male technique, to genetically modified organisms, to *Wolbachia*. These are powerful new tools we need to integrate into mosquito control. We must strive to be open to new technologies, expand academic-operational cooperation and support innovative research. Education, training, research and collaboration are vital to deliver new, sustainable vector control strategies.

Acknowledgements

I would also like to thank Rutgers University's Center for Vector Biology and my dearest mentor Dr Randy Gaugler for making opportunities available to me, because I am a true believer in research and collaboration between academia and mosquito control agencies. I would also like to thank my colleagues Dr Ary Faraji, Nicholas Indelicato and Dr Ilia Rochlin for making every challenge I encountered a pleasure and a valuable learning experience. Last but certainly not least, I would like to extend my deepest gratitude to Kurt Neinstedt for helping me review the literature on ATSB and SIT sections of this review.

References

1. Unlu I, Faraji A, Indelicato N, Fonseca DM. The hidden world of Asian tiger mosquitoes: immature *Aedes albopictus* (Skuse) dominate in rainwater corrugated extension spouts. Transactions of the Royal Society of Tropical Medicine and Hygiene 2014;108:699–705. doi: 10.1093/trstmh/tru139.
2. Unlu I, Farajollahi A, Strickman D, Fonseca DM. Crouching tiger, hidden trouble: urban sources of *Aedes albopictus* (Diptera: Culicidae) refractory to source-reduction. PLoS One 2013;8:e77999.
3. Crans WJ. A classification system for mosquito life cycles: life cycle types for mosquitoes of the northeastern United States. Journal of Vector Ecology 2004;29:1–10.
4. Wheeler AS, Petrie WD, Malone D, Allen F. Introduction, control, and spread of *Aedes albopictus* on Grand Cayman Island, 1997–2001. Journal of the American Mosquito Control Association 2009;25:251–9.
5. Fonseca DM, Unlu I, Crepeau T, Farajollahi A, Healy SP, Bartlett-Healy K, *et al.* Area-wide management of *Aedes albopictus*. Part 2: gauging the efficacy of traditional integrated pest control measures against urban container mosquitoes. Pest Management Science 2013;69:1351–61.
6. Preechaporn W, Jaroensutasinee M, Jaroensutasinee K. The larval ecology of *Aedes aegypti* and *Ae. albopictus* in three topographical areas of Southern Thailand. Dengue Bulletin 2006;30:204.
7. Richards SL, Ghosh SK, Zeichner BC, Apperson CS. Impact of source reduction on the spatial distribution of larvae and pupae of *Aedes albopictus* (Diptera: Culicidae) in suburban neighborhoods of a Piedmont community in North Carolina. Journal of Medical Entomology 2008;45:617–28.
8. Caputo B, Ienco A, Cianci D, Pombi M, Petrarca V, Baseggio A, *et al.* The 'auto-dissemination' approach: a novel concept to fight *Aedes albopictus* in urban areas. PLoS Neglected Tropical Diseases 2012;6:e1793.

9. Richards SL, Apperson CS, Ghosh SK, Cheshire HM, Zeichner BC. Spatial analysis of *Aedes albopictus* (Diptera: Culicidae) oviposition in suburban neighborhoods of a Piedmont community in North Carolina. Journal of Medical Entomology 2006;43:976–89.
10. Rochlin I, Ninivaggi DV, Hutchinson ML, Farajollahi A. Climate change and range expansion of the Asian tiger mosquito (*Aedes albopictus*) in Northeastern USA: implications for public health practitioners. PLoS One 2013;8:e60874.
11. Bartlett-Healy K, Hamilton G, Healy S, Crepeau T, Unlu I, Farajollahi A, *et al.* Source reduction behavior as an independent measurement of the impact of a public health education campaign in an integrated vector management programme for the Asian tiger mosquito. International Journal of Environmental Research and Public Health 2011;8:1358–67.
12. Ladeau SL, Leisnham PT, Biehler D, Bodner D. Higher mosquito production in low-income neighborhoods of Baltimore and Washington, DC: understanding ecological drivers and mosquito-borne disease risk in temperate cities. International Journal of Environmental Research and Public Health 2013;10:1505–26.
13. Unlu I, Klingler K, Indelicato N, Faraji A, Strickman D. Suppression of *Aedes albopictus*, the Asian tiger mosquito, using a 'hot spot' approach. Pest Management Science 2016;72:1427–32.
14. Benedict MQ, Levine RS, Hawley WA, Lounibos LP. Spread of the tiger: global risk of invasion by the mosquito *Aedes albopictus*. Vector-Borne and Zoonotic Diseases 2007;7:76–85.
15. Unlu I, Farajollahi A. Vectors without borders: imminent arrival, establishment, and public health implications of the Asian bush (*Aedes japonicus*) and Asian tiger (*Aedes albopictus*) mosquitoes in Turkey. Hacettepe Journal of Biology and Chemistry 2012;40:23–36.
16. Gubler DJ, Rosen L. Variation among geographic strains of *Aedes albopictus* in susceptibility to infection with dengue viruses. The American Journal of Tropical Medicine and Hygiene 1976;25:318–25.
17. Effler PV, Pang L, Kitsutani P, Vorndam V, Nakata M, Ayers T, *et al.* Dengue fever, Hawaii, 2001–2002. Emerging Infectious Diseases 2005;11:742–9.
18. Turell MJ, Beaman JR, Tammariello RF. Susceptibility of selected strains of *Aedes aegypti* and *Aedes albopictus* (Diptera: Culicidae) to chikungunya virus. Journal of Medical Entomology 1992;29:49–53.
19. Mctighe SP, Vaidyanathan R. Vector competence of *Aedes albopictus* from Virginia and Georgia for Chikungunya virus isolated in the Comoros Islands, 2005. Vector-Borne and Zoonotic Diseases 2012;12:867–71.
20. Azar SR, Roundy CM, Rossi SL, Huang JH, Leal G, Yun R, *et al.* Differential vector competency of *Aedes albopictus* populations from the Americas for Zika virus. The American Journal of Tropical Medicine and Hygiene 2017;97:330–9.
21. Vanlandingham DL, Higgs S, Huang Y-JS. *Aedes albopictus* (Diptera: Culicidae) and mosquito-borne viruses in the United States. Journal of Medical Entomology 2016;53:1024–8.
22. Paupy C, Delatte H, Bagny L, Corbel V, Fontenille D. *Aedes albopictus*, an arbovirus vector: from the darkness to the light. Microbes and Infection 2009;11:1177–85.
23. Ponlawat A, Harrington LC. Blood feeding patterns of *Aedes aegypti* and *Aedes albopictus* in Thailand. Journal of Medical Entomology 2005;42:844–9.
24. Faraji A, Egizi A, Fonseca DM, Unlu I, Crepeau T, Healy SP, *et al.* Comparative host feeding patterns of the Asian tiger mosquito, *Aedes albopictus*, in urban and suburban Northeastern USA and implications for disease transmission. PLoS Neglected Tropical Diseases 2014;8:e3037.
25. Westbrook CJ. Larval ecology and adult vector competence of invasive mosquitoes *Aedes albopictus* and *Aedes aegypti* for Chikungunya virus 2010;72(2).
26. Juliano SA, Philip Lounibos L. Ecology of invasive mosquitoes: effects on resident species and on human health. Ecology Letters 2005;8:558–74.
27. Gratz NG. Critical review of the vector status of *Aedes albopictus*. Medical and Veterinary Entomology 2004;18:215–27.
28. Shragai T, Tesla B, Murdock C, Harrington LC. Zika and chikungunya: mosquito-borne viruses in a changing world. Annals of the New York Academy of Sciences 2017;1399:61–77.
29. CDC. Chikungunya in the Caribbean; 2014. Available from: URL: http://wwwnc.cdc.gov/travel/notices/watch/chikungunya-saint-martin.
30. Sabin AB. Research on dengue during World War II. The American Journal of Tropical Medicine and Hygiene 1952;1:30–50.
31. Xu G, Dong H, Shi N, Liu S, Zhou A, Cheng Z, *et al.* An outbreak of dengue virus serotype 1 infection in Cixi, Ningbo, People's Republic of China, 2004, associated with a traveler from Thailand

and high density of *Aedes albopictus*. The American Journal of Tropical Medicine and Hygiene 2007;76:1182–8.

32. Vazeille M, Moutailler S, Coudrier D, Rousseaux C, Khun H, Huerre M, *et al.* Two Chikungunya isolates from the outbreak of La Reunion (Indian Ocean) exhibit different patterns of infection in the mosquito, *Aedes albopictus*. PLoS One 2007;2:e1168.
33. Lambrechts L, Scott TW, Gubler DJ. Consequences of the expanding global distribution of *Aedes albopictus* for dengue virus transmission. PLoS Neglected Tropical Diseases 2010;4:e646.
34. Paupy C, Ollomo B, Kamgang B, Moutailler S, Rousset D, Demanou M, *et al.* Comparative role of *Aedes albopictus* and *Aedes aegypti* in the emergence of Dengue and Chikungunya in Central Africa. Vector-Borne and Zoonotic Diseases 2010;10:259–66.
35. Enserink M. Infectious diseases. Chikungunya: no longer a third world disease. Science 2007;318:1860–1.
36. Pagès F, Peyrefitte CN, Mve MT, Jarjaval F, Brisse S, Iteman I, *et al.* *Aedes albopictus* mosquito: the main vector of the 2007 Chikungunya outbreak in Gabon. PLoS One 2009;4:e4691.
37. De Lamballerie X, Leroy E, Charrel R, Ttsetsarkin K, Higgs S, Gould E. Chikungunya virus adapts to tiger mosquito via evolutionary convergence: a sign of things to come? Virology Journal 2008;5:33.
38. Tsetsarkin KA, Vanlandingham DL, Mcgee CE, Higgs S. A single mutation in chikungunya virus affects vector specificity and epidemic potential. PLoS Pathogens 2007;3:e201.
39. Tsetsarkin KA, Chen R, Weaver SC. Interspecies transmission and chikungunya virus emergence. Current Opinion in Virology 2016;16:143–50.
40. Tsetsarkin KA, Weaver SC. Sequential adaptive mutations enhance efficient vector switching by chikungunya virus and its epidemic emergence. PLoS Pathogens 2011;7:e1002412.
41. Dick G, Kitchen S, Haddow A. Zika virus (I). Isolations and serological specificity. Transactions of the Royal Society of Tropical Medicine and Hygiene 1952;46:509–20.
42. Faye O, Freire CC, Iamarino A, Faye O, De Oliveira JVC, Diallo M, *et al.* Molecular evolution of Zika virus during its emergence in the 20th century. PLoS Neglected Tropical Diseases 2014;8:e2636.
43. Volk SM, Chen R, Tsetsarkin KA, Adams AP, Garcia TI, Sall AA, *et al.* Genome-scale phylogenetic analyses of chikungunya virus reveal independent emergences of recent epidemics and various evolutionary rates. Journal of Virology 2010;84:6497–504.
44. Haddow AD, Schuh AJ, Yasuda CY, Kasper MR, Heang V, Huy R, *et al.* Genetic characterization of Zika virus strains: geographic expansion of the Asian lineage. PLoS Neglected Tropical Diseases 2012;6:e1477.
45. Duffy MR, Chen T-H, Hancock WT, Powers AM, Kool JL, Lanciotti RS, *et al.* Zika virus outbreak on Yap Island, federated states of Micronesia. New England Journal of Medicine 2009;360:2536–43.
46. Calvet G, Aguiar RS, Melo AS, Sampaio SA, De Filippis I, Fabri A, *et al.* Detection and sequencing of Zika virus from amniotic fluid of fetuses with microcephaly in Brazil: a case study. The Lancet Infectious Diseases 2016;16:653–60.
47. Wong P-SJ, Li M-ZI, Chong C-S, Ng L C, Tan C-H. *Aedes* (Stegomyia) *albopictus* (skuse): a potential vector of Zika virus in Singapore. PLoS Neglected Tropical Diseases 2013;7:e2348.
48. Di Luca M, Severini F, Toma L, Boccolini D, Romi R, Remoli ME, *et al.* Experimental studies of susceptibility of Italian *Aedes albopictus* to Zika virus. Eurosurveillance 2016;21(18):pii=30223. https://doi.org/10.2807/1560-7917.ES.2016.21.18.30223.
49. Mogi M, Armbruster P, Fonseca DM. Analyses of the northern distributional limit of *Aedes albopictus* (Diptera: Culicidae) with a simple thermal index. Journal of Medical Entomology 2014;49:1233–43.
50. Delatte H, Toty C, Boyer S, Bouetard A, Bastien F, Fontenille D. Evidence of habitat structuring *Aedes albopictus* populations in Réunion Island. PLoS Neglected Tropical Diseases 2013;7: e2111.
51. Campos GS, Bandeira AC, Sardi SI. Zika virus outbreak, Bahia, Brazil. Emerging Infectious Diseases 2015;21:1885–6.
52. Dinger J, Schüffner W, Snijders E, Swellen-Grebel NH. Yellow Fever Research in Holland. (First Communication). 1929;28. https://www.cabdirect.org/cabdirect/abstract/19292902285.
53. Johnson BW, Chambers TV, Crabtree MB, Filippis AM, Vilarinhos PT, Resende MC, *et al.* Vector competence of Brazilian *Aedes aegypti* and *Ae. albopictus* for a Brazilian yellow fever virus isolate. Transactions of the Royal Society of Tropical Medicine and Hygiene 2002;96:611–3
54. Holick J, Kyle A, Ferraro W, Delaney RR, Iwaseczko M. Discovery of *Aedes albopictus* infected with West Nile virus in southeastern Pennsylvania. Journal of the American Mosquito Control Association 2002;18:131.

55. Unlu I, Mackay AJ, Roy A, Yates MM, Foil LD. Evidence of vertical transmission of West Nile virus in field-collected mosquitoes. Journal of Vector Ecology 2010;35:95–9.
56. Turell MJ, O'Guinn ML, Dohm DJ, Jones JW. Vector competence of North American mosquitoes (Diptera: Culicidae) for West Nile virus. Journal of Medical Entomology 2001;38:130–4.
57. Sardelis MR, Turell MJ, O'Guinn ML, Andre RG, Roberts DR. Vector competence of three North American strains of *Aedes albopictus* for West Nile virus. Journal of the American Mosquito Control Association 2002;18:284–9.
58. Kogan M. Integrated pest management: historical perspectives and contemporary developments. Annual Review of Entomology 1998;43:243–70.
59. Brogdon WG, Mcallister JC. Insecticide resistance and vector control. Emerging Infectious Diseases 1998;4:605.
60. Patterson G. The Mosquito Crusades: A History of the American Anti-Mosquito Movement From the Reed Commission to the First Earth Day. Rutgers University Press, New Brunswick, NJ; 2009.
61. Palmisano CT, Taylor V, Caillouet K, Byrd B, Wesson DM. Impact of West Nile virus outbreak upon St. Tammany Parish mosquito abatement district. Journal of the American Mosquito Control Association 2005;21:33–8.
62. Garcia R, Colwell AE, Voigt WG, Woodward DL. Fay-Prince trap baited with CO_2 for monitoring adult abundance of *Aedes sierrensis* (Diptera: Culicidae). Journal of Medical Entomology 1989;26:327–31.
63. Gillies M. The role of carbon dioxide in host-finding by mosquitoes (Diptera: Culicidae): a review. Bulletin of Entomological Research 1980;70:525–32.
64. Davis TJ, Kaufman PE, Tatem AJ, Hogsette JA, Kline DL. Development and evaluation of an attractive self-marking ovitrap to measure dispersal and determine skip oviposition in *Aedes albopictus* (Diptera: Culicidae) field populations. Journal of Medical Entomology 2015;tjv170.
65. Service MW. Mosquito Ecology. Field Sampling Methods. Applied Science Publishers Ltd, London; 1976.
66. Nicolas G, Sillans D. Immediate and latent effects of carbon dioxide on insects. Annual Review of Entomology 1989;34:97–116.
67. Headlee TJ. The Mosquitoes of New Jersey and their Control. Rutgers University Press, New Brunswick, New Jersey; 2003, pp. 106–128.
68. Slaff M, Crans WJ, Mccuiston J. A comparison of three mosquito sampling techniques in northwestern New Jersey. Mosquito News 1983;43:287–290.
69. Sudia W, Chamberlain R. Battery operated light trap, an improved model. Mosquito News 1962;22:126–9.
70. Kilpatrick AM, Kramer LD, Campbell SR, Alleyne EO, Dobson AP, Daszak P. West Nile virus risk assessment and the bridge vector paradigm. Emerging Infectious Diseases 2005;11:425.
71. Muirhead-Thompson R. Trap Responses of Flying Insects: the Influence of Trap Design on Capture Efficiency. Academic Press, London; 2012.
72. Lothrop HD, Reisen WK. Landscape affects the host-seeking patterns of *Culex tarsalis* (Diptera: Culicidae) in the Coachella Valley of California. Journal of Medical Entomology 2001;38:325–32.
73. Burkett DA, Lee WJ, Lee KW, Kim HC, Lee HI, Lee JS, *et al.* Light, carbon dioxide, and octenol-baited mosquito trap and host-seeking activity evaluations for mosquitoes in a malarious area of the Republic of Korea. Journal of the American Mosquito Control Association 2001;17:196–205.
74. Reiter P. A portable battery-powered trap for collecting gravid *Culex* mosquitoes. Mosquito News 1983;43:496–8.
75. Unlu I, Farajollahi A. A multiyear surveillance for *Aedes albopictus* with Biogents Sentinel trap counts for males and species composition of other mosquito species. Journal of the American Mosquito Control Association 2014;30:122–5.
76. Farajollahi A, Fonseca DM, Kramer LD, Marm Kilpatrick A. 'Bird biting' mosquitoes and human disease: a review of the role of *Culex pipiens* complex mosquitoes in epidemiology. Infection, Genetics and Evolution 2011;11:1577–85.
77. Farajollahi A, Kesavaraju B, Price DC, Williams GM, Healy SP, Gaugler R, *et al.* Field efficacy of BG-Sentinel and industry-standard traps for *Aedes albopictus* (Diptera: Culicidae) and West Nile virus surveillance. Journal of Medical Entomology 2009;46:919–25.
78. Robertson RC, Hu SMK. The tiger mosquito in Shanghai. China Journal 1935;23:299–306.
79. Lampman RL, Novak RJ. Attraction of *Aedes albopictus* adults in sod infusion. Journal of the American Mosquito Control Association 1996;12:119–24.

80. Reiter P, Amador MA, Colon N. Enhancement of the CDC ovitrap with hay infusions for daily monitoring of *Aedes aegypti* populations. Journal of the American Mosquito Control Association 1991;7:52–5.
81. Gopalakrishnan R, Das M, Baruah I, Veer V, Dutta P. Studies on the ovitraps baited with hay and leaf infusions for the surveillance of dengue vector, *Aedes albopictus* in northeastern India. Tropical Biomedicine 2012;29:598–604.
82. Trexler JD, Apperson CS, Schal C. Laboratory and field evaluations of oviposition responses of *Aedes albopictus* and *Aedes triseriatus* (Diptera: Culicidae) to oak leaf infusions. Journal of Medical Entomology 1998;35:967–76.
83. CDC. Surveillance and control of *Aedes aegypti* and *Aedes albopictus* in the United States. CDC; 2016. Available from: URL: http://www.cdc.gov/chikungunya/resources/vector-control.html (last accessed 28 September 2018, 7 September 2016).
84. Mackay AJ, Amador M, Barrera R. An improved autocidal gravid ovitrap for the control and surveillance of *Aedes aegypti*. Parasites & Vectors 2013;6:225.
85. Focks DA. A Review of Entomological Sampling Methods and Indicators for Dengue Vectors. WHO, Geneva; 2003.
86. Connelly C, Carlson D. Florida Mosquito Control: The state of the mission as defined by mosquito controllers, regulators, and environmental managers. Florida Coordinating Council on Mosquito Control. University of Florida, Institute of Food and Agricultural Sciences. Florida Medical Entomology Laboratory, Vero Beach; 2009.
87. Fonseca D, Kaplan L, Heiry R, Strickman D. Density-dependent oviposition by female *Aedes albopictus* (Diptera: Culicidae) spreads eggs among containers during the summer but accumulates them in the fall. Journal of Medical Entomology 2015;52:705–12.
88. Fay R, Eliason DA. A preferred oviposition site as a surveillance method for *Aedes aegypti*. Mosquito News 1966;26:531–4.
89. Johnson BJ, Hurst T, Quoc HL, Unlu I, Freebairn C, Faraji A, *et al.* Field comparisons of the *Gravid Aedes* Trap (GAT) and BG-sentinel trap for monitoring *Aedes albopictus* (Diptera: Culicidae) populations and notes on indoor GAT collections in Vietnam. Journal of Medical Entomology 2016;54(2):340–8.
90. CDC. West Nile Virus in the United States: Guidelines for Surveillance, Prevention, and Control. 4th ed. Centers for Disease Control and Prevention. Division of Vector-Borne Diseases, Fort Collins, Colorado; 2013.
91. Tun-Lin W, Kay B, Barnes A, Forsyth S. Critical examination of *Aedes aegypti* indices: correlations with abundance. The American Journal of Tropical Medicine and Hygiene 1996;54:543–7.
92. Focks DA, Chadee DD. Pupal survey: an epidemiologically significant surveillance method for *Aedes aegypti*: an example using data from Trinidad. The American Journal of Tropical Medicine and Hygiene 1997;56:159–67.
93. Boyer S, Foray C, Dehecq J-S. Spatial and temporal heterogeneities of *Aedes albopictus* density in La Reunion Island: rise and weakness of entomological indices. PLoS One 2014;9:e91170.
94. Getis A, Morrison AC, Gray K, Scott TW. Characteristics of the spatial pattern of the dengue vector, *Aedes aegypti*, in Iquitos, Peru. The American Journal of Tropical Medicine and Hygiene 2003;69:494–505.
95. Meeraus WH, Armistead JS, Arias JR. Field comparison of novel and gold standard traps for collecting *Aedes albopictus* in Northern Virginia. Journal of the American Mosquito Control Association 2008;24:244–8.
96. Krockel U, Rose A, Eiras AE, Geier M. New tools for surveillance of adult yellow fever mosquitoes: comparison of trap catches with human landing rates in an urban environment. Journal of the American Mosquito Control Association 2006;22:229–38.
97. Unlu I, Farajollahi A, Healy SP, Crepeau T, Bartlett-Healy K, Williges E, *et al.* Area-wide management of *Aedes albopictus*: choice of study sites based on geospatial characteristics, socioeconomic factors and mosquito populations. Pest Management Science 2011;67:965–74.
98. Unlu I, Faraji A, Morganti M, Vaeth R, Akaratovic K, Kiser J, *et al.* Reduced performance of a PVC-coated Biogents Sentinel prototype in comparison to the original Biogents Sentinel for monitoring the Asian tiger mosquito, *Aedes albopictus*, in temperate North America. PLoS One 2017;12:e0172963.
99. Akaratovic KI, Kiser JP, Gordon S, Abadam CF. Evaluation of the trapping performance of four Biogents AG traps and two lures for the surveillance of *Aedes albopictus* and other host-seeking mosquitoes. Journal of the American Mosquito Control Association 2017;33:108–15.

100. Arimoto H, Harwood JF, Nunn PJ, Richardson AG, Gordon S, Obenauer PJ. Comparison of trapping performance between the original BG-Sentinel® trap and BG-Sentinel 2® trap 1. Journal of the American Mosquito Control Association 2015;31:384–7.
101. Li Y, Kamara F, Zhou G, Puthiyakunnon S, Li C, Liu Y, *et al.* Urbanization increases *Aedes albopictus* larval habitats and accelerates mosquito development and survivorship. PLoS Neglected Tropical Diseases 2014;8:e3301.
102. Williams GM, Unlu I, Rochlin I, Wang Y, Gaugler R. Timing of lambda-cyhalothrin and pyriproxyfen barrier treatments for *Aedes albopictus* (Diptera: Culicidae) management. Journal of Economic Entomology 2019, toz020, https://doi.org/10.1093/jee/toz020.
103. Crepeau TN, Unlu I, Healy SP, Farajollahi A, Fonseca DM. Experiences with the large-scale operation of the Biogents Sentinel trap. Journal of the American Mosquito Control Association 2013;29:177–80.
104. Unlu I, Baker M. Comparison of BG-Sentinel prototype, BG-Sentinel-1, and BG-Sentinel-2: better results with modification of earlier design. Journal of the American Mosquito Control Association 2018;34:237–9.
105. Unlu I, Suman DS, Wang Y, Klingler K, Faraji A, Gaugler R. Effectiveness of autodissemination stations containing pyriproxyfen in reducing immature *Aedes albopictus* populations. Parasites & Vectors 2017;10:139.
106. Unlu I, Williams GM, Rochlin I, Suman D, Wang Y, Chandel K, Gaugler R. Evaluation of lambda-cyhalothrin and pyriproxyfen barrier treatments for *Aedes albopictus* (Diptera: Culicidae) management in urbanized areas of New Jersey. Journal of Medical Entomology 2017;55(2):472–6.
107. Rochlin I, Kawalkowski M, Ninivaggi DV. Comparison of mosquito magnet and Biogents Sentinel traps for operational surveillance of container-inhabiting *Aedes* (Diptera: Culicidae) species. Journal of Medical Entomology 2016;53:454–9.
108. Eiras AE, Buhagiar TS, Ritchie SA. Development of the gravid *Aedes* trap for the capture of adult female container-exploiting mosquitoes (Diptera: Culicidae). Journal of Medical Entomology 2014;51:200–9.
109. Ritchie SA, Buhagiar TS, Townsend M, Hoffmann A, Van Den Hurk AF, Mcmahon JL, *et al.* Field validation of the gravid *Aedes* trap (GAT) for collection of *Aedes aegypti* (Diptera: Culicidae). Journal of Medical Entomology 2014;51:210–9.
110. Ritchie SA, Rapley L, Williams C, Johnson P, Larkman M, Silcock R, *et al.* A lethal ovitrap-based mass trapping scheme for dengue control in Australia: I. Public acceptability and performance of lethal ovitraps. Medical and Veterinary Entomology 2009;23:295–302.
111. John KH, Stoll JR, Olson J. An economic assessment of the benefits of mosquito abatement in an organized mosquito control district. Journal of the American Mosquito Control Association 1987;3:8–14.
112. Morris CD, Clanton KB. Significant associations between mosquito control service requests and mosquito populations. Journal of the American Mosquito Control Association 1989;5:36–41.
113. Unlu I, Baker MA, Indelicato N, Drews D, Zeng Z, Vaidyanathan R. Nighttime applications of two formulations of pyrethroids are effective against diurnal *Aedes albopictus*. Journal of the American Mosquito Control Association 2018;34(2):158–62.
114. Unlu I, Farajollahi A, Rochlin I, Crepeau TN, Strickman D, Gaugler R. Differences in male–female ratios of *Aedes albopictus* (Diptera: Culicidae) following ultra-low volume adulticide applications. Acta Tropica 2014;137:201–5.
115. Farajollahi A, Healy SP, Unlu I, Gaugler R, Fonseca DM. Effectiveness of ultra-low volume nighttime applications of an adulticide against diurnal *Aedes albopictus*, a critical vector of dengue and chikungunya viruses. PLoS One 2012;7:e49181.
116. Knipling E. Possibilities of insect control or eradication through the use of sexually sterile males. Journal of Economic Entomology 1955;48:459–62.
117. Dame DA, Curtis CF, Benedict MQ, Robinson AS, Knols BG. Historical applications of induced sterilisation in field populations of mosquitoes. Malaria Journal 2009;8:S2.
118. Brelsfoard CL, Séchan Y, Dobson SL. Interspecific hybridization yields strategy for South Pacific filariasis vector elimination. PLoS Neglected Tropical Diseases 2008;2:e129.
119. Zabalou S, Apostolaki A, Livadaras I, Franz G, Robinson A, Savakis C, *et al.* Incompatible insect technique: incompatible males from a Ceratitis capitata genetic sexing strain. Entomologia Experimentalis et Applicata 2009;132:232–40.
120. Dobson SL, Marsland EJ, Rattanadechakul W. Mutualistic Wolbachia infection in *Aedes albopictus*: accelerating cytoplasmic drive. Genetics 2002;160:1087–94.

121. Dutra HLC, Rocha MN, Dias FBS, Mansur SB, Caragata EP, Moreira LA. Wolbachia blocks currently circulating Zika virus isolates in Brazilian *Aedes aegypti* mosquitoes. Cell Host & Microbe 2016;19:771–4.
122. Blagrove MS, Arias-Goeta C, Di Genua C, Failloux A-B, Sinkins SP. AWolbachia wMel transinfection in *Aedes albopictus* is not detrimental to host fitness and inhibits Chikungunya virus. PLoS Neglected Tropical Diseases 2013;7:e2152.
123. Xi Z, Khoo CC, Dobson SL. Wolbachia establishment and invasion in an *Aedes aegypti* laboratory population. Science 2005;310:326–8.
124. Klasson L, Kambris Z, Cook PE, Walker T, Sinkins SP. Horizontal gene transfer betweenWolbachia and the mosquito *Aedes aegypti*. BMC Genomics 2009;10:33.
125. Brelsfoard CL, Dobson SL. Wolbachia-based strategies to control insect pests and disease vectors. Asia Pacific Journal of Molecular Biology and Biotechnology 2009;17:55–63.
126. Thomas DD, Donnelly CA, Wood RJ, Alphey LS. Insect population control using a dominant, repressible, lethal genetic system. Science 2000;287:2474–6.
127. Labbé GM, Scaife S, Morgan SA, Curtis ZH, Alphey L. Female-specific flightless (fsRIDL) phenotype for control of *Aedes albopictus*. PLoS Neglected Tropical Diseases 2012;6:e1724.
128. Alphey L. Genetic control of mosquitoes. Annual Review of Entomology 2014;59:205–224.
129. Rozeboom LE, Rosen L, Ikeda J. Observations on oviposition by *Aedes* (s.) *albopictus* skuse and A.(S.) *polynesiensis* marks in nature. Journal of Medical Entomology 1973;10:397–9.
130. Lacour G, Chanaud L, L'Ambert G, Hance T. Seasonal synchronization of diapause phases in *Aedes albopictus* (Diptera: Culicidae). PLoS One 2015;10:e0145311.
131. Edgerly J, Mcfarland M, Morgan P, Livdahl T. A seasonal shift in egg-laying behaviour in response to cues of future competition in a treehole mosquito. Journal of Animal Ecology 1998;67:805–18.
132. Chandel K, Suman DS, Wang Y, Unlu I, Williges E, Williams GM, *et al.* Targeting a hidden enemy: Pyriproxyfen autodissemination strategy for the control of the container mosquito *Aedes albopictus* in cryptic habitats. PLoS Neglected Tropical Diseases 2016;10:e0005235.
133. Sun D, Williges E, Unlu I, Healy S, Williams GM, Obenauer P, *et al.* Taming a tiger in the city: comparison of motorized backpack applications and source reduction against the Asian tiger mosquito, *Aedes albopictus*. Journal of the American Mosquito Control Association 2014;30:99–105.
134. Kartzinel MA, Alto BW, Deblasio MW, Burkett-Cadena ND. Testing of visual and chemical attractants in correlation with the development and field evaluation of an autodissemination station for the suppression of *Aedes aegypti* and *Aedes albopictus* in Florida. Journal of the American Mosquito Control Association 2016;32:194–202.
135. Marina CF, Bond JG, Casas M, Muñoz J, Orozco A, Valle J, *et al.* Spinosad as an effective larvicide for control of *Aedes albopictus* and *Aedes aegypti*, vectors of dengue in southern Mexico. Pest Management Science 2011;67:114–21.
136. Nelder M, Kesavaraju B, Farajollahi A, Healy S, Unlu I, Crepeau T, *et al.* Suppressing *Aedes albopictus*, an emerging vector of dengue and chikungunya viruses, by a novel combination of a monomolecular film and an insect-growth regulator. American Journal of Tropical Medicine and Hygiene 2010;82:831–7.
137. Farajollahi A, Williams GM, Condon GC, Kesavaraju B, Unlu I, Gaugler R. Assessment of a direct application of two Bacillus thuringiensis israelensis formulations for immediate and residual control of *Aedes albopictus*. Journal of the American Mosquito Control Association 2013;29:385–8.
138. Abramides GC, Roiz D, Guitart R, Quintana S, Guerrero I, Giménez N. Effectiveness of multiple intervention strategy for the control of the tiger mosquito (*Aedes albopictus*) in Spain. Transactions of the Royal Society of Tropical Medicine and Hygiene 2011;105:281–8.
139. Becnel J, Garcia J, Johnson M. Effects of three larvicides on the production of *Aedes albopictus* based on removal of pupal exuviae. Journal of the American Mosquito Control Association 1996;12:499–502.
140. Boyce R, Lenhart A, Kroeger A, Velayudhan R, Roberts B, Horstick O. *Bacillus thuringiensis israelensis* (*Bti*) for the control of dengue vectors: systematic literature review. Tropical Medicine & International Health 2013;18:564–77.
141. Nayar JK, Ali A, Zaim M. Effectiveness and residual activity comparison of granular formulations of insect growth regulators pyriproxyfen and s-methoprene against Florida mosquitoes in laboratory and outdoor conditions. Journal of the American Mosquito Control Association 2002;18:196–201.

142. Toma L, Severini F, Bella A, Romi R. A semifield evaluation of Vectobac DT (ABG-6499), a new formulation of *Bacillus thuringiensis israelensis* for control of *Aedes albopictus*. Journal of the American Mosquito Control Association 2003;19:424–9.
143. Nishiura JT, Ho P, Ray K. Methoprene interferes with mosquito midgut remodeling during metamorphosis. Journal of Medical Entomology 2003;40:498–507.
144. Levy R, Nichols M, Miner T. Jr Control of immature mosquitoes with single-, joint-, or multi-action polymer-base insecticide delivery system. Proceedings, International Symposium on Controlled Release of Bioactive Materials, Basel, Switzerland, 1998.
145. Nasci R, Wright G, Willis F. Control of *Aedes albopictus* larvae using time-release larvicide formulations in Louisiana. Journal of the American Mosquito Control Association 1994;10:1–6.
146. Furutani S, Arita-Tsutsumi L. Use of *Bacillus thuringiensis israelensis* and methoprene to control Asian tiger mosquito, *Aedes albopictus* (Skuse) (Diptera: Culicidae), in non-circulating hydroponics tanks; 2001.
147. Ravensberg WJ. A Roadmap to the Successful Development and Commercialization of Microbial Pest Control Products for Control of Arthropods. Dordrecht: Springer Science & Business Media; 2011, pp. 1–53.
148. Cohen E. Chitin biochemistry: synthesis and inhibition. Annual Review of Entomology 1987;32:71–93.
149. Mulla MS, Thavara U, Tawatsin A, Chompoosri J, Zaim M, Su T. Laboratory and field evaluation of novaluron, a new acylurea insect growth regulator, against *Aedes aegypti* (Diptera: Culicidae). Journal of Vector Ecology 2003;28:241–54.
150. Arredondo-Jimenez JI, Valdez-Delgado KM. Effect of Novaluron (Rimon® 10 EC) on the mosquitoes Anopheles albimanus, Anopheles pseudopunctipennis, *Aedes aegypti*, *Aedes albopictus* and *Culex quinquefasciatus* from Chiapas, Mexico. Medical and Veterinary Entomology 2006;20:377–87.
151. Farnesi LC, Brito JM, Linss JG, Pelajo-Machado M, Valle D, Rezende GL. Physiological and morphological aspects of *Aedes aegypti* developing larvae: effects of the chitin synthesis inhibitor novaluron. PLoS One 2012;7:e30363.
152. Hirano M, Hatakoshi M, Kawada H, Takimoto Y. Pyriproxyfen and other juvenile hormone analogues. Reviews in Toxicology 1998;2:357–94.
153. Doud CW, Hanley AM, Chalaire KC, Richardson AG, Britch SC, Xue R-D. Truck-mounted area-wide application of pyriproxyfen targeting *Aedes aegypti* and *Aedes albopictus* in Northeast Florida. Journal of the American Mosquito Control Association 2014;30:291–7.
154. Unlu I, Faraji A, Williams GM, Marcombe S, Fonseca DM, Gaugler R. Truck-mounted area-wide applications of larvicides and adulticides for extended suppression of adult *Aedes albopictus*. Pest Management Science 2019;75(4):1115–22.
155. Seccacini E, Lucia A, Harburguer L, Zerba E, Licastro S, Masuh H. Effectiveness of pyriproxyfen and diflubenzuron formulations as larvicides against *Aedes aegypti*. Journal of the American Mosquito Control Association 2008;24:398–403.
156. Consuitation I. Report of the WHO; 1998.
157. Sullivan J, Kean SG. Environmental Fate of Pyriproxyfen. Journal of Pesticide Science 2008;33(4):339–50.
158. Schaefer C, Mulligan F. III Potential for resistance to pyriproxyfen: a promising new mosquito larvicide. Journal of the American Mosquito Control Association 1991;7:409–11.
159. Yapabandara A, Curtis C. Laboratory and field comparisons of pyriproxyfen, polystyrene beads and other larvicidal methods against malaria vectors in Sri Lanka. Acta Tropica 2002;81:211–23.
160. Vythilingam I, Luz BM, Hanni R, Beng TS, Huat TC. Laboratory and field evaluation of the insect growth regulator pyriproxyfen (Sumilarv 0.5 G) against dengue vectors. Journal of the American Mosquito Control Association 2005;21:296–300.
161. Dame D, Wichterman G, Hornby J. Mosquito (*Aedes taeniorhynchus*) resistance to methoprene in an isolated habitat. Journal of the American Mosquito Control Association 1998;14:200–3.
162. Sihuincha M, Zamora-Perea E, Orellana-Rios W, Stancil JD, Lopez-Sifuentes V, Vidal-Ore C, *et al.* Potential use of pyriproxyfen for control of *Aedes aegypti* (Diptera: Culicidae) in Iquitos, Peru. Journal of Medical Entomology 2005;42:620–30.
163. Gaugler R, Suman D, Wang Y. An autodissemination station for the transfer of an insect growth regulator to mosquito oviposition sites. Medical and Veterinary Entomology 2012;26:37–45.
164. Wang Y, Suman DS, Bertrand J, Dong L, Gaugler R. Dual-treatment autodissemination station with enhanced transfer of an insect growth regulator to mosquito oviposition sites. Pest Management Science 2014;70:1299–304.

165. Mains JW, Brelsfoard CL, Dobson SL. Male mosquitoes as vehicles for insecticide. PLoS Neglected Tropical Diseases 2015;9:e0003406–6.
166. Itoh T. Utilization of blood fed females of *Aedes aegypti* as a vehicle for the transfer of the insect growth regulator, pyriproxyfen, to larval habitats. Tropical Medicine 1994;36:243–8.
167. Snetselaar J, Andriessen R, Suer RA, Osinga AJ, Knols BG, Farenhorst M. Development and evaluation of a novel contamination device that targets multiple life-stages of *Aedes aegypti*. Parasites & Vectors 2014;7:200.
168. Kawada H, Shono Y, Ito T, Abe Y. Laboratory evaluation of insect growth regulators against several species of anopheline mosquitoes. Japanese Journal of Sanitary Zoology 1993;44:349–53.
169. Abad-Franch F, Zamora-Perea E, Ferraz G, Padilla-Torres SD, Luz SL. Mosquito-disseminated pyriproxyfen yields high breeding-site coverage and boosts juvenile mosquito mortality at the neighborhood scale. PLoS Neglected Tropical Diseases 2015;9(4):e0003702.
170. Lam PHY, Boon CS, Yng NY, Benjamin S. *Aedes albopictus* control with spray application of *Bacillus thuringiensis israelensis*, strain AM 65-52. Southeast Asian Journal of Tropical Medicine and Public Health 2010;41:1071.
171. Fiorenzano JM, Koehler PG, Xue R-D. Attractive toxic sugar bait (ATSB) for control of mosquitoes and its impact on non-target organisms: A review. International Journal of Environmental Research and Public Health 2017;14:398.
172. Lea A. Sugar-baited insecticide residues against mosquitoes. Mosquito News 1965;25:65–6.
173. Xue R-D, Kline DL, Ali A, Barnard DR. Application of boric acid baits to plant foliage for adult mosquito control. Journal of the American Mosquito Control Association 2006;22:497–500.
174. Beier JC, Müller G C, Gu W, Arheart KL, Schlein Y. Attractive toxic sugar bait (ATSB) methods decimate populations of Anopheles malaria vectors in arid environments regardless of the local availability of favoured sugar-source blossoms. Malaria Journal 2012;11(1):31.
175. Naranjo DP, Qualls WA, Müller GC, Samson DM, Roque D, Alimi T, *et al.* Evaluation of boric acid sugar baits against *Aedes albopictus* (Diptera: Culicidae) in tropical environments. Parasitology Research 2013;112:1583–7.
176. Revay EE, Müller GC, Qualls WA, Kline DL, Naranjo DP, Arheart KL, *et al.* Control of *Aedes albopictus* with attractive toxic sugar baits (ATSB) and potential impact on non-target organisms in St. Augustine, Florida. Parasitology Research 2014;113:73–9.
177. Britch SC, Linthicum KJ, Wynn WW, Walker TW, Farooq M, Smith VL, *et al.* Evaluation of barrier treatments on native vegetation in a southern California desert habitat. Journal of the American Mosquito Control Association 2009,25:184–93.
178. Trout R, Brown G, Potter M, Hubbard J. Efficacy of two pyrethroid insecticides applied as barrier treatments for managing mosquito (Diptera: Culicidae) populations in suburban residential properties. Journal of Medical Entomology 2007;44:470–7.
179. Doyle MA, Kline DL, Allan SA, Kaufman PE. Efficacy of residual bifenthrin applied to landscape vegetation against *Aedes albopictus*. Journal of the American Mosquito Control Association 2009;25:179–83.
180. Li C-X, Wang Z-M, Dong Y-D, Yan T, Zhang Y-M, Guo X-X, *et al.* Evaluation of lambda-cyhalothrin barrier spray on vegetation for control of *Aedes albopictus* in China. Journal of the American Mosquito Control Association 2010;26:346–8.
181. Muzari OM, Adamczyk R, Davis J, Ritchie S, Devine G. Residual effectiveness of λ-cyhalothrin harbourage sprays against foliage-resting mosquitoes in North Queensland. Journal of Medical Entomology 2014;51:444–9.
182. Cilek J, Hallmon C. Residual effectiveness of pyrethroid-treated foliage against adult *Aedes albopictus* and *Culex quinquefasciatus* in screened field cages. Journal of the American Mosquito Control Association 2006;22:725–31.
183. Lothrop HD, Lothrop BB, Gomsi DE, Reisen WK. Intensive early season adulticide applications decrease arbovirus transmission throughout the Coachella Valley, Riverside County, California. Vector-Borne and Zoonotic Diseases 2008;8:475–90.
184. Macedo PA, Schleier III JJ, Reed M, Kelley K, Goodman GW, Brown DA, *et al.* Evaluation of efficacy and human health risk of aerial ultra-low volume applications of pyrethrins and piperonyl butoxide for adult mosquito management in response to West Nile virus activity in Sacramento County, California. Journal of the American Mosquito Control Association 2010;26:57–66.

185. Mount GA. A critical review of ultralow-volume aerosols of insecticide applied with vehicle-mounted generators for adult mosquito control. Journal of the American Mosquito Control Association 1998;14:305–34.
186. Reiter P, Gubler DJ, Ooi EE, Vasudevan S, Farrar J. Surveillance and control of urban dengue vectors. Dengue and Dengue Hemorrhagic Fever, 2nd edition. Wallingford, UK: CAB International; 2014, pp. 481–518.
187. Focks D, Kloter K, Carmichael G. The impact of sequential ultra-low volume ground aerosol applications of malathion on the population dynamics of *Aedes aegypti* (L.). The American Journal of Tropical Medicine and Hygiene 1987;36:639–47.
188. Gubler DJ, Clark GG. Community involvement in the control of *Aedes aegypti*. Acta Tropica 1996;61:169–79.
189. Dave DC. An evaluation of malathion ULV spraying against caged and natural populations of *Aedes aegypti* in Trinidad, West Indies; 1985.
190. Kline D, Lemire G. Evaluation of attractant-baited traps/targets for mosquito management on Key Island, Florida, USA. Journal of Vector Ecology: Journal of the Society for Vector Ecology 1998;23:171–85.
191. Vale G, Hargrove J, Cockbill G, Phelps R. Field trials of baits to control populations of *Glossina morsitans morsitans* Westwood and *G. pallidipes* austen (Diptera: Glossinidae). Bulletin of Entomological Research 1986;76:179–93.
192. Takken V, Oladunmade M, Dengwat L, Feldmann H, Onah J, Tenabe S, *et al.* The eradication of *Glossina palpalis palpalis* (Robineau-Desvoidy) (Diptera: Glossinidae) using traps, insecticide-impregnated targets and the sterile insect technique in central Nigeria. Bulletin of Entomological Research 1986;76:275–86.
193. Kline DL. Evaluation of various models of propane-powered mosquito traps. Journal of Vector Ecology 2002;27:1–7.
194. Barrera R, Amador M, Acevedo V, Hemme RR, Félix G. Sustained, area-wide control of *Aedes aegypti* using CDC autocidal gravid ovitraps. The American Journal of Tropical Medicine and Hygiene 2014;91:1269–76.
195. Johnson BJ, Brosch D, Christiansen A, Wells E, Wells M, Bhandoola AF, *et al.* Neighbors help neighbors control urban mosquitoes. Scientific Reports 2018;8:15797.
196. Heringer L, Johnson BJ, Fikrig K, Oliveira BA, Silva RD, Townsend M, *et al.* Evaluation of alternative killing agents for *Aedes aegypti* (Diptera: Culicidae) in the gravid *Aedes* trap (GAT). Journal of Medical Entomology 2016;53:873–79.
197. Marten G. A survey of cyclopoid copepods for control of *Aedes albopictus* larvae. Bulletin of the Society of Vector Ecologists 1989;14:232–6.
198. Marten GG. Evaluation of cyclopoid copepods for *Aedes albopictus* control in tires. Journal of the American Mosquito Control Association 1990;6:681–8.
199. Marten GG, Bordes ES, Nguyen M. Use of cyclopoid copepods for mosquito control. In: Ecology and Morphology of Copepods. Springer, Dordrecht; 1994, pp. 491–496.
200. Miyagi I, Toma T, Mogi M. Biological control of container-breeding mosquitoes, *Aedes albopictus* and *Culex quinquefasciatus*, in a Japanese island by release of *Toxorhynchites splendens* adults. Medical and Veterinary Entomology 1992;6:290–300.
201. Focks DA, Sackett SR, Bailey DL. Field experiments on the control of *Aedes aegypti* and *Culex quinquefasciatus* by *Toxorhynchites rutilus rutilus* (Diptera: Culicidae). Journal of Medical Entomology 1982;19:336–9.
202. Nauen R. Insecticide resistance in disease vectors of public health importance. Pest Management Science 2007;63(7):628–33.
203. Kamgang B, Marcombe S, Chandre F, Nchoutpouen E, Nwane P, Etang J, *et al.* Insecticide susceptibility of *Aedes aegypti* and *Aedes albopictus* in Central Africa. Parasites & Vectors 2011;4:79.
204. Harris AF, Rajatileka S, Ranson H. Pyrethroid resistance in *Aedes aegypti* from Grand Cayman. The American Journal of Tropical Medicine and Hygiene 2010;83:277–84.
205. Marcombe S, Farajollahi A, Healy SP, Clark GG, Fonseca DM. Insecticide resistance status of United States populations of *Aedes albopictus* and mechanisms involved. PLoS One 2014;9:e101992.
206. Kay B, Nam VS. New strategy against *Aedes aegypti* in Vietnam. The Lancet 2005;365:613–7.
207. Dowling Z, Armbruster P, Ladeau SL, Decotiis M, Mottley J, Leisnham PT. Linking mosquito infestation to resident socioeconomic status, knowledge, and source reduction practices in suburban Washington, DC. EcoHealth 2013;10:36–47.

208. Sanchez L, Perez D, Perez T, Sosa T, Cruz G, Kouri G, *et al.* Intersectoral coordination in *Aedes aegypti* control. A pilot project in Havana City, Cuba. Tropical Medicine & International Health 2005;10:82–91.
209. Schreiber E, Morris C. Evaluation of public information packets for mosquito source reduction in two Florida cities. Journal of the American Mosquito Control Association-Mosquito News 1995;11:186–90.
210. Healy K, Hamilton G, Crepeau T, Healy S, Unlu I, Farajollahi A, *et al.* Integrating the public in mosquito management: active education by community peers can lead to significant reduction in peridomestic container mosquito habitats. PLos One 2014;9:e108504.
211. Carrieri M, Bellini R, Maccaferri S, Gallo L, Maini S, Celli G. Tolerance thresholds for *Aedes albopictus* and *Aedes caspius* in Italian urban areas. Journal of the American Mosquito Control Association 2008;24:377–86.
212. Corrêa PRL, França E, Bogutchi TF. *Aedes aegypti* infestation and occurrence of dengue in the city of Belo Horizonte, Brazil. Revista de Saude Publica 2005;39:33–40.
213. Sanchez L, Cortinas J, Pelaez O, Gutierrez H, Concepción D, Van Der Stuyft P. Breteau Index threshold levels indicating risk for dengue transmission in areas with low *Aedes* infestation. Tropical Medicine & International Health 2010;15:173–5.
214. Thammapalo S, Chongsuvivatwong V, Geater A, Dueravee M. Environmental factors and incidence of dengue fever and dengue haemorrhagic fever in an urban area, Southern Thailand. Epidemiology & Infection 2008;136:135–43.
215. Chadee D. Dengue cases and *Aedes aegypti* indices in Trinidad, West Indies. Acta Tropica 2009;112:174–80.
216. Maciel-De-Freitas R, Lourenco-De-Oliveira R. Does targeting key-containers effectively reduce *Aedes aegypti* population density? Tropical Medicine & International Health 2011;16:965–73.
217. Tun-Lin W, Lenhart A, Nam V, Rebollar-Téllez E, Morrison A, Barbazan P, *et al.* Reducing costs and operational constraints of dengue vector control by targeting productive breeding places: a multi-country non-inferiority cluster randomized trial. Tropical Medicine & International Health 2009;14:1143–53.
218. Faraji A, Unlu I. The eye of the tiger, the thrill of the fight: effective larval and adult control measures against the Asian tiger mosquito, *Aedes albopictus* (Diptera: Culicidae), in North America. Journal of Medical Entomology 2016;53:1029–47.
219. Suter TT, Flacio E, Fariña BF, Engeler L, Tonolla M, Regis LN, *et al.* Surveillance and control of *Aedes albopictus* in the Swiss-Italian border region: differences in egg densities between intervention and non-intervention areas. PLoS Neglected Tropical Diseases 2016;10(1):e0004315.
220. Facchinelli L, Valerio L, Pombi M, Reiter P, Costantini C, Della Torre A. Development of a novel sticky trap for container-breeding mosquitoes and evaluation of its sampling properties to monitor urban populations of *Aedes albopictus*. Medical and Veterinary Entomology 2007;21:183–95.
221. Carrieri M, Angelini P, Venturelli C, Maccagnani B, Bellini R. *Aedes albopictus* (Diptera: Culicidae) population size survey in the 2007 Chikungunya outbreak area in Italy. I. Characterization of breeding sites and evaluation of sampling methodologies. Journal of Medical Entomology 2011;48:1214–25.
222. Unlu I, Roy AF, Yates M, Garrett D, Bell H, Harden T, *et al.* Evaluation of surveillance methods for detection of West Nile virus activity in East Baton Rouge Parish, Louisiana, 2004–2006. Journal of the American Mosquito Control Association 2009;25:126–33.
223. Koppenhöfer HS, Gaugler R. Entomopathogenic nematode and bacteria mutualism. In: White JF, Torres MS, editors. Defensive Mutualism in Microbial Symbiosis, CRC Press, Boca Raton, FL; 2009, pp. 99–116.
224. Cho C-H, Whang KS, Gaugler R, Yoo SK. Submerged monoxenic culture medium development for *Heterorhabditis bacteriophora* and its symbiotic bacterium *Photorhabdus luminescens*: protein sources. Journal of Microbiology and Biotechnology 2011;21:869–73.
225. Gaugler R. Entomopathogenic Nematology. CABI; 2002.
226. Poinar GO Jr, Kaul H. Parasitism of the mosquito *Culex pipiens* by the *nematode Heterorhabditis bacteriophora*. Journal of Invertebrate Pathology 1982;39:382–7.
227. Cardoso DDO, Gomes VM, Dolinski C, Souza RM, Idalino WSS. Efficacy of *Heterorhabditis indica* LPP35 against *Aedes aegypti* larvae in human-generated containers and bromeliads. Nematoda 2016;3.
228. Gaugler R, Lewis E, Stuart RJ. Ecology in the service of biological control: the case of entomopathogenic nematodes. Oecologia 1997;109:483–9.

229. Welch H, Bronskill JF. Parasitism of mosquito larvae by the nematode, DD136 (Nematoda: Neoaplectanidae). Canadian Journal of Zoology 1962;40:1263–8.
230. Pérez-Pacheco R, Rodríguez-Hernández C, Lara-Reyna J, Montes-Belmont R, Ruiz-Vega J. Control of the mosquito *Anopheles pseudopunctipennis* (Diptera: Culicidae) with *Romanomermis iyengari* (Nematoda: Mermithidae) in Oaxaca, Mexico. Biological Control 2005;32:137–42.
231. Abagli AZ, Alavo TB, Platzer EG. Efficacy of the insect parasitic nematode, *Romanomermis iyengari*, for malaria vector control in BeninWest Africa. Malaria Journal 2012;11:5.
232. Petersen J, Willis Q. Procedures for the mass rearing of a mermithid parasite of mosquitoes. Mosquito News 1972;32:226–30.
233. Dickinson JL, Zuckerberg B, Bonter DN. Citizen science as an ecological research tool: challenges and benefits. Annual Review of Ecology, Evolution, and Systematics 2010;41:149–72.
234. Cohnstaedt LW, Ladner J, Campbell LR, Busch N, Barrera R. Determining mosquito distribution from egg data: The role of the citizen scientist. The American Biology Teacher 2016;78:317–22.
235. Jordan RC, Sorensen AE, Ladeau S. Citizen science as a tool for mosquito control. Journal of the American Mosquito Control Association 2017;33:241–5.
236. Bartlett-Healy K, Unlu I, Obenauer P, Hughes T, Healy S, Crepeau T, *et al.* Larval mosquito habitat utilization and community dynamics of *Aedes albopictus* and *Aedes japonicus* (Diptera: Culicidae). Journal of Medical Entomology 2012;49:813–24.

10 The threat of the larger grain borer, *Prostephanus truncatus* (Coleoptera: Bostrichidae) and practical control options for the pest

Bernardo L. Muatinte[1*], Nickolas G. Kavallieratos[2], Maria C. Boukouvala[2,3], Silverio García-Lara[4], L. Margarita López-Castillo[4] and Brighton M. Mvumi[5]
[1]Department of Biological Sciences, Faculty of Sciences, Eduardo Mondlane University, P.O. Box 257, Maputo, Mozambique; [2]Laboratory of Agricultural Zoology and Entomology, Department of Crop Science, Agricultural University of Athens, 75 Iera Odos str., 11855, Athens, Attica, Greece; [3]Laboratory of Organic Chemistry, Department of Chemistry, University of Ioannina, Panepistimioupolis, 45110, Ioannina, Greece; [4]Biotechnology Center, Tecnológico de Monterrey, Campus Monterrey, C.P. 64849, Monterrey, N.L., Mexico; [5]Department of Soil Science and Agricultural Engineering, Faculty of Agriculture, University of Zimbabwe, P.O. Box MP167, Mt Pleasant, Harare, Zimbabwe

Abstract

It is about 40 years since *Prostephanus truncatus*, a devastating pest of stored maize and dried cassava, was accidentally introduced into Africa from meso-America. Losses of stored maize at least doubled because of the destructive nature of the pest. *Prostephanus truncatus* management measures generated and recommended then worked at that time. However, the pest remains a serious threat to food security in developing tropical countries as it continues to spread to other geographical areas. This paper reviews the status of scientific knowledge of the pest and practical control options in the context of reductions in synthetic pesticide use; pesticide resistance management; and consumer, worker and environmental safety, and in the process, identifies research and development gaps. We present a scenario of the threat of *P. truncatus* on stored-maize grain and dry cassava chips and discuss, critically analyse the economic feasibility of control methods that have been generated and applied through scientific research. This review reveals a clear paucity of reliable data on economic analysis of most interventions, a critical factor for farmer adoption. A gap is also identified in the uptake of some of the developed technologies; mainly ascribed to affordability issues and a lack of evidence-based economic analysis under the user's circumstances. There is a lack of reliable national postharvest loss data based on the measurement and currently, the most robust loss estimates for sub-Saharan Africa are available on the African Postharvest Loss Information System (APHLIS) online platform. Any emerging *P. truncatus* control methods should be evaluated for efficacy and cost-effectiveness under realistic circumstances together with the target end-users of the methods to assess appropriateness.

*Corresponding author: muatinteb@uem.mz, muatinteb@yahoo.com

Keywords: Economic analysis, Postharvest food loss, Chemical and non-chemical control, Geographical distribution, Integrated pest management

Review Methodology

Published and unpublished documents were sought from online databases and search engines using the terms "*Prostephanus truncatus* and/or larger grain borer (LGB)" either in the title and/or abstract. The documents were screened according to content relevance, and scientific and methodological consistency. Furthermore, documents were critically analysed for: (a) quantitative and economic loss caused by *P. truncatus* in dried maize and cassava chips; and (b) efficacy and economics of available control measures for *P. truncatus*.

Introduction

The larger grain borer, *Prostephanus truncatus* (Horn) (Coleoptera: Bostrichidae), is native to Central and South America [1] and it was first recorded in Tanzania at the end of the 1970s [2–4]. The theory is that it was accidentally introduced from Central America, where it was considered an occasional pest of stored maize [5, 6], into several countries of Africa [6, 7]. Data on the presence of *P. truncatus* in Asia (China) [8] and in North America (Arizona and California in the USA) [7], where it has a likely restricted distribution, require more details. From Tanzania, the pest spread to east and central Africa [9–11], west Africa [12–16] and southern Africa [17–21].

Prostephanus truncatus is a pest of economic importance in stored maize [16, 21–23] and dried cassava roots [24–27], particularly in Africa, where it reduces the storage period and availability of these commodities as a staple food and as income sources, mostly affecting small-scale farmers. *Prostephanus truncatus* is capable of infesting and damaging stored timber and timber products [28]. However, levels of infestation, damage and losses caused by *P. truncatus* in stored commodities are still underestimated [6, 19, 22, 29, 30] and assessment of the economic effect is often missing in most scientific publications or reports on pest control measures.

Management measures for *P. truncatus* worldwide have focused on suppressing population densities. Studies on chemical control [31–34], use of inert dust [35, 36], genetic resistance of host commodities [37–39] and use of plant-derived products [40–43] have been conducted to develop technologies to control *P. truncatus*. Moreover, research into the biological control of *P. truncatus* has also been conducted with strategies such as the use of the predator beetle, *Teretrius nigrescens* Lewis (Coleoptera: Histeridae) [22, 30, 44–46], parasitoids [47, 48] and entomopathogenic fungi [49–51]. The use of semiochemicals, particularly in mass trapping has recently attracted the attention of researchers working on storedproduct insect control, including *P. truncatus* [52, 53].

Integrated pest management (IPM), an ecosystem-based strategy that comprises a combination of several control techniques and measures, is an effective approach, applied for long-term prevention of pests, or their damage [21, 54]. IPM measures against *P. truncatus* include the application of combinations of the afore-mentioned technologies such as combinations of diatomaceous earth (DE) with spinosad or pyrethroids [55–57], wood ash with conidia of the fungus *Beauveria bassiana* (Balsamo) Vuillemin (Ascomycota: Hypocreales) [58], hermetic bags [59, 60] and metal silos [38].

However, reports on the economic impact and practical feasibility and efficacy of results of research tests and application of almost all control methods against *P. truncatus* are scarce. Hence, the objective of this review was to systematize and interrogate the impact of *P. truncatus* threat expressed as stored maize and dry cassava infestation and damage, in conjunction with practical control methods of this pest. For that purpose, we critically reviewed the relevant globally available scientific publications that deal with this pest. With the aim of illustrating the economic impact of *P. truncatus* threat, we collated and systematized the published results on damage and losses of stored-maize grain and dried cassava chips expressed in monetary values. We then analysed the threat of *P. truncatus* on maize grain and dry cassava chips and the practical feasibility of control methods that have been generated and applied. The present review could be a useful resource when designing effective management strategies for *P. truncatus*.

Materials and Methods

The study involved literature screening which was conducted in two parts:

(1) Information screening to identify relevant literature and studies, published and unpublished (grey literature) that included either in the title or abstract the terms '*Prostephanus truncatus* or/and larger grain borer (LGB)' plus:
   (a) Occurrence, distribution, infestation, damage, population dynamics
   (b) Pest control, insect pest management, chemical control, bio-control, integrated insect pest management, biotechnological pest control, use of DE for pest control,
   (c) Biology and life traits and
   (d) Stored products,maize and/or cassava chips storage.

   The terms '*Prostephanus truncatus* or/and larger grain borer (LGB)' were combined either in pair or with several of those mentioned in (a) to (d) terms. The literature and studies screened were coded, e.g. as Th (threat), BC (biocontrol), DE (diatomaceous earth), Oc (occurrence), etc., and afterwards they were deleted from the list as they were counted. This procedure minimized the risk of double counting the literature.

   The published literature screened included papers found in scientific journals and in grey literature including M.Sc. and Ph.D. theses, technical reports, working papers, project reports, evaluation reports, conference presentations and abstracts.

(2) Screening of the relevant available literature based on sound study methodological procedures. Studies with severe methodological imperfections, e.g. no replication of treatments or seasons, were not considered but they were used only to count the relative number of literature published on *P. truncatus*. Published and unpublished information was sought from online databases such as CAB direct (Life Science reference Database) of The Centre for Agriculture and Bioscience International (CABI), Access to Global Online Research in Agriculture (AGORA), Scopus/Science direct, Agricultural On-Line Access (AGRICOLA)/Elton B. Stephens Co. (EBSCO), Journal Storage (JSTOR), Agricultural Information System (AGRIS)/ United Nations for Food and Agriculture Organization (UN-FAO) and the African Postharvest Losses Information System (APHLIS). Research Gate, Google scholar, Google search engines were used to access online published and unpublished literature from publishing houses as well as records from government agencies, universities, research institutions, non-governmental organizations and development partners [The Regional Universities Forum for Capacity Building in Agriculture (RUFORUM), the German Corporation for International Cooperation (GIZ), Federal Ministry of Economic Cooperation and Development (Germany) (BMZ), The Department for International Development, UK (DFID), International Fund for Agricultural Development (IFAD), World Food Programme (WFP), Alliance for a Green Revolution in Africa (AGRA), the French Agricultural Research Centre for International Development (CIRAD), United States Agency for International Development (USAID), Denmark's Development Cooperation (DANIDA), Swiss Agency for Development and Cooperation (SDC), Natural Resource Institute (NRI) (UK), the International Institute of Tropical Agriculture (IITA), and International Maize and Wheat Improvement Center (CIMMYT), The International Food Policy Research Institute (IFPRI)].

Finally, at the synthesis stage, an analytical and critical review report was written, where the percentage loss and monetary values from infestation and damage caused by *P. truncatus* in maize, cassava chips or in other host products were recorded, systematized and presented in tables or graphs. Quantitative values associated with the use of several types of control measures for *P. truncatus* were also recorded. The pest management technique cited in (1b) was viewed as practical for the control of *P. truncatus* if it was reported as cost-effective [14, 29, 32–36], efficacious [29, 38] and environmentally sustainable [21, 34–36, 44, 45]. For this purpose, the control methods cited in (1b) were categorized into cultural and hygiene, chemical, bio-control, IPM, biotechnological, control by inert dust and miscellaneous methods. The miscellaneous category comprised a set of mostly emerging and under research control methods, such as the potential of propolis [61] and symbiotic bacteriosomes [62] for the management of *P. truncatus*. Further on, five original papers published on each of the category (type) of control method were selected and were rated on a scale from 1 to 5, where 1 – method was cited as not effective; 2 – method was cited as fairly cost-effective and efficacious; 3 – method was cited as cost-effective but environmentally unsustainable; 4 – method was referred to as cost-effective but needs much future research; 5 – method was unanimously cited as cost-effective, needs more research and is environmentally sustainable. The papers were selected if they presented a consistent citation of one of the contents mentioned on scales 1–5 and if they followed sound methodological procedures. Moreover, the criteria adopted by Cugala *et al.* [63] were followed to select the papers.

Results and Discussion

The threat posed by Prostephanus truncatus to stored products

Analysis of the threat from *P. truncatus* can be done by considering the magnitude of losses caused at an economic level, the socio-economic environment (whether the economy is developed or developing), the type of host commodities and forest plant species, and the methodological liability of estimated losses caused by this pest. Most of the data on quantitative losses arising from *P. truncatus* either in stored maize [22, 30, 64] or dried cassava chips [24, 27, 65] have been expressed as a percentage of infestation or damage or weight loss of dry matter rather than as economic losses. Reports on the percentage of infestation, damage and weight loss varied within and between countries depending on the period, length of storage season and geographical location (Table 10.1). For instance, first measures of stored-maize weight losses due to *P. truncatus* attack in western Tanzania were estimated at 9–34% over a 6-month stored period [3], while in Arusha (northeastern Tanzania), dry weight loss of this cereal increased to 30% after 9 months of storage [66]. Cugala *et al.* [19] reported up to 62% of maize grain damage recorded from smallholder farmer granaries in the Manica Province, Mozambique. Twelve years later, Muatinte *et al.* [64] estimated 80–100% stored-maize damaged grain in granaries in the Manica Province (Mozambique). Records on stored-cassava losses followed similar patterns as illustrated in Nigeria [28].

Concern about the serious and destructive effect of *P. truncatus* in stored products, such as dried maize and dry cassava chips, was recorded for the first time in the Tabora region, Tanzania [2] seven decades after it was intercepted for the first time in the USA [85]. Fisher [86] recorded *P. truncatus* when taxonomically revising the family Bostrichidae of the USA, supporting, therefore, the findings of Chittenden [85]. The serious concerns regarding the destructive nature of *P. truncatus* can be evaluated by the empirically high number of about 1234 reports released that dealt with the threat, biology and control measures of the pest (Figure 10.1).

The spatial and temporal increase of infestation and damage of commodities by *P. truncatus* is supported by the number of papers published worldwide per decade (Figure 10.2). The data suggest that the publication activity was at a peak in the 1990s and thereafter subsided. This is in accord with the oscillating pattern of interest in sub-Saharan Africa postharvest losses (PHLs) reported by Mvumi and Stathers [87]. The oscillations have largely been linked to food price hikes over the decades. According to Mvumi and Stathers [87], the renewed interest of the international development community in reducing PHLs caused by *P. truncatus* seems to be contradictory to the decline in scientific

Table 10.1. Percentage of maize grain weight losses caused by the larger grain borer, *Prostephanus truncatus*, worldwide

Commodity	(%) loss	Length of storage period (months)	Region/country	References
	15–25	Not specified	Mexico	Najera [67]
Maize	23–77	Not specified	Benin	Holst *et al.* [68], Borgemeister, Holst *et al.* [69], Meikle *et al.* [70], Edoh Ognakossan *et al.* [71]
	13–25	Not specified	Nigeria	Holst *et al.* [68]
	30–45	8	Togo	Richter *et al.* [44], Golob [72], Bell *et al.* [73] Pantennius [74]
	30–40	Not specified	Ghana	Mao and Enderson [75] Anankware *et al.* [76]
	9–48	6	Kenya	Nang'ayo *et al.* [77], Tefera *et al.* [38]
	34–80	6–8	Tanzania	Henckes [78], Keil [66], Nang'ayo *et al.* [77], Bell *et al.* [73], Boxall [5]
	>62	8	Mozambique	Cugala *et al.* [19], Muatinte *et al.* [64]
	40–100	3–6	Malawi	Denning *et al.* [79], Kamanula *et al.* [80]
Cassava	70–80	Not specified	Nigeria	Wright [28], Mukanga *et al.* [81]
	20	7	Togo	Wright [82]
	30	6	Ghana	Magrath *et al.* [83], Bell [73]
	19–30	6–9	Western and Eastern Africa	Markham *et al.* [4], Espinal *et al.* [84], Pantennius [74], Osipitan *et al.* [39]

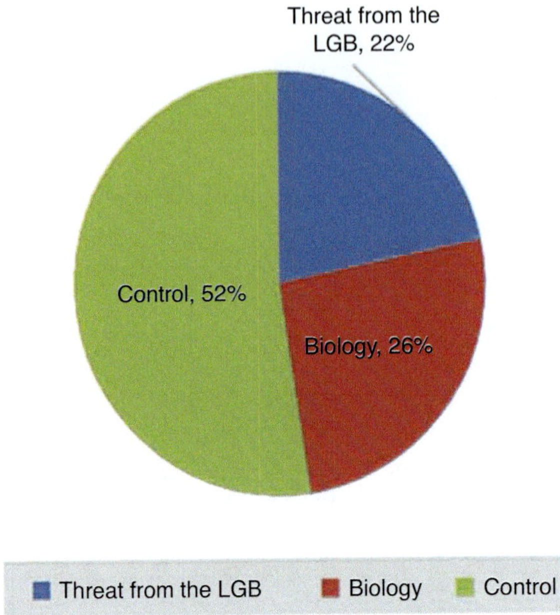

Fig. 10.1. Relative percentage of published literature worldwide on different aspects related to *Prostephanus truncatus* (1911–2018) (Source: Authors' compilations from various sources; 2018).

publication recorded on the threat of the pest (Figure 10.2). Records on the monetary effect of the threat, particularly of infestation, damage and weight loss that the beetle poses, mostly in stored-maize grain and dried cassava chips are scarce. Farrell and Schulten [6] reported a 4-year project of about US$5 million, aimed to suppress *P. truncatus* outbreak to economically acceptable levels in 1982, in Tanzania. The authors estimated that the total disbursement on all *P. truncatus* work in Tanzania was about £6.4 million in 1998 prices. However, the gross benefit was £21.5 million, giving a net benefit of £15.1 million.

Krause and Mück [88], based on the assumption of the absence of the predator, *T. nigrescens* and FAO production statistics for 1994, estimated the annual maize loss, potentially caused by *P. truncatus* of about US$7 million in Benin, Togo and Tanzania. Thirty per cent of annual stored-cassava chip losses of smallholder farmers, corresponding to about US$10.5 million in 1997, was recorded in Eastern Ghana [83]. Moreover, through a developed model under the framework of a long-term *P. truncatus* control project, it has been deduced that if no control measures were undertaken, the pest could cause losses of about US$4 billion in Ghana [89]. Although, the monetary expression of the above losses is out-of-date, and no new data have been published in the last two decades on this aspect. However, it still gives an idea of the potential economic loss.

Furthermore, the threat from *P. truncatus* can be measured by means of commercial losses based on stored-maize trade quarantine status. Therefore, losses may result from complete cessation or ban of maize trade or demand for pesticide treatment of the commodity before exporting. In 1987, the Somalian authorities refused a shipment of maize grain that was exported from Malawi via Tanzania without being fumigated or milled to reduce the risk of infestation by *P. truncatus* [5]. The shipment of maize was also rejected among other countries including by Ethiopia and Mozambique. The costs for fumigation and milling of that maize, the quantity of which was not reported were around US$700 000 [5, 90]. Losses can also be attributed to additional shipping costs and maize re-sell at a lower price [90]. Boxall [5] reported additional direct costs expressed as losses for

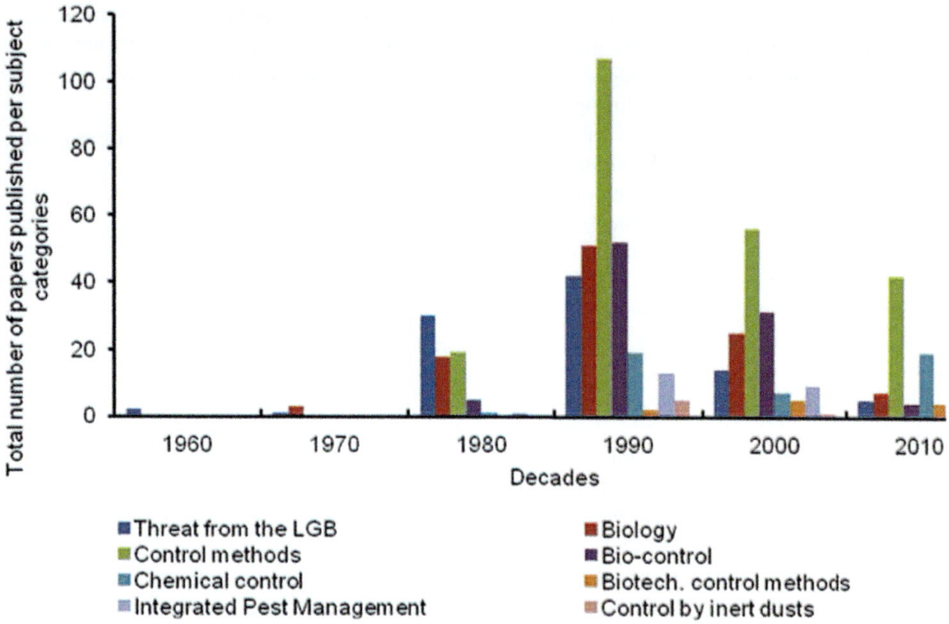

Fig. 10.2. Relative number of papers published worldwide on different control methods related to *Prostephanus truncatus* (1960–2018) (Source: Authors' compilations from various sources; 2018).

both exporting and importing countries and agencies, with consequent and obvious loss of foreign exchange, especially for the exporter. Lugg and Ndibalema [90] addressed additional costs to maize traders, state agencies, loss of income to producers and traders and costs associated with possible disincentives for maize production at national or international level.

Investments in research, stakeholder training, administrative and logistical costs worsen threats from *P. truncatus*. These investments can emerge from costs of planning and conducting scientific studies, organizing and conducting training courses, national and international workshops, conferences, seminars and extension service support activities. Moreover, refurbishment, construction and introduction of new storage facilities that aim to minimize the negative effect of the occurrence of *P. truncatus* also constitute investments that aggravate threats from the pest. In the absence of this notorious pest, these expenses could be channelled towards other developmental investments, such as building of health, education, commercial, administration, justice and many other infrastructural facilities and services.

Threats from *P. truncatus* vary according to the host commodity. However, the host range of this pest is an issue of on-going research often yielding contradictory results. For instance, *P. truncatus* can damage several dried commodities such as maize, wheat, rice, cassava, cowpea, cocoa, coffee, haricot, nuts, yam and sweet potato [3, 5, 91] but its development is favoured by maize and cassava [5, 91, 92]. Under laboratory conditions, *P. truncatus* has been cultured and successfully developed from egg to adult on sorghum, millet, barley and oats [93, 94]. In contrast, the beetle did not cause damage on rice, sorghum and beans [3] and it could not develop on cowpea, haricot beans, cocoa, coffee beans, hard winter wheat and on rough rice [95].

The potential of *P. truncatus* to grow on different grain and amylaceous commodities has recently been studied [96]. Results of the latter studies reinforced reports on the favourable development of *P. truncatus* on maize, mostly cracked kernels and flour. Thus, the condition of stored maize merits attention when it is planned to be stored for long periods or transported within or between countries to prevent the risk of infestation, damage and losses caused by *P. truncatus*. Moreover, unsorted grain and co-existence of various types of bagged infested and uninfested stored products during transshipment can result in high risk of infestation and damage of commodities by *P. truncatus* since this pest can perforate storage bags either by direct penetration or through already defective bags by insects or by improper sealing or even by accidental tearing [97–104].

The threat from *P. truncatus* is worsened by the fact that the beetle can easily move within and between locations and colonize different habitats and grain lots [92, 95, 104] due to its good ability to fly [105]. Therefore, attention should be given to commodities that can be potential hosts of *P. truncatus*, a species of high quarantine concern. Further investigation is required to establish *P. truncatus* host suitability and dispersal patterns in order to design more effective pest management strategies for the pest.

The threat of Prostephanus truncatus arising from its capacity to thrive on forest plant species

The threat from *P. truncatus* can arise from the ability of this beetle to survive, develop and breed in the wood of several forest plant species. The beetle can infest and damage stored-timber and timber-derived products and forest plant species [77]. Reports have been published on the capability of *P. truncatus* to survive and breed in the wood of 20 plant species including the cultivated trees *Anacardium occidentale* (Anacardiaceae), *Mangifera indica* (Anacardiaceae) and *Azadirachta indica* (Meliaceae) [22] and of 27 native and agroforestry trees and shrubs in Kenya [77]. Besides the findings on survival and breeding of *P. truncatus* through forest plant species [106–108], the pest was successfully reared under laboratory conditions in the wood of six plant species sold as firewood [21] in Mozambique.

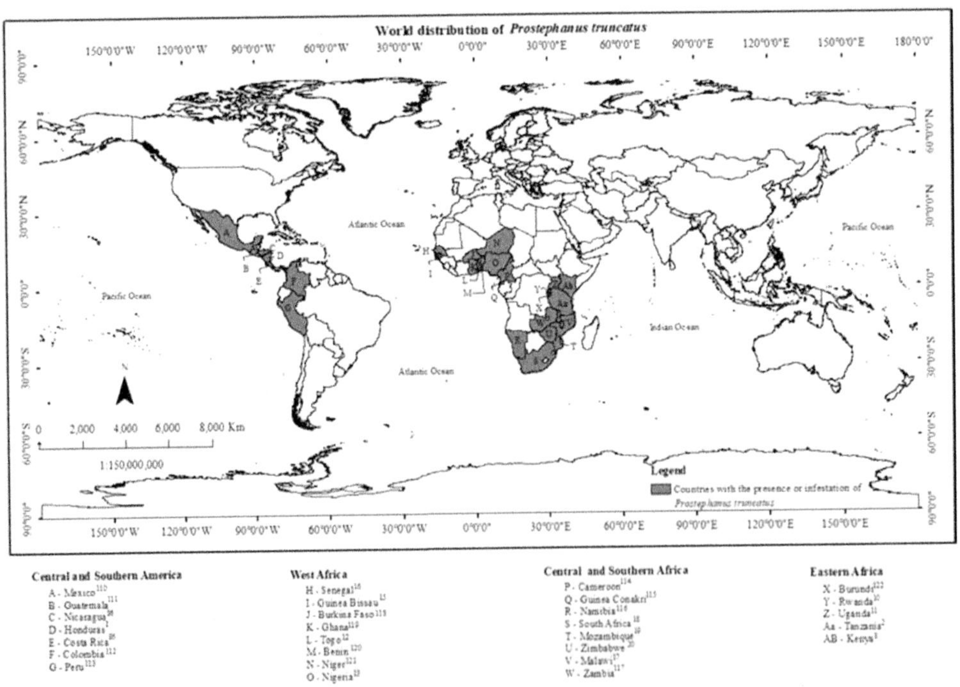

Fig. 10.3. Current world distribution of *Prostephanus truncatus* (Map drawn by Mr Joaquim Campira, SECOSUD II Italia-Mozambique Cooperation Project, Eduardo Mondlane University, Mozambique. ArcGIS 10.2/2013).

Survival, development and breeding of *P. truncatus* in forest plant species enable the quantitative multiplication of beetle individuals, particularly when the preferred commodities are in short supply. In-country and regional transport and trade of such suitable plant species may enable the spatial and temporal dispersal of the pest, increasing the capacity of the beetle to invade and colonize new habitats, infest, damage and cause losses to maize and dry cassava globally where the environmental conditions are suitable. The use of most of these plant species for constructing postharvest structures such as granaries and cribs and homestead fences is an additional factor that contributes to *P. truncatus* breeding and dispersal, especially in Africa. The latter factors may contribute to, and increase, the threat from *P. truncatus*. Hence for people who rely on maize and/or cassava as sources of food and income, their food and nutrition security can be significantly threatened. However, threats that result from the survival and breeding of *P. truncatus* on forest plant species is still unknown and needs further investigation.

The threat from *P. truncatus* is related to its spatial and temporal dispersal, colonization, damage and high losses of stored maize grain and dry cassava in the nearest future. Current global distribution shows that the beetle has already spread to about 21 African countries with Ethiopia, Somalia and the Kingdom of Eswatini in danger of *P. truncatus* invasion or already hosting the beetle but not yet officially reported (Figure 10.3). Despite the estimates of potential spatial expansion range of *P. truncatus* based on the regional similarity of climate variables such as temperature, humidity and precipitation [109], countries with either intensive export or import of maize or bordering countries should be concerned about the risk of *P. truncatus* occurrence. Since dispersal, establishment, colonization of new habitats, infestation, damage caused by the pest mostly on stored dry maize and cassava depend on climate, availability of host commodities [65, 91, 94, 96], the presence of forest host plants [21, 77], the extensiveness of intra-country or international trade and the effectiveness of applied control measures [55–60, 64, 69–71]. *Prostephanus truncatus* now occurs in approximately 40% of the African continent. However, data on pest's status are available and up-to-date from the southern [20, 21] and eastern [29, 63] Africa (Figure 10.3). Recently, the pest has a restricted distribution in most West African countries [7]. *Prostephanus truncatus* has been intercepted in the past in Iraq [92, 110] and Israel [92, 111] and in France [92, 112]. Moreover, the presence of *P. truncatus* has also been reported in numerous countries such as Peru, Columbia, Nicaragua, Costa Rica, Honduras and Brazil. However, there is no up-to-date information on pest's status in these countries, an issue that may imprecise the current knowledge about the worldwide distribution of *P. truncatus* (Figure 10.3).

The phytosanitary classification of *P. truncatus* is A1 [absent from the European Plant Protection Organization (EPPO) region] or A2 (present in the EPPO region) [7] while it is a quarantine organism in Israel, Jordan and New Zealand [7]. *Prostephanus truncatus* can also be found infesting maize in the field before harvest [21, 113] but its presence cannot be easily detected during this period, an issue that commonly leads to the introduction of the infested maize into storage facilities [113–115]. The extensive spatial distribution of *P. truncatus* [7] indicates that it has a very high invasive potential. Therefore, the development of cost-effective, practical and environmentally sustainable management tools for the pest is needed to prevent its further expansion.

Chemical control based on pyrethroids and organophosphorus compounds

Most farmers have tended to rely on synthetic pesticides as their first line of protecting stored maize. However, one of the most important issues regarding the chemical control of *P. truncatus* is the development of insecticide resistance [116, 117]. Therefore, several research efforts on *P. truncatus* have been conducted in the last decades to optimize the management of this species by using single or combined insecticidal treatments under different treatment scenarios (Table 10.2). Suppression of *P. truncatus* progeny production has been achieved in synthetic pesticide-treated maize grain stored in bins or polypropylene bags for a period of 9 months [118]. The maize grain was treated with 0.5 mg/kg of deltamethrin combined with either 8 mg/

kg of pirimiphos-methyl or chlorpyrifos-methyl. More recently, *P. truncatus* has been restricted to <50 adults/kg of maize treated with a mixture of fenitrothion 1% and deltamethrin 0.13% or pirimiphos-methyl 1.6% and permethrin 0.4% for a period of 40 weeks of storage in polypropylene bags in Zimbabwe [117]. Based on the current local market prices in southern Africa, the cost of application of the former mixture is US$7.5/1000 kg of maize grain and the latter is US$12.5/1000 kg of maize at the label dose [117, 118].

Single treatments with spinosad, spinetoram, pyrrole derivatives, insect growth regulators (IGRs), pyrethroids, organophosphorus, neonicotinoid insecticides as grain protectants are also capable of effectively controlling *P. truncatus*. Athanassiou *et al.* [119] suggested that 0.5 ppm of a liquid formulation of spinosad [an insecticide that is based on the secondary metabolites spinosyns A and D of the actinomycete *Saccharopolyspora spinosa* Mertz and Yao (Actinomycetales: Pseudonocardiaceae)] [51, 130] was adequate to generate >90% mortality of *P. truncatus* adults on maize at 25 or 30 °C and 55 or 75% relative humidity even after 7 days of exposure. When 0.1 ppm of liquid spinetoram, an insecticide that is based on the secondary metabolites spinosyn J and L of *S. spinosa* [131–133], was applied on maize grain, it killed 100% of *P. truncatus* adults after 14 days of exposure at 25 °C [122]. Similarly, almost all *P. truncatus* adults died after 7 days of exposure on maize treated with 1 ppm of liquid spinetoram or liquid spinosad at a range of 20–30 °C [124]. In the same study, several combinations of spinosad with spinetoram did not alter the very high effectiveness of single treatments of both insecticides against *P. truncatus*.

For the pyrrole chlorfenapyr, Kavallieratos *et al.* [120] reported that mortality of *P. truncatus* adults reached 98.3% in maize treated with 1 ppm of this active ingredient after 14 days of exposure. The application 5 ppm of the IGR diflubenzuron in maize completely suppressed (100%) the progeny production of *P. truncatus* 60 days post-treatment under different temperature levels (i.e. 20–30 °C) [121]. Pozidi Metaxa and Athanassiou [123] showed that 0.5 ppm deltamethrin or 4 ppm pirimiphosmethyl caused >98% mortality to *P. truncatus* adults after 14 days of exposure on treated maize while no progeny production was recorded 60 days post-treatment with deltamethrin at 20, 25 and 30 °C. When the neonicotinoid insecticide thiamethoxam was applied in maize at 1 or 10 ppm, it produced 98.3% delayed mortality or 96.7% direct mortality to *P. truncatus* adults after 4 days of exposure, respectively [125].

More recently, additional research efforts have been applied to control *P. truncatus*. For instance, 700 g of the gaseous insecticide phosphine at low pressure (525 mm Hg absolute pressure) in a chamber that contained dried figs caused 89.5% mortality of *P. truncatus* adults [126]. Moreover, the application of thiamethoxam in plastic or concrete surfaces and α-cypermethrin in plastic surfaces generated 100% delayed mortality to *P. truncatus* adults after 7 days of exposure [127, 128]. However, it must be appreciated that in the typical grain storage ecosystem, *P. truncatus* does not occur alone. The pest shares its ecological niche with many other storage pest species [21, 27, 63]. Thus any chemical control measure will need to be able to control the whole pest spectrum that occurs in that particular stored-grain ecosystem.

Another method applied to control stored-product insects deals with the impregnation of insecticides into the fabrics of storage bags [e.g. ZeroFly polypropylene polymer storage bags (Vestergaard S.A., Lausanne, Switzerland)] or with the direct treatment of packaging material (e.g. paper, polyethylene, polypropylene) [103]. Kavallieratos *et al.* [101] reported 94.4% direct mortality of *P. truncatus* adults that had been exposed to ZeroFly storage bags after 5 days of exposure. Moreover, the application of α-cypermethrin, chlorfanapyr and pirimiphos-methyl in polypropylene storage bags can kill >91% of *P. truncatus* adults after 5 days of exposure while α-cypermethrin almost completely suppressed the damage (i.e. holes/bites) that the insect individuals cause to the surface of the storage bag [127].

Furthermore, it was found that 100% of exposed *P. truncatus* adults to ZeroFly storage bags were dead [129]. Adults could not remain upright and exhibited an uncoordinated movement for five or more seconds. They were turned on their backs and displayed loss of appendage control and tremors, or only exhibited movement in legs, antennae or mouthparts when probed by a fine-hair brush after 30 min. The use of

Table 10.2. Effect of insecticides against the larger grain borer, *Prostephanus truncatus*

Insecticides	Country	Lab or field study	Outcome	References
Deltamethrin with pirimiphos-methyl or deltamethrin with chlorpyrifos-methyl	Tanzania	Field study	Suppressed progeny production of *P. truncatus* in maize stored in bins or sacks for a period of 9 months	Dales and Golob [118]
Liquid spinosad	Greece	Lab study	0.5 ppm of a liquid formulation of spinosad was adequate to generate >90% mortality of *P. truncatus* adults on maize at 25 or 30 °C and 55 or 75% relative humidity even after 7 days of exposure	Athanassiou *et al.* [119]
Chlorfenapyr	Greece	Lab study	Mortality of *P. truncatus* adults reached 98.3% in maize treated with 1 ppm of chlorfenapyr after 14 days of exposure	Kavallieratos *et al.* [120]
Diflubenzuron	Greece	Lab study	5 ppm of diflubenzuron in maize completely suppressed (100%) the progeny production of *P. truncatus* 60 days post-treatment under different temperature levels (i.e. 20–30 °C)	Kavallieratos *et al.* [121]
Liquid spinetoram	Greece	Lab study	0.1 ppm of liquid spinetoram was applied on maize grain, it killed all *P. truncatus* adults after 14 days of exposure at 25 °C	Vassilakos *et al.* [122]
Deltamethrin, pirimiphos-methyl	Greece	Lab study	0.5 ppm deltamethrin or 4 ppm pirimiphos-methyl caused >98% mortality to *P. truncatus* adults after 14 days of exposure on treated maize while no progeny production was recorded 60 days post-treatment with deltamethrin at 20, 25 and 30 °C	Pozidi Metaxa and Athanassiou [123]
Spinosad, spinetoram or spinosad with spinetoram	Greece	Lab study	Almost all exposed *P. truncatus* adults died after 7 days of exposure on maize treated with 1 ppm of liquid spinetoram or liquid spinosad at a range of 20–30 ° C and the several combinations of spinosad with spinetoram did not alter the very high effectiveness of single treatments of both insecticides against *P. truncatus*	Athanassiou and Kavallieratos [124]

Continued

Table 10.2. Continued

Insecticides	Country	Lab or field study	Outcome	References
Thiamethoxam	Greece	Lab study	When thiamethoxam applied in maize at 1 ppm or 10 ppm it produced 98.3% delayed mortality or 96.7% direct mortality to *P. truncatus* adults after 4 days of exposure, respectively	Tsaganou *et al.* [125]
Phosphine	Greece	Lab study	700 g of the gaseous insecticide phosphine at low pressure (525 mm Hg absolute pressure) in a chamber that contained dried figs caused 89.5% mortality of *P. truncatus* adults	Athanassiou *et al.* [126]
Deltamethrin-treated woven polypropylene storage bag material (ZeroFly storage bags)	USA	Lab study	The direct mortality of *P. truncatus* adults that had been exposed to ZeroFly storage bags was 94.4%, after 5 days of exposure	Kavallieratos *et al.* [101]
α-cypermethrin, chlorfanapyr and pirimiphos-methyl	Greece	Lab study	The application of α-cypermethrin, chlorfanapyr and pirimiphos-methyl in polypropylene storage bags can kill >91% of *P. truncatus* adults after 5 days of exposure while α-cypermethrin almost completely suppressed the damage (i.e. holes/bites) that the insect individuals cause to the surface of the storage bag	Kavallieratos *et al.* [127]
Thiamethoxam, α-cypermethrin	Greece	Lab study	The application of thiamethoxam in plastic or concrete surfaces and α-cypermethrin in plastic surfaces generated 100% delayed mortality to *P. truncatus* adults after 7 days of exposure	Doganay *et al.* [128]
Fenitrothion with deltamethrin or pirimiphos-methyl with permethrin	Zimbabwe	Field study	Keep *P. truncatus* <50 individuals/kg of treated maize for a storage period of 40 weeks in polypropylene bags in different areas of Zimbabwe	Mlambo *et al.* [117]
Deltamethrin-treated woven polypropylene storage bag material (ZeroFly storage bags)	USA	Lab study	All exposed *P. truncatus* adults to ZeroFly storage bags were affected (adults could not remain upright and exhibited uncoordinated movement for five or more seconds, were turned on their backs and displayed loss of appendage control and tremors, or only exhibited movement in legs, antennae or mouthparts when probed by a fine-hair brush) after 30 min	Scheff *et al.* [129]

ZeroFly bags is strongly associated with the pre-fumigation of the grain commodities that are to be bagged to suppress internal insect infestations [101]. This fact could be considered as a disadvantage of the use of ZeroFly bags by smallholders that, due to low income gained, cannot cover the cost of fumigation and purchasing the bags [134]. Therefore, the enhanced ZeroFly storage bags, which include an inner hermetic design, have been proposed as an additional tool to address the aforementioned disadvantage of the nonhermetic ZeroFly bags [134].

Use of diatomaceous earths

One other category of grain protectants that have been successfully tested for the management of *P. truncatus* are enhanced DEs (Table 10.3). The DEs Protect-It (Hedley Technologies Inc., Mississauga, Ontario), containing 83.7% SiO_2 with 10% silica aerogel and PyriSec (Agrinova Gmbh, Obrigheim/Mühleim), which consists of 88% SiO_2 and 1.2% natural pyrethrum (25%), 3.1% piperonyl butoxide were found not effective against *P. truncatus* between 20 and 30 °C or 55 and 75% relative humidity at 500–1500 ppm [128]. Contrary to these DEs, the novel enhanced DE, DEA-P (Research and Consulting Inc., Toronto, ON, Canada), has been reported to be highly effective against *P. truncatus*, causing 98–100% mortality of the pest adults [135]. The combination of DEs with other insecticidal active ingredients can be a promising approach towards the effective control of *P. truncatus* [135, 136]. Combinations of the DEs Protect-It or Chemutsi (a raw DE from Zimbabwe that contains 86.7% total silica) with spinosad or deltamethrin in low doses suppressed 100% *P. truncatus* adults after 14 days of exposure [57].

The fact that some active ingredients, such as methyl bromide (CH_3Br), have been withdrawn or may have been locally or globally phased out of the market immediately generates a gap of the available management tools. However, the aim of using DEs is to provide a safer replacement option for the highly toxic insecticides. On the other hand, a major problem

Table 10.3. Effect of diatomaceous earths against *Prostephanus truncatus*

Diatomaceous earths	Country	Laboratory or field study	Outcome	References
Protect-It, PyriSec and DEA-P (mixture of abamectin and freshwater DE that contains 92.4% SiO_2)	Greece	Lab study	The diatomaceous earths Protect-It and PyriSec were not effective between 20 and 30 °C or 55 and 75% relative humidity at 500–1500 ppm against *P. truncatus* while the novel enhanced DE, DEA-P exhibited elevated effectiveness (i.e. 99–100% mortality) against adults of *P. truncatus*, while this efficacy was not affected by temperature or relative humidity at the low doses of 75 or 150 ppm	Athanassiou *et al.* [135]
Protect-It, Chemutsi, spinosad, deltamethrin	Zimbabwe	Lab study	The majority of all combinations of the DEs Protect-It or Chemutsi with spinosad or deltamethrin in low doses completely suppressed (100%) *P. truncatus* adults after 14 days of exposure	Machekano *et al.* [57]
DEA-P (mixture of abamectin and freshwater DE that contains 92.4% SiO_2)	Greece	Lab study	DEA-P was highly effective at 75 ppm by killing 98–99% of *P. truncatus* adults and suppressing their progeny production in five maize hybrids	Kavallieratos *et al.* [136]

is the use of the same active ingredients which leads to increasing resistance of *P. truncatus* towards these substances and finally to their ineffective use. Some methods require expensive equipment that is not always available, especially in developing countries. A possible solution is the rotation of the already known insecticides by applying them in label doses in order to decelerate the emergence of resistance. At this point, novel active ingredients should be developed and exhaustively tested before they become commercially available, bearing in mind that they should be safe for workers, consumers, animals and the broader environment. An interesting field of research is the possible combinations of novel insecticidal substances with low-risk local natural insecticides (e.g. DEs or other inert dust) aiming at enhancing the direct or delayed mortality of the exposed insects and/or increasing the time which is needed to kill the pest from the moment it will contact the toxicant. These efforts can be combined with the development of storage bags whose fabric has synthetic pesticide incorporated into it as an additional tool against *P. truncatus*.

Biological control with the use of predators

Multiple and diverse control measures need to be included in, and complemented with, the IPM package against *P. truncatus* [137–139], particularly the biocontrol with the use of predators. One of the natural enemies of *P. truncatus* is the predatory beetle *T. nigrescens*. This predator attacks a wide variety of appropriately sized prey encountered in its environment [69] and is found commonly associated with *P. truncatus* in mesoAmerica [70]. Studies of the biology of *T. nigrescens* demonstrated that this predator exhibits a strong preference for feeding on *P. truncatus* in choice tests indicating a relatively close association between the predator and its prey [140]. Adults and larvae of *T. nigrescens* feed voraciously on *P. truncatus* eggs and larvae and develop at a comparable rate on both the beetle stages. *Teretrius nigrescens* has been released in Togo [141], Ghana [142], Kenya [143], Zambia [144] and Guinea Conakry [145], Tanzania [2] and in Mozambique [19]. However, releases of the predator in these countries have shown variable success [146]. For instance, there was no impact of the predator found on *P. truncatus* after 2 years of its release in the field in the Manica and Barue districts, Mozambique, in 2007 [147]. However, augmented releases of the predator provided successful control of *P. truncatus* in Benin, given that there were reductions between 5 and 11% of cassava chip losses after 1 and 5 months of storage, respectively [46]. Inundated and augmented releases of *T. nigrescens* can increase its abundance, followed by increased efficacy against *P. truncatus* in natural environments. Reduction of maize losses at 3.2 and 8.6% with or without the predator, respectively, and in two seasons (1992–1993) was achieved after 8 months of storage in Togo [148]. It must be emphasized that under practical farmer storage systems, *T. nigrescens* is only appropriate in stores where no synthetic pesticides have been applied. The predator is more suitable for natural forests to suppress the numbers of *P. truncatus* flying to maize maturing in the field or to drying or stored maize at the homestead.

The use of the predator, *T. nigrescens* can provide the reduction of maize losses and increment monetary income. A decline in losses of 28 000 tonnes of maize with a value of US$5 million was estimated as macro-economic benefits from the use of *T. nigrescens* in the coastal region of West Africa [72, 73]. These estimates were also conducted in Benin, Togo and Guinea Conakry for a 30-year period by considering several issues such as costs of development and/implementation of the biological control programme, predator's mass rearing, release and monitoring. An analysis of additional cost-effective integrated postharvest management techniques indicated a further reduction in costs of about US$5 million [73].

The establishment of a biological control agent in the field is a major challenge in any classical biological control programme. Therefore, due to spatial and temporal variations of climatic conditions, integration of other control measures, particularly sanitary and hygienic, structural (the type of storage infrastructures), mass trapping and sustainable use of registered pesticides is recommended. Since data on the economic cost-effectiveness of the use of *T. nigrescens* as a biocontrol agent in the field or

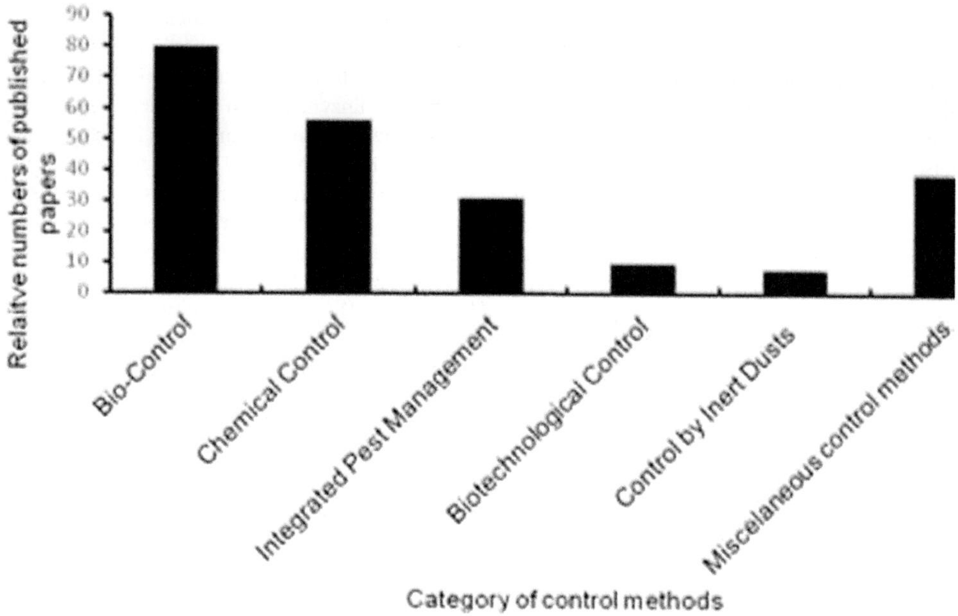

Fig. 10.4. Distribution of relative numbers of published papers per categories of *Prostephanus truncatus* control methods, worldwide (1980–2018) (Source: authors' compilation, 2018).

granaries are still scarce, further research is needed in this direction. Therefore, it can be deduced from Figure 10.4 that research on biocontrol followed by chemical control received much attention of scientists and pest management workers. However, the long-term effect and sustainability of biocontrol is required, especially *T. nigrescens*.

Biotechnological/genetic control: use of varietal plant resistance

Host plant resistance (HPR) has been defined as the capacity of plant species to avoid, tolerate or recover from injuries caused by insect populations, limiting insect feeding and/or reproduction [135]. HPR is determined by genetic, biophysical and biochemical characteristics of the plant, which influence insect development and behaviour [149–154]. Depending on the impact of insect damage, HPR could be manifested as immunity, high resistance, moderate resistance, low resistance or susceptibility [150, 151].

Resistance against *P. truncatus* has been attributed mainly to pericarp thickness/toughness and to the action of biochemical factors, including α-amylase inhibitors [155] and allelochemicals, localized in the endosperm aleurone layer, such as phenolic acid amides [152, 156]. However, the knowledge of the specific mechanisms by which these traits contribute to resistance against *P. truncatus* remains unknown [149–152]. Several maize varieties and native landraces from Mexico and Central America such as Cónico or Nal-Tel have been identified as natural sources of resistance to *P. truncatus* [157–162]. Resistant varieties displayed a reduction of grain weight losses ranging from 18 to 33% in comparison with susceptible genotypes [163]. However, *P. truncatus*-resistant landraces often display poor agronomic characteristics [149, 163].

Therefore, international breeding programmes have made important efforts to develop maize varieties resistant to *P. truncatus*, using Caribbean insect-resistant accessions (such as the afore-mentioned) as sources of resistance [163, 164]. Recently, significant gains in storage insect pest resistance have been achieved through recurrent selection. An example is P84c3, which is the result of a S_3 intra-population improvement [165], which led to a dramatic reduction (up to 75%) of grain weight losses caused by *Sitophilus zeamais* Motschulsky (Coleoptera: Curculionidae) and up to 30% of losses caused by *P. truncatus* [150, 162]. In Africa, open pollinated varieties and hybrids resistant to *P. truncatus* have been developed [152]. However, the lack of knowledge mechanisms and bases of *P. truncatus* resistance leads to slow progress in trait transference into elite germplasm [149]. Thus, in order to accelerate the development of *P. truncatus*-resistant maize varieties, the identification of additional *P. truncatus* resistance traits (biophysical and/or biochemical) should become the main concern for breeding programmes.

Considering the success of the implementation of Bt-maize for the control of pests such as the European corn borer, *Ostrinia nubilalis* (Hübner) (Lepidoptera: Crambidae) [166], genetic modification technology has been explored as an alternative for the acceleration of the development of *P. truncatus*-resistant maize varieties. Transgenic avidin maizewas considered as a promising alternative for the reduction of insect post-harvest losses [167]. This GMO displayed toxic effects against diverse storage pests, including *S. zeamais* and *Sitotroga cerealella* (Lepidoptera: Gelechiidae). However, *P. truncatus* displayed tolerance to avidin and was not affected [167]. Genes associated with *P. truncatus* resistance, such as α-amylase inhibitors [155], could be considered for this technology. However, rigid international regulations and low social acceptance of GMO hinder the application of this technology for the development of *P. truncatus*-resistant maize [168, 169].

Thus, strategies based on genetic selection, such as marker-assisted selection, could be explored as an alternative for the identification of resistance traits. Molecular markers are used to accelerate the transfer of genes or genetic regions associated with resistance in modern pre-breeding and breeding programmes [170, 171]. In terms of maize postharvest pest resistance, molecular breeding efforts have focused on the identification of genomic areas and genes involved in resistance to the major pests. The principal quantitative trait loci (QTL) and specific associated genes in maize for lepidopteran insect resistance have been elucidated [172–176]. By this strategy, genes such as cellulose synthases [177] have been associated with biophysical resistance. Enzymes such as chalcone synthase [178], phenylalanine ammonium lyase, caffeate *O*-methyltransferase [179], arginine decarboxylase, ornithine decarboxylase, spermidine synthase [180] and peroxidases [150, 153, 162] have been identified as biochemical bases of resistance to *S. zeamais* by QTL mapping [175, 176]. However, in the case of *P. truncatus*, resistance traits and resistance-associated genes remain undiscovered. QTL-based strategies or novel genome-wide association studies, which permit the association of resistance traits to single nucleotide polymorphisms [181], could be suitable for the identification of traits associated with *P. truncatus* resistance.

Identification of resistance traits should become a priority for breeding programmes through which it will be possible to develop efficient and insect-resistant maize lines in a shorter time, taking advantage of maize genetic variation [138, 182–184]. In economic terms, the monetary impact of biotechnological and genetic control of *P. truncatus* remains unreported. Economic impact of the liberation and promotion of insect-resistant varieties needs to be evaluated in terms of market price, quality of the grains, agronomic yields, costs associated with grain production and acceptability of these varieties to farmers and consumers. Thus, studies associated with the economic impact of *P. truncatus*-resistant varieties should be performed in the near future.

Use of plant-derived products

Investigation on the potential of the plant-derived products for the control of stored-product insect pests such as *P. truncatus* on stored maize and dry cassava chips has been widely studied globally [40–43, 184–189]. In Africa, the traditional use of available plant materials for pest control has been reported in many countries, e.g. Mozambique [185], Nigeria [186], Botswana [187] and in Ghana [188] (Table 10.4). The presence of a mixture of phytochemicals, apparently with different mechanisms of insecticidal activity of the tested plant extracts may delay the development of resistance by insect pests relative to the synthetic insecticide.

The effectiveness of plant-derived products, such as barks, seed powders or essential oils has been evaluated by means of the quantity in weight or volume per unit mass, enough to cause 50–100% mortality of *P. truncatus* adult individuals at 1–24 h exposure periods. In Ghana, it was demonstrated that only 1000 μL of leaf extract essential oil of *Ocimum suave* (Labiatae) were necessary to treat one tonne of maize grain for effective control of *P. truncatus* [40]. Nukenine *et al.* [42] reported that nearly 9 kg of the leaf powder was needed to treat one tonne of maize grain in Cameroon. These results question the amount of raw plant material (barks, leaves, roots or seeds) and consequently, plant populations necessary to support wider-scale pest control in a community over human generations. Rapid plant propagation becomes a key for the sustainable use of the plant-derived product at a larger scale in pest management. For example, *Eucalyptus globulus* (Myrtaceae) and *A. indica* (Meliaceae) are perennial plants which require not <5–10 years to grow and produce suitable plant extracts for use in IPM [193]. The cost-benefit analysis of using plant-derived products in stored-product IPM remains unreported.

For the evaluation of the sustainability of the use of plant-derived products in IPM for insect pest control with a focus on *P. truncatus*, it is necessary to consider the expenses involved in procuring the plant materials, bio-chemical or chemical reagents, processing and logistic costs compared to conventional practice. However, other benefits such as environmental and health implications remain priceless. Plant species also suffer the negative effect of veldt fire, human deforestation, animal over-grazing and over-browsing, leading to their spatial and temporal reduction or ultimately, extinction. Besides the ecological issues raised above, research questions about dosage effectiveness dependence, particularly of essential plant oils, spatial and temporal variation of plant extract concentrations and means of enhancing the persistence of active ingredients on the surface of commodities [40, 184–189] are of concern and deserve further comprehensive investigation. Selection and mass propagation of annual multigenerational shrub and grass species and integration with other control measures could be the optional ways to increase the sustainability and practical use of plant-derived products in pest management for *P. truncatus*. Some of the tree species such as *Bobgunnia madagascariensis* (Fabaceae) are difficult to propagate due to poor and erratic germination of its seeds and slow growth of its seedlings [193]. Hence, research is required to improve the propagation of such species.

Understanding the plant chemistry and its spatiotemporal variability is critical in assessing the potential of plants for pest management as well as the factors that underpin these differences [193–198]. For example, it has recently been demonstrated that two morphologically distinct varieties of *Lippia javanica* (Verbenaceae), a commonly used pesticidal plant including grain protection in Southern Africa, occur as *L. javanica* var. *javanica* and *L. javanica* var. *whytei*, and these correspond to two distinct chemotypes [199]. The two varieties showed variations in the chemical amount and composition of their essential oils with respect to the time of harvesting. Hence, both factors will influence the success with which farmers benefit from them as a pest control technology. Correct identification of insecticidal plant species is also critical but is not always done correctly in practice [200]. For example, Ndakidemi *et al.* [201] report that perhaps one of the most widely used pesticidal plant material in Africa, *Tephrosia vogelii* (Fabaceae), is being incorrectly identified and promoted for use by farmers as *T. candida* which can be highly confusing.

To ensure recommendations on the use of pesticidal plant materials are robust; further evaluations of the mammalian toxicity and potential side-effects to non-target organisms such as predators and parasitoids are required [201] despite the common perceptions of these materials as safe to humans and the environment [202]. Some plants are, in fact, well known to be extremely toxic in nature [202, 203]. Because plant materials kill animals such as insects, it means caution must be exercised in their use, especially considering that they are applied directly on stored foods to control the storage insect pests. Internationally, published empirical evidence of vertebrate toxicity of African pesticidal plants is scarce which is probably ascribed

Table 10.4. Reports of plant species with insecticidal effect and with potential to protect stored maize grain or dry cassava chips against *Prostephanus truncatus*

Plant species	Family	Parts to use	Activity	Effectiveness	Countries	References
Ocimum kenyense Ayob. ex A.J.Paton	Lamiaceae	Dry ground leaves and essential oils	Repellency Toxicity	LD_{100} 0.5–10 μl/insect	Kenya	Obeng-Ofori *et al.* [189]
Ocimum suave Willd.		Essential oil	Repellency	LD_{80-100} 1 μl/kg grain	Ghana	Obeng-Ofori and Reichmuth [40]
Ocimum kilimandscharicum Guerke				LD_{100} 1 mg/250 g grain		Obeng-Ofori *et al.* [190]
Azadirachta indica A. Juss	Meliaceae			LD_{50-100}		Osipitan *et al.* [191]
Cola millennia K. Schum	Sterculiaceae	Bark extracts		50 ml/kg grain		
Delonix regia (Boj. ex Hook.) Raf.	Fabaceae					
Vernonia Amygdalina Delile	Asteraceae	Leaf extracts				
Chenopodium ambrosioides L.	Chenopodiaceae	Dry leaf powder and essential oils	Contact toxicity	LD_{50-100} 0.05–0.80 wt/wt 0.2 μl/cm^2 oil	Cameroon	Tapondjou *et al.* [192]
Clausena anisata (Willd.) Hook. f. ex Benth.	Rutaceae	Leaf powder Essential oil	Contact toxicity	LD_{50} 0.9–1.9 g/100 kg grain	Cameroon	Nukenine *et al.* [42]

to the fact that the costs of commercializing pesticidal plants are prohibitive, so there are often no requirements to test them officially [203]. Therefore, farmers also need capacity-building for sustainable harvesting of pesticidal plants to avoid over-harvesting and possible subsequent extinction.

Use of hermetic containers

Use of hermetic bags

The chemical control of *P. truncatus* may meet the severe problem of resistance to pyrethroid and organophosphorous active ingredients which are extensively used in storage facilities [116, 117, 204]. Since resistance can easily be induced to *P. truncatus* populations [116], alternative technologies for the effective management of this species should be developed that will also prolong the storage period with no or minimal grain losses and being safe to individuals who are involved in any step of the postharvest processes, consumers, animals and the environment. One of these technologies is the use of hermetic grain bags [71] in which the air replenishment is minimal when loaded. Therefore, the respiration and metabolic activities of insects, fungi and grain diminish the level of O_2 and increase the level of CO_2, resulting in the death of insects, resident inside the hermetic bags [205, 206]. Recent research efforts have shown that hermetic grain bags are a highly viable tool against *P. truncatus*. For example, Edoh Ognakossan *et al.* [71] reported that hermetic grain bags were highly suitable for the storage of maize that was artificially infested by *P. truncatus* compared to woven polypropylene bags. The authors found that hermetic grain bags significantly diminished the population growth or increased the mortality of the pest compared to woven polypropylene bags for a storage period between 90 and 150 days. Simultaneously, maize losses were kept <6.0% in hermetic grain bags versus 27.1% in woven polypropylene bags. However, the damage of hermetic grain bags by *P. truncatus* was not totally avoided since the insect made holes on the material of the hermetic grain bags.

In contrast, García-Lara *et al.* [207] found that the enhanced hermetic storage bags SuperGrainbags SGB-IV-R liners (consisting of tough plastic material) (GrainPro Inc., Concord, MA, USA) provided 100% mortality of *P. truncatus* in pre-infested maize after 4 weeks of storage in Mexico. However, these bags, though extra tough and perforation-resistant, were finally damaged by incoming *P. truncatus* adult infestation after 3 weeks of storage. Super Grain bags (SGBs) (consisting of high-density polyethylene liner and polypropylene bag) offered a high level of storage of pre-infested or naturally infested maize by *P. truncatus*. The SGBs recovered <6 adult insect individuals and ≤2.5 g of frass/kg of maize were produced by insects' activity after 4 months of storage in two locations of Zimbabwe [208]. SGBs performed significantly better when filled with preor non-infested maize that was treated with a mixture of pirimiphos-methyl 2.5% with deltamethrin 0.5%. Purdue Improved Crop Storage (PICS) hermetic bags have also been reported as suitable for the safe storage of preinfested maize since they kept low population growth of *P. truncatus* [209]. The use of PICS resulted in only 2.0% weight loss of maize grain after 6 months of storage in Kenya. Moreover, PICS bags caused 95–100% mortality of *P. truncatus* after 6.5 months of storage while suppressing damage and weight loss of naturally infested kernels in a field trial in Benin, Burkina Faso and Ghana [210].

PICS bags demonstrated their efficacy in controlling losses caused by storage pests after 35 weeks of storage under participatory on-farm trials conducted in rural villages in eastern Kenya [210] and in protecting several cereals, such as wheat, from stored-product pests other than *P. truncatus* [211]. However, for maize, tests were only done with *S. zeamais* under controlled conditions [212] or under on-farm locations [209]. Perforation of storage bags by *P. truncatus* adults either from inside or outside the bags is a major issue that threatens the use of these facilities for commodity storage [213]. The perforation action considerably shortens the active life of bags [213]. The potential solution to this problem could be the treatment of the bags with insecticides [101]. Although PICS bags seemed an adequate tool for storage of maize, they were not suitable for the storage of cassava chips since after 8 months of storage the chips sustained high levels of *P. truncatus* infestation and damage either when the chips were artificially infested. This was probably due to large pieces of cassava chips that leave gaps with O_2 supply for the developing insects inside bags, contrary to closely packed maize kernels [209].

Hermetic storage is also effective in preventing the growth of mould-producing aflatoxins, a major public health hazard [214–216]. Hermetic flexible liners and large flexible containers have been used worldwide for the storage of maize [215] because they reduce losses. Recent evidence demonstrated that hermetic plastic bags have also been used for maize storage against *P. truncatus*. However, the erratic effectiveness of this technology depends on the type of plastic and manufacturer [61, 217].

Use of metal silos

Metal silos have lately been the focus on the use of hermetic storage as an alternative grain storage option in America [208, 218, 219]. Metal silos are based on a similar design to those promoted in South America under the POSTCOSECHA programme in the last century [220, 221] with an airtight and cylindrical shape and adapted to the local agro-ecological environment by small changes [222]. The functional principle of metal silos is like that described by Navarro [222] regarding ordinary hermetic bags. Metal silos have been extensively tested and found to be highly effective in reducing grain storage losses under smallholder farming conditions, even when *P. truncatus* occurs and therefore providing a pesticide-free alternative to use synthetic pesticides [220–222]. However, several issues remain unsolved: (1) air tightness only work for long-term storage, while for short storage period hermetic seal does not work (farmers open the silo very often breaking the process of oxygen depletion); (2) the weld with lead could contaminate the grain and increase the risk of chronic diseases; (3) do not prevent aflatoxin contamination if the grain is inadequately dry or if there is a high temperature fluctuation, leading to the development of 'hotspots'; humid condensation generates hotspot of fungal-toxic contamination; (4) lack of quality control in silo manufacturing can result in severe problems of iron oxidation; (5) hard to clean or preserve and is extremely difficult to extract the grain after the reductions of the content to <50%; (6) very expensive, difficult to fix and ineffective after the loss of airtightness [220, 222]. A cost-benefit economic analysis conducted by SDC [222] concluded that the use of metal silos generated rural business or employment and provided farmers with savings of about 50 000 tonnes of grain loss per year (i.e. US$12 million) in Honduras, Guatemala, Nicaragua and El Salvador, in the 1980s under the POSTCOSECHA programme.

Existing hermetic solutions have notable shortcomings, either in terms of the expense of metal silos or effectiveness of plastic bags [38, 223]. Tefera *et al.* [38] reported that prices of metal silos of 120, 500 and 1800 kg of capacity varied between US$12–23, 30–70 and 50–187, respectively. These prices were selected from 11 countries including Mozambique, Malawi, Senegal, Burkina Faso and Bolivia. Metal silos with capacity ranging from 2000 to 3000 kg cost between US$320 and 480 [38]. The ineffectiveness and/or the fact that metal silos could be unaffordable to smallholder farmers are other limitations of the hermetic storage technology, especially when *P. truncatus* is present in stored maize as it happens in many developing countries of Africa, Central and Southern America and Asia [38, 223]. The integration of hermetic storage methods with mass trapping, sanitary, hygienic, use of local claymade silos and sustainable use of chemical methods could be the best way to minimize the effect of the aforementioned limitations.

Use of semiochemicals: mass trapping for the control of Prostephanus truncatus

Semiochemicals, particularly sex and aggregation pheromones, deal mostly with the monitoring of spatial and temporal flight activity of insects in stores and in the environment [53, 224, 225], male and female intra-specific communication [224] and prey–predator interactions [226, 227]. However, the cost-effectiveness of the use of mass trapping as a tool in the management of *P. truncatus* is limited. Phillips and Throne [228] and Muatinte *et al.* [147] remarked that semiochemical-based control measures against insect-stored pests are cost-effective and friendly to the environment. Field trials conducted in Mozambique revealed that mass trapping using pheromone baited

Uni-traps was effective in reducing population numbers of *P. truncatus* in granaries of smallholder farmers over a 6-month period [64]. The method achieved a 40% reduction of damaged maize kernels in granaries in relation to those without traps. Previous studies have shown that mass trapping is efficient against several noxious insects. For instance, Schlyter *et al.* [229] trapped 0.5–1.7 million individuals of the bark beetle, *Ipis duplicatus* (Sahlberg) (Coleoptera: Scolytidae)/per year in Mongolia, China, resulting in a significant reduction of tree mortality in a 3-year period. Mass trapping was found effective in reducing numbers of the palm weevil, *Rhynchophorus ferrugineus* Olivier (Coleoptera: Rhynchophoridae) in Saudi Arabia [230] and love bugs, *Plecia nearctica* Hardy (Diptera: Bibionidae) [231] in the USA. Records of significant reduction of insect numbers through mass trapping either in granaries or in a forest environment indicate that this method could be potentially used as an additional tool in IPM. However, the capture of natural enemies of the target pest [64] could be a limitation of this method, particularly when using sticky pheromone-baited traps. Therefore, non-sticky pheromone baited traps like Uni-traps should be used instead. Although the former catch of *T. nigrescens*, collecting and releasing it back to the environment is allowed. This procedure enhances the population number of the natural enemy contrary to sticky traps that will kill it. Analysis of the cost-effectiveness and the sustainability of mass trapping in IPM should be focused on monetary values of infestation, damage and weight loss reduction of commodities. Costs of the traps and pheromones per area and period should also be considered. Further research on the origin of trapped individuals of *P. truncatus*, the statistical model that defines the relationships between trap catches and reduction of infestation, damage and weight loss of maize grain should be conducted. The knowledge of the influence of the type and quantity of traps and their exposure in the environment may increase the effectiveness of the mass trapping method [64, 212, 217]. Despite the outstanding limitations, mass trapping is a promising tool that can be integrated with other control methods to suppress stored-product insects, such as *P. truncatus*, to economically acceptable levels, particularly in granaries of smallholder farmers and in the forest environment.

Successes and limitations in Prostephanus truncatus management

The advantages of *P. truncatus* management lies on the numerical reduction of the pest with concomitant reduction of commodity damage and weight losses resulting in improved household food security of households, especially regarding dried stored maize and cassava. Considerable reduction of losses has been noticed in Kenya [6] as a result of the extensive implementation of control measures in the frame of a programme that was funded by the German Technical Cooperation Agency (GTZ). The implementation of IPM programmes, mainly in West African countries, resulted in the generation of employment and income-generating opportunities in rural areas, decrease in maize grain market prices, stimulation of farmers' self-capacity, development of human resources, increase in commodity export and decrease in deficit of trade. Nevertheless, reports and activities on successful control of the pest are scarce [73] and where available the scientific information is very old [6, 11, 14, 18] and probably outdated. The main contributions on *P. truncatus* deal with scientific research that were related to the effectiveness of chemical pesticides [120–126, 128, 133], (enhanced) DE [135, 136] and biological control using the predator, *T. nigrescens* [19, 22, 30, 44, 69, 70, 140–143]. Moreover, the success of pest management can result from the current research on the potential of using semiochemicals, with emphasis on mass trapping [63], hermetic storage [38, 59, 71, 205–210], HPR [37, 149–155] and application of plant-derived products [40–43, 168, 169].

The enhancement of collaboration of governmental (e.g. Universities, Ministries and Research Institutes) and development partners [e.g. FAO, GIZ, United States Agency for International Development (USAID), OXFAM International, CAB International, Denmark's Development Cooperation (DANIDA), International Fund for Agricultural Development (IFAD), WFP, SDC, Federal Ministry of Economic Cooperation and Development (Germany)

(BMZ)], international research organizations [NRI, IITA CIMMYT and IFPRI] constitute an effective and active network that has optimized and publicized global management guidelines for *P. truncatus*. Also, new knowledge on *P. truncatus* and its natural enemies has been generated through numerous (>20) Ph.D. studies [21, 78, 82, 232–247]

Constraints in the management of *P. truncatus* are attributed to different methodological approaches that have been adopted during PHL assessments [63, 73] or due to seasonal, national and regional variations of infestation [59, 60, 144, 145, 232]. Empirical approaches, simplifications and methodological generalizations in the assessment of PHL caused by *P. truncatus* may hinder the success of management of the pest. Fluctuations of the devastating effect of *P. truncatus* may also depend on the regional differences of climate, diversity and abundance of potential forest host plants or magnitude of national and international maize trade [20]. Finally, the degree of the adoption of new control methods against *P. truncatus* is dependent on whether the country is developed or developing. IPM of *P. truncatus* takes into account the pest's seasonal population dynamics, at national or international levels and reinforcement of quarantine measures. Effective control of *P. truncatus* within countries and involvement of local farmers in evaluating the cost-effectiveness of new IPM technologies against *P. truncatus* have the potential to positively contribute to the reduction of the pest to economically acceptable levels.

Concluding remarks and recommendations

The threat from *P. truncatus* results from (1) very high infestation, damage and weight loss (30–100%) of dried stored maize grain and cassava chips in developing countries, (2) losses from regional and international maize grain trade opportunities because of quarantine limitations and requirements, (3) reduction of opportunities and income generation for socio-economic development due to re-direction of investments for the control of *P. truncatus*, (4) continuous increase of the spatio-temporal dispersal and distribution of *P. truncatus* through the trade of maize with consequent habitat invasion, colonization, commodity infestation, weight loss, worldwide with emphasis on developing countries.

Chemical, hermetic storage, biological and biotechnological management approaches in conjunction with the use of host plant-resistant varieties seem to be the most effective and practical control options of *P. truncatus*. However, the effectiveness of each method varies according to the socio-economic context among countries either developed or developing, and more research work on the costs involved for the implementation of particularly biological and biotechnological use of host plant-resistant varieties is needed.

The potential use of mass trapping could be a valuable tool in the overall management of *P. truncatus*. Any emerging control methods should be evaluated in terms of their efficacy against *P. truncatus* and their cost-effectiveness under farmer circumstances and together with the target end-users. This is an important goal since any positive research outcome could be introduced in developing countries, where post-harvest products are mostly kept by smallholder resource-poor farmers.

Overall, in order to obtain positive results regarding the management of *P. truncatus*, adequate knowledge is required about *P. truncatus* biology, behaviour, detection coupled with the safe use of synthetic insecticides given that not all postharvest stakeholders are familiar with the potential devastation of this pest and/or how it can be safely and effectively controlled. Continuous farmer and extension staff training is a key to the successful management of pest as these key stakeholders need to be constantly appraised of the changes in biology, the ecology of the pest and management strategies, as well as the climate.

Acknowledgements

We express our profound acknowledgement to Dr David Hemmings and the CAB International for inviting us to write this review. We thank Dr Harriet Muyinza, National Agricultural Research Laboratories (NARL) and National Agriculture Research Organization (NARO), Uganda for the information on the actual threat of *P. truncatus* in Uganda. We thank Mr Joaquim Campira (MSc), SECOSUD II Italia-Mozambique Cooperation Project for the updated *P. truncatus* world distribution map.

References

1. Hoppe T. Storage insects of basic food grains in Honduras. Tropical Sciences 1986;26:25–8.
2. Dunstan WR and Magazini IA. Outbreaks and new records, United Republic of Tanzania. The larger grain borer on stored products. FAO Plant Protection Bulletin 1981;29:80–1.
3. Hodges RJ, Dunstan WR, Magazini I, Golob P. An outbreak of *Prostephanus truncatus* (Horn) (Coleoptera: Bostrichidae) in East Africa. Protection Ecology 1983;5:183–94.
4. Markham RH, Wright VF, Rios Ibarra RM. A selective review of research on *Prostephanus truncatus* (Coleoptera: Bostrichidae) with an annotated and updated bibliography. Ceiba 1991;32:1–90.
5. Boxall RA. Damage and loss caused by the larger grain borer *Prostephanus truncatus*. International Pest Management Review 2002;7:105–21.
6. Farrell G, Schulten GGM. Larger grain borer in Africa: a history of efforts to limit its impact. Integrated Pest Management Reviews 2002;7:67–4.
7. EPPO (European and Mediterranean Plant Protection Organization). EPPO global data base. *Prostephanus truncatus*. 2019. Available from: URL: https://gd.eppo.int/taxon/PROETR/distribution (last accessed 13 May 2019).
8. APPPC (The Asia-Pacific Plant Protection Commission). Appendixes A (covering revised List A1 and A2) and B of the Plant Protection Agreement for the Asia and Pacific region. Technical Document no. 139/1991, FAO Regional Office for Asia and the Pacific, Bangkok (TH), 19 pp; 1991. Available from: URL: https://gd.eppo.int/taxon/PROETR/distribution/CN.
9. Kega VK, Warui CW. *Prostephanus truncatus* in Coast Province, Kenya. Tropical Stored Products Information 1983;46:2.
10. Bonzi SM, Ntambabazi C. Primier cas de grand capuchin du maïs (*Prostephanus truncatus*) signalé au Rwanda. FAO Plant Protection Bulletin 1993;41:204–5.
11. Opolot O, Odong M. Overview of phytosanitary quarantine services in Uganda. In: Nahdy SM, Agona JA, editors. Proceedings of the 1st Conference on Stored-Product Insect Pests. Kampala, Uganda, 1999. p. 24–29. 29 November-1 December. Available from: URL: https://tspace.library.utoronto.ca/html/1807/21892/cs00036.html (last accessed 21 June 2019).
12. Harnisch R, Krall S. Further distribution of the larger grain borer in Africa. FAO Plant Protection Bulletin 1984;32:13–114.
13. Pike V, Akinnabagbe JJA, Bosque-Perez NA. Larger grain borer (*Prostephanus truncatus*) outbreak in western Nigeria. FAO Plant Protection Bulletin 1992;40:170–73.
14. Tyler PS, Hodges RJ. Phytosanitary measures against the larger grain borer *Prostephanus truncatus* (Horn) (Coleoptera: Bostrichidae), in international trade. Integrated Pest Management Reviews 2002;7:279–89.
15. Aman OB, Schulthess F, Masiga D, Van den Berg J. On the diversity of Teretrius nigrescens, predator of the larger grain borer. 17ème Conférence de l'Association des Entomologistes Africains et de la Société Entomologique du Sénégal 2007. 11–15 June 2007. Dakar, Senegal; 2007.
16. Gueye MT, Goergen G, Badiane D, Hell K, Lamboni L. First report of the larger grain borer *Prostephanus truncatus* (Horn) (Coleoptera: Bostrichidae) occurrence in Senegal. African Entomology 2008;16:309–11.
17. Munthali SCM. The larger grain borer, *Prostephanus truncatus* in Malawi: current status. *Plant Protection Workshop*, Lilongwe, Malawi. 1992. 1–5 June.
18. Roux PWJ. Larger grain borer: further developments. Plant protection news (South Africa). Plant Protection Research Institute 1999;55:3–4.

19. Cugala D, Sidumo A, Santos L, Mariquele B, Cumba V, Bulha M. Assessment of status, distribution and weight lost due to *Prostephanus truncatus* (horn) (Coleoptera: Bostrichidae) in Mozambique. African Crop Science Conference Proceedings 2007;8:975–79. ISSN 1023-070X/2007. El-Minia, Egypt.
20. Nyagwaya LDM, Mvumi BM, Saunyama IGM. Occurrence and distribution of *Prostephanus truncatus* (Horn) (Coleoptera: Bostrichidae) in Zimbabwe. International Journal of Tropical Insect Science 2010;30:221–31.
21. Muatinte BL. Population dynamics and integrated management of *Prostephanus truncatus* (Coleoptera: Bostrichidae) in Manica Province, Mozambique'. Thesis submitted in fulfilment of the requirements for the award of the Degree Doctor of Philosophy in Environmental Sciences and Management at the North-West University (Potchefstroom Campus). North-West University, South Africa. 2017. Available from: URL: http://dspace.nwu.ac.za/bitstream/handle/10394/25594/Muatinte_BL_2017.pdf?sequence=1&isAllowed=y (last accessed 21 May 2018).
22. Helbig J. The Ecology of *Prostephanus truncatus* in Togo with Particular Emphasis on Interaction with the Predator *Teretriosoma nigrescens*. GTZ, Eschborn, Germany; 1995. 111 pp.
23. Giliomee JH. Recent establishment of many alien insects in South Africa – a cause of concern. African Crop Science Journal 2011;19:151–5.
24. Hodges RJ, Meikle J, Denton H. Infestation of dry cassava (Manihot esculenta Crantz) by *Prostephanus truncatus* (Horn) (Coleoptera: Bostrichidae). Journal of Stored Products Research 1985;20:205–13.
25. Stumpf E. Post-harvest loss due to pests in dry cassava chips and comparative methods for its assessment. A case study on small-scale farm household in Ghana. 1998. GTZ, 172 pp. Available from: URL: http://www2.gtz.de/postharvest/documents/new_5426e/x5426e00.htm (last accessed 30 September 2013).
26. Schäfer K, Gorgen G, Borgemeister C. An illustrated identification key to four different species of adult Dinoderus (coleoptera: Bostrichidae), commonly attacking dried cassava chips in West Africa. Journal of Stored Products Research 2000;36:245–52.
27. Gnonlonfin GJB, Hell K, Siame AB, Fandohan P. Infestation and population dynamics of insects on stored cassava and yams chips in Benin, West Africa. Journal of Economic Entomology 2008;101:1967–73.
28. Wright VF. World distribution of *Prostephanus truncatus*. Proceedings of GASGAWorkshop on the Larger grain borer *Prostephanus truncatus*. Slough. Publ. 1984. GTZ, Eschborn. p. 11–16.
29. Kimenju SC, De Groote H. Economic Analysis of Alternative Maize Storage Technologies in Kenya. Joint 3rd African Association of Agricultural Economists (AAAE) and 48th Agricultural Economists Association of South Africa (AEASA) Conference, Cape Town, South Africa, September 19–23. 2010. Available from: URL: https://ageconsearch.umn.edu/bitstream/96419/2/169.%20Economic%20Analysis%20of%20Alternative%20Maize%20Storage%20Technologies%20in%20Kenya.pdf (last accessed 28 June 2018).
30. Muatinte BL, Cugala DR. Monitoring the establishment and dispersal of Teretrius nigrescens Lewis (Coleoptera: Histeridae), a predator of *Prostephanus truncatus* (Horn) (Coleoptera: Bostrichidae) in the Manica Province, Mozambique. African Entomology 2015;23:250–54.
31. Wong-Corral FJ, Ramirez-Martinez M, Cortez-Rocha MO, Borboa-Flores J, Leos-Martinez J. Presence of *Prostephanus truncatus* (Horn) (Coleoptera: Bostrichidae) in Sonora, Mexico, first report. Southwestern Entomologist 2001;26:151–58.
32. Meikle WG, Degbey P, Oussou R, Holst N, Nansen C, Markham RH. Pesticide use in grain stores: an evaluation based on survey data from Benin. PhAction News 1999;1:5–9. Available from: URL: http://www.iita.org/info/phnews/ph-res4.hmt (last accessed 19 September 2018).
33. Kavallieratos NG, Athanassiou CG, Vayias BJ, Betsi PCC. Insecticidal efficacy of fipronil against four stored-product insect pests: influence of commodity, dose, exposure interval, relative humidity and temperature. Pest Management Science 2010;66:640–49.
34. Richter J, Biliwa A, Henning-Helbig S. Efficacy of dust formulated insecticides in traditional maize stores in West Africa. Journal of Stored Products Research 1998;34:181–87.
35. Golob P. Current status and future perspectives for inert dust for control of stored product insects. Journal of Stored Products Research 1997;33:69–9.
36. Stathers TE, Riwa W, Mvumi BM, Mosha R, Kitandu L, Mngara K, *et al.* Do diatomaceous earths have potential as grain protectants for small-holder farmers in sub-Saharan Africa? The case of Tanzania. Crop Protection 2008;27:44–70.

37. Arnason JT, Gale J, Conilh-de-Eyssac B, Sen A, Miller SS, Philogene JR, *et al.* Role of phenolics in resistance of maize grain to the stored grain insects *Prostephanus truncatus* (Horn) and *Sitophilus zeamais* (Motscholski). Journal of Stored Products Research 1992;28:119–26.
38. Tefera T, Kanampiu F, De Groote H, Hellin J, Mugo S, Kimenju S, *et al.* The metal silo: an effective grain storage technology for reducing post-harvest insect and pathogen losses in maize while improving smallholder farmers' food security in developing countries. Crop Protection 2011;30:240–5.
39. Osipitan AA, Sangowusi VT, Omoniyi I, Lawal OI, Popoola KO. Correlation of chemical compositions of cassava varieties to their resistance to *Prostephanus truncatus* Horn (Coleoptera: Bostrichidae). Journal of Insect Science 2015;15:1–7. Available from: URL: https://doi.org/10.1093/jisesa/ieu173 (last accessed 21 June 2019).
40. Obeng-Ofori D, Reichmuth CH. Plant oils as potential agents of monoterpenes for protection of stored grains against damage by stored product beetle pests. International Journal of Pest Management 1999;45:155–9.
41. Rajendran S, Sriranjini V. Plant products as fumigants for stored-product insect control. Journal of Stored Products Research 2008;44:126–35.
42. Nukenine EN, Adler C, Reichmuth C. Efficacy of Clausena anisata and *Plectranthus glandulosus* leaf powder against *Prostephanus truncatus* (Coleoptera: Bostrichidae) and two strains of *Sitophilus zeamais* (Coleoptera: Curculionidae) on maize. Journal of Pesticide Sciences 2010;83:181–90.
43. Silva GN, Faroni LRA, Sousa AH, Freitas RS. Bioactivity of *Jatrofa curcas* L. to insect pests of stored products. Journal of Stored Products Research 2012;48:111–13.
44. Richter J, Biliwa A, Helbig J, Henning-Helbig S. Impact of *Teretriosoma nigrescens* Lewis (Coleoptera: Histeridae) on *Prostephanus truncatus* (Horn) (Coleoptera: Bostrichidae) and losses in traditional maize stores in southern Togo. Journal of Stored Products Research 1997;33:137–42.
45. Holst N, Meikle W. *Teretrius nigrescens* against larger grain borer *Prostephanus truncatus* in African maize stores: biological control at work. Journal of Applied Ecology 2003;40:307–19.
46. Hell K, Lamboni Y, Houndekon T, Alidou GM. Augmented release of *Teretrius nigrescens* Lewis (Coleoptera: Histeridae) for the control of *Prostephanus truncatus* (Horn) (Coleoptera: Bostrichidae) in stored cassava chips. Journal of Stored Products Research 2006;42:367–76.
47. Helbig J. Ability of natural occurring parasitoids to suppress the introduced pest *Prostephanus truncatus* (Horn) (Coleoptera: Bostrichidae) in traditional maize stores in Togo. Journal of Stored Products Research 1998;34:287–95.
48. Bajpai NK, Ballal CR, Rao NS, Singh SP, Bhaskaran TV. Competitive interaction between two ichneu monid parasitoids of *Spodoptera litura*. BioControl 2006;51:419–38.
49. Smith SM, Moore D, Karanja LW, Chandi EA. Formulation of vegetable fat pellets with pheromone and *Beauveria bassiana* to control the larger grain borer, *Prostephanus truncatus* (Horn). Pesticide Sciences 1999;5:711–18.
50. Kassa A, Zimmermann G, Stephan D, Vidal S. Susceptibility of *Sitophilus zeamais* (Motsch.) (Coleoptera: Curculionidae) and *Prostephanus truncatus* (Horn) (Coleoptera: Bostrichidae) to entomopathogenic fungi from Ethiopia. Biocontrol Science and Technology 2002;12:727–36.
51. Hertlein MB, Thompson GD, Subramanyam Bh, Athanassiou CG. Spinosad: a new natural product for stored grain protection. Journal of Stored Products Research 2011;47:131–46.
52. Cox PD. Potential for using semiochemicals to protect stored products from insect infestation. Journal of Stored Products Research 2004;40:1–25.
53. El-Sayed AM, Suckling DM, Wearing CH, Byers JA. Potential of mass trapping for long-term pest management and eradication of invasive species. Journal of Economic Entomology 2006;99:155–56.
54. Adda C, Borgemeister C, Biliwa A, Meikle WG, Markham RH, Poehling HM. Integrated pest management in post-harvest maize: a case study from the Republic of Togo (West Africa). Agriculture, Ecosystems and Environment 2002;93:305–21.
55. Vayias BJ, Athanassiou G, Buchelos CT. Effectiveness of spinosad combined with diatomaceous earth against different European strains of *Tribolium confusum* du Val. (Coleoptera: Tenebrionidae): influence of commodity and temperature. Journal of Stored Products Research 2009;45:165–76.
56. Kavallieratos NG, Athanasiou CG, Vayias BJ, Kotzamanidis S, Synodis SD. Efficacy and adherence ratio of diatomaceous earth and spinosad in three wheat varieties against three stored product insect pests. Journal of Stored Products Research 2010;46:73–80. Available from: URL: http://ir.lib.uth.gr/handle/11615/29318 (last accessed 21 June 2019).

57. Machekano H, Mvumi BM, Chinwada P, Richardson Kageler SJ, Rwafa R. Efficacy of diatomaceous earths and their low-dose combinations with spinosad of deltamethrin against three beetle pests of stored-maize. Journal of Stored Products Research 2017;72:128–37.
58. Smith SM, Moore D, Oduor GI, Wright DJ, Chandi EA, Agano JO. Effect of wood ash and conidia of *Beauveria bassiana* (Balsamo) Vuillemin on mortality of *Prostephanus truncatus* (Horn). Journal of Stored Products Research 2006;42:357–66.
59. Quezada MY, Moreno J, Vázque ME, Mendoza M, Méndez-Albores A, Moreno-Martínez E. Hermetic storage system preventing the proliferation of *Prostephanus truncatus* Horn and storage fungi in maize with different moisture contents. Postharvest Biology and Technology 2006;39:321–6.
60. De Groote H, Kimenju SC, Likhayo P, Kanampiu F, Tefera T, Hellin J. Effectiveness of hermetic systems in controlling maize storage pests in Kenya. Journal of Stored Products Research 2013;53:27–36.
61. Adedoyin OA, Ayotunde OL, Gbotemi AI, Ibrahim AD. Propolis production by honey bee, *Apis mellifera* (Hymenoptera: Apidae) and its potential for the management of the larger grain borer, *Prostephanus truncatus* (Horn) (Coleoptera: Bostrichidae) on maize grains. Journal of Plant Protection Research 2010;50:61–66.
62. Kleespies RG, Nansen C, Adouhoun T, Huger AM. Ultrastructure of bacteriomes and their sensitivity to ambient temperatures in *Prostephanus truncatus* (Horn). Biocontrol Science and Technology 2001;1:217–32.
63. Cugala D, Tostao E, Affognon H, Mutungi C. Postharvest losses in Africa – analytical review and synthesis: the case of Mozambique. 2012. Available from: URL: https://www.researchgate.net/publication/285594414 (last accessed 21 August 2018).
64. Muatinte BL, Santos LA, Van den Berg J. The use of mass trapping to suppress population numbers of *Prostephanus truncatus* (Horn) (Coleoptera: Bostrichidae) in small-scale farmer granaries in Mozambique. African Entomology 2018;26:301–16.
65. Wright MAP, Akou-edi D, Stabrawa A. Larger grain borer Project Togo. Infestation of dried cassava and maize by *Prostephanus truncatus*: entomological and socio-economic assessments for the development of loss reduction strategies. *NRI Report* R1941. 141 pp. 1993.
66. Keil H. Losses caused by the larger grain borer in farm stored maize in the Arusha Region of Tanzania. In: Schulten GGM, Toet AJ, editors. Proceedings of the Workshop on the Containment and Control of the Larger Grain Borer. Arusha, Tanzania, 16–21 May 1988. FAO, Rome, Italy; 1988, p. 28–52. Report 2, 209 pp.
67. Najera RMB. Ecology and control of the larger grain borer, *Prostephanus truncatus* in the central areas of Jalisco. 1991. Available from: URL: http://agris.fao.org/agris-search/search.do?recordID =MX19930037639 (last accessed 22 August 2018).
68. Holst N, Meikle WG, Markham RH. Grain injury models for *Prostephanus truncatus* (Coleoptera: Bostrichidae) and *Sitophilus zeamais* (Coleoptera: Curculionidae) in rural maize stores in West Africa. Journal of Economic Entomology 2000;93:1338–46.
69. Borgemeister C, Holst N, Hodges RJ. Biological control and other pest management options for the larger grain borer *Prostephanus truncatus*. In: Neuenschwander P, Borgemeister C, Langewald J, editors. Biological Control in IPM Systems in Africa. CAB international, Wallingford, 2003:311–328. Available from: URL: https://pure.au.dk/portal/en/persons/niels-holst(76beb126-8903-4355-b81bec79cc6a86d7)/publications/biological-control-and-other-pest-management-options-for-larger-grain-borerprostephanus-truncatus(9bc96380-0d90-11db-aa3e-000ea68e967b).html (last accessed 21 June 2019).
70. Meikle WG, Rees D, Markham RH. Biological control of the larger grain borer, *Prostephanus truncatus* (Horn) (Coleoptera: Bostrichidae). Integrated Pest Management Reviews 2002;7:123–38.
71. Edoh Ognakossan KL, Tounou AK, Lamboni Y, Hell K. Post-harvest insect infestation in maize grain stored in woven polypropylene and in hermetic bags. International Journal of Tropical Insect Science 2013;33:71–81.
72. Golob P. Current status of the larger grain borer *Prostephanus truncatus* (Horn) in Africa. Insect Science and its Application 1988;9:737–45.
73. Bell A, Mück O, Mutlu P, Schneider H. Integrated post-harvest protection is worth its money. Deutsche Gesellschaft für Technische Zusammenarbeit (GTZ) GmbH. 1999. Available from: URL: http://citeseerx.ist.psu.edu/viewdoc/download? doi=10.1.1.456.4419&rep=rep1&type=pdf (last accessed 21 June 2018).
74. Pantennius CU. Storage losses in traditional maize granaries in Togo. Insect Science and its Application 1988;9:725–35.

75. Mao L, Henderson G. Evaluation of potential use of nootkatone against maize weevil, *Sitophilus zeamais* Motschulsky and rice weevil, *S. oryzae* (l.) (Coleoptera: Curculionidae). Journal of Stored Products Research 2010;46:72–5.
76. Anankware PJ, Fatunbi AO, Afreh -Nuamah K, Obeng-Ofori D, Ansah AF. Efficacy of the multiple-layer hermetic storage bag for biorational management of primary beetle pests of stored maize. Academic Journal of Entomology 2012;5:47–53.
77. Nang'ayo FLO, Hill MG, Chandi EA, Nzeve NV, Obiero J. The natural environment as a reservoir for the larger grain borer *Prostephanus truncatus* (Horn) (Coleoptera: Bostrichidae) in Kenya. African Crop Science Journal 1993;1:39–47.
78. Henckes C. Investigations into insect population dynamics, damage and losses of stored maize - an approach to IPM on small farms in Tanzania with special reference to *Prostephanus truncatus* (Horn) [Ph.D. thesis]. Technische Universität Berlin, Germany; 1992.
79. Denning G, Kabambe P, Sanchez P, Malik A, Flor R, Harawa R, *et al.* Input subsidies to improve smallholder maize productivity in Malawi: toward an African green revolution. PLoS Biology 2009;7:1–9. Available from: URL: https://journals.plos.org/plosbiology/article/file?id=10.1371/journal.pbio.1000023&type=printable (last accessed 21 June 2019).
80. Kamanula J, Sileshi GW, Belmain SR, Sola P, Mvumi BM, Nyirenda GKC, *et al.* Farmers' insect pest management practices and pesticide plant use in the protection of stored maize and beans in Southern Africa. International Journal of Pest Management 2011;57:41–9.
81. Mukanga M, Deedat Y, Mwangala FS. Toxic effects of five plant extracts against the larger grain borer, *Prostephanus truncatus*. African Journal of Agricultural Research 2010;5:3369–78.
82. Wright MAP. Contributions on the ecology and control of *Prostephanus truncatus* (Horn) in stored cassava [PhD thesis]. University of Reading, UK; 1993.
83. Magrath PA, Compton JAF, Motte FF, Awuku M. Coping with A new Storage Insect Pest: The Impact of the Larger Grain Borer in Eastern Ghana. Natural Resources Institute, Chatham, UK; 1996.
84. Espinal R, Markham RH, Wright VF. Honduras summary of research activities on larger grain borer and storage pest status in meso-America. In: Greathead AHM, Hill G, Kibata GN, editors. Proceedings Management of farm storage pests in East and Central Africa proceedings of the East and Central African Storage Pest Management Workshop held in Naivasha, Kenya. International Institute of Biological Control, Ascot, Berks, UK. 1996. 255 pp. https://cgspace.cgiar.org/handle/10568/48786 (last accessed 21 June 2019).
85. Chittenden F.H. Papers on insects affecting stored products. The lesser grain borer. The larger grain borer, U.S. Department of Agriculture. Bulletin of the Bureau of Entomology 1911;96:48–2.
86. Fisher WC. A revision of the North American species of beetles belonging to the family Bostrichidae. *Miscellaneous Publications of the U.S. Department of Agriculture*, 1950; 698.
87. Mvumi BM, Stathers T. The food security challenges in sub-Saharan Africa: the potential contribution of postharvest skills and technology in closing the gap. In: Arthur FH, Kengkanpanich R, Chayaprasert W, Suthisut D, editors. 11th International Working Conference on Stored Product Protection. Chiang Mai, Thailand; 2015. p. 32–43. November 24–28. Available from: URL: https://www.semanticscholar.org/paper/Food-security-challenges-in-Sub-Saharan-Africa%3A-The-Mvumi-Stathers/55e2c08dfe7e48923133174bae0dce923be768ad (last accessed 25 June 2019)
88. Krause G, Mück O. Bericht über die Projektfortschrittskontrolle zum Projekt 93.2504.4 Integrierte Bekämpfung des Großen Kornbohrers und assoziierter Schadinsekten in bäuerlichen Speichern Überregional/Afrika. Eschborn, Germany; 1996.
89. Boxall RA, Compton JAF. Ghana larger grain borer project. Final technical report. Chatham: Natural Resources Institute, Chatham Maritime, Kent, ME4 4TB, U.K; 1996.
90. Lugg DG, Ndibalema VRJ. Effects of larger grain borer on inter-African grain trade. In: Schulten GGM, Toet AJ, editors. Technical Papers. Workshop on the Containment and Control of the Larger Grain Borer. 16–21 May 1988. Pest Control Services, Ministry of Agriculture and Livestock Development/Food and Agriculture Organization of the United Nations, Arusha, Tanzania; 1988:180–192. Available from: URL: https://revistas.zamorano.edu/index.php/CEIBA/article/viewFile/1006/950 (last accessed 21 June 2019).
91. Shires SW. Ability of *Prostephanus truncatus* (Horn) (Coleoptera: Bostrichidae) to damage and breed on several food commodities. Journal of Stored Products Research 1977;13:205–8.
92. Nansen C, Meikle WG. The biology of the larger grain borer, *Prostephanus truncatus* (Horn) (Coleoptera: Bostrichidae). International Journal of Pest Management Review 2002;7:91–4.

93. Howard DC. The Population biology of the greater grain borer, *Prostephanus truncatus* (Horn) [PhD thesis]. University of ReadingUK; 1983.
94. Mailafiya DM, Ayertey JN, Cudjoe AR. Suitability of sorghum grain for the development of the larger grain borer *Prostephanus truncatus* (Horn) (Coleoptera: Bostrichidae). Science World Journal 2007;2. Available from URL: http://www.scienceworldjournal.org/article/view/2467/2002 (last accessed 21 September 2018).
95. Hodges RJ. The biology and control of *Prostephanus truncatus* (Horn) (Coleoptera: Bostrichidae) – a destructive storage pest with increasing range. Journal of Stored Products Research 1986;22:1–14.
96. Athanassiou CG, Kavallieratos NG, Boukouvala MC, Nika EP. Influence of commodity on the population growth of the larger grain borer, *Prostephanus truncatus* (Horn) (Coleoptera: Bostrichidae). Journal of Stored Products Research 2017;73:129–34.
97. Danho M, Haubruge E, Gaspar C, Lognay G. Selection of grain-hosts by *Prostephanus truncatus* (Coleoptera, Bostrichidae) in the presence of *Sitophilus zeamais* (Coleoptera, Curculionidae) on previously infested grains. Belgian Journal of Zoology 2000;130:3–9.
98. Mullen MA, Vardeman JM, Bagwell J. Insect-resistant packaging. In: Hagstrum DW, Phillips TW, Cuperus G, editors. Stored Product Protection. Kansas State University, Manhattan, KS; 2012. p. 135–42.
99. Navarro S. The use of modified and controlled atmospheres for the disinfestation of stored products. Journal of Pest Sciences 2012;85:301–22.
100. Costa SJ. Reducing Food Losses in sub-Saharan Africa: an "Action Research" Evaluation Trial From Uganda and Burkina Faso. UN World Food Programme, Kampala, Uganda; 2014.
101. Kavallieratos NG, Athanassiou CG, Arthur FH. Effectiveness of insecticide-incorporated bags to control stored-product beetles. Journal of Stored Products Research 2017;70:18–24.
102. Stejskal V, Bostlova M, Nesvorna M, Volek V, Dolezal V, Hubert J. Comparison of the resistance of mono- and multiplayer packaging films to stored-product insects in a laboratory test. Food Control 2017;73:566–73.
103. Kavallieratos NG, Boukouvala MC. Efficacy of four insecticides on different types of storage bags for the management of *Trogoderma granarium* Everts (Coleoptera: Dermestidae) adults and larvae. Journal of Stored Products Research 2018;78:50–8.
104. Pike V. The behaviour of *Prostephanus truncatus* in relation to pheromone-baited crevice traps and flight mill performance in Mexico. Report R1914. Natural Resources Institute, Chatham; 1993.
105. Fadamiro HY. Free flight capacity determination in a sustained flight tunnel: effects of age and sexual state on the flight duration of *Prostephanus truncatus*. Physiological Entomology 1997;22:29–36.
106. Makundi RH. The spread of *Prostephanus truncatus* (Horn) in Africa and measures for its control. FAO Plant Protection Bulletin 1987;35:121–6.
107. Nang'ayo FLO, Hill MG, Wright DJ. Potential hosts of *Prostephanus truncatus* (Coleoptera: Bostrichidae) among native and agroforestry trees in Kenya. Bulletin of Entomological Research 2002;92:499–06.
108. Borgemeister C, Goergen G, Tchabi A, Awande S, Markham RH, Scholz D. Exploitation of a wood host plant and Cerambycid associated volatiles as host find cues by the larger grain borer (Horn) (Coleoptera: Bostrichidae). Annals of the Entomological Society of America 1998;91:741–7.
109. Arthur FH, William R, Morrison WR, Morey AC. Modeling the potential range expansion of larger grain borer, *Prostephanus truncatus* (Coleoptera: Bostrichidae). Scientific reports 2019;9:6862. Available from: URL: https://doi.org/10.1038/s41598-019-42974-5. Available from: URL: https://www.researchgate.net/publication/332847474_Modeling_the_potential_range_expansion_of_larger_grain_borer_Prostephanus_truncatus_Coleoptera_Bostrichidae/references (last accessed 2 May 2019).
110. Al-Sousi EJ, El-Haidari H, Al-Ani JN. Outbreak and new records. FAO Plant Protection Bulletin 1970;18:92–3.
111. Calderon M, Donahaye E. First record of *Prostephanus truncatus* in stored grain. FAO Plant Protection Bulletin 1962;10:43–4.
112. Lesne P. Revision des col´eopt`eres de la famille des Bostrychides. Annales de la Société Entomologique de France 1897;66:319–50.
113. Jia F, Toews MG, Campbell J, Ramaswamy SB. Survival and reproduction of the lesser grain borer, *Rhyzopertha dominica* (F.) (Coleoptera: Bostrichidae) on flora associated with native habitats in Kansas. Journal of Stored Products Research 2008;44:366–72.

114. Hill MG, Borgemeister C, Nansen C. Ecological studies on the larger grain borer, *Prostephanus truncatus* (Horn) (Coleoptera: Bostrichidae) and their implications on integrated pest management. Integrated Pest Management Reviews 2002;7:201–21.
115. Borgemeister C, Adda C, Dlomamou B, Degbay P, Agbaka A, Djossou F, *et al.* The effect of maize cob selection and the impact of field infestation on stored maize losses by the larger grain borer *Prostephanus truncatus* (Horn) (Coleoptera: Bostrichidae) and associated pests. In: Highley E, Wright EJ, Banks HJ, Champ BR, editors. Proceedings of the 6th International Working Conference on Stored Product Protection. CABI, Wallingford, UK; 1994, p. 906–09. April 17–23, Canberra, Australia.
116. Golob P. Chemical, physical and cultural control of *Prostephanus truncatus*. Integrated Pest Management Reviews 2002;7:245–77.
117. Mlambo S, Mvumi BM, Stathers TE, Mubayiwa M, Nyabako T. Field efficacy and persistence of synthetic pesticidal dusts on stored maize grain under contrasting agro-climatic conditions. Journal of Stored Products Research 2018;76:129–39.
118. Dales MJ, Golob P. The protection of maize against *Prostephanus truncatus* (Horn), using insecticide sprays in Tanzania. International Journal of Pest Management 1997;43:39–43.
119. Athanassiou CG, Kavallieratos NG, Yiatilis AE, Vayias BJ, Mavrotas CS, Tomanović Ž. Influence of temperature and humidity on the efficacy of spinosad against four stored grain beetle species. Journal of Insect Sciences 2008;8:60.
120. Kavallieratos NG, Athanassiou CG, Hatzikonstantinou AN, Kavallieratou HN. Abiotic and biotic factors affect efficacy of chlorfenapyr for control of stored-product insect pests. Journal of Food Protection 2011;74:1288–99.
121. Kavallieratos NG, Athanassiou CG, Vayias BJ, Tomanović Ž. Efficacy of insect growth regulators as grain protectants against two stored-product pests in wheat and maize. Journal of Food Protection 2012;75:942–50.
122. Vassilakos TN, Athanassiou CG, Saglam O, Chloridis AS, Dripps JE. Insecticidal effect of spinetoram against six major stored grain insect species. Journal of Stored Products Research 2012;51:69–73.
123. Pozidi Metaxa E, Athanassiou CG. Comparison of spinosad with three traditional grain protectants against *Prostephanus truncatus* (Horn) and *Ephestia kuehniella* (Zeller) at different temperatures. Journal of Pesticide Sciences 2013;86:203–10.
124. Athanassiou CG, Kavallieratos NG. Evaluation of spinetoram and spinosad for control of *Prostephanus truncatus*, *Rhyzopertha dominica*, *Sitophilus oryzae*, and *Tribolium confusum* on stored grains under laboratory conditions. Journal of Pesticide Sciences 2014;87:469–83.
125. Tsaganou FC, Vassilakos TN, Athanassiou CG. Knockdown and mortality of five stored product beetle species after short exposures of thiamethoxam. Journal of Economical Entomology 2014;107:2222–28.
126. Athanassiou CG., Rumbos CI, Sakka M, Sotiroudas V. Insecticidal efficacy of phosphine fumigation at low pressure against major stored-product insect species in a commercial dried fig processing facility. Crop Protection 2016;90:177–85.
127. Kavallieratos NG, Athanassiou CG, Nika EP, Boukouvala MC. Efficacy of alpha-cypermethrin, chlorfenapyr and pirimiphos-methyl for the control of *Prostephanus truncatus* (Horn), *Rhyzopertha dominica* (F.) and *Sitophilus oryzae* (L.). Journal of Stored Products Research 2017;73:54–61.
128. Doganay I, Agrafioti P, Isikber AA, Saglam O, Athanassiou CG. Immediate and delayed mortality of the larger grain borer, *Prostephanus truncatus* (Horn), on different surfaces treated with thiamethoxam and alpha-cypermethrin. Journal Stored Products Research 2018;76:1–6.
129. Scheff DS, Arthur FH, Campbell JF. Insecticide treated packaging for the control of stored product insects. In: Adler CS, Opit G, Fórstenau B, Móller-Blenkle C, Kern P, Arthur FH, *et al.* editors. Proceedings of the 12th International Working Conference on Stored Product Protection (IWCSPP). Berlin, Germany; 2018, p. 920–24. Available from: URL: https://www.cabdirect.org/?target=%2fcabdirect%2fsearch%2f%3fq%3dpb%253a%2522Julius%2bK%25c3%25bchn%2bInstitut%252c%2bBundesforschungsinstitut%2bf%25c3%25bcr%2bKulturpflanzen%2522 (last accessed 21 June 2019).
130. Thompson GD, Michel KH, Yao RC, Mynderse JS, Mosburg CT, Worden TV, *et al.* The discovery of *Saccharopolyspora spinosa* and a new class of insect control products. Down to Earth 1997,52:1–5.
131. Sparks TC, Crouse GD, Dripps JE, Anzeveno P, Martynow J, Deamicis CV, *et al.* Neural network-based QSAR and insecticide discovery: spinetoram. Journal of Computer-Aided Molecular Design 2008;22:393–401.

132. Jones MM, Robertson JL, Weinzierl RA. Susceptibility of oriental fruit moth (Lepidoptera: Tortricidae) larvae to selected reduced-risk insecticides. Journal of Economic Entomology 2010;103:1815–20.
133. Sparks TC, Dripps JE, Watson GB, Paroonagian D. Resistance and cross-resistance to the spinosyns – a review and analysis. Pesticide Biochemistry and Physiology, 2012;102:1–10.
134. Mlambo S, Mvumi BM, Stathers TE, Mubayiwa M, Nyabako T. Field efficacy of hermetic and other maize grain storage options under smallholder farmer management. Crop Protection 2017;98:198–10.
135. Athanassiou CG, Kavallieratos NG., Peteinatos GG, Petrou SE, Boukouvala MC, Tomanović Ž. Influence of temperature and humidity on insecticidal effect of three diatomaceous earth formulations against larger grain borer (Coleoptera: Bostrichidae). Journal of Economic Entomology 2007;100:599–03.
136. Kavallieratos NG, Athanassiou CG, Peteinatos GG, Boukouvala MC, Benelli G. Insecticidal effect and impact of fitness of three diatomaceous earths on different maize hybrids for the eco-friendly control of the invasive stored-product pest *Prostephanus truncatus* (Horn). Environmental Science Pollution Research 2018;25:1407–17.
137. Bergvinson DJ. Opportunities and challenges for IPM in developing countries. In: Koul O, Dhaliwal GS, Cuperus GW, editors. Integrated Pest Management Potential, Constraints and Challenges. CAB International; 2004, p. 281–12. Available from: URL: https://knowledgecenter.cimmyt.org/cgi-bin/koha/opac-detail.pl?biblionumber=5814&shelfbrowse_itemnumber=5890 (last accessed 21 June 2018).
138. Bergvinson D, García-Lara S. Genetic approaches to reducing losses of stored grain to insects and diseases. Current Opinion in Plant Biology 2004;7:480–5. doi: 10.1016/ j.pbi.2004.05.001.
139. García-Lara S, Bergvinson DJ. Integral program to reduce post-harvest losses in maize. Agricultura Técnica en México 2007;33:181–9.
140. Ayertey JN, Meikle WG, Borgemeister C, Camara M, Markham RH. Studies on predation of *Prostephanus truncatus* (Horn) (Coleoptera., Bostrichidae) and *Sitophilus zeamais* Mots. (Coleoptera: Curculionidae) at different densities on maize by *Teretriosoma nigrescens* Lewis (Coleoptera: Histeridae). Journal of Applied Entomology 1999;123:265–71.
141. Biliwa A, Böye J, Fischer HU, Helbig J. Stratégie de lâchers et études de suivi de *Teretriosoma nigrescens* au Togo. In: Böye J, Wright M, Laborius G-A, editors. Implementation of and Further Research on Biological Control of the Larger Grain Borer. Proceedings of an FAO/GTZ Co-ordination Meeting. FAO, Rome, Italy, and GTZ, Eschborn, Germany. FAO, Rome, Italy; 1992, p. 138–42.
142. Compton JAF, Ofosu A. Biological control of the larger grain borer with *Teretriosoma nigrescens*. In: Compton JAF, editor. Quarterly Report 4/94, Ghana Larger Grain Borer Project (Research Programme Volta Region). Ghana Ministry of Food and Agriculture, Accra, Ghana and United Kingdom Overseas Development Administration, London; 1994, p. 1–8.
143. Giles PH, Hill MG, Nang'ayo FLO, Farrell G, Kibata GN. Release and establishment of the predator *Teretriosoma nigrescens* Lewis for the biological control of *Prostephanus truncatus* (Horn) in Kenya. African Crop Science Journal 1996;4:325–37.
144. Milimo JM, Munene V. Technical report of larger grain borer monitoring survey, Northern Province, Zambia, 16–29 September *Tour report*, Mt. Makulu Research Station, Zambia. Mudsprings Geographers, 2001. AWhere-ACT 3.5. Mudsprings Geographers, Inc. USAID/CIMMYT; 1993.
145. Kalivogui K, Müch O. Larger grain borer *Prostephanus truncatus*, in Guinea. FAO Plant Protection Bulletin 1989;39:43.
146. Schneider H, Borgemeister C, Se´tamou M, Affognon H, Bell A, Zweigert ME, *et al.* Biological control of the larger grain borer *Prostephanus truncatus* (Horn) (Coleoptera: Histeridae) in Togo and Benin. Biological Control 2004;30:241–55.
147. Muatinte BL, Van den Berg J, Santos LA. *Prostephanus truncatus* in Africa: a review of biological trends and perspectives on future pest management strategies. African Crop Science Journal 2014;22:237–56.
148. Mutlu P. Ability of the predator *Teretriosoma nigrescens* Lewis (coleoptera: Histeridae) to control larger grain borer (*Prostephanus truncatus*) (Horn) (Coleoptera: Bostrichidae) under rural storage conditions in the southern region of Togo. In: Highley E, Wright EJ, Banks HJ, Champ BR, editors. Proceedings of the Sixth International Working Conference on Stored-Product Protection. Canberra, Australia; 1994, p. 1116–21. Available from: URL: http://spiru.cgahr.ksu.edu/proj/iwcspp/iwcspp6.html (last accessed 21 June 2019).
149. Sharma HC, Ortiz R. Host plant resistance to insects: an eco-friendly approach for pest management and environment conservation. Journal of Environmental Biology 2002;23:111–35.

150. Tefera T, Mugo S, Beyene Y. Developing and deploying insect resistant maize varieties to reduce pre-and post-harvest food losses in Africa. Food Security 2016;8:211–20. doi: 10.1007/s12571-015-0537-7.
151. López-Castillo LM, Silva-Fernández SE, Winkler R, Bergvinson DJ, Arnason JT, García-Lara S. Postharvest insect resistance in maize. Journal of Stored Products Research 2018;77:66–76. doi: 10.1016/j.jspr.2018.03.004.
152. Bruce TJA. Interplay between insects and plants: dynamic and complex interactions that have coevolved over millions of years but act in milliseconds. Journal of Experimental Botany 2015;66:455–65. doi: 10.1093/jxb/eru391.
153. Wiseman BR. Types and mechanisms of host plant resistance to insect attack. Insect Science and its Application 1985;6:239–42. doi: 10.1017/S1742758400004483.
154. Saulnier L, Thibault JF. Ferulic acid and diferulic acids as components of sugar-beet pectins and maize branheteroxylans. Journal of the Science of Food and Agriculture 1999;79:396–402. doi: 10.1002/(SICI)1097-0010 (19990301)79:3<396: AID-JSFA262>3.0.CO;2-B.
155. Sen A, Bergvinson D, Miller SS, Atkinson J, Fulcher RG, Arnason JT. Distribution and microchemical detection of phenolic acids, flavonoids, and phenolic acid amides in maize kernels. Journal of Agricultural and Food Chemistry 1994;42:1879–83. doi: 10.1021/jf00045a009.
156. García-Lara S, Arnason JT, Díaz-Pontones D, Gonzalez E, Bergvinson DJ. Soluble peroxidase activity in maize endosperm associated with maize weevil resistance. Crop Science 2007;47:1125–30. doi: 10.2135/cropsci2006.10.0687.
157. Akpodiete ON, Lale NES, Umeozor OC, Zakka U. Role of physical characteristics of the seed on the stability of resistance of maize varieties to maize weevil (*Sitophilus zeamais* Motschulsky). IOSR Journal of Environmental Science Toxicology and Food Technology 2015;9:2319–99. doi: 10.9790/2402-09226066.
158. Blanco-Labra A, Chagolla-López A, Martínez-Gallardo NA, Valdes-Rodríguez S. Further characterization of the 12 kDa protease/alpha amylase inhibitor present in maize seeds. Journal of Food Biochemistry 1995;19:27–41.
159. Arnason JT, Conilh de Beyssac B, Philogène BJR, Bergvinson D, Serratos JA, Mihm JA. Mechanism of resistance in maize grain to the maize weevil and the larger grain borer. In: Mihm JA, editor. Insect Resistance Maize: Recent Advances and Utilization; Proceeding of an International Symposium Held at CIMMYT. CIMMYT, Mexico, DF, Mexico; 1997:91–95. Available from: URL: https://www.researchgate.net/publication/41667756_Role_of_Dehydrodiferulates_in_Maize_Resistance_to_Pests_and_Diseases (last accessed 25 June 2019).
160. Arnason JT, Gale J, Conilh de Beyssac B, Sen A, Miller SS, Philogène BJR, *et al.* Role of phenolics in resistance of maize grain to the stored grain insects. Journal of Stored Products Research 1992;28:119–26.
161. García-Lara S, Bergvinson DJ. Identification of maize landraces with high level of resistance to storage pests *Sitophilus zeamais* motschulsky and *Prostephanus truncatus* (Horn) in Latin America. Revista Fitotecnia Mexicana 2013;36:347–56.
162. Kumar H. Resistance in maize to the larger grain borer, *Prostephanus truncatus* (Horn) (Coleoptera: Bostrichidae). Journal of Stored Products Research 2002;38:267–80.
163. García-Lara S, Bergvinson DJ. Phytochemical and nutraceutical changes during recurrent selection for storage pest resistance in tropical maize. Crop Sciences 2014;54:2423. doi:10.2135/cropsci2014.03.0223.
164. Bergvinson DJ, García-Lara S. Synergistic effects of insect-resistant maize and *Teretrius nigrescens* on the reduction of grain losses caused by *Prostephanus truncatus* (Horn.). Journal of Stored Products Research 2011;47:95–00. doi: 10.1016/j.jspr.2011.01.003.
165. Winkler R, García-Lara S. Activity-directed identification of maize kernel peroxidases associated with postharvest insect resistance. Molecular BioSystems 2010;6:1810–12. doi: 10.1039/c003112 g.
166. Koziel MG, Beland GL, Bowman C, Carozzi NB, Crenshaw R, Crossland L, *et al.* Field performance of elite transgenic maize plants expressing an insecticidal protein derived from *Bacillus thuringiensis*. Nature Biotechnology 1993;11:194–00. doi: 10.1038/nbt0293-194.
167. Kramer KJ, Morgan TD, Throne JE, Dowell FE, Bailey M, Howard J. Transgenic avidin maize is resistant to storage insect pests. Nature Biotechnology 2000;18:670–4. doi: 10.1038/76531.
168. Araki M, Ishii T. Towards social acceptance of plant breeding by genome editing'. Trends in Plant Science 2015;20:145–9. doi: 10.1016/j.tplants.2015.01.010.

169. Azadi H, Taube F, Taheri F. Co-existence of GM, conventional and organic crops in developing countries: main debates and concerns. Critical Reviews in Food Science and Nutrition 2017;8398:1–12. doi: 10.1080/10408398.2017.1322553.
170. Sakiyama NS, Ramos HCC, Caixeta ET, Pereira MG. Plant breeding with marker-assisted selection in Brazil. Crop Breeding and Applied Biotechnology 2014;14:54–60. Available from: URL: http://www.scielo.br/pdf/cbab/v14n1/09.pdf (last accessed 21 June 2019).
171. Gorjanc G, Jenko J, Hearne SJ, Hickey JM. Initiating maize pre-breeding programs using genomic selection to harness polygenic variation from landrace populations. BMC Genomics 2016;17:1–15. Available from: URL: https://www.ncbi.nlm.nih.gov/pmc/articles/PMC4702314/ (last accessed 25 June 2019).
172. Yencho GC, Cohen MB, Byrne PF. Applications of tagging and mapping insect resistance loci in plants. Annual Review of Entomology 2000;45:393–422.
173. Cardinal AJ, Lee M. Genetic relationships between resistance to stalk-tunneling by the European corn borer and cell-wall components in maize population B73xB52. Theoretical and Applied Genetics 2005;111:1–7.
174. Krakowsky MD, Lee M, Holland JB. Genotypic correlation and multivariate QTL analyses for cell wall components and resistance to stalk tunneling by the European corn borer in maize. Crop Science 2007;47:485–90.
175. García-Lara S, Burt AJ, Arnason JT, Bergvinson DJ. QTL mapping of tropical maize grain components associated with maize weevil resistance. Crop Science 2010;50:815–25. doi: 10.2135/cropsci2009.07.0415.
176. Castro-Álvarez FF, William M, Bergvinson DJ, García-Lara S. Genetic mapping of QTL for maize weevil resistance in a RIL population of tropical maize. Theoretical and Applied Genetics 2015;128:411–19. doi: 10.1007/s00122-014-2440-6.
177. Holland N, Holland D, Helentjaris T, Dhugga KS, Xoconostle-Cazares B, Delmer DP. A comparative analysis of the plant cellulose synthase (*CesA*) gene family. Plant Physiology 2000;123:1313–24.
178. Meyer JDF, Snook ME, Houchins KE, Rector BG, Widstrom NW, McMullen MD. Quantitative trait loci for maysin synthesis in maize (*Zea mays* L.) lines selected for high silk maysin content. Theoretical and Applied Genetics 2007;115:119–28.
179. Davis GL, McMullen MD, Baysdorfer C, Musket T, Grant D, Staebell M, *et al.* A maize map standard with sequenced core markers, grass genome reference points and 932 expressed sequence tagged sites (ESTs) in a 1736-locus map. Genetics 1999;152:1137–72.
180. Rodríguez-Kessler M, Alpuche-Solís AG, Ruiz OA, Jiménez-Bremont JF. Effect of salt stress on the regulation of maize (*Zea mays* L.) genes involved in polyamine biosynthesis. Plant Growth Regulation 2006;48:175–85.
181. Samayoa LF, Malvar RA, Olukolu BA, Holland JB, Butrón A. Genome-wide association study reveals a set of genes associated with resistance to the Mediterranean corn borer (*Sesamia nonagrioides l.*) in a maize diversity panel. BMC Plant Biology 2015;15:1–15. Available from: URL: https://bmcplantbiol.biomedcentral.com/articles?tab=keyword&searchType=journalSearch&sort=PubDate&volume=15&page=6 (last accessed 25 June 2019)
182. Badji A, Otim M, Machida L, Odong TL, Kyamanywa S, Rubaihayo P. Maize resistance to stem borers and storage pests: the need for new genetic and functional genomics approaches in future research. African Journal of Rural Development 2017;2:467–80.
183. Bouda H, Tapondjou LA, Fontem DD, Gumedzoe MYD. Effect of essential oils from leaves of *Agerantum conyzoides*, *Lantana camara* and *Chromolaema odorata* on the mortality of *Sitophilus zeamais* (Coleoptera: Curculionidae). Journal of Stored Products Research 2001;37:103–9.
184. Tapondjou AL, Adler C, Fontem DA, Bouda H, Reichmuth C. Bioactivity of cymol and essential oils of *Cupressus sempervirens* and *Eucalyptus saligna* against *Sitophilus zeamais* Motscholski and *Tribolium confusum* du Val. Journal of Stored Products Research 2005;41:91–02.
185. Segeren P, Van den Oever R, Compton J. Principais doenças e pragas. Relatório de ensaios realizados no regadio do Chókwe, 1986/1990. INIA, Série Investigação 1993;10:359.
186. Poswal MAT, Akpa AD. Current trends in the use of traditional and organic methods for the control of crop pests and diseases in Nigeria. Tropical Pest Management 1991;37:329–33.
187. Jackai LE, Dust R. Insect pests of cowpeas. Annual Review of Entomology 1986;31:95–9.
188. Owusu EO. Effect of some Ghanaian plant components on control of two stored-product insect pests of cereals. Journal of Stored Products Research 2001;37:85–1. doi: 43:1001-10.

189. Obeng-Ofori D, Reichmuth CH. Bioactivity of euginol, a major component of essential oil of *Ocimum suave* (Wild.) against four species of stored-product Coleoptera. International Journal of Pest Management 1997;43:89–94.
190. Obeng-Ofori D, Reichmuth CH, Bekele AJ, Hassanali A. Toxicity and protectant potential of camphor, a major component of essential oil of *Ocimum kilimandscharicum*, against four stored product beetles. International Journal of Pest Management 1998;44:203–9.
191. Osipitan AA, Odebiyi JA, Lawal OA, Somade AA. Evaluation of the extracts of some tropical plants in the management of *Prostephanus truncatus* (Horn) infestation in maize (Zea mays L.). Archives of Phytopathology and Plant Protection 2010;43:1001–10.
192. Tapondjou LA, Adler C, Bouda H, Fontem DA. Efficacy of powder and essential oil from *Chenopodium ambrosioides* leaves as post-harvest grain protectants against six-stored products beetles. Journal of Stored Product Research 2002;38:395–402.
193. Van Wyk B, Van Wyk P. Field Guide to Trees of Southern Africa. Struik Publishers (Pty) Ltd. Cape Town, South Africa. 536 pp. 1997. https://www.amazon.co.uk/Field-Southern-Africa-Struik-Publishers/dp/1770079114. (last accessed 21 June 2019).
194. Amoabeng BW, Gurr GM, Gitau CW, Stevenson PC. Cost-benefit analysis of botanical insecticide use in cabbage: implications for smallholder farmers in developing countries. Crop Protection 2014;57:71–6.
195. Sosef MSM, Dauby G, Blach-Overgaard A, van der Burgt X, Catarino L, Damen TH, *et al.* Exploring the floristic diversity of tropical Africa. BMC Biology 2017;15:15. doi: 10.1186/s12915-017-0356-8.
196. Belmain SR, Amoah BA, Nyirenda SP, Kamanula JF, Stevenson PC. Highly variable insect control efficacy of Tephrosia vogelii chemotypes. Journal of Agricultural and Food Chemistry 2012;60:1055–63.
197. Stevenson PC, Nyirenda SP, Mvumi BM, Sola P, Kamanula JF, Sileshi G, *et al.* Pesticidal plants: a viable alternative insect pest management approach for resource-poor farming in Africa. In: Koul O, Dhaliwal GS, Khokhar S, Singh R, editors. Biopesticides in Environment and Food Security: Issuers and Strategies. Scientific Publishers, Jodhpur, India; 2012, p. 212–38.
198. Stevenson PC, Arnold SEJ, Belmain SR. Pesticidal plants for stored product pest in smallholder farming in Africa. In: Singh D, editor. Advances in Plant Biopesticides Chapter: Pesticidal Plants for Stored Product Pests on Small-holder Farms in Africa. Springer India. 2014. p. 212–38. DOI: 10.1007/978-81-322-2006-0_9. https://www.researchgate.net/publication/268684590_Pesticidal_Plants_for_Stored_Product_Pests_on_Smallholder_Farms_in_Africa. (last accessed 21 June 2019).
199. Kamanula JF, Belmain SR, Hall DR, Farman DI, Goyderc DJ, Mvumi BM, *et al.* Chemical variation and insecticidal activity of Lippia javanica (Burm. f.) spreng essential oil against Sitophilus zeamais Motschulsky. Industrial Crops and Products 2017;110:75–82. Available from: URL: https://doi org/10.1016/j.indcrop.2017.06.036.
200. Mazhawidza E, Mvumi B. Field evaluation of aqueous indigenous plant extracts against the diamondback moth, Plutella xylostella L. and the rape aphid, Brevicoryne brassicae L. in brassica production. Industrial Crops and Products 2017;110:34–44. doi: 10.1016/j.indcrop.2017.09.053.
201. Ndakidemi B, Mtei K, Ndakidemi PA. Impacts of synthetic and botanical pesticides on beneficial insects. Agricultural Sciences 2016;7:364–72. doi: 104236/as.201676038. Available from URL: http://file.scirp.org/pdf/AS_2016061715061431.pdf (last accessed 21 September 2018).
202. Thokozani BLK, Zulu D, Sileshi GW, Gondwe D, Sarasan V, Stevenson PhC. Seed germination and in vitro regeneration of the African medicinal and pesticidal plant, Bobgunnia madagascariensis. African Journal of Biotechnology 2011;10:5959–96.
203. Sola P, Mvumi BM, Ogendo JO, Mponda O, Kamanula JF, Nyirenda SP, *et al.* Botanical pesticide production, trade and regulatory mechanisms in sub-Saharan Africa: making a case for plant-based pesticidal products. Food Security 2014;6:369–84.
204. Rumbos CI, Dutton AC, Athanassiou CG. Comparison of two pirimiphos-methyl formulations against major stored-product insect species. Journal of Stored Products Research 2013;55:106–15.
205. Varnava A, Navarro S, Donahaye E. Long-term hermetic storage of barley in PVC-covered concrete platforms under Mediterranean conditions. Postharvest Bio Technology 1995;6:177–86.
206. Murdock LL, Margam V, Baoua I, Bakfe S, Shade RE. Death by desiccation: effects of hermetic storage on cowpea bruchids. Journal of Stored Products Research 2012;49:166–70.
207. García Lara S, Ortíz-Islas S, Villers P. Portable hermetic storage bag resistant to *Prostephanus truncatus*, *Rhyzopertha dominica*, and *Callosobruchus maculatus*. Journal of Storage Products Research 2013;54:23–5. doi: 10.1016/j.jspr.2013.04.001.
208. Carson JW, Holmes T. Silo failures: why do they happen? Task Quarterly 2003;7:499–12.

209. Njoroge AW, Affognon HD, Mutungi CM, Manono J, Lamuka PO, Murdock LL. Triple bag hermetic storage delivers a lethal punch to *Prostephanus truncatus* (Horn) (Coleoptera: Bostrichidae) in stored maize. Journal of Stored Products Research 2014;58:12–9.
210. Hell K, Edoh Ognakossan K, Lamboni Y. PICS hermetic storage bags ineffective in controlling infestations of *Prostephanus truncatus* and *Dinoderus* spp. in traditional cassava chips. Journal of Stored Products Research 2014;58:53–8. doi: https://dx.doi.org/10.1016/j.jspr.2014.03.003.
211. De Bruin T, Villers P, Wagh A, Navarro S. Worldwide use of hermetic storage for the preservation of agricultural products. *9th International Controlled Atmosphere & Fumigation Conference*. Antalya, Turkey, October 2012; 2012, p. 15–19.
212. Martin DT, Baributsa D, Huesing JE,Williams SB, Murdock LL. PICS bags protect wheat grain, *Triticum aestivum* (L.), against rice weevil, *Sitophilus oryzae* (L.) (Coleoptera: Curculionidae). Journal of Stored Products Research 2015;63:22–30.
213. Villers P, De Bruin T, Navarro S. Development and applications of the hermetic storage technology. In: Lorini I, *et al.*, editors. Proceedings of 9th the International Work Conference on Stored Products Protection. ABRAPOS, Campinas, Sao Paulo, Brazil; 2006, p. 719–29.
214. Ng'ang'a J, Mutungi C, Imathiu SM, Affognon H. Low permeability triple-layer plastic bags prevent losses of maize caused by insects in rural on-farm stores. Food Security 2016;8:621–33.
215. Williams SB, Murdock LL, Baributsa D. Storage of maize in Purdue improved crop storage (PICS) Bags. PLoS ONE 2017:12:1–12. Available from: URL: https://www.ncbi.nlm.nih.gov/pmc/articles/PMC5225016/pdf/pone.0168624.pdf (last accessed 21 June 2019).
216. Abass AB, Fischler M, Schneider K, Daudi S, Gaspar A, Rüst J, *et al.* On-farm comparison of different postharvest storage technologies in a maize farming system of Tanzania Central Corridor. Journal of Stored Products Research 2018;77:55–5.
217. Coffi H, Nyabicha J, Ouma JO. The use of hermetic bags for on farm storage of grains and pulses against insect pests. Outlooks on Pest Management 2016;27:231–5.
218. Bravo-Martinez J. Metal silos and food security. Lessons learned from a successful central American postharvest program. Ibarra, Ecuador: Work- Print; 2009.
219. Chigoverah AA, Mvumi BM. Efficacy of metal silos and hermetic bags against stored-maize insect pests under simulated smallholder farmer conditions. Journal of Stored Products Research 2016;69:179–89.
220. Dogangun A, Karaca Z, Durmus A, Sezen H. Cause of damage and failures in silo structures. Journal of Performance of Constructed Facilities 2009;23:65–1.
221. García-Lara S. Postharvest technologies, past and present experiences in Central Mexico. *Technical Report*. Tecnologico de Monterrey. MTY, Mexico. 13 pp; 2013.
222. SDC (Swiss Agency for Development and Cooperation). Metal silos. Manual for Manufacturing Metal Silos for Grain Storage. 2nd ed. Berne, Switzerland; 2008, p. 58–118. Available from: URL: https://www.shareweb.ch/site/Agriculture-and-Food-Security/focusareas/Documents/phm_postcosecha_manual_silos_e.pdf (last accessed 21 June 2019).
223. SDC (Swiss Agency for Development and Cooperation). Agriculture and food security network; 2013. Available from: URL: http://www.sdc-foodsecurity.ch/en/Home/Focus_areas/Post_harvest (last accessed 22 September 2018).
224. Omondi BA, Jiang N, Van den Berg J, Schulthess F. The flight activity of *Prostephanus truncatus* (Horn) (Coleoptera: Bostrichidae) and *Teretrius nigrescens* Lewis (Coleoptera: Histeridae) in Kenya. Journal of Stored Products Research 2011;47:13–9.
225. Fadamiro HY. Flight and landing behaviour of *Prostephanus truncatus* (Coleoptera: Bostrichidae) in relation to wind speed. Journal of Stored Products Research 1996;32:233–8.
226. Steward-Jones A, Hodges RJ, Farman DI, Hall DA. Solvent extraction of cues in the dust and frass of *Prostephanus truncatus* and analysis of behavioural mechanisms leading to arrestment of the predator *Teretrius nigrescens*. Physiological Entomology 2006;31:63–72.
227. Camara M, Borgemeister C, Markham RH, Poehling HM. Electrophoretic analysis of the prey spectrum of *Teretrius nigrescens* (Lewis) (Coleoptera: Histeridae), a predator of *Prostephanus truncatus* (Horn) (Coleoptera: Bostrichidae), in Mexico, Honduras and Benin. Journal of Applied Entomology 2003;127:360–8.
228. Phillips TW, Throne JE. Bio-rational approaches to managing stored-product insects. Annual Review of Entomology 2010;55:375–97.
229. Schlyter F, Zhang Q, Liu G, Ji L. A successful case of pheromone mass trapping of the bark beetle, *Ips duplicatus* in a forest island, analyzed by 20-years' time-series data. Integrated Pest Management Reviews 2001;6:185–96.

230. Vidyasagar PSPV, Hagi M, Abozuhairah RA, Al-Mohanna OE, Al-Saihati AA. Impact of mass pheromone trapping on red palm weevil: adult population and infestation level in date palm gardens of Saudi Arabia. Planter 2000;76:347–55.
231. Arthurs P, Morales-Reyes C, Cherry RH. Trap design for Lovebugs, Plecia nearctica (Diptera: Bibionidae). Florida Entomologist 2015;98:892–8.
232. Böye J. Autoecological investigations on the behaviour of the larger grain borer *Prostephanus truncatus* (Horn) (Coleoptera: Bostrichidae) in Costa Rica [PhD thesis]. Institute for Phytopathology, Christian Albrechts University Kiel; 1988.
233. Wekesa PW. Field and store ecology of the larger grain borer, *Prostephanus truncatus* (Horn) (Coleoptera: Bostrichidae), in Kenya [PhD thesis]. University of Leicester UK; 1994.
234. Fadamiro HY. Flight behaviour and pheromone communication of the larger grain borer, *Prostephanus truncatus* (Horn) (Coleoptera: Bostrichidae) [PhD thesis]. Green College Oxford, UK; 1995.
235. Nang'ayo FLO. Ecological studies on Larger grain borer in savannah woodlands of Kenya [PhD thesis]. Imperial College London, UK; 1996.
236. Scholz D. Dispersal and host-finding behaviour of *Prostephanus truncatus* (Horn) (Coleoptera: Bostricidae) [PhD. Dissertation]. Hannover University Germany; 1997.
237. Schulthess F. Impact Assessment of *Teretriosoma nigrescens* LEWIS (Coleoptera: Histeridae), a predator of the larger grain borer *Prostephanus truncatus* (Horn) (Coleoptera: Bostrichidae) [PhD dissertation]. Vom Fachbereich Gartenbau der Universit Hannover Alemanha; 1999.
238. Nansen C. Spatial Distribution and Potential Hosts of the larger grain borer, *Prostephanus truncatus* (Horn) (Coleoptera: Bostrichidae), in a forest in Benin, West Africa [PhD thesis]. Royal Veterinary and Agricultural University Copenhagen, Denmark; 2002.
239. Osipitan AA. Distribution and control of the larger grain borer-*Prostephanus truncatus* (Horn) (Coleoptera: Bostrichidae) in Ogun state [PhD dissertation]. University of IbadanIbadan, Oyo State, Nigeria; 2005.
240. Omondi BAO. Ecology and population genetic structure of strains of *Teretrius nigrescens* (Coleoptera: Histeridae), predator of *Prostephanus truncatus* (Coleoptera: Bostrichidae). Thesis submitted in fulfillment of the requirements for the award of the degree Doctor of Philosophy in Environmental Sciences at the North West University (Potchefstroom Campus). South Africa; 2009.
241. Stewart-Jones A. Investigation of close-range behaviours and chemical cues used by *Teretrius nigrescens* (Lewis) (Coleoptera: Histeridae) in finding its prey *Prostephanus truncatus* (Horn) (Coleoptera: Bostrichidae) [PhD thesis]. Natural Resources Institute, University of Greenwich; 2002.
242. Birkinshaw LA. Mate choice in *Prostephanus truncatus* (Horn) (Coleoptera: Bostrichidae): the role of male produced aggregation pheromone [PhD thesis]. University of Leicester United Kingdom; 1998.
243. Leliveldt B. Natural enemy complex of *Prostephanus truncatus* (Horn) (Coleoptera Bostrichidae) [PhD thesis]. Freie University Berlin; 1990.
244. Guntrip J. Variation and covariation in and between life-variation in resource acquisition between individuals or history traits of the larger grain borer, *Prostephanus truncatus* groups of individuals may often obscure underlying phenotypic (Coleoptera, Bostrichidae), and the Cowpea weevil, *Callosobruchus maculatus* (Coleoptera, Bruchidae), with particular reference to [PhD thesis]. University of Reading UK; 1994.
245. Li L. Behavioral ecology and life history evolution in the larger grain borer, *Prostephanus truncatus* (Horn) [PhD thesis]. University of Reading; 1988.
246. Pöschko M. Biologie and Wirtsspezifitä't von *Teretriosoma nigrescens* Lewis (Coleoptera: Histeridae) [PhD thesis]. Technische Universität Berlin Germany; 1993.
247. Schneider H. Impact assessment of *Teretriosoma nigrescens* Lewis (Coleoptera: Histeridae), a predator of the larger grain borer *Prostephanus truncatus* (Horn) (Coleoptera: Bostrichidae) [PhD dissertation]. Hannover University Hannover, Germany; 1999.

11 Invasive *Pomacea* snails: actual and potential environmental impacts and their underlying mechanisms

P.R. Martín[1,2*], S. Burela[1,2], M.E. Seuffert[1,2], N.E. Tamburi[1,3] and L. Saveanu[1,3]

[1] Instituto de Ciencias Biológicas y Biomédicas del Sur (INBIOSUR, UNS-CONICET), San Juan 671, Bahía Blanca, Argentina. [2] Departamento de Biología, Bioquímica y Farmacia, Universidad Nacional del Sur, San Juan 670, Bahía Blanca, Argentina. [3] Departamento de Matemática, Universidad Nacional del Sur, Alem 1250, Bahía Blanca, Argentina

Abstract

Apple snails are large freshwater snails belonging to the family Ampullariidae that inhabit tropical to temperate areas. The South American apple snails *Pomacea canaliculata* and *Pomacea maculata* have been introduced to other continents where they have successfully established and spread. Our review aims to analyse the mechanisms of the impacts that these invasive *Pomacea* provoke or may provoke. Nine basic mechanisms were identified: grazing/herbivory/browsing, competition, predation, disease transmission, hybridisation with native species, poisoning/toxicity, interaction with other invasive species, promotion of collateral damage of control methods on non-target species and when acting as prey. The most important impacts are those related to their grazing on aquatic macrophytes, algae and rice and their competition and predation on other aquatic animals, mostly macroinvertebrates, including other apple snails. Invasive *Pomacea* are also responsible for outbreaks of an emergent parasitic disease (human eosinophilic meningitis). Their great abundance in invaded areas, their bioaccumulation of pollutants and their natural toxicity may impact on their predators and on trophic webs through apparent competition, trophic cascades and biomagnification. The biota from man-managed and natural wetlands may be unintentionally affected by mechanical, chemical and biological control against invasive *Pomacea*. Their capacity to hybridize may affect the distinctiveness and ecological traits of native *Pomacea* in invaded regions of America. Established populations of these invaders may either facilitate or resist the establishment of other exotic species. Field surveys and more realistic experimental approaches with multiple interacting species are needed to better understand the environmental impacts of invasive *Pomacea* and their underlying mechanisms.

Keywords: Golden apple snail, Biological invasion, Pest, Plague, Risk

Review Methodology

We searched the following databases with 31 December 2018 as deadline: Scopus, Scielo, PubMed, Agris and CAB Abstracts. The following combination of terms was used for searching in tittles, abstracts and keywords: ((pomacea) OR (apple AND snail) OR (ampullaria) OR (ampullariidae) OR (ampullariid) OR (ampularid) OR (ampullarid) OR (caracol AND manzana) OR (ampularia) AND (impact) OR (risk) OR (threat) OR (menace) OR (damage) OR (danger)). Wildcards were used when necessary (e.g. damag*). A total of 430 articles were retrieved

*Corresponding author: pablorafaelmartin@gmail.com

using the search options mentioned above. A few of these (53) did not bear a relationship with apple snails, being the result of coincidental homonymy, or lack an available abstract. Many articles (210) mentioned impact-related terms in their title, abstract and keywords but upon closer examination they probed to contain no new evidence or knowledge about impacts. The remaining 167 articles, even if the authors did not aim to evaluate impacts explicitly, contained information that was considered relevant to the aims of the present review. Most of these articles belong to studies performed in Southeastern Asia (58.7%), mainly China (32.7%), Philippines (18.4%) and Japan (8.2%), followed by studies in the Americas (32.9%) and Europe (2.4%). Except for most papers from Argentina and Brazil, the native range of invasive *Pomacea*, these numbers reflect the distribution of present impacts and the concerns about future impacts around the world. Most relevant articles dealt with *P. canaliculata* (55.7%) or *P. maculata* (5.4%) and a few with two or more *Pomacea* spp. (21.6%). Most studies were performed in rice fields (31.7%) followed by laboratory microcosms (22.2%) and those based on field experiments or mesocosms (16.8%).

Introduction

Apple snails (Ampullariidae) are a diverse and widespread group of freshwater snails with very peculiar features such as an aerial–aquatic respiratory system, the largest sized shells among freshwater snails and, in some genera, cleidoic eggs with calcareous eggshells [1, 2]. A few species of this family, notably the South American *Pomacea canaliculata* and *Pomacea maculata*, have been transported intra- and inter-continentally and attracted much basic and applied scientific interest, mostly due to their invasion and impacts in wetlands and aquatic crops [3–8]. The present review aims to analyse the diverse mechanisms of the multiple impacts that *P. canaliculata* and *P. maculata* (from now on referred as invasive *Pomacea*) have provoked or may provoke in temperate to tropical regions around the world. Most information belongs to *P. canaliculata* but our conclusions may be extensive to *P. maculata* and perhaps to other species of the genus.

Mechanisms Underlying Impacts of Invasive *Pomacea*

The present review will focus on actual and potential environmental impacts, i.e. significant changes of ecological properties or processes, regardless of actual or perceived value to humans [9, 10], including impacts on human-managed ecosystems and human beings but not those on human society and economy. About a dozen different basic mechanisms are considered responsible for most of the impacts of invasive species [10]: grazing/herbivory/browsing, competition, predation, disease transmission, hybridisation with native species, poisoning/toxicity, interaction with other invasive species, parasitism, bio-fouling, increase in vegetation flammability, rooting/digging and trampling (disturbances on stones, logs and soil or damage to plants and animals caused by the activities or footsteps of big tetrapods). It is noteworthy that except for the last five, three of which are intrinsically terrestrial, all other mechanisms can be related to the impacts of invasive *Pomacea*. In addition, there are other mechanisms which also seem important in the case of invasive *Pomacea*, for example, when acting passively as prey or as promoters of the collateral damage of control methods on non-target species. The information retrieved from the databases will be presented in this framework as well as the bioecological traits related to each mechanism and their main outcomes. The magnitude of the impacts of invasive *Pomacea* has been related to their biomass [7, 11], which may be high due to their large sizes and high densities.

Grazing/herbivory/browsing

The most widely recognized impacts of invasive *Pomacea* are related to their ability to scrape and graze on live aquatic plants, including vascular ones, an ability that is uncommon among freshwater snails [12]. They have strong jaws and radular teeth and a muscular and cuticularized gizzard [13–15]. *Pomacea canaliculata* has also powerful sets of digestive enzymes, including autogenously and endosymbiotically synthetized proteases and cellulases [16–19]. This apple snail is able to detect the preferred aquatic macrophytes from a distance [20] and it can use ingested lithic particles to increase the mechanical digestion of food and its growth efficiency [21]. Invasive *Pomacea* also have higher feeding rates than other non-invasive snails [22–24].

Many studies reported their ability to ingest various species and types of plants, from aquatic to semi-aquatic or even terrestrial, but also showed great differences in their palatability and nutritive quality for apple snails [22, 24–33], even among plants with the same life form (e.g. submersed macrophytes [34]). Palatability is inversely related to physical and chemical defences and low nutrient content which impair apple snail fitness, reducing growth, fecundity and survival [28, 30, 35]. Different parts or organs of each macrophyte may have different palatability and can be affected differentially by *Pomacea* spp. [26, 36, 37]. The best predictor of macrophyte palatability to *P. canaliculata* is the ratio between nitrogen and phenolic content [33], although other factors may be important. For instance, *P. canaliculata* grazing induces macrophyte chemical defences that depress its feeding and growth [38] and the palatability and nutritive quality of macrophytes can increase, decrease or remain unchanged after their senescence [30]. Moreover, the actual feeding rate on an aquatic macrophyte, and hence the impact on it, depends also on the relative palatability of the other macrophytes available in the same waterbody [31, 34]. The most preferred macrophytes are consumed and eradicated first but then the apple snails feed on the less preferred ones [25, 34]. Field studies have shown that the abundance of invasive *Pomacea* is negatively correlated to the abundance and richness of preferred types of aquatic macrophytes [39, 40]. The feeding preferences of invasive *Pomacea* for different aquatic macrophytes need to be considered in restoration projects as they may affect their success [27, 35, 41].

Although *P. canaliculata* has been usually and generally regarded as macrophytophagous due to its strong impacts on aquatic vegetation and crops, it also consumes other diverse trophic resources [42, 43]. Benthic, filamentous and periphytic algae can be heavily depleted by *P. canaliculata* grazing [26, 34, 36]. This species can also survive and grow collecting organic material from the water surface, including floating plants and other neustonic organisms, with its forefoot [44, 45]. It heavily consumes vegetal detritus [46–49], usually the most important trophic item in its digestive contents in waterbodies from both the invaded and native range (but for exceptions see [50]). The detrital biomass from submersed macrophytes can be reduced to almost one-third (36%) by *P. canaliculata*, a reduction even stronger than the 42% observed for live biomass [34]. In rice fields, *P. canaliculata* is able to grow consuming only rice straw and acts as an active decomposer that accelerates nutrient cycling [51, 52] and, like other *Pomacea* species [53], may also contribute to the breakdown of leaf litter from riparian vegetation.

The herbivory of invasive *Pomacea* releases nutrients to water [54] that may result in eutrophication due to its uptake by phytoplankton [34, 39]; their role as detritus decomposers may also contribute to eutrophication [51, 52]. Even if there is no increase in productivity, phytoplankton-specific composition can be strongly affected [36]. Eradication of submersed macrophytes may also facilitate sediment resuspension and a consequent increase in water turbidity that lessens their recovery chances and promotes a transition from clear- to turbid-water states [39].

Invasive *Pomacea*, especially *P. canaliculata*, are regarded as major pests of aquatic crops such as rice, taro and lotus [3, 4, 6, 55–57]. Although many of these plants are quite unpalatable, in monocultures they are almost the only trophic option. In rice, damage is mostly related to water management and cultural practices that allow the snails to attack the most vulnerable stages, as seedlings and very young plants [57–62]. *Pomacea canaliculata* can also deplete rice field weeds and farmers promote it as a weed biocontrol agent through cultural and water management practices that impede at least in part the attack to rice [63–66].

Competition

Several traits of invasive *Pomacea* such as their large size [2], high trophic flexibility [42, 43], high feeding rate [22, 67] and high fecundity

[68–71] indicate that they may be strong competitors of other snail species [72]. Their tolerance to abiotic stressors such as hypoxia, high salinity, desiccation and extreme temperatures [73–80], their life-cycle plasticity [81–84] and their endurance to starvation [85] may also contribute to their competitive dominance. Invasive *Pomacea* have been associated with the decline of native apple snails in Southeastern Asia [86–88] since many localities formerly inhabited by *Pila* spp. are now occupied only by *P. canaliculata*. In laboratory experiments, *P. maculata* depresses the growth of the Florida apple snail *Pomacea paludosa* [89–91], probably through depletion of shared trophic resources combined with behavioural or chemical interference. *Pomacea canaliculata* impairs growth and survival of the viviparid snail *Bellamya ferrugineus* due to an alteration of water physicochemical quality [54], whereas avoidance of contacts with it, or its mucous trails and faeces, restrains habitat use in three small snails [72]. Besides, *P. canaliculata* grazing induces chemical defences in aquatic macrophytes [38] which could reduce their palatability to other macroinvertebrates feeding on them.

Recent studies showed that detritus is usually the most important trophic item in the digestive contents of *P. canaliculata* from natural wetlands in its invaded and native range [46, 47, 49]. Even in the presence of palatable submersed macrophytes, its intense detritivory can heavily deplete this trophic resource [34] and hence could result in strong resource competition with other detritivorous macroinvertebrates, mostly arthropods and other snails, and promote bottom-up trophic cascades (PRM, unpub. results). *Pomacea* spp. also feed on fresh and partially rotten carrion from vertebrates and invertebrates [42, 43, 92] and may compete with macroinvertebrate scavengers, particularly insects, which are not very specialized in this feeding mode [93].

Predation

Invasive *Pomacea* are opportunistic predators that will prey on any animal too small or too slow to avoid contact, capture and ingestion. Even though they seem unable to detect them from a distance, their wide foot, long cephalic and labial tentacles and their relatively high crawling velocity probably facilitate the encounters with this kind of prey [94]. Their prey are usually other invertebrates but predation on amphibian egg masses has been reported in China and North America [95, 96]. Among invertebrates, bryozoan colonies [97], oligochaete worms [92] and other snails have been reported. Egg masses, hatchlings and adults of snails, especially pulmonates, are usually predated by *Pomacea* spp. in laboratory settings [72, 92, 94, 98, 99], constituting a case of intraguild predation (i.e. a combination of predation and competition). Biocontrol field experiments showed strong declines and even extirpation of schistosome-bearing snails after *Pomacea* spp. introduction [100–102] but these effects may be a combination of intraguild predation and habitat alteration [101, 102] (PRM, unpub. results).

Disease transmission

Although in their native range, *Pomacea* spp. harbour many commensals, parasites and pathogens [103, 104], none of them seems to have been reported in invaded areas. This suggests that they may have been transported mostly as eggs, the most axenic life stage, and that enemy release, especially from pathogens and parasites, may explain in part their success in invaded areas. Probably due to their complex immunological defences, which include molecular and cellular components and as well as barrier organs [105–107], invasive *Pomacea* are susceptible to infestation by only a few parasites from invaded areas, notably the rat lungworm *Angiostrongylus cantonensis*. *Pomacea* spp. and other snails host the third stage larvae, which may be acquired through ingestion by rats, the definitive host, but also by people eating raw or insufficiently cooked snails [108, 109]. *Pomacea canaliculata* recently become one of the main vectors responsible for an emergent parasitic disease (human eosinophilic meningitis) caused by this worm in Southeastern Asia [109–111]. Native apple snails like *Pila polita* are more susceptible to the rat lungworm than the invasive *Pomacea* [112] but the recent decline in the former and the increase in the latter are probably boosting the importance of exotic apple snails in this parasite transmission [110]. The rat lungworm has also reached other regions of the world and the native or introduced *Pomacea* spp. may contribute to the emergence of human eosinophilic meningitis in regions like Brazil and North America [106, 113, 114]. *Pomacea canaliculata* can also act as a facultative or paratenic host of another nematode (*Gnathostoma spinigerum*), which parasitize cats, dogs and humans in Thailand and Japan [115]. On the contrary, invasive *Pomacea* may be able to disrupt the life cycles of some human parasites in invaded areas. *Pomacea* spp. can act as decoys for schistosome miracidia and hence they may reduce the infestation rates of *Biomphalaria* snails with these parasites, whose adults cause schistosomiasis or bilharziasis in humans [107, 116–118]. The high densities and large sizes of invasive *Pomacea* may be especially relevant because the decoy effect increases with the number and individual body surface of non-target apple snails [116].

Hybridisation with native species

Even though *P. canaliculata* and *P. maculata* belong to different clades and are not closely related within the genus *Pomacea* [119], their hybridisation has been probed in laboratory crossings and molecular genetic evidence was also found in natural populations from the native and invaded range [120, 121]. More than 96 *Pomacea* species inhabit the Neotropical region [2] and hence the probability of co-occurrence of native and invasive *Pomacea* in the same waterbody is high. Genetic exchange may blur the distinctiveness of natives but also enhance their tolerance and biotic potential [121]. The heterospecific copulations may likewise result in a decrease of the fecundity of natives due to partial sterility or to low viability of hybrid progeny [120].

Poisoning/toxicity

The aerial eggs of *Pomacea* spp. are regarded as unpalatable to most predators and are seldom eaten by vertebrates or invertebrates, either in their native or invaded range [122, 123]. The perivitelline fluid of invasive *Pomacea* eggs shows neurotoxic, antidigestive and antinutritive properties to terrestrial and aquatic vertebrates [124–128]. Moreover, female reproductive organs involved in perivitellin synthesis have been recently shown to share these egg properties [129]. The most specialized avian predators of *Pomacea* along its native range, the Snail Kite *Rostrhamus sociabilis* and the Limpkin *Aramus guarauna*, are known to discard these organs despite the significant loss of captured biomass [130]. Some predators such as rats, raccoons and storks learn to discard these organs in the invaded range [131–133], but perhaps other naïve native predators from invaded areas may be unable to identify the toxic organs and may be affected negatively by this very unusual trait.

Invasive *Pomacea* are especially tolerant to diverse pollution types [134–136] and are able to strongly bioconcentrate several toxic chemicals that naturally or anthropogenically occur in their habitats, ranging from heavy metals [135, 137, 138] and synthetic compounds [139–141] to uranium [137] and cianotoxins [142]. The potential dangers of these bioaccumulated chemicals to humans and aquaculture or farm animals have been already put forward [137, 143] but there are also potential effects on wild predators attracted by the abundant invasive *Pomacea*. For instance, a cyanotoxin accumulated by *P. maculata* may produce avian vacuolar myelinopathy in the already endangered Snail Kite in Florida [142].

Pesticides used against invasive apple snails may also result in negative effects on organisms feeding on them. For instance, organic pesticides used against *P. canaliculata* in rice fields reduced the number, biomass and diversity of scavenger flies emerging from apple snail corpses with likely consequences in other trophic levels [144].

Interaction with other invasive species

Multiple invasions are increasingly common and facilitation among invaders may enhance the chances of establishment, spread and impacts, provoking invasional meltdown [145, 146]. For instance, the differential grazing of the invasive *P. maculata* may favour the growth of an invasive aquatic macrophyte over native ones in North America, although without beneficial effects for the apple snail [147]. The populations of the native *P. paludosa* were associated with native macrophytes in a Florida lake, whereas *P. maculata* ones were associated with the invasive

Hydrilla verticillata, suggesting at least a unilateral facilitation through enhanced propagule dispersal and grazing on native competing macrophytes [148]. Conversely, an established invader may help to impede or slow the establishment of new potential invaders [146] and contribute to the biotic resistance of the community. Such a role has been suggested for apple snails as they seem to prefer exotic over native macrophytes [31]. The grazing of *P. canaliculata* and of an unidentified *Pomacea* species in Brazil reduces the growth and propagule production of *H. verticillata* and may exert biotic resistance [37, 149], since at least *P. canaliculata* seems to prefer the invasive macrophyte over natives [150, 151]. The biotic resistance potential of *P. canaliculata* against other snails depends on their reproductive and morphological traits: small pulmonate snails that lay gelatinous egg masses are the most affected whereas big snails that give birth to their offspring are barely affected, irrespective of their native or exotic status [72, 94].

Collateral damage on non-target species

The attempts to control invasive *Pomacea* spp. are mostly limited to rice fields and other aquatic crops (but for exceptions see [152, 153]). Most control measures, mechanical, chemical or biological, may also have effects on non-target species inhabiting or visiting the fields or irrigation channels and the last two may also impact nearby natural wetlands [59, 144, 154]. Control measures like inundation of rice fields with seawater may affect both their production and biodiversity [155]. Several chemical molluscicides are effective against apple snails but not very specific and may affect invertebrates and vertebrates as well [156–159]; this prompted the development of new environmentally-friendly molluscicides with low or no toxic effects on non-target freshwater biota [160]. Biocontrol agents tried against invasive apple snails may also be effective but are not very specific either and may significantly impact on other aquatic fauna and flora [161, 162]; biotechnologically developed all-male predatory prawns used for inundative biocontrol of apple snails may still have undesired impacts but are not expected to persist and will be localized [163]. Even hand-picking, one of the most environmentally-friendly control methods, may impact on native apple snails if farmers are not able to distinguish them from invasive ones [154].

As prey

Pomacea spp. are a valuable trophic resource for many predators [2] and the density and biomass of invasive *Pomacea* may reach higher values than those of their native counterparts [148, 164]. In the invaded range, many native and exotic predators take advantage of *Pomacea* snails. Generalist predators soon learn to capture, handle and eat the novel and abundant trophic resource [39, 165–169] which may alter trophic webs provoking bottom-up cascades. Diverse changes in the populations of more specialized predators are also to be expected. For instance, the establishment of dense populations of invasive *Pomacea* led to range expansion and use of new habitats as foraging and nesting grounds by Snail Kites and Limpkins in South and North America [164, 170–172]. In Southeastern Asia, the malacophagous Open-bill stork (*Anastomus oscitans*) has similarly modified its migratory routes and foraging sites after the establishment of *P. canaliculata* populations [132, 173]. However, some invaded wetlands may become ecological traps for specialized predators if an abundant invasive prey differs in subtle ways from native staple prey [174, 175]. Notwithstanding, at least in the case of the endangered Snail Kite in Florida wetlands, the survival and breeding success increases when foraging on *P. maculata*, as its greater size and abundance provides a higher nutritional reward than the native *P. paludosa* [164]. On the other hand, the increase in geographical range or abundance of these specialized predators due to the presence of invasive *Pomacea* may contribute to the decline of other potential native prey, for instance *P. paludosa* in Florida and *Pila* spp. in Southeastern Asia, due to apparent competition (i.e. negative effects due to a shared predator).

Conclusions

Some of the multiple recognized impacts of invasive *Pomacea* have been studied frequently but others remain to be confirmed and explored with more detail. With the exception of studies in rice fields, many impacts of invasive *Pomacea* may be regarded as potential as most studies were performed in laboratory microcosms with only one other species at a time to interact. Field surveys and more realistic experimental settings like mesocosms or enclosures with several interacting species are needed to obtain sound evidence about the environmental impacts of invasive *Pomacea* and their underlying mechanisms.

Acknowledgements

This work was funded with a grant from UNS ('Universidad Nacional del Sur', PGI 24/B232). All the authors are researchers in CONICET ('Consejo Nacional de Investigaciones Científicas y Técnicas'). The authors wish to thank two anonymous reviewers for their helpful criticisms.

References

1. Hayes KA, Cowie RH, Thiengo SC. A global phylogeny of apple snails: Gondwanan origin, generic relationships, and the influence of outgroup choice (Caenogastropoda: Ampullariidae). Biological Journal of the Linnean Society 2009;98(1):61–76.
2. Hayes KA, Burks RL, Castro-Vazquez A, Darby PC, Heras H, Martín PR, *et al.* Insights from an integrated view of the biology of apple snails (Caenogastropoda: Ampullariidae). Malacologia 2015;58(1–2):245–302.
3. Cowie RH. Apple snails (Ampullariidae) as agricultural pests: their biology, impacts and management. In: Barker GM, editor. Molluscs as Crop Pests. New Zealand CABI Publishing, Wallingford; 2002. p. 145–92.
4. Joshi R, Sebastian SS. Global Advances in Ecology and Management of Golden Apple Snails. Philippines: Philippine Rice Research Institute; 2006. p. 600.
5. Horgan FG, Stuart AM, Kudavidanage EP. Impact of invasive apple snails on the functioning and services of natural and managed wetlands. Acta Oecologica 2014;54:90–100.
6. Joshi RC, Cowie RH, Sebastian LS. Biology and management of invasive apple snails. Maligaya, Science City of Muñoz, Nueva Ecija 3119: Philippine Rice Research Institute (PhilRice); 2017.
7. Gilioli G, Schrader G, Carlsson N, van Donk E, van Leeuwen CHA, Martín PR, *et al.* Environmental risk assessment for invasive alien species: a case study of apple snails affecting ecosystem services in Europe. Environmental Impact Assessment Review 2017;65:1–11.
8. Yang Q, Liu S, He C, Cowie RH, Yu X, Hayes KA. Invisible apple snail invasions: importance of continued vigilance and rigorous taxonomic assessments. Pest Management Science 2019;75(5):1277–86.
9. Simberloff D, Martin JL, Genovesi P, Maris V, Wardle DA, Aronson J, *et al.* Impacts of biological invasions: what's what and the way forward. Trends in Ecology & Evolution 2013;28(1):58–66.
10. Blackburn TM, Essl F, Evans T, Hulme PE, Jeschke JM, Kühn I, *et al.* A unified classification of alien species based on the magnitude of their environmental impacts. PLoS Biology 2014;12(5):e1001850.
11. Gilioli G, Pasquali S, Martín PR, Carlsson N, Mariani L. A temperature-dependent physiologically based model for the invasive apple snail *Pomacea canaliculata*. International Journal of Biometeorology 2017;61(11):1899–911.
12. Dillon RT. The Ecology of Freshwater Molluscs. Cambridge University Press, Cambridge, UK; 2000.
13. Andrews EB (editor). The functional anatomy of the gut of the prosobranch gastropod *Pomacea canaliculata* and of some other pilids. Proceedings of the Zoological Society of London; 1965: Wiley Online Library.
14. Moretto HJA, Nahabedian DE. On the radulae of three species of Neotropical *Pomacea* (Mollusca: Mesogastropoda: Pilidae). Walkerana 1989;3(10):195–202.
15. Martín SM, Negrete LHL. Radular ultrastructure of South American Ampullariidae (Gastropoda: Prosobranchia). Brazilian Journal of Biology 2007;67(4):721–6.
16. Godoy MS, Castro-Vazquez A, Vega IA. Endosymbiotic and host Proteases in the digestive tract of the invasive snail *Pomacea canaliculata*: diversity, origin and characterization. PLoS ONE 2013;8(6):e66689.

17. Luo MZ, Zhang JE, Hu JL, Zhao BL. Comparative study on the digestive enzyme activities between golden apple snails (*Pomacea canaliculata*) and local field snails (*Cipangopaludina chinensis*). Shengtai Xuebao/Acta Ecologica Sinica 2015;35(11):3580–7.
18. Yang L, Cheng TY, Zhao FY. Comparative profiling of hepatopancreas transcriptomes in satiated and starving *Pomacea canaliculata*. BMC Genetics 2017;18(1):18.
19. Gong M, Guan Q, Ou X, Chen X, Xie Z. Screening and production of cellulase of bacteria isolated from intestinal tract of *Pomacea canaliculata*. Journal of Biobased Materials and Bioenergy 2017;11(4):278–81.
20. Estebenet AL. Food and feeding in *Pomacea canaliculata* (Gastropoda: Ampullariidae). Veliger 1995;38:277–83.
21. Manara E, Saveanu L, Martín PR. Effects of ingestion of lithic particles on growth of the apple snail *Pomacea canaliculata* (Caenogastropoda, Ampullariidae). Malacologia 2016;59(2):211–23.
22. Morrison WE, Hay ME. Feeding and growth of native, invasive and non-invasive alien apple snails (Ampullariidae) in the United States: invasives eat more and grow more. Biological Invasions 2011;13(4):945–55.
23. Monette D, Ewe S, Markwith SH. Effects of the consumption behavior of adult *Pomacea maculata* and *Pomacea paludosa* on *Vallisneria americana*. Southeastern Naturalist 2016;15(4):689–96.
24. Xu M, Mu X, Dick JTA, Fang M, Gu D, Luo D, *et al.* Comparative functional responses predict the invasiveness and ecological impacts of alien herbivorous snails. PLoS ONE 2016;11(1):e0147017.
25. Carlsson NOL, Lacoursiére JO. Herbivory on aquatic vascular plants by the introduced golden apple snail (*Pomacea canaliculata*) in Lao PDR. Biological Invasions 2005;7:233–41.
26. Carlsson NOL, Brönmark C. Size dependent effects of an invasive herbivorous snail (*Pomacea canaliculata*) on macrophytes and periphyton in Asian wetlands. Freshwater Biology 2006;51(4):695–704.
27. Burlakova LE, Karatayev AY, Padilla DK, Cartwright LD, Hollas DN. Wetland restoration and invasive species: apple snail (*Pomacea insularum*) feeding on native and invasive aquatic plants. Restoration Ecology 2008;17(3):433–40.
28. Qiu JW, Kwong KL. Effects of macrophytes on feeding and life-history traits of the invasive apple snail *Pomacea canaliculata*. Freshwater Biology 2009;54(8):1720–30.
29. Wong PK, Liang Y, Liu NY, Qiu JW. Palatability of macrophytes to the invasive freshwater snail *Pomacea canaliculata*: Differential effects of multiple plant traits. Freshwater Biology 2010;55(10):2023–31.
30. Qiu JW, Chan MT, Kwong KL, Sun J. Consumption, survival and growth in the invasive freshwater snail *Pomacea canaliculata*: Does food freshness matter? Journal of Molluscan Studies 2011;77(2):189–95.
31. Morrison WE, Hay ME. Herbivore preference for native vs. exotic plants: generalist herbivores from multiple continents prefer exotic plants that are evolutionarily naïve. PLoS ONE 2011;6(3):e17227.
32. Low L, Anderson CJ. The threat of a nonnative, invasive apple snail to oligohaline marshes along the Northern Gulf of Mexico. Journal of Coastal Research 2017;33(6):1376–82.
33. Grutters BMC, Roijendijk YOA, Verberk WCEP, Bakker ES. Plant traits and plant biogeography control the biotic resistance provided by generalist herbivores. Functional Ecology 2017;31(6):1184–92.
34. Manara E, Maldonado MA, Martín PR. The role of an invader in its native range: could differential grazing by apple snails structure the submersed macrophytes assemblages in Southern Pampas (Argentina)? Hydrobiologia 2019;828(1):229–42.
35. YamR, Fan Y-T,Wang T-T. Importance of macrophyte quality in determining life-history traits of the apple snails *Pomacea canaliculata*: Implications for bottom-up management of an invasive herbivorous pest in constructed wetlands. International Journal of Environmental Research and Public Health 2016;13(3):248.
36. Fang L, Wong PK, Lin L, Lan C, Qiu JW. Impact of invasive apple snails in Hong Kong on wetland macrophytes, nutrients, phytoplankton and filamentous algae. Freshwater Biology 2010;55(6):1191–204.
37. Ribas LGDS, Cunha ER, Vitule JRS, Mormul RP, Thomaz SM, Padial AA. Biotic resistance by snails and fish to an exotic invasive aquatic plant. Freshwater Biology 2017;62(7):1266–75.
38. Morrison WE, Hay ME. Induced chemical defenses in a freshwater macrophyte suppress herbivore fitness and the growth of associated microbes. Oecologia 2011;165(2):427–36.
39. Carlsson N, Kestrup Å, Mårtensson M, Nyström P. Lethal and non-lethal effects of multiple indigenous predators on the invasive golden apple snail (*Pomacea canaliculata*). Freshwater Biology 2004;49(10):1269–79.
40. Smith C, Boughton EH, Pierre S. *Pomacea maculata* (Island Apple Snail) invasion in seasonal wetlands on Florida Ranchland: association with plant-community structure and aquatic-predator abundance. Southeastern Naturalist 2015;14(3):561–76.

41. Wang Z, Pei Y. Ecological risk resulting from invasive species: a lesson from riparian wetland rehabilitation. Procedia Environmental Sciences 2012;13:1798–808.
42. Cazzaniga NJ, Estebenet AL. Revisión y notas sobre los hábitos alimentarios de los Ampullariidae (Gastropoda). Historia Natural 1984;4(22):213–24.
43. Saveanu L, Manara E, Martín PR. Carrion consumption and its importance in a freshwater trophic generalist: the invasive apple snail *Pomacea canaliculata*. Marine and Freshwater Research 2017;68(4):752–9.
44. Saveanu L, Martín PR. Pedal surface collecting as an alternative feeding mechanism of the invasive apple snail *Pomacea canaliculata* (Caenogastropoda: Ampullariidae). Journal of Molluscan Studies 2013;79(1):11–8.
45. Saveanu L, Martín PR. Neuston: a relevant trophic resource for apple snails? Limnologica 2015;52:75–82.
46. Ocón C, López-van Oosterom MV, Muñoz MI, Rodrigues-Capítulo A. Macroinvertebrate trophic responses to nutrient addition in a temperate stream in South America. Fundamental and Applied Limnology/Archiv für Hydrobiologie 2013;182(1):17–30.
47. López van Oosterom MV, Ocón CS, Brancolini F, Maroñas ME, Sendra ED, Rodrigues Capítulo A. Trophic relationships between macroinvertebrates and fish in a pampean lowland stream (Argentina). Iheringia Série Zoologia 2013;103(1):57–65.
48. López van Oosterom V, Ocón C, Ferreira AC, Rodrigues Capítulo A. The diet of *Pomacea canaliculata* (Gastropoda: Ampullariidae) in its native habitat based on gut content and stable isotopes analysis. Intropica 2016;11:73–83.
49. Kwong KL, Dudgeon D, Wong PK, Qiu JW. Secondary production and diet of an invasive snail in freshwater wetlands: implications for resource utilization and competition. Biological Invasions 2010;12(5):1153–64.
50. Saigo M, Marchese M, Wantzen KM. A closer look at the main actors of Neotropical floodplain food webs: functional classification and niche overlap of dominant benthic invertebrates in a floodplain lake of Paraná River. Iheringia Série Zoologia 2016;106:e2016004.
51. Panteleit J, Horgan FG, Türke M, Schmidt A, Schädler M, Bacht M, *et al.* Effects of detritivorous invertebrates on the decomposition of rice straw: evidence from a microcosm experiment. Paddy and Water Environment 2018;16(2):279–86.
52. Türke M, Schmidt A, Schädler M, Hotes S, Weisser WW. Are invasive apple snails important neglected decomposers of rice straw in paddy fields? Biological Agriculture & Horticulture 2018;34(4):245–57.
53. Tanaka MO, Ribas ACA, de Souza ALT. Macroinvertebrate succession during leaf litter breakdown in a perennial karstic river in Western Brazil. Hydrobiologia 2006;568(1):493–8.
54. Fang M, Xu M, Luo D, Gu DE, Yang YX, Mu XD, *et al.* Impact of aquaculture liquid from invasive species apple snails (*Pomacea canaliculata*) on water quality and indigenous species *Bellamya aeruginosa*. Chinese Journal of Ecology 2016;35(7):1879–88.
55. Halwart M. The golden apple snail *Pomacea canaliculata* in Asian rice farming systems: present impact and future threat. International Journal of Pest Management 1994;40(2):199–206.
56. Horgan FG, Nogues Palenzuela A, Stuart AM, Naredo AI, Ramal AF, Bernal CC, *et al.* Effects of silicon soil amendments and nitrogen fertilizer on apple snail (Ampullariidae) damage to rice seedlings. Crop Protection 2017;91:123–31.
57. Horgan FG, Yanes Figueroa J, Almazan MLP. Seedling broadcasting as a potential method to reduce apple snail damage to rice. Crop Protection 2014;64:168–76.
58. Teo SS. Damage potential of the golden apple snail *Pomacea canaliculata* (Lamarck) in irrigated rice and its control by cultural approaches. International Journal of Pest Management 2003;49(1):49–55.
59. Wada T. Strategies for controlling the apple snail *Pomacea canaliculata* (Lamarck) (Gastropoda: Ampullariidae) in Japanese direct-sown paddy fields. Japan Agricultural Research Quarterly 2004;38(2):75–80.
60. Litsinger JA, Estano DB. Management of the golden apple snail *Pomacea canaliculata* (Lamarck) in rice. Crop Protection 1993;12(5):363–70.
61. Yanes Figueroa J, Almazan MLP, Horgan FG. Reducing seed-densities in rice seedbeds improves the cultural control of apple snail damage. Crop Protection 2014;62:23–31.
62. Arfan AG, Muhamad R, Omar D, Nor Azwady AA, Manjeri G. Sowing methods and water levels influence apple snail damage to rice and its yield in peninsular Malaysia. Experimental Agriculture 2017;53(4):528–38.
63. Okuma M, Tanaka K, Sudo S. Weed control method using apple snail (*Pomacea canaliculata*) in paddy fields. Weed Research (Tokyo) 1994;39(2):114–9.

64. Park Y-S, Bae M-J, Kwon Y-S. Effects of global warming on the distribution of overwintering *Pomacea canaliculata* (Gastropoda: Ampullariidae) in Korea. Korean Journal of Ecology and Environment 2012;45(4):453–8.
65. Arfan AG, Muhamad R, Omar D, Aziz NAA, Gnanasegaram M. Foes can be friends: laboratory trials on invasive apple snails, *Pomacea* spp. Preference to invasive weed, *Limnocharis flava* (L.) Buchenau compared to rice, *Oryza sativa* L. Pakistan. Journal of Zoology 2016;48(3):673–9.
66. Lee SB, Lee SM, Park CB, Lee CR, Ko BG, Park KL, *et al.* The environmental adaptability of *Pomacea canaliculata* used for weed control in wet rice paddies and crop damage caused by overwintered golden apple snails. Korean Journal of Environmental Agriculture 2019;38(1):23–33.
67. Tamburi NE, Martín PR. Feeding rates and food conversion efficiencies in the apple snail *Pomacea canaliculata* (Caenogastropoda: Ampullariidae). Malacologia 2009;51(2):221–32.
68. Martín PR, Estebenet AL. Interpopulation variation in life-history traits of *Pomacea canaliculata* (Gastropoda: Ampullariidae) in southwestern Buenos Aires Province, Argentina. Malacologia 2002;44(1):153–63.
69. Barnes MA, Fordham RK, Burks RL. Fecundity of the exotic apple snail, *Pomacea insularum*. Journal of Natural American Benthology Society 2008;27(3):738–45.
70. Burela S, Martín PR. Evolutionary and functional significance of lengthy copulations in a promiscuous apple snail, *Pomacea canaliculata* (Caenogastropoda: Ampullariidae). Journal of Molluscan Studies 2011;77(1):54–64.
71. Kyle CHS, Plantz AL, Shelton T, Burks RL. Count your eggs before they invade: identifying and quantifying egg clutches of two invasive apple snail species (*Pomacea*). PLoS ONE 2013;8(10):e77736.
72. Maldonado MA, Martín PR. Dealing with a hyper-successful neighbor: effects of the invasive apple snail *Pomacea canaliculata* on exotic and native snails in South America. Current Zoology 2018;14(10):4296–308.
73. Mu H, Sun J, Fang L, Luan T, Williams GA, Cheung SG, *et al.* Genetic basis of differential heat resistance between two species of congeneric freshwater snails: insights from quantitative proteomics and base substitution rate analysis. Journal of Proteome Research 2015;14(10):4296–308.
74. Bernatis JL, McGaw IJ, Cross CL. Abiotic tolerances in different life stages of apple snails *Pomacea canaliculata* and *Pomacea maculata* and the implications for distribution. Journal of Shellfish Research 2016;35(4):1013–25.
75. Glasheen PM, Calvo C, Meerhoff M, Hayes KA, Burks RL. Survival, recovery, and reproduction of apple snails (*Pomacea* spp.) following exposure to drought conditions. Freshwater Science 2017;36(2):316–24.
76. Liu C, Zhang Y, Ren Y, Wang H, Li S, Jiang F, *et al.* The genome of the golden apple snail *Pomacea canaliculata* provides insight into stress tolerance and invasive adaptation. GigaScience 2018;7(9):giy101.
77. Mu H, Sun J, Cheung SG, Fang L, Zhou H, Luan T, *et al.* Comparative proteomics and codon substitution analysis reveal mechanisms of differential resistance to hypoxia in congeneric snails. Journal of Proteomics 2018;172:36–48.
78. Mueck K, Deaton LE, Lee A, Guilbeaux T. Physiology of the apple snail *Pomacea maculata*: Aestivation and overland dispersal. Biological Bulletin 2018;235(1):43–51.
79. Yang S, Zhong JR, Zhao LL, Wu H, Du ZJ, Liu Q, *et al.* The salinity tolerance of the invasive golden apple snail (*Pomacea canaliculata*). Molluscan Research 2018;38(2):90–8.
80. Underwood EB, Darden TL, Knott DM, Kingsley-Smith PR. Determining the salinity tolerance of invasive island apple snail *Pomacea maculata* (Perry, 1810) hatchlings to assess the invasion probability in Estuarine Habitats in South Carolina, USA. Journal of Shellfish Research 2019;38(1):177–82.
81. Estoy GF, Yusa Y, Wada T, Sakurai H, Tsuchida K. Effects of food availability and age on the reproductive effort of the apple snail, *Pomacea canaliculata* (Lamark) (Gastropoda: Ampullariidae). Applied Entomology and Zoology 2002;37(4):543–50.
82. Estoy JGF, Yusa Y, Wada T, Sakurai H, Tsuchida K. Size and age at first copulation and spawning of the apple snail, *Pomacea canaliculata* (Gastropoda: Ampullariidae). Applied Entomology and Zoology 2002;37(1):199–205.
83. Tamburi NE, Martín PR. Reaction norms of size and age at maturity of *Pomacea canaliculata* (Gastropoda: Ampullariidae) under a gradient of food deprivation. Journal of Molluscan Studies 2009;75(1):19–26.
84. Tamburi NE, Martín PR. Effects of food availability on reproductive output, offspring quality and reproductive efficiency in the apple snail *Pomacea canaliculata*. Biological Invasions 2011;13(10):2351–60.

85. Tamburi NE, Martín PR. Effects of absolute fasting on reproduction and survival of the invasive apple snail *Pomacea canaliculata* in its native range. Current Zoology 2016;62(4):369–75.
86. Tan SK, Lee YL, Ng TH. The status of the apple snail, *Pila scutata* (Gastropoda: Ampullariidae) in Singapore. Nature in Singapore 2013;6:135–41.
87. Chaichana R, Sumpan T. The potential ecological impact of the exotic snail *Pomacea canaliculata* on the Thai native snail *Pila scutata*. Science Asia 2014;40:11–5.
88. Ng TH, Tan SK, Ahmad A, Van Tu D, Joshi RC,Wang WY, *et al.* Not in the least concern: anthropogenic influences on a South-east Asian apple snail *Pila scutata* (Ampullariidae). Oryx 2018;53(2):230–8.
89. Conner SL, Pomory CM, Darby PC. Density effects of native and exotic snails on growth in juvenile apple snails *Pomacea paludosa* (Gastropoda: Ampullariidae): a laboratory experiment. Journal of Molluscan Studies 2008;00:1–8.
90. Posch H, Garr AL, Reynolds E. The presence of an exotic snail, *Pomacea maculata*, inhibits growth of juvenile Florida apple snails, *Pomacea paludosa*. Journal of Molluscan Studies 2013;79(4):383–5.
91. Monette D, Ewe S, Markwith SH. Interspecies competition and growth response of *Pomacea maculata* and *Pomacea paludosa*. Florida Scientist 2017;80(1):1–10.
92. Aditya G, Raut SK. Food of the snail *Pomacea bridgesi*, introduced in India. Current Science 2001;80(8):919–21.
93. Fenoglio S, Merritt RW, Cummins KW. Why do no specialized necrophagous species exist among aquatic insects? Freshwater Science 2014;33(3):711–5.
94. Kwong KL, Chan RKY, Qiu JW. The potential of the invasive snail *Pomacea canaliculata* as a predator of various life-stages of five species of freshwater snails. Malacologia 2009;51:343–56.
95. Karraker NE, Dudgeon D. Invasive apple snails (*Pomacea canaliculata*) are predators of amphibians in South China. Biological Invasions 2014;16(9):1785–9.
96. Carter J, Johnson D, Merino S. Exotic invasive *Pomacea maculata* (giant apple snail) will depredate eggs of frog and toad species of the Southeastern US. Southeastern Naturalist 2018;17(3):470–5.
97. Wood T, Anurakpongsatorn P, Chaichana R, Mahujchariyawong J, Satapanajaru T. Predation on freshwater bryozoans by the apple snail, *Pomacea canaliculata*, Ampulariidae, an invasive species in Southeast Asia: a summary report. Denisia 2005;16:283–6.
98. Cazzaniga NJ. Predation of *Pomacea canaliculata* (Ampullariidae) on adult *Biomphalaria peregrina* (Planorbidae). Annals of Tropical Medicine and Parasitology 1990;84(1):97–100.
99. Aditya G, Raut SK. Destruction of *Indoplanorbis exustus* (Planorbidae) eggs by *Pomacea bridgesi* (Ampullariidae). Molluscan Research 2002;22(1):87–90.
100. Milward-de-Andrade R, Carvalho OS. Colonização de *Pomacea haustrum* (Reeve, 1856) em localidade com esquistossomose mansoni: Baldim, MG (Brasil). (Prosobranchia, Pilidae). Revista de Saúde Pública 1979;13:92–107.
101. Pointier JP, Théron A, Imbert-Establet D, Borel G. Eradication of a sylvatic focus of *Schistosoma mansoni* using biological control by competitor snails. Biological Control 1991;1(3):244–7.
102. Pointier JP, David P. Biological control of *Biomphalaria glabrata*, the intermediate host of schistosomes, by *Marisa cornuarietis* in ponds of Guadeloupe: long-term impact on the local snail fauna and aquatic flora. Biological Control 2004;29(1):81–9.
103. Damborenea MC, Brusa F, Negrete J. Symbionts and diseases associated with invasive apple snails. In: Joshi RC, Cowie RH, Sebastian LS, editors. Biology and Management of Invasive Apple Snails. Philippine Rice Research Institute; 2017. p. 73–97.
104. Vega IA, Damborenea MC, Gamarra-Luques C, Koch E, Cueto JA, Castro-Vázquez A. Facultative and obligate symbiotic associations of *Pomacea canaliculata* (Caenogastropoda, Ampullariidae). Biocell (Mendoza) 2006;30(2):367–75.
105. Cueto JA, Rodriguez C, Vega IA, Castro-Vazquez A. Immune defenses of the invasive apple snail *Pomacea canaliculata* (Caenogastropoda, Ampullariidae): phagocytic hemocytes in the circulation and the kidney. PLoS ONE 2015;10(4): e0123964.
106. Rodriguez C, Prieto GI, Vega IA, Castro-Vazquez A. Assessment of the kidney and lung as immune barriers and hematopoietic sites in the invasive apple snail *Pomacea canaliculata*. PeerJ 2018;6:e5789.
107. Silva BBTN, Alves LC, Brayner FA, Junior NCP, Tenório FCAM, Lira ACS, *et al.* Immunological parameters of the *Pomacea lineata* Spix, 1827 (Mollusca: Caenogastropoda) exposed to *Schistosoma mansoni* Sambon, 1907. Acta Parasitologica 2019;64(1):31–43.
108. Yang T-B, Wu Z-D, Lun Z-R. The apple snail *Pomacea canaliculata*, a novel vector of the rat lungworm, *Angiostrongylus cantonensis:* its introduction, spread, and control in China. Hawai'i Journal of Medicine & Public Health: a Journal of Asia Pacific Medicine & Public Health 2013;72(6):23–5.

109. Lv S, Zhang Y, Steinmann P, Yang GJ, Yang K, Zhou XN, *et al.* The emergence of angiostrongyliasis in the People's Republic of China: the interplay between invasive snails, climate change and transmission dynamics. Freshwater Biology 2011;56(4):717–34.
110. Lv S, Guo Y-H, Nguyen HM, Sinuon M, Sayasone S, Lo NC, *et al.* Invasive *Pomacea* snails as important intermediate hosts of *Angiostrongylus cantonensis* in Laos, Cambodia and Vietnam: implications for outbreaks of eosinophilic meningitis. Acta Tropica 2018;183:32–5.
111. Song L, Wang XH, Yang Z, Lv Z, Wu Z. *Angiostrongylus cantonensis* in the vector snails *Pomacea canaliculata* and *Achatina fulica* in China: a meta-analysis. Parasitology Research 2016;115(3): 913–23.
112. Tesana S, Srisawangwong T, Sithithaworn P, Laha T. *Angiostrongylus cantonensis*: experimental study on the susceptibility of apple snails, *Pomacea canaliculata* compared to *Pila polita*. Experimental Parasitology 2008;118:531–5.
113. Teem JL, Qvarnstrom Y, Bishop HS, da Silva AJ, Carter J, White-McLean J, *et al.* The occurrence of the rat lungworm, *Angiostrongylus cantonensis*, in nonindigenous snails in the Gulf of Mexico region of the United States. Hawai'i Journal of Medicine & Public Health: a Journal of Asia Pacific Medicine & Public Health 2013;72(6):11–4.
114. Thiengo SC, Pinto HA, de Mattos AC, Fernandez MA. Helminths parasitizing species of Pomacea in South America (Caenogastropoda; Ampullariidae). In: Sant'Anna BS, Hattori GY, editors. Amazonian Apple Snails. New York: Nova Science Publishers; 2017. p. 177–96.
115. Komalamisra C, Nuamtanong S, Dekumyoy P. *Pila ampullacea* and *Pomacea canaliculata*, as new paratenic hosts of *Gnathostoma spinigerum*. Southeast Asian Journal of Tropical Medicine and Public Health 2009;40(2):243–6.
116. Upatham ES, Sturrock RF. Field investigations on the effect of other aquatic animals on the infection of *Biomphalaria glabrata* by *Schistosoma mansoni* miracidia. The Journal of Parasitology 1973:448–53.
117. Laracuente A, Brown RA, Jobin WR. Comparison of four species of snails as potential decoys to intercept schistosome miracidia. The American Journal of Tropical Medicine and Hygiene 1979;28(1):99–105.
118. Milward-De-Andrade R, Souza CP. *Pomacea haustrum* (Reeve, 1856) as a decoy to infection of *Biomphalaria glabrata* (Say, 1818) with *Schistosoma mansoni* Sambon, 1907. Revista Brasileira de Malariologia e Doencas Tropicais Publicacoes Avulsas 1979;31:59.
119. Hayes KA, Cowie RH, Thiengo SC, Strong EE. Comparing apples with apples: clarifying the identities of two highly invasive Neotropical Ampullariidae (Caenogastropoda). Zoological Journal of the Linnean Society 2012;166(4):723–53.
120. Matsukura K, Okuda M, Cazzaniga NJ, Wada T. Genetic exchange between two freshwater apple snails, *Pomacea canaliculata* and *Pomacea maculata* invading East and Southeast Asia. Biological Invasions 2013;15(9):2039–48.
121. Matsukura K, Izumi Y, Yoshida K, Wada T. Cold tolerance of invasive freshwater snails, *Pomacea canaliculata*, *P. maculata*, and their hybrids helps explain their different distributions. Freshwater Biology 2015;61(1):80–7.
122. Yusa Y. Predation on eggs of the apple snail *Pomacea canaliculata* (Gastropoda: Ampullariidae) by the fire ant *Solenopsis geminata*. Journal of Molluscan Studies 2001;67(3):275–9.
123. dos Santos-Neto CR. Record of Egg Predation of the Gastropods of Genus Pomacea. In: Sant'Anna BS, Hattori GY, editors. Amazonian Apple Snails. New York: Nova Science Publishers; 2017. p. 145–54.
124. Pasquevich MY, Dreon MS, Qiu J-W, Mu H, Heras H. Convergent evolution of plant and animal embryo defences by hyperstable non-digestible storage proteins. Scientific Reports 2017;7(1):15848.
125. Lajmanovich RC, Attademo AM, Peltzer PM, Junges CM, Martinuzzi CS. Acute toxicity of apple snail *Pomacea canaliculata*'s eggs on *Rhinella arenarum* tadpoles. Toxin Reviews 2017;36(1):45–51.
126. Giglio ML, Ituarte S, Pasquevich MY, Heras H. The eggs of the apple snail *Pomacea maculata* are defended by indigestible polysaccharides and toxic proteins. Canadian Journal of Zoology 2016;94(11):777–85.
127. Dreón MS, Frassa MV, Ceolín M, Ituarte S, Qiu J-W, Sun J, *et al.* Novel animal defenses against predation: a snail egg neurotoxin combining lectin and pore-forming chains that resembles plant defense and bacteria attack toxins. PLoS ONE 2013;8(5):e63782.
128. Heras H, Frassa MV, Fernández PE, Galosi CM, Gimeno EJ, Dreon MS. First egg protein with a neurotoxic effect on mice. Toxicon 2008;52(3):481–8.

129. Cadierno MP, Dreon MS, Heras H. Apple snail perivitellin precursor properties help explain predators' feeding behavior. Physiological and Biochemical Zoology 2017;90(4):461–70.
130. Cadierno MP, Burela S, Dreon MS, Martín PR, Heras H. The influence of energy, nutritional value and noxiousness of prey in sex- and size-biased predation by Snail Kites in southern South America. Emu – Austral Ornithology 2017;117(4):382–7.
131. Yusa Y, Sugiura N, Ichinose K. Predation on the apple snail, *Pomacea canaliculata* (Ampullariidae), by the Norway rat, *Rattus norvegicus*, in the field. Veliger 2000;43(4):349–53.
132. Sawangproh W, Round PD, Poonswad P. Asian openbill stork *Anastomus oscitans* as a predator of the invasive alien gastropod *Pomacea canaliculata* in Thailand. Iberus 2012;30(1):111–7.
133. Carter J, Merino S, Prejean D, LaFleur Jr G. Observations of raccoon (*Procyon lotor*) predation on the invasive maculata apple snail (*Pomacea maculata*) in Southern Louisiana. Southeastern Naturalist 2017;16(3):N14–N8.
134. Chaichana R, Sumpan T. Environmental tolerance of invasive golden apple snails (*Pomacea canaliculata* (Lamarck, 1822)) and Thai native apple snails (*Pila scutata*,(Mousson, 1848)). Tropical Ecology 2015;56(3):347–55.
135. Huang F, Peng L, Zhang J, Lin W, Chen S. Cadmium bioaccumulation and antioxidant enzyme activity in hepatopancreas, kidney, and stomach of invasive apple snail *Pomacea canaliculata*. Environmental Science and Pollution Research 2018;25:18682–92.
136. Deng PY, Shu WS, Lan CY, Liu WH. Metal contamination in the sediment, pondweed, and snails of a stream receiving effluent from a lead/zinc mine in southern China. Bulletin of Environmental Contamination and Toxicology 2008;81(1):69–74.
137. Vega IA, Arribére MA, Almonacid AV, Guevara SR, Castro-Vazquez A. Apple snails and their endosymbionts bioconcentrate heavy metals and uranium from contaminated drinking water. Environmental Science and Pollution Research 2012;19(8):3307–16.
138. Campoy-Diaz AD, Arribére MA, Guevara SR, Vega IA. Bioindication of mercury, arsenic and uranium in the apple snail *Pomacea canaliculata* (Caenogastropoda, Ampullariidae): Bioconcentration and depuration in tissues and symbiotic corpuscles. Chemosphere 2018;196:196–205.
139. Martínez ML, Piol MN, Sbarbati Nudelman N, Verrengia Guerrero NR. Tributyltin bioaccumulation and toxic effects in freshwater gastropods *Pomacea canaliculata* after a chronic exposure: field and laboratory studies. Ecotoxicology 2017;26(5):691–701.
140. Wu JP, Chen XY, Wu SK, Tao L, She YZ, Luo XJ, *et al.* Polychlorinated biphenyls in apple snails from an abandoned e-waste recycling site, 2010–2016: a temporal snapshot after the regulatory efforts and the bioaccumulation characteristics. Science of the Total Environment 2019;650:779–85.
141. Koch E, Altamirano JC, Covaci A, Lana NB, Ciocco NF. Should apple snail *Pomacea canaliculata* (Caenogastropoda, Ampullariidae) be used as bioindicator for BDE-209? Environmental Science and Pollution Research 2014;21(1):761–5.
142. Dodd SR, Haynie RS, Williams SM, Wilde SB. Alternate food-chain transfer of the toxin linked to avian vacuolar myelinopathy and implications for the endangered Florida snail kite (*Rostrhamus sociabilis*). Journal of Wildlife Diseases 2016;52(2):335–44.
143. Berry JP, Lind O. First evidence of 'paralytic shellfish toxins' and cylindrospermopsin in a Mexican freshwater system, Lago Catemaco, and apparent bioaccumulation of the toxins in 'tegogolo' snails (*Pomacea patula catemacensis*). Toxicon 2010;55(5):930–8.
144. Horgan FG. The ecophysiology of apple snails in rice: implications for crop management and policy. Annals of Applied Biology 2018;172(3):245–67.
145. Simberloff D. Invasional meltdown 6 years later: important phenomenon, unfortunate metaphor, or both? Ecology Letters 2006;9(8):912–9.
146. Kuebbing SE, Nuñez MA. Negative, neutral, and positive interactions among nonnative plants: patterns, processes, and management implications. Global Change Biology 2015;21(2):926–34.
147. Meza-Lopez MM, Siemann E. Experimental test of the Invasional Meltdown Hypothesis: an exotic herbivore facilitates an exotic plant, but the plant does not reciprocally facilitate the herbivore. Freshwater Biology 2015;60(7):1475–82.
148. Monette D, Ewe S, Dinkins JM, Markwith SH. Interactions of exotic and native *Pomacea* with wetland vegetation structure in the Greater Everglades, Florida, USA. Fundamental and Applied Limnology 2017;189(4):291–9.
149. Calvo C, Mormul RP, Figueiredo BRS, Cunha ER, Thomaz SM, Meerhoff M. Herbivory can mitigate, but not counteract, the positive effects of warming on the establishment of the invasive macrophyte *Hydrilla verticillata*. Biological Invasions 2019;21(1):59–66.

150. Cruz C, Silva AF, Venturini FP, Garlich N, Pitelli RLCM, Pitelli RA. Food preference and consumption of aquatic macrophytes submerged by snail *Pomacea canaliculata*. Planta Daninha 2015;33(3):433–9.
151. Oliveira MV, Dainez M, Bertoncin AP, Muniz C, Meurer T, Figueiredo B, *et al.* Native snails choose an invasive over a native macrophyte as a food resource. Canadian Journal of Zoology 2018;97(4):362–7.
152. Martin CW, Bayha KM, Valentine JF. Establishment of the invasive island apple snail *Pomacea insularum* (Gastropoda: Ampullaridae) and eradication efforts in Mobile, Alabama, USA. Gulf of Mexico Science 2012;30(1–2):30–8.
153. Bernatis JL, Warren GL. Effectiveness of a hand removal program for management of nonindigenous apple snails in an urban pond. Southeastern Naturalist 2014;13(3):607–18.
154. Schneiker J, Weisser WW, Settele J, Bustamante JV, Marquez L, Villareal S, *et al.* Is there hope for sustainable management of golden apple snails, a major invasive pest in irrigated rice? NJAS-Wageningen Journal of Life Sciences 2016;79:11–21.
155. EFSA. Scientific opinion on the environmental risk assessment of the apple snail for the EU. EFSA Journal 2014;12(4):3641–97.
156. Chen RS, Wang KL, Wu CY. Assessment of the camellia seed meal impact on loaches in paddy fields. Paddy and Water Environment 2012;10(4):291–300.
157. Attademo AM, Lajmanovich RC, Peltzer PM, Junges CM. Acute Toxicity of metaldehyde in the invasive rice snail *Pomacea canaliculata* and sublethal effects on tadpoles of a non-target species (*Rhinella arenarum*). Water, Air, and Soil Pollution 2016;227(11):400.
158. Olivier HM, Jenkins JA, Berhow M, Carter J. A pilot study testing a natural and a synthetic molluscicide for controlling invasive apple snails (*Pomacea maculata*). Bulletin of Environmental Contamination and Toxicology 2016;96(3):289–94.
159. Horgan FG, Bernal CC, Letana S, Naredo AI, Ramp D, Almazan MLP. Reduced efficiency of tropical flies (Diptera) in the decomposition of snail cadavers following molluscicide poisoning. Applied Soil Ecology 2018;129:61–71.
160. Castillo-Ruiz M, Cañon-Jones H, Schlotterbeck T, Lopez MA, Tomas A, San Martín R. Safety and efficacy of quinoa (*Chenopodium quinoa*) saponins derived molluscicide to control of *Pomacea maculata* in rice fields in the Ebro Delta, Spain. Crop Protection 2018;111:42–9.
161. Wong PK, Kwong KL, Qiu JW. Complex interactions among fish, snails and macrophytes: implications for biological control of an invasive snail. Biological Invasions 2009;11(10):2223–32.
162. Ip KKL, Liang Y, Lin L,Wu H, Xue J, Qiu J-W. Biological control of invasive apple snails by two species of carp: effects on non-target species matter. Biological Control 2014;71:16–22.
163. Savaya-Alkalay A, Ovadia O, Barki A, Sagi A. Size-selective predation by all-male prawns: implications for sustainable biocontrol of snail invasions. Biological Invasions 2018;20(1):137–49.
164. Cattau CE, Fletcher Jr RJ, Reichert BE, Kitchens WM. Counteracting effects of a non-native prey on the demography of a native predator culminate in positive population growth. Ecological Applications 2016;26(7):1952–68.
165. Yusa Y, Sugiura N, Wada T. Predatory potential of freshwater animals on an invasive agricultural pest, the apple snail *Pomacea canaliculata* (Gastropoda: Ampullariidae), in southern Japan. Biological Invasions 2006;8(2):137–47.
166. Yamanishi Y, Yoshida K, Fujimori N, Yusa Y. Predator-driven biotic resistance and propagule pressure regulate the invasive apple snail *Pomacea canaliculata* in Japan. Biological Invasions 2012;14(7):1343–52.
167. Dorn NJ, Hafsadi M. Native crayfish consume more non-native than native apple snails. Biological Invasions 2016;18(1):159–67.
168. Bertolero A, Navarro J. A native bird as a predator for the invasive apple snail, a novel rice field invader in Europe. Aquatic Conservation: Marine and Freshwater Ecosystems 2018;28(5):1099–104.
169. Works AJ, Olson DH. Diets of two nonnative freshwater turtle species (*Trachemys scripta* and *Pelodiscus sinensis*) in Kawai Nui Marsh, Hawaii. Journal of Herpetology 2018;52(4):445–53.
170. Horgan FG, Felix MI, Portalanza DE, Sánchez L, Moya Rios WM, Farah SE, *et al.* Responses by farmers to the apple snail invasion of Ecuador's rice fields and attitudes toward predatory snail kites. Crop Protection 2014;62:135–43.
171. Pias KE, Fletcher Jr RJ, Kitchens WM. Assessing the value of novel habitats to snail kites through foraging behavior and nest survival. Journal of Fish and Wildlife Management 2016;7(2):449–60.
172. Marzolf N, Smith C, Golladay S. Limpkin (*Aramus guarauna*) establishment following recent increase in nonnative prey availability in Lake Seminole, Georgia. The Wilson Journal of Ornithology 2019;131(1):179–84.

173. Ratanakorn P, Suwanpakdee S, Wiriyarat W, Eiamampai K, Chaichoune K, Wiratsudakul A, *et al.* Satellite telemetry tracks flyways of Asian Openbill storks in relation to H5N1 avian influenza spread and ecological change. BMC Veterinary Research 2018;14(1):349.
174. Cattau CE, Martin J, Kitchens WM. Effects of an exotic prey species on a native specialist: example of the snail kite. Biological Conservation 2010;143(2):513–20.
175. Wilcox RC, Fletcher Jr RJ. Experimental test of preferences for an invasive prey by an endangered predator: implications for conservation. PLoS ONE 2016;11(11):e0165427.

12 An overview of ash (*Fraxinus* spp.) and the ash dieback disease in Europe

Rasmus Enderle[1]*, Jan Stenlid[2] and Rimvydas Vasaitis[2]

[1] Institute for Plant Protection in Horticulture and Forests, Julius Kuehn Institute, 38104 Braunschweig, Germany. [2] Department of Forest Mycology and Plant Pathology, Swedish University of Agricultural Sciences, 75007 Uppsala, Sweden

Abstract

Ash dieback, caused by the invasive fungal pathogen *Hymenoscyphus fraxineus*, has become a serious threat to ash trees (*Fraxinus* spp.) and ash-related ecosystems in Europe. The fast emergence and expansion of this new disease has called for an intensive investigation, which resulted in a large number of research projects in Europe. Recently, as a result of the European cooperation and exchange programme FRAXBACK, numerous reports containing detailed information about the situation of ash dieback in different countries and related research have been published. For this review, we performed a systematic analysis of these country reports. We focussed on differences and similarities between European regions and countries regarding the importance, genetics and resistance of ash, the spread, monitoring, impact and management of the disease and on factors that have an influence on its severity. By including most recent scientific literature, this review provides a concise yet substantial overview about ash and ash dieback in Europe.

Keywords: Ash dieback, *Fraxinus*, *Hymenoscyphus fraxineus*, Invasive species, Emerging disease, Forest pathology

Review Methodology

The main source of information for this review is the FRAXBACK book [1] and references therein. Further information was gathered from a special issue on ash dieback of the journal Baltic Forestry [2] and by online searches (e.g. Google, Google Scholar and Web of Science).

Introduction

Ash dieback is a lethal disease of ash trees (*Fraxinus* spp.) in Europe. It is caused by an invasive ascomycete, *Hymenoscyphus fraxineus* Baral et al. (syn. *H. pseudoalbidus* Queloz et al., anamorph: *Chalara fraxinea* T. Kowalski), which is native to Far East Asia [3–5]. First symptoms of ash dieback have been observed in the early 1990s in north-eastern Poland [6, 7 and references therein, 8], but a genetic study suggests that the introduction of the pathogen to Europe had occurred one to three decades earlier [3]. The disease spreads fast and is today present in most part of the ash distribution range in Europe, causing a dramatic decline of native trees that belong to the genus *Fraxinus*. Leaves and petioles are infected by airborne ascospores that can be transported over long distances [5, 7]. The fungus can spread to woody tissues through the petiole–shoot junction [9], where it causes bark necroses and locally kills shoots and twigs. The

*Corresponding author: rasmus.enderle@julius-kuehn.de

successive crown decline caused by every year occurring infections often is lethal to the trees. Another important yet little understood factor of mortality is the symptom of root collar necrosis, which is most likely caused primarily by *H. fraxineus* and often is followed by root and but rot. In this regard, wood-decay fungi, particularly fungi of the genus *Armillaria*, play a presumably secondary yet decisively aggravating role [10–15].

The European dimension of the ash dieback epidemic did soon demand for an international approach to address challenging research questions. Within the European Cooperation in Science and Technology (COST) programme, an action called FP1103 FRAXBACK was conducted from 2012 to 2016, providing a platform for more than 200 scientists from 38 countries and different disciplines to exchange and interact. As a main outcome of this action, a comprehensive book (299 pages) [1] containing detailed country reports about the situation and research regarding ash dieback was published in 2017, along with a special issue of the scientific journal Baltic Forestry [2]. This review article can be understood as a short summary of the book with a focus on differences and similarities between European countries and regions. It deals with the importance, genetics and resistance of ash, the spread, monitoring, impact and management of the disease and with factors that have an influence on its severity. While the book is the main source of information for this review, it is placed in the wider context of other literature including most recent publications on the topic.

The exchange that was triggered by the FRAXBACK action was also a basis for a review on silvicultural strategies [16]. There are other reviews that cover evolutionary aspects regarding ash dieback [17], the possible pathways of the pathogen's introduction to Europe [18], the morphology and ecology of *H. fraxineus* (and some related species) [4], the quantitative genetics of disease resistance [19], general knowledge about *H. fraxineus* [5] and possible ecologic consequences of the disease [20].

Importance of Ash in Europe

There are four species of ash native to Europe. Common ash (*Fraxinus excelsior* L.) is the most widely distributed species and present in all European countries except Island, Malta and Portugal, but rather scarce in the Mediterranean. Here, narrow-leafed ash (*Fraxinus angustifolia* Vahl) and manna ash (*Fraxinus ornus* L.), which are restricted to more southern parts of Europe, are of higher importance. *F. ornus* is not very susceptible to ash dieback [21] and a rather small tree, although of economic importance in southern Europe [22]. The fourth species, *Fraxinus pallisiae* Wilmott ex Pallis, grows rarely in steppe and steppe-forest zones in south-eastern Europe, as does a subspecies of common ash, *F. excelsior* ssp. *coriariaefolia* [23]. The susceptibility of the latter two taxa to ash dieback is unknown. This review focuses on *F. excelsior* and *F. angustifolia*.

In most European countries, ash trees account for 1–2% of the forest area or standing volume stock [8, 23–28]. At the northern distribution range, the share of ash trees is even smaller [29–31], while ash trees hold a slightly higher proportion in forests of central Europe [21, 32–34]. Of particular importance is ash in England and Wales, where the proportion of this tree species in woodland ranges from 4.5 to 12.7% [35], and on the Island of Ireland, where ash is an omnipresent landscape component as a major part of hedgerows and woodland patches [36].

The economical importance of ash in Europe was much greater in ancient times and before the onset of the industrialization, when its wood was indispensable because of its toughness and bending properties and was used widely, e.g. for tools, arms and transport [37]. As a pollarding tree, ash has also been a source of fodder for livestock [29, 37]. In the last century, the use of ash timber has often been replaced by steel or other modern materials. However, ash timber is still a favourite for furniture, certain sports equipment, tool handles and some other goods (e.g. [25, 33]). An example in this respect is the hurl (used in traditional Gaelic sports hurley and camogie), which is crafted from ash wood [36]. Overall, the economic contribution of ash to forestry is considered significant in many regions of Europe, as is indicated in almost all country reports of the book [1]. Moreover, ash is very important from an ecological point of view. It provides critical ecological services, such as wildlife habitat and niche, stabilization of water balance and stream banks, production of high quality litter and as a component in slowing zones for forest fires [24, 25, 35, 38]. A large number of organisms are associated with or even entirely dependent on ash [21, 29, 30, 35, 39]. In forestry, ash was popular also because of its robustness, its drought tolerance, fast growth (on rich sites) and prolific regeneration. In the decades before the outbreak of the epidemic, forestry deliberately increased the proportion of ash in many regions [25, 26, 33].

The importance of ash is also demonstrated by a British study, which identified a wide range of national stakeholders for ash from government, the private sector and civil society that either have an influence on or a special interest in the ash dieback epidemic [40]. However, stakeholders in Romania tend to underestimate the importance of ash and the implications of ash dieback, as was concluded from a survey by Drăgoi et al. [41].

Gene Flow, Genetic Structure and Diversity of *H. fraxineus*

Several population studies demonstrated that in Europe there is low allelic diversity and little to no genetic differentiation between *H. fraxineus* populations [26, 29, 31, 34, 35, 42, 43]. Genetic differentiation was even small between long-established populations (from Lithuania) and populations from the epidemic front (in Switzerland), and there was no reduced genetic diversity in the front populations [26, 34]. Also between populations in the UK and Continental Europe there was only little genetic differentiation [43]. The uniform genetic structure throughout Europe indicates high-gene flow and a recent bottleneck due to a very small founder population. In fact, analyses of the coding regions of core eukaryotic genes in Asian and European isolates suggest that the European population was founded by two divergent haploid individuals [44]. This finding is supported by comparisons of European and eastern Asian populations, where much higher genetic diversity and similarly high-gene flow was observed [3, 34, 35]. An ancestry analysis indicates that the Russian Far East is the source of the European populations [3].

It can be assumed from expansion rates and gene flow analyses that viable ascospores are transmitted by wind over long distances. However, on a small scale, along a tree species mixture gradient in Germany, some slight population differentiation was observed, which may have reflected recent colonization history [45]. Moreover, an increase of genetic diversity of *H. fraxineus* was observed along an altitude gradient in Poland [8, 46]. Population studies also indicate that the fungus is outcrossing and high-genotypic diversity was detected within the European populations [26, 29, 34, 35, 42]. These results may explain part of the variations in virulence, growth rates, temperature optima for growth and exoenzyme profiles, which have been observed among isolates [47–49].

Gene Flow, Genetic Structure and Diversity of Ash in Europe

The genetic structure of European *F. excelsior* populations is still shaped by the ice ages, despite the predominantly outcrossing nature of the species. After the last ice age, the present range was recolonized from glacial refugia in Iberia, Italy, the eastern Alps and the Balkan Peninsula, as was clearly shown by analyses of chloroplast DNA (cpDNA) variation of 62 populations throughout Europe [50]. In western and central Europe, admixture of populations from different refugia and high-gene flow during the recolonization process led to relatively high homogeneity, allelic richness and genetic diversity [36, 51]. For example, 96% of the variation was detected within Irish *F. excelsior* populations with little evidence of population structuring [36]. Results of a recent reciprocal transplanting experiment along a transect from Scotland to the Pyrenees led to an assumption that there is little local adaption in western Europe, as some populations grew well at all sites while local populations never performed best [35]. On the other hand, restricted postglacial gene flow caused by a rapid population expansion in the mountainous topography has likely caused rather strong genetic differentiation over short distances and less genetic diversity within populations in south-eastern Europe, as was suggested by a microsatellite study [51]. There is evidence that the present northern range of common ash was colonized from south-eastern Europe and similarly is characterized by rather limited gene flow and low-allelic richness [29, 31, 50, 52]. It is feared that these more fragmented populations are particularly vulnerable to presumable genetic effects of ash dieback such as loss of allelic structure and inbreeding [31]. In Lithuania, for example, it is very likely that the density of flowering trees and thus the effective population size has already decreased to a degree that compromises the remaining level of genetic diversity and possibilities for successful regeneration of ash populations [26].

When compared with other central European populations, higher genetic variation was detected in south-eastern Germany, where phylogenetic lineages from different refugia meet [33]. The transgression zone between the central and eastern European lineage along Hungary, Slovakia, Czech Republic and Poland could be detected in a corridor of 36 km width when analysing maternally inherited cpDNA, whereas the analysis of bi-parentally inherited nuclear gene markers detected an average width of the transgression zone of 275 km [27]. Another study in Germany confirmed that, while seed dispersal is rather restricted, there is considerable pollen flow at the landscape level [53]. In former times it was believed that *F. excelsior* trees from floodplain forest and from dry, calcareous sites form two ecotypes, but such distinction could never be proved by reciprocal transplants or progeny trials [21, 30]. There are inconsistent results concerning differences in observed heterozygosity between groups of differently susceptible *F. excelsior* trees in Germany and Poland [54, 55].

Similarly to *F. excelsior*, cpDNA analyses of *F. angustifolia* and *F. ornus* populations indicate distinct glacial refugia in southern Europe, with the south-eastern populations being relatively polymorphic [56]. Overall cpDNA diversity in these two species was lower than in *F. excelsior*. Generally, just like *F. excelsior*, *F. angustifolia* is a species with high intrapopulation genetic variation and low differentiation among populations [22]. Only little is known about local inbreeding or outcrossing rates and their impact on the genetic structure of *F. angustifolia* and *F. ornus* [38].

Genetic Resistance of Ash to *H. fraxineus*

Throughout Europe, there is a remarkable consistency in reports of genotypic differences in susceptibility of *F. excelsior* and *F. angustifolia* to ash dieback. Typically, about 1% of trees show no or only insignificant symptoms (Figure 12.1) and are considered resistant (or tolerant, depending on definition). Remarkably consistent is also the high degree of genetic control of this quantitative resistance, as was indicated by heritability values in a number of studies from several countries (Table 12.1). Heritability is defined as the proportion of the total variance due to genetic effects (e.g. [57]). Further evidence for genetic control of the resistance is provided from clonal seed orchards in Austria, the Netherlands, Croatia and Slovakia for both *F. excelsior* [21, 58] and *F. angustifolia* [21, 58–60]. There is also indication that development and severity of ash collar necroses are under considerable genetic control (h^2 (narrow-sense heritability) = 0.49 for prevalence and h^2 = 0.42 for necrosis width) [61]. Based on these studies it can be assumed with very high certainty that throughout the European populations there is a small proportion of ash trees with genetically controlled resistance. However, there was no significant correlation of damage data between parent ash trees and their offspring in natural regeneration in two heterogenic Austrian ash stands, indicating that environmental conditions and the interaction of genotypes with these play an important role *in situ* [61, 62].

Although some part of the variation in the health status of different genotypes is probably due to disease escape or phenological mismatch (see section 'Influencing factors'), inoculation experiments clearly indicate that an active defence is involved [66]. The mechanisms that play a role in this defence are still virtually unknown. Different gene expression markers and a coding single-nucleotide polymorphism were associated with resistance to ash dieback and indicate a role of MADS box transcription factors (named after genesMCM1, AGAMOUS, DEFICIENS, and SRF) and iridoid glycoside [74, 75]. Fourier-transform infrared spectroscopy of phenolic extracts from bark tissue proved to be

Fig. 12.1. This picture (taken in August 2018) shows neighbouring crowns of two *F. excelsior* trees in Gotland (Sweden), where the ash dieback pathogen arrived probably 2001 or 2002 [73]. Only minor parts of the left crown are still alive, because epicormic shoots, which were produced to compensate for dead twigs, gradually became diseased as well. In contrast, despite similar infection pressure for much more than a decade, the right tree is barely affected by ash dieback. It can be assumed that it possesses a relatively high degree of genetic resistance to ash dieback.

Table 12.1. Heritability values for traits related to ash dieback severity (crown symptoms) as a measure for genetic control of resistance to ash dieback

Country	Trial type	Number of sites	Year of establishment	Time of data collection	Species	Heritability value(s)	Reference(s)
Denmark	Progeny	1	2003	2009–2014	*F. excelsior*	h^2: 0.42–0.53	[63]
Denmark	Progeny	2	2004	2008–2010	*F. excelsior*	h^2: 0.37–0.52	[64]
Denmark	Clonal	2	1998	2007–2009	*F. excelsior*	H^2: 0.25–0.54	[65, 66]
France	Progeny	1	1995	2010–2014	*F. excelsior*	h^2: 0.42	[61]
Germany	Clonal	4	1991–1995	2012–2013	*F. excelsior*	H^2: 0.48–0.59	[67]
Lithuania	Progeny	2	2005	2010	*F. excelsior*	h^2: 0.40–0.49	[26, 68]
Lithuania	Clonal	7	2012	2012–2013	*F. excelsior*	H^2: 0.21–0.65	[26, 69]
Slovenia	Clonal	1	1989	2009–2012	*F. angustifolia*	H^2: 0.23–0.50	[70]
Sweden	Clonal	2	1992–1995	2006–2011	*F. excelsior*	H^2: 0.10–0.42	[71, 72]

H^2, broad-sense heritability; h^2, narrow-sense heritability.
The values are based on studies with different settings and plant materials.

a relatively effective method to identify resistant individuals of ash, although on its own is not indicative regarding underlying resistance mechanisms [76]. However, such technologies can significantly advance efforts in selection and breeding for resistance.

Spread and Monitoring of Ash Dieback

According to Hietala et al. [77], the strong saprobic competence of *H. fraxineus* in defending the sporulation niche against abiotic stress and microbial competition, which enables production of a large amount of sexual spores, may be a main reason for the successful and invasive spread of this pathogen in Europe. The authors further assume that the ability to produce conidia may be connected to its invasion success, although the biological role of these asexual spores needs further investigation.

Our present knowledge of the spatiotemporal spread of the epidemic throughout Europe is based on dates of observations of symptoms and pathogen records, which is, however, biased by the spatiotemporally uneven presence of forest pathology experts and their individual awareness of the disease. For example, *H. fraxineus* was first detected in England in 2012 on imported nursery stock, but mortality dating of ash trees indicated that ash dieback was already active in England in 2004 or 2005 [78]. A rough overview of the historical spread of ash dieback throughout Europe was given by McKinney et al. [19], showing a more or less concentric expansion from the epicentre of the disease, which is believed to be located in north-eastern Poland [8]. It is quite likely that the disease in its early stage can be overlooked and that the first introduction remained localized for a prolonged period. The subsequent concentric range expansion is documented in some countries on a regional scale, e.g. in Norway, Switzerland, France and Italy, typically with continuous steps of about 30–70 km per year [31, 34, 79, 80]. It is believed that this pattern of disease spread is mainly due to wind dispersal of airborne ascospores. In contrast, irregular or even inverse disease spread occurred in numerous other countries (e.g. [21, 27, 30, 33]), and there is broad agreement that this is an effect of anthropogenic movement of plant material, i.e. the trade of ash plants. It is probable that even to the UK, despite its isolated geographic situation, *H. fraxineus* was introduced by both wind dispersal and imports of infected plant material [43].

Large parts of central and eastern Europe have been infested before the first description of the causal agent in 2006 [6] and even larger areas before the correct identification of its teleomorph in 2011 [81]. This is why in mainland Europe there were no official attempts to limit the spread of the disease. *H. fraxineus* had been listed in the EPPO Altert List from 2007 to 2014, but has never been listed as a regulated organism (A1 or A2 List) in the Plant Health Directive of the European Union [82]. Only on the British Isles, legislations were changed in order to limit or prevent further introductions from mainland Europe by control of importation of plant material. Moreover, ash plants on infested sites were removed and destroyed [35, 36]. Although soon it became clear that there is no hope that these measures can lead to eradication of *H. fraxineus*, they were continued in Ireland in order to slow spread and build-up of the disease within the island [36]. Measures to prevent possible outbreaks of ash dieback were also taken in the northern Apennines, Italy, by the Regional Phytosanitary Services of Tuscany and Emilia Romagna [79]. There are concerns that introduction of further pathogen genetic diversity to Europe may cause increased disease severity. It is thus recommended to regulate imports of live ash plants and material that may contain ash leave debris from eastern Asia [44].

Ash dieback and its effects have been monitored in many local trials and forests throughout Europe. An extensive list of studies and sites was provided by Coker et al. [83]. These studies generated much data, which are, however, difficult to compare, because different scoring systems were used and the scoring of ash dieback and its severity is intricate and, in visual assessments, to some degree subjective. Kopinga and de Vries [84] compared different methods of ash dieback assessment in a large-scale survey. Examples for the assessed data are the degree of dieback, defoliation and the presence of epicormic shoots in percentage or ordinal scales. All evaluated specific aspects of ash dieback increased the accuracy of the health status assessments of trees and were significantly correlated, but a certain combination of aspects that is the most distinctive in monitoring of ash dieback has not yet been identified. Mortality is unambiguous and was recorded in many studies. It can serve as a measure of ash dieback severity, although it can have other causes. Mortality rates were reviewed and modelled including many different data sources in two recent papers, one using data from studies throughout Europe [83], another using data from Belgium and France [80]. Both studies suggest that the increase of annual mortality rates tend to stagnate at a level that depends on the forest stand after about 6–8 years of strong infection pressure. According to theory, future decrease of mortality rates due to a decrease of susceptibility in the remaining population and due to a dilution effect is likely, but extrapolation of mortality data with regards to ash dieback is still a matter of great uncertainty [14, 80, 83]. Also, detailed studies on the expansion of the pathogen within diseased trees indicated that a significant proportion of individual twig and branch infections stop expanding and eventually die out [85, 86]. This suggests that the dynamics of new and old infections within a tree are crucial for ash survival.

Influencing Factors

Severity of ash dieback in a stand depends on the time of pathogen arrival, the age of ash trees in the stand and several other factors. In numerous monitoring studies a number of factors that were connected to the severity of ash dieback symptoms and the extent of disease-induced mortality have been identified. An overview of factors that were associated with ash dieback is given in Table 12.2.

As shown there, most of these factors were features of individual trees. Characteristics that are related to age and size of trees are often negatively correlated with the severity of symptoms (i.e. the disease often progresses more slowly on larger or older trees). However, this result depends on the set of investigated trees, which is why such relationship had not been detected in every study (see Table 12.2 for reference). Moreover, the severity of symptoms is connected to the host tree species, since *F. angustifolia* seems

Table 12.2. Factors that are associated with symptom severity or mortality due to ash dieback according to references in the FRAXBACK book [1]

Factor	Symptoms are more severe on/at ...	Reference(s)
Characteristics of individual trees		
Host species	... *F. excelsior* compared with *F. angustifolia* (depending on site)	[27, 84]
Gender	... male trees	[58]
Provenance	Not applicable	[25–27, 33, 66]
Cultivar	Not applicable	[25, 58, 84]
Leaf yellowing/leaf shed	... later shedding trees	[66]
Spring phenology	... later flushing trees	[25, 26, 29, 87]
Diameter at breast height (DBH)	... (inconsistently) trees of smaller DBH	[26, 29, 39, 80]
Crown surface projection	... trees with smaller crowns	[8]
Age	... (inconsistently) younger trees	[25, 33, 88]
Seedling versus sprouts	... sprouts	[26]
Armillaria root rot	... *Armillaria*-infected trees	[26]
Site conditions		
Altitude	... lower altitudes	[25]
Shade, woodland, open, grazed	... (inconsistently) shaded, forested sites	[21, 39, 84]
Temperature	... low to moderate temperature	[28, 79]
Drought, arid climate	... less dry sites	[21, 23, 28, 79]
Moisture, humidity	... more humid sites	[21, 79]
Soil type	... not specified	[27]
Management		
Pollarding, pruning	Inconsistent results	[39, 58]
Thinning	Not significant	[29]

to be slightly less susceptible than *F. excelsior* (and on *F. ornus* only minor symptoms occur on leaves). In a Dutch study, predominantly male ash trees performed better than trees with higher proportions of female flowers. Relatively strong differences in susceptibility were also reported for certain cultivars. Differences in susceptibility between central and western European provenances are only small, but Slovakian provenances apparently differ more, which possibly is connected to higher population differentiation (see the section 'Gene flow, genetic structure and diversity of ash in Europe'). In Lithuania, autochthonous provenances performed better than foreign ones, which may be due to already commenced natural selection in this long-time infested country, or due to genotype – environment interactions, which were found significant in some trials. Processes of natural selection may also be the reason for young seedlings performing better than regeneration that emerged by sprouting. Trees that are infected by *Armillaria* at the root collar are more likely to be severely damaged in the crown.

The severity of ash dieback symptoms also depends on environmental factors. There is a consensus that moisture and humidity is conducive for the disease. These factors seem to be especially important for the symptom of collar lesion [13]. It was speculated that the formation of apothecia on shoots and roots, which generally occurs only very rarely, may be of higher importance in the wet, maritime climate of Ireland [36]. Although it was not confirmed in a study on veteran trees in Sweden [39], it is a general observation that trees in forests are more severely affected than trees in the open landscape [21], which may be due to higher humidity and infection pressure. On the other hand, there is evidence that drought and high temperature are unfavourable for ash dieback [89]. The disease risk is considered to be low in warm areas of southern Europe and global warming is expected to alleviate the disease to some extent in the future [89, 90]. Thinning can change the microclimate and possibly alter conditions of infection and reduce disease severity. However, in young even-aged ash stands in Denmark, disease severity was the worst in unthinned plots, but otherwise unrelated to stand density [91]. Pruning of diseased branches reduces the probability of stem infections, although such control tactic may not be economical [92]. Interestingly, old trees that had been pollarded some decades ago are less affected than other more recently or not at all pollarded veteran trees [39]. Although some fungal ash endophytes with high *in vitro* antagonistic activity against *H. fraxineus* have been identified [93–96], an influence of endophytic mycobiomes on the health status of ash trees (i.e. the susceptibility to ash dieback) or on the establishment of the pathogen has not yet been demonstrated [77, 94, 96, 97].

Impact of Ash Dieback on Timber Logging

An obvious reaction to ash dieback is increased timber logging due to salvage cuttings. Published data on this increase, however, are relatively rare and often of rather indirect nature. Swedish national forest inventory data reveal a notable decline of the standing timber stock since 2009, which is probably due to increased cuttings in stands damaged by ash dieback [29]. In Lithuania, where symptoms of ash dieback were first observed in 1996, heavy salvage cutting (selective felling and clearcutting of more than 1500 ha of ash stands) commenced in 2001 [26]. Annual salvage cutting reached its maximum in 2006 (about 3500 ha) and smoothly decreased down to about 1000 ha in 2015. At this time, 18 years after the first observation of symptoms, the former area of ash stands was reduced by more than 50% [26]. Clearcutting was also carried out in other countries. For example, as an ultimate option, immature but heavily devastated ash stands often have been clearcut in Austria [21, 98]. In Grafenegg (Austria), where ash dieback arrived about 2005, logging of ash more than doubled by 2011, quadrupled by 2014 and sextupled by 2015 when compared with 2008. A similar exponential increase in timber logging is reported from state owned forests in Baden-Württemberg, Germany [98], and from Denmark [19, 98]. In Wallonia (Belgium), a very sharp increase in the number of ash timber lots greater than 50 m^3 was recorded in 2015 [32]. In summary, it can be said that sharp increases of ash timber logging commenced 5–8 years after ash dieback symptoms were detected in a region or country for the first time.

Management of Ash Dieback

Silvicultural strategies in response to ash dieback were reviewed comprehensively by Skovsgaard et al. [16], who concluded with the main advice to retain tolerant-appearing ash trees wherever reasonable within the context of management objectives and overall forest structure in order to preserve these genetic resources (i.e. heritable resistance). However, there are reports from numerous countries that forest owners do not always follow this advice, which may be due to the lack of information [39], economic reasons [26] or general loss of interest and expectation regarding forestry with ash [21, 25, 26, 33]. The latter often is connected to the experience of observing declining trees that appeared resistant in the early stages of the disease.

Generally, official recommendations for disease management in the ash dieback-affected countries are in line with the strategies that are proposed in the silviculture review [16], but there are regional adaptations and emphases of certain aspects. For example, measures to prevent disease spread and inoculum build-up (i.e. destruction of infected plants, plant material and leaf litter) were carried out extensively in the British Isles, although with low impact [35, 36]. In contrast, in continental Europe there were only a few local attempts to slow down the spread of the disease, e.g. in order to prevent outbreaks in the Apennines in Italy [79]. Awareness of the risks rose too late for efficient quarantine measures in large parts of Europe, where the disease spread before fundamental knowledge was available.

In most infested countries nursing and planting of new ash trees is not recommended and has decreased sharply since the onset of the disease [24, 25, 29, 32, 33, 58, 66]. In Poland, the production of ash plant material even was banned completely [8]. On the other hand, large numbers of ash plants were still produced in Slovakia (up to 873 000 plants in 2015), where 150 000 to 175 000 trees were annually planted in the forests [27]. In 2017, planting of ash was still subsidized in Flanders, Ireland and Greece [22, 24, 36]. At the northern limit of the range of *F. excelsior*, planting of healthy seedlings from local sources is considered to be the best option to prevent loss of genetic diversity on the one hand and loss of local adaption on the other [31], which both is crucial in these areas (see the section 'Gene flow, genetic structure and diversity of ash in Europe'). A similar strategy is also proposed in so far not infected areas, i.e. in Portugal, with the aim to increase among- and within-population genetic diversity and hence prepare ash populations for invasive pathogens such as *H. fraxineus* [38].

Tending and early thinning is recommended in Poland in order to increase tree vigour, which helps ash trees to survive the disease, by fostering the growth of crowns and root systems [8]. In Germany, such early promotion is recommended only for resistant appearing ash trees [33], whereas in Lithuania pre-commercial thinning is recommended only in heavily damaged stands with low proportions of ash [26].

For the management of middle-aged stands there are different concepts that depend on the severity of the disease and the proportion of ash. Generally, stands should be thinned as usual to promote canopy development while focusing on retaining resistant appearing trees and species other than ash. Foresters in Sweden are advised to mark the healthiest ash trees in order to avoid cutting these in error and to enable somewhat objective observations of the development of the disease over time [29]. In Norway it was recommended to preferably remove susceptible ash trees in order to lower the local overall infection pressure and to reduce their genetic influence on the next tree generation [31]. However, new research demonstrated that trees with higher levels of crown damage produce much less amounts of propagules than healthier trees [99]. This relationship is stronger for seeds than for pollen, so it seems recommendable to focus particularly on diseased male trees when thinning. Valuable timber should be harvested before wood quality deteriorates. If, however, it can be expected that the damage will become too large to allow for silvicultural control of stand development, salvage felling of larger parts of the stand or even clearcutting may be the only options.

Replacement of ash trees that have died or have been felled due to the disease can be achieved by either natural forest regeneration or planting. Natural regeneration of ash can be used and even fostered, e.g. by retaining some relatively healthy ash individuals as seed trees, as is recommended in Lithuania and Estonia [18, 26]. Since planting of new ash trees is, if at all, recommended only exceptionally (see above), the choice of alternative tree species needs to be considered carefully. Especially in monocultural ash woodlands, which often is the case in Ireland, it is suggested to aim for a transition to more mixed species forests, in order to build a greater degree of resilience [100]. There is a consensus that replacement in forests by non-native *Fraxinus* species is not a preferable option, even if these species are more resistant to ash dieback [16, 18, 21, 23, 25, 30, 33]. Instead, recommendations for alternative tree species are mainly based on specific site adaption and thus differ between regions and countries. In Great Britain, the ability of tree species and species mixtures to substitute the ecosystem functions of ash is considered as an additional important criterion [35, 101].

Conclusion

Ash species native to Europe provide important ecosystem services and are considered of high value both as ornamental trees and as a forest component. In many places, particularly in central Europe and the British Isles, wood production with *F. excelsior* or *F. angustifolia* contributes significantly to local economy. There is little genetic differentiation of ash populations in western and central Europe, but some population differentiation is characteristic to its eastern and, due to local adaptation, particularly to its very northern range. The European population of *H. fraxineus* is characterized by extremely low genetic variation and local differentiation, which is a result of its recent introduction, its subsequent rapid spread and effective sexual outcrossing. The disease spread both by wind dispersal of ascospores and by human-mediated transportation of infected plant material. Large parts of the continent were infested before the identification of *H. fraxineus* as the causal agent of ash dieback, and sharp increases of ash timber logging typically commenced 5–8 years after ash dieback symptoms were detected for the first time in a region.

Severity of ash dieback is influenced by individual tree characteristics (e.g. tree age) and site conditions (e.g. moisture or humidity), but there is rather little and inconsistent evidence for an influence of forest management practices. Strategies for disease management, which differ slightly between countries, thus aim mainly for the retention and propagation of resistant or tolerant trees. Heritability values for ash crown damage and other disease-related traits from numerous clonal trials and progeny trials were consistently high, regardless of their location in Europe. This is an indication for strong genetic control of these traits, although it was not yet possible to demonstrate a connection of tree health between parent trees and natural regeneration under heterogeneous conditions.

Acknowledgements

The authors would like to thank the COST Action FP1103 FRAXBACK, which founded the basis of this review article with its final publications. R. Vasaitis acknowledges the support from The Organisation for Economic Co-operation and Development (OECD) Programme on Biological Resource Management for Sustainable Agricultural Systems, UK Government Department for Environment, Food and Rural Affairs (DEFRA) within EUPHRESCO project 'Risk-based strategies to prepare for and manage invasive tree borers' (PREPSYS).

References

1. Vasaitis R, Enderle R, editors. Dieback of European ash (*Fraxinus* spp.) – Consequences and Guidelines for Sustainable Management. Uppsala, Sweden: Swedish University of Agricultural Sciences; 2017. 299 p. Available from: URL: http://www.slu.se/globalassets/ew/org/inst/mykopat/forskning/stenlid/dieback-of-european-ash.pdf.

2. Enderle R, Pliura A, Vasaitis R, editors. Advances in ash dieback research – and some other invasive diseases of trees. Baltic Forestry 2017;23(1):1–333. Available from: URL: https://www.balticforestry.mi.lt/bf/index.php?option=com_content&view=article&catid=14&id=482.
3. Sønstebø JH, Vivian-Smith A, Adamson K, Drenkhan R, Solheim H, Hietala A. Genome-wide population diversity in *Hymenoscyphus fraxineus* points to an eastern Russian origin of European ash dieback. bioRxiv 2017:154492. doi: 10.1101/154492.
4. Baral H-O, Bemmann M. *Hymenoscyphus fraxineus* vs. *Hymenoscyphus albidus* – A comparative light microscopic study on the causal agent of European ash dieback and related foliicolous, stroma-forming species. Mycology 2014;5(4):228–90. doi: 10.1080/21501203.2014.963720. PubMed PMID: 25544935.
5. Gross A, Holdenrieder O, Pautasso M, Queloz V, Sieber TN. *Hymenoscyphus pseudoalbidus*, the causal agent of European ash dieback. Molecular Plant Pathology 2014;15(1):5–21. doi: 10.1111/mpp.12073. PubMed PMID: 24118686.
6. Kowalski T. *Chalara fraxinea* sp. nov. associated with dieback of ash (*Fraxinus excelsior*) in Poland. Forest Pathology 2006;36(4):264–70. doi: 10.1111/j.1439-0329.2006.00453.x.
7. Timmermann V, Børja I, Hietala AM, Kirisits T, Solheim H. Ash dieback: pathogen spread and diurnal patterns of ascospore dispersal, with special emphasis on Norway. EPPO Bulletin 2011;41(1):14–20. doi: 10.1111/j.1365-2338.2010.02429.x.
8. GilW, Kowalski T, KrajW, Zachara T, Łukaszewicz J, Paluch R, *et al.* Ash dieback in Poland – History of the phenomenon and possibilities of its limitation. In: Vasaitis R, Enderle R, editors. Dieback of European ash (*Fraxinus* spp.) – Consequences and Guidelines for Sustainable Management. Uppsala, Sweden: Swedish University of Agricultural Sciences; 2017. p. 176–84.
9. Haňáčková Z, Koukol O, Čmoková A, Zahradník D, Havrdová L. Direct evidence of *Hymenoscyphus fraxineus* infection pathway through the petiole-shoot junction. Forest Pathology 2017;47(6):e12370. doi: 10.1111/efp.12370.
10. Husson C, Caël O, Grandjean JP, Nageleisen LM, Marçais B. Occurrence of *Hymenoscyphus pseudoalbidus* on infected ash logs. Plant Pathology 2012;61(5):889–95. doi: 10.1111/j.1365-3059.2011.02578.x.
11. Chandelier A, Gerarts F, San Martin G, Herman M, Delahaye L. Temporal evolution of collar lesions associated with ash dieback and the occurrence of Armillaria in Belgian forests. Forest Pathology 2016;46(4):289–97. doi: 10.1111/efp.12258.
12. Enderle R, Peters F, Nakou A, Metzler B. Temporal development of ash dieback symptoms and spatial distribution of collar rots in a provenance trial of *Fraxinus excelsior*. European Journal of Forest Research 2013;132(5–6):865–76. doi: 10.1007/s10342-013-0717-y.
13. Marçais B, Husson C, Godart L, Caël O. Influence of site and stand factors on *Hymenoscyphus fraxineus*-induced basal lesions. Plant Pathology 2016;65(9):1452–61. doi: 10.1111/ppa.12542.
14. Enderle R, Sander F, Metzler B. Temporal development of collar necroses and butt rot in association with ash dieback. iForest 2017;10(3):529–36. doi: 10.3832/ifor2407-010.
15. Langer G. Collar rots in forests of Northwest Germany affected by ash dieback. Baltic Forestry 2017;23(1):4–19.
16. Skovsgaard JP, Wilhelm GJ, Thomsen IM, Metzler B, Kirisits T, Havrdová L, *et al.* Silvicultural strategies for *Fraxinus excelsior* in response to dieback caused by *Hymenoscyphus fraxineus*. Forestry: An International Journal of Forest Research 2017;90(4):455–72. doi: 10.1093/forestry/cpx012.
17. Landolt J, Gross A, Holdenrieder O, Pautasso M. Ash dieback due to *Hymenoscyphus fraxineus*: what can be learnt from evolutionary ecology? Plant Pathology 2016;65(7):1056–70. doi: 10.1111/ppa.12539.
18. Drenkhan R, Sander H, Hanso M. Introduction of Mandshurian ash (*Fraxinus mandshurica* Rupr.) to Estonia: Is it related to the current epidemic on European ash (*F. excelsior* l.)? European Journal of Forest Research 2014;133(5):769–81. doi: 10.1007/s10342-014-0811-9.
19. McKinney LV, Nielsen LR, Collinge DB, Thomsen IM, Hansen JK, Kjaer ED. The ash dieback crisis: genetic variation in resistance can prove a long-term solution. Plant Pathology 2014;63(3):485–99. doi: 10.1111/ppa.12196.
20. Pautasso M, Aas G, Queloz V, Holdenrieder O. European ash (*Fraxinus excelsior*) dieback – A conservation biology challenge. Biological Conservation 2013;158:37–49. doi: 10.1016/j.biocon.2012.08.026.
21. Heinze B, Tiefenbacher H, Litschauer R, Kirisits T. Ash dieback in Austria – history, current situation and outlook. In: Vasaitis R, Enderle R, editors. Dieback of European ash (*Fraxinus* spp.) – Consequences

and Guidelines for Sustainable Management. Uppsala, Sweden: Swedish University of Agricultural Sciences; 2017. p. 33–52.
22. Spanos K, Gaitanis D. An overview of ash species in Greece: ecology, biology and taxonomy, silviculture, genetics and health status. In: Vasaitis R, Enderle R, editors. Dieback of European ash (*Fraxinus* spp.) – Consequences and Guidelines for Sustainable Management. Uppsala, Sweden: Swedish University of Agricultural Sciences; 2017. p. 273–83.
23. Chira D, Chira F, Tăut I, Popovici O, Blada I, Doniţă N, *et al.* Evolution of ash dieback in Romania. In: Vasaitis R, Enderle R, editors. Dieback of European ash (*Fraxinus* spp.) – Consequences and Guidelines for Sustainable Management. Uppsala, Sweden: Swedish University of Agricultural Sciences; 2017. p. 185–94.
24. Sioen G, Roskams P, de Cuyper B, Steenackers M. Ash dieback in Flanders (Belgium): research on disease development, resistance & management. In: Vasaitis R, Enderle R, editors. Dieback of European ash (*Fraxinus* spp.) – Consequences and Guidelines for Sustainable Management. Uppsala, Sweden: Swedish University of Agricultural Sciences; 2017. p. 61–67.
25. Rozsypálek J, Dvořák M, Longauer R, Botella L, Prouza M, Palovčíková D, *et al.* Ash and ash dieback in the Czech Republic. In: Vasaitis R, Enderle R, editors. Dieback of European ash (*Fraxinus* spp.) – Consequences and Guidelines for Sustainable Management. Uppsala, Sweden: Swedish University of Agricultural Sciences; 2017. p. 79–88.
26. Pliūra A, Bakys R, Suchockas V, Marčiulynienė D, Gustienė A, Verbyla V, *et al.* Ash dieback in Lithuania: disease history, research on impact and genetic variation in disease resistance, tree breeding and options for forest management. In: Vasaitis R, Enderle R, editors. Dieback of European ash (*Fraxinus* spp.) – Consequences and Guidelines for Sustainable Management. Uppsala, Sweden: Swedish University of Agricultural Sciences; 2017. p. 150–66.
27. Longauerová V, Kunca A, Longauer R, Maľová M, Leontovyč R. The ash and ash dieback in Slovakia. In: Vasaitis R, Enderle R, editors. Dieback of European ash (*Fraxinus* spp.) – Consequences and Guidelines for Sustainable Management. Uppsala, Sweden: Swedish University of Agricultural Sciences; 2017. p. 209–19.
28. Davydenko K, Meshkova V. The current situation concerning severity and causes of ash dieback in Ukraine caused by *Hymenoscyphus fraxineus*. In: Vasaitis R, Enderle R, editors. Dieback of European ash (*Fraxinus* spp.) – Consequences and Guidelines for Sustainable Management. Uppsala, Sweden: Swedish University of Agricultural Sciences; 2017. p. 220–27.
29. Cleary M, Nguyen D, Stener DG, Stenlid J, Skovsgaard JP. Ash and ash dieback in Sweden: A review of disease history, current status, pathogen and host dynamics, host tolerance and management options in forests and landscapes. In: Vasaitis R, Enderle R, editors. Dieback of European ash (*Fraxinus* spp.) – Consequences and Guidelines for Sustainable Management. Uppsala, Sweden: Swedish University of Agricultural Sciences; 2017. p. 195–208.
30. Drenkhan R, Agan A, Palm K, Rosenvald R, Jürisoo L, Maaten T, *et al.* Overview of ash and ash dieback in Estonia. In: Vasaitis R, Enderle R, editors. Dieback of European ash (*Fraxinus* spp.) – Consequences and Guidelines for Sustainable Management. Uppsala, Sweden: Swedish University of Agricultural Sciences; 2017. p. 115–24.
31. Børja I, Timmermann V, Hietala AM, Tollefsrud MM, Nagy NE, Vivian-Smith A, *et al.* Ash dieback in Norway – current situation. In: Vasaitis R, Enderle R, editors. Dieback of European ash (*Fraxinus* spp.) – Consequences and Guidelines for Sustainable Management. Uppsala, Sweden: Swedish University of Agricultural Sciences; 2017. p. 166–75.
32. Chandelier A, Delahaye L, Claessens H, Lassois L. Ash dieback in Wallonia, southern Belgium: research on disease development, resistance & management options. In: Vasaitis R, Enderle R, editors. Dieback of European ash (*Fraxinus* spp.) – Consequences and Guidelines for Sustainable Management. Uppsala, Sweden: Swedish University of Agricultural Sciences; 2017. p. 53–60.
33. Enderle R, Fussi B, Lenz HD, Langer G, Nagel R, Metzler B. Ash dieback in Germany: research on disease development, resistance and management options. In: Vasaitis R, Enderle R, editors. Dieback of European ash (*Fraxinus* spp.) – Consequences and Guidelines for Sustainable Management. Uppsala, Sweden: Swedish University of Agricultural Sciences; 2017. p. 89–105.
34. Queloz V, Hopf S, Schoebel CN, Rigling D, Gross A. Ash dieback in Switzerland: history and scientific achievements. In: Vasaitis R, Enderle R, editors. Dieback of European ash (*Fraxinus* spp.) – Consequences and Guidelines for Sustainable Management. Uppsala, Sweden: Swedish University of Agricultural Sciences; 2017. p. 68–78.

35. Clark J, Webber J. The ash resource and the response to ash dieback in Great Britain. In: Vasaitis R, Enderle R, editors. Dieback of European ash (*Fraxinus* spp.) – Consequences and Guidelines for Sustainable Management. Uppsala, Sweden: Swedish University of Agricultural Sciences; 2017. p. 228–37.
36. McCracken AR, Douglas GC, Ryan C, Destefanis M, Cooke LR. Ash dieback on the island of Ireland. In: Vasaitis R, Enderle R, editors. Dieback of European ash (*Fraxinus* spp.) – Consequences and Guidelines for Sustainable Management. Uppsala, Sweden: Swedish University of Agricultural Sciences; 2017. p. 125–39.
37. Pratt J. Management and use of ash in Britain from the prehistoric to the present: some implications for its preservation. In: Vasaitis R, Enderle R, editors. Dieback of European ash (*Fraxinus* spp.) – Consequences and Guidelines for Sustainable Management. Uppsala, Sweden: Swedish University of Agricultural Sciences; 2017. p. 1–14.
38. Bragança H, Varela M. C. *Fraxinus angustifolia* and forest health in Portugal – an overview. In: Vasaitis R, Enderle R, editors. Dieback of European ash (*Fraxinus* spp.) – Consequences and Guidelines for Sustainable Management. Uppsala, Sweden: Swedish University of Agricultural Sciences; 2017. p. 284–87.
39. Bengtsson V, Senström A. Ash dieback – a continuing threat to veteran ash trees? In: Vasaitis R, Enderle R, editors. Dieback of European ash (*Fraxinus* spp.) – Consequences and Guidelines for Sustainable Management. Uppsala, Sweden: Swedish University of Agricultural Sciences; 2017. p. 262–72.
40. Dandy N, Marzano M, Porth EF, Urquhart J, Potter C. Who has a stake in ash dieback? A conceptual framework for the identification and categorisation of tree health stakeholders. In: Vasaitis R, Enderle R, editors. Dieback of European ash (*Fraxinus* spp.) – Consequences and Guidelines for Sustainable Management. Uppsala, Sweden: Swedish University of Agricultural Sciences; 2017. p. 15–26.
41. Drăgoi M, Chira D, Popovici O. Socio-economic outcomes of ash-dieback in Romania. In: Vasaitis R, Enderle R, editors. Dieback of European ash (*Fraxinus* spp.) – Consequences and Guidelines for Sustainable Management. Uppsala, Sweden: Swedish University of Agricultural Sciences; 2017. p. 27–32.
42. Bengtsson SBK, Vasaitis R, Kirisits T, Solheim H, Stenlid J. Population structure of *Hymenoscyphus pseudoalbidus* and its genetic relationship to Hymenoscyphus albidus. Fungal Ecology 2012;5(2):147–53. doi: 10.1016/j.funeco.2011.10.004.
43. Orton ES, Brasier CM, Bilham LJ, Bansal A, Webber JF, Brown JKM. Population structure of the ash dieback pathogen, *Hymenoscyphus fraxineus*, in relation to its mode of arrival in the UK. Plant Pathology 2018;67(2):255–64. doi: 10.1111/ppa.12762. PubMed PMID: 29527064.
44. McMullan M, Rafiqi M, Kaithakottil G, Clavijo BJ, Bilham L, Orton E, *et al.* The ash dieback invasion of Europe was founded by two genetically divergent individuals. Nature Ecology and Evolution 2018;2(6):1000–8. doi: 10.1038/s41559-018-0548-9. PubMed PMID: 29686237.
45. Nguyen DT, Cleary MR, Enderle R, Berlin A, Stenlid J. Analyses of the ash dieback pathogen, *Hymenoscyphus fraxineus*, suggest role of tree species diversity on colonization and population structure differentiation. Forest Pathology 2016;46(1):82–4. doi: 10.1111/efp.12236.
46. Kraj W, Zarek M, Kowalski T. Genetic variability of *Chalara fraxinea*, dieback cause of European ash (*Fraxinus excelsior* L.). Mycological Progress 2012;11(1):37–45. doi: 10.1007/s11557-010-0724-z.
47. Kowalski T, Bartnik C. Morphological variation in colonies of *Chalara fraxinea* isolated from ash (*Fraxinus excelsior* L.) stems with symptoms of dieback and effects of temperature on colony growth and structure. Acta Agrobotanica 2010;63(1):99–106.
48. Junker C, de Vries J, Eickhorst C, Schulz B. Each isolate of *Hymenoscyphus fraxineus* is an individualist as shown by exoenzyme and growth rate profiles. Baltic Forestry 2017;23(1):25–40.
49. Lygis V, Prospero S, Burokiene D, Schoebel CN, Marčiulynienė D, Norkute G, *et al.* Virulence of the invasive ash pathogen *Hymenoscyphus fraxineus* in old and recently established populations. Plant Pathology 2017;66(5):783–91. doi: 10.1111/ppa.12635.
50. Heuertz M, Fineschi S, Anzidei M, Pastorelli R, Salvini D, Paule L, *et al.* Chloroplast DNA variation and postglacial recolonization of common ash (*Fraxinus excelsior* L.) in Europe. Molecular Ecology 2004;13(11):3437–52. doi: 10.1111/j.1365-294X.2004.02333.x. PubMed PMID: 15488002.
51. Heuertz M, Hausman J-F, Hardy OJ, Vendramin GG, Frascaria-Lacoste N, Vekemans X. Nuclear microsatellites reveal contrasting patterns of genetic structure between western and southeastern European populations of the Common ash (*Fraxinus excelsior* L.). Evolution 2004;58(5):976–88. doi: 10.1111/j.0014-3820.2004.tb00432.x.

52. Tollefsrud MM, Myking T, Sønstebø JH, Lygis V, Hietala AM, Heuertz M. Genetic structure in the northern range margins of Common ash, *Fraxinus excelsior* L. PLoS One 2016;11(12): e0167104. doi: 10.1371/journal.pone.0167104. PubMed PMID: 27907032.
53. Semizer-Cuming D, Kjær ED, Finkeldey R. Gene flow of common ash (*Fraxinus excelsior* L.) in a fragmented landscape. PLoS One 2017;12(10):e0186757. doi: 10.1371/journal.pone.0186757. PubMed PMID: 29053740.
54. Fussi B, Konnert M. Genetic analysis of European common ash (*Fraxinus excelsior* L.) populations affected by ash dieback. Silvae Genetica 2014;63(5):198–212.
55. Pacia A, Nowakowska JA, Tkaczyk M, Sikora K, Tereba A, Borys M, *et al.* Common ash stand affected by ash dieback in the Wolica Nature Reserve in Poland. Baltic Forestry 2017;23(1):183–97.
56. Heuertz M, Carnevale S, Fineschi S, Sebastiani F, Hausman JF, Paule L, *et al.* Chloroplast DNA phylogeography of European ashes, *Fraxinus* sp. (Oleaceae): roles of hybridization and life history traits. Molecular Ecology 2006;15(8):2131–40. doi: 10.1111/j.1365-294X.2006.02897.x. PubMed PMID: 16780430.
57. Burton GW, Devane EH. Estimating heritability in tall fescue (*Festuca arundinacea*) from replicated clonal material. Agronomy Journal 1953;45(10):478–81.
58. de Vries SMG, Kopinga J. Differences in susceptibility to *Hymenoscyphus fraxineus* (twig dieback of ash) of selections of Common ash (*Fraxinus excelsior*) in The Netherlands – report of the observations and results of 2012 and 2015. In: Vasaitis R, Enderle R, editors. Dieback of European ash (*Fraxinus* spp.) – Consequences and Guidelines for Sustainable Management. Uppsala, Sweden: Swedish University of Agricultural Sciences; 2017. p. 238–48.
59. Diminić D, Kajba D, Milotić M, Andrić I, Kranjec J. Susceptibility of *Fraxinus angustifolia* clones to *Hymenoscyphus fraxineus* in lowland Croatia. Baltic Forestry 2017;23(1):233–43.
60. Adamčíková K, Pažitný J, Pastirčáková K. Individual resistance of *Fraxinus angustifolia* and *F. excelsior* clones to *Hymenoscyphus fraxineus*. Journal of Plant Protection Research 2018;58(3):227–33.
61. Muñoz F, Marçais B, Dufour J, Dowkiw A. Rising out of the ashes: additive genetic variation for crown and collar resistance to *Hymenoscyphus fraxineus* in *Fraxinus excelsior*. Phytopathology 2016;106(12):1535–43. doi: 10.1094/PHYTO-11-15-0284-R. PubMed PMID: 27349738.
62. Wohlmuth A, Essl F, Heinze B. Genetic analysis of inherited reduced susceptibility of *Fraxinus excelsior* L. seedlings in Austria to ash dieback. Forestry: An International Journal of Forest Research 2018;91(4):514–25. doi: 10.1093/forestry/cpy012.
63. Lobo A, Hansen JK, McKinney LV, Nielsen LR, Kjær ED. Genetic variation in dieback resistance: growth and survival of *Fraxinus excelsior* under the influence of *Hymenoscyphus pseudoalbidus*. Scandinavian Journal of Forest Research 2014;29(6):519–26. doi: 10.1080/02827581.2014.950603.
64. Kjær ED, McKinney LV, Nielsen LR, Hansen LN, Hansen JK. Adaptive potential of ash (*Fraxinus excelsior*) populations against the novel emerging pathogen *Hymenoscyphus pseudoalbidus*. Evolutionary Applications 2012;5(3):219–28. doi: 10.1111/j.1752-4571.2011.00222.x. PubMed PMID: 25568043.
65. McKinney LV, Nielsen LR, Hansen JK, Kjær ED. Presence of natural genetic resistance in *Fraxinus excelsior* (Oleraceae) to *Chalara fraxinea* (Ascomycota): an emerging infectious disease. Heredity (Edinb) 2011;106(5):788–97. doi: 10.1038/hdy.2010.119. PubMed PMID: 20823903.
66. Kjær ED, McKinney LV, Hansen LN, Olrik DC, Lobo A, Thomsen IM, *et al.* Genetics of ash dieback resistance in a restoration context – experiences from Denmark. In: Vasaitis R, Enderle R, editors. Dieback of European ash (*Fraxinus* spp.) – Consequences and Guidelines for Sustainable Management. Uppsala, Sweden: Swedish University of Agricultural Sciences; 2017. p. 106–14.
67. Enderle R, Nakou A, Thomas K, Metzler B. Susceptibility of autochthonous German *Fraxinus excelsior* clones to *Hymenoscyphus pseudoalbidus* is genetically determined. Annals of Forest Science 2015;72(2):183–93. doi: 10.1007/s13595-014-0413-1.
68. Pliura A, Lygis V, Suchockas V, Bartkevièius E. Performance of twenty four European *Fraxinus excelsior* populations in three Lithuanian progeny trials with a special emphasis on resistance to *Chalara fraxinea*. Baltic Forestry 2011;17(1):17–34.
69. Pliura A, Marèiulynienë D, Bakys R, Suchockas V. Dynamics of genetic resistance to *Hymenoscyphus pseudoalbidus* in juvenile *Fraxinus excelsior* clones. Baltic Forestry 2014;20(1):10–27.
70. Hauptman T, Ogris N, de Groot M, Piškur B, Jurc D. Individual resistance of *Fraxinus angustifolia* clones to ash dieback. Forest Pathology 2016;46(4):269–80. doi: 10.1111/efp.12253.
71. Stener L-G. Clonal differences in susceptibility to the dieback of *Fraxinus excelsior* in southern Sweden. Scandinavian Journal of Forest Research 2013;28(3):205–16. doi: 10.1080/02827581.2012.735699.

72. Stener L-G. Genetic evaluation of damage caused by ash dieback with emphasis on selection stability over time. Forest Ecology and Management 2018;409:584–92. doi: 10.1016/j.foreco.2017.11.049.
73. Östbrant I-L, Vasaitis R, Stenlid J, Pliūra A, Menkis A. Natura 2000 habitats dominated by ash and elm, invaded by alien invasive fungi on the Gotland Island of Sweden: an overview. Baltic Forestry 2017;23(1):264–9.
74. Harper AL, McKinney LV, Nielsen LR, Havlickova L, Li Y, Trick M, *et al.* Molecular markers for tolerance of European ash (*Fraxinus excelsior*) to dieback disease identified using Associative Transcriptomics. Scientific Reports 2016;6:19335. doi: 10.1038/srep19335. PubMed PMID: 26757823.
75. Sollars ESA, Harper AL, Kelly LJ, Sambles CM, Ramirez-Gonzalez RH, Swarbreck D, *et al.* Genome sequence and genetic diversity of European ash trees. Nature 2017;541(7636):212–6. doi: 10.1038/nature20786. PubMed PMID: 28024298.
76. Villari C, Dowkiw A, Enderle R, Ghasemkhani M, Kirisits T, Kjær ED, *et al.* Advanced spectroscopy-based phenotyping offers a potential solution to the ash dieback epidemic. Scientific Reports; 2018.
77. Hietala AM, Børja I, Cross H, Nagy NE, Solheim H, Timmermann V, *et al.* Dieback of European ash: what can we learn from the microbial community and species-specific traits of endophytic fungi associated with ash? In: Pirttilä AM, Frank AC, editors. Endophytes of Forest Trees: Biology and Applications. Forestry Sciences. Cham, Switzerland: Springer International Publishing; 2018. p. 229–58.
78. Wylder B, Biddle M, King K, Baden R, Webber J. Evidence from mortality dating of *Fraxinus excelsior* indicates ash dieback (*Hymenoscyphus fraxineus*) was active in England in 2004–2005. Forestry: An International Journal of Forest Research 2018;91(4):434–43. doi: 10.1093/forestry/cpx059.
79. Ghelardini L, Migliorini D, Santini A, Pepori AL, Maresi G, Vai N, *et al.* From the Alps to the Apennines: possible spread of ash dieback in Mediterranean areas. In: Vasaitis R, Enderle R, editors. Dieback of European ash (*Fraxinus* spp.) – Consequences and Guidelines for Sustainable Management. Uppsala, Sweden: Swedish University of Agricultural Sciences; 2017. p. 140–49.
80. Marçais B, Husson C, Caël O, Dowkiw A, Saintonge F-X, Delahaye L, *et al.* Estimation of ash Mortality Induced by *Hymenoscyphus fraxineus* in France and Belgium. Baltic Forestry 2017;23(1):159–67.
81. Queloz V, Grünig CR, Berndt R, Kowalski T, Sieber TN, Holdenrieder O. Cryptic speciation in *Hymenoscyphus albidus*. Forest Pathology 2011;41(2):133–42. doi: 10.1111/j.1439-0329.2010.00645.x.
82. European Plant Protection Organisation (EPPO). Mini data sheet on *Chalara fraxinea* [Internet]; 2014 [cited 2019 Feb 12]. Available from: URL: https://gd.eppo.int/download/doc/991_minids_CHAAFR.pdf.
83. Coker TLR, Rozsypálek J, Edwards A, Harwood TP, Butfoy L, Buggs RJA. Estimating mortality rates of European ash (*Fraxinus excelsior*) under the ash dieback (*Hymenoscyphus fraxineus*) epidemic. Plants, People, Planet 2019;1(1):48–58. doi: 10.1002/ppp3.11.
84. Kopinga J, de Vries SMG. Ash DieBack (ADB) in amenity trees in the city of Amsterdam – development of a monitoring system and the first results of a large-scale survey in 2015. In: Vasaitis R, Enderle R, editors. Dieback of European ash (*Fraxinus* spp.) – Consequences and Guidelines for Sustainable Management. Uppsala, Sweden: Swedish University of Agricultural Sciences; 2017. p. 249–61.
85. Matisone I, Matisons R, Kenigsvalde K, Gaitnieks T, Burneviča N. Seasonal development of lesions caused by *Hymenoscyphus fraxineus* on young *Fraxinus excelsior* trees in Latvia. iForest 2018;11(1):17–23. doi: 10.3832/ifor2283-010.
86. Bengtsson SBK, Barklund P, von Brömssen C, Stenlid J. Seasonal pattern of lesion development in diseased *Fraxinus excelsior* infected by *Hymenoscyphus pseudoalbidus*. PLoS One 2014;9(4):e76429. doi: 10.1371/journal.pone.0076429. PubMed PMID: 24759550.
87. Nielsen LR, McKinney LV, Kjær ED. Host phenological stage potentially affects dieback severity after *Hymenoscyphus fraxineus* infection in *Fraxinus excelsior* seedlings. Baltic Forestry 2017;23(1):229–32.
88. Enderle R, Metzler B, Riemer U, Kaendler G. Ash dieback on sample points of the National Forest Inventory in south-western Germany. Forests 2018;9(1):1–13. doi: 10.3390/f9010025.
89. Grosdidier M, Ioos R, Marçais B. Do higher summer temperatures restrict the dissemination of *Hymenoscyphus fraxineus* in France? Forest Pathology 2018;48(4):e12426. doi: 10.1111/efp.12426.
90. Goberville E, Hautekèete N-C, Kirby RR, Piquot Y, Luczak C, Beaugrand G. Climate change and the ash dieback crisis. Scientific Reports 2016;6:35303. doi: 10.1038/srep35303. PubMed PMID: 27739483.
91. Bakys R, Vasaitis R, Skovsgaard JP. Patterns and severity of crown dieback in young even-aged stands of European ash (*Fraxinus excelsior* L.) in relation to stand density, bud flushing phenotype, and season. Plant Protection Science 2013;49 (No. 3):120–6. doi: 10.17221/70/2012-PPS.

92. Marciulyniene D, Davydenko K, Stenlid J, Cleary M. Can pruning help maintain vitality of ash trees affected by ash dieback in urban landscapes? Urban Forestry & Urban Greening 2017;27:69–75. doi: 10.1016/j.ufug.2017.06.017.
93. Kosawang C, Amby DB, Bussaban B, McKinney LV, Xu J, Kjær ED, *et al.* Fungal communities associated with species of *Fraxinus* tolerant to ash dieback, and their potential for biological control. Fungal Biology 2018;122(2–3):110–20. doi: 10.1016/j.funbio.2017.11.002. PubMed PMID: 29458714.
94. Haňáčková Z, Havrdová L, Černý K, Zahradník D, Koukol O. Fungal endophytes in ash shoots – diversity and inhibition of *Hymenoscyphus fraxineus*. Baltic Forestry 2017;23(1):89–106.
95. Schulz B, Haas S, Junker C, Andrée N, Schobert M. Fungal endophytes are involved in multiple balanced antagonisms. Current Science 2015;109(1):39–45.
96. Schlegel M, Dubach V, von Buol L, Sieber TN. Effects of endophytic fungi on the ash dieback pathogen. FEMS Microbiology Ecology 2016;92(9):1–8. doi: 10.1093/femsec/fiw142. PubMed PMID: 27364360.
97. Schlegel M, Queloz V, Sieber TN. The endophytic mycobiome of European ash and sycamore maple leaves – geographic patterns, host specificity and influence of ash dieback. Frontiers in Microbiology 2018;9:2345. doi: 10.3389/fmicb.2018.02345. PubMed PMID: 30405540.
98. ForstBW. Herausforderung Eschentriebsterben: Waldbauliche Behandlung geschädigter Bestände. [Internet]. Stuttgart; 2018 [cited 2018 Jul 18]. Available from: URL: http://www.forstbw.de/fileadmin/forstbw_infothek/forstbw_praxis/ForstBW_Praxis_Eschentriebsterben_20180327.pdf.
99. Semizer-Cuming D, Finkeldey R, Nielsen LR, Kjær ED. Negative correlation between ash dieback susceptibility and reproductive success: good news for European ash forests. Annals of Forest Science 2019;76(1):219. doi: 10.1007/s13595-019-0799-x.
100. Short I, Hawe J. Ash dieback in Ireland – A review of European management options and case studies in remedial silviculture. Irish Forestry [Internet] 2018 [cited 2019 Feb 15];75:44–72. Available from: URL: https://t-stor.teagasc.ie/handle/11019/1654.
101. Broome A, Ray D, Mitchell R, Harmer R. Responding to ash dieback (*Hymenoscyphus fraxineus*) in the UK: woodland composition and replacement tree species. Forestry: An International Journal of Forest Research 2019;92(1):108–19. doi: 10.1093/forestry/cpy040.

13 Species distribution models (SDM): applications, benefits and challenges in invasive species management

Vivek Srivastava, Valentine Lafond and Verena C. Griess*

Department of Forest Resources Management, Faculty of Forestry, University of British Columbia, Forest Sciences Centre, 2424 Main Mall, Vancouver, British Columbia V6T1Z4, Canada

Abstract

The use of species distribution models (SDM) is of increasing popularity when studying biological invasions, e.g. to assess the impact of climate change on invasive species, to prioritize conservation measures, or to study invasive evolutionary biology. SDM correlate known occurrences of species with environmental variables and predict a species' potential distribution on other geographies over space and time. Today, SDM are widely used to produce invasion risk maps by delineating probable risk areas based on climatic suitability for a species. These maps can guide early detection and rapid response measures. While recent developments in modelling approaches and wider availability of environment datasets have helped to create better and more accurate SDM, in many cases the used models ignore the associated underlying ecological processes for e.g. dispersal and biotic interactions and thus provide only an incomplete picture of invasion risks. In this paper, we present a review of the most common applications of SDM in invasive species management and provide an overview of possible solutions to various challenges like uncertainty and transferability associated with them. Based on our review we conclude that ways towards more accurate model outputs include fitting models with existing ecological knowledge (hybrid models), address uncertainty and biotic interactions and link species dispersal traits with projections of species distributions. In the future, work is required on developing more hybrid approaches and models that addresses both local diffusion and long distance movement of alien species.

Keywords: SDM, MaxEnt, Habitat suitability maps, Predictive biogeography, Invasive species, Biological invasion

Review Methodology

We conducted a literature search using Clarivate Analytic's web of science search platform using the search phrases 'Species Distribution Modelling', 'Niche Modelling', 'Habitat Predictive Modelling', 'Habitat Mapping', Invasive Species Niche Modelling', and 'Invasive Species Risk Mapping'. We used these search phrases as published articles in this field often contains these phrases in the title of the article. Hence, we sought papers that contained these search phrases in the title. The search phrases were entered separately with an 'OR' separator on the basic search function, which yielded a total of 2753 articles published between 2000 and 2017 (Figure 13.1). We first restricted this selection to papers published between 2000 and 2017, to focus the review on recent SDM applications and methodological advancements along with modelling challenges. Later, we refined our search based on the research areas in which they were published: environmental sciences ecology, biodiversity conservation, zoology, plant sciences, forestry and entomology. We did this to analyse the trend of using SDM in

*Corresponding author: verena.griess@ubc.ca

different areas of ecology and biogeography. We then selected 100 scientific papers centered on invasive species distribution modelling. Papers were chosen based on the relevancy to the topic. We also analysed the temporal trend of using SDM in the fields of ecology, biodiversity conservation, environmental sciences, forestry and plant sciences, evolutionary biology and multidisciplinary sciences (Figures 13.1 and 13.2). This was accomplished by using Web of Science research categories. Considering these resources, we provide a brief review of the various aspects related to creating an effective distribution model for invasive species, along with the benefits, challenges and good practices associated with correlative SDM.

Introduction

Biological invasions are becoming one of the main causes of the global loss of biodiversity and the extinction of species [1–5]. They are also responsible for substantial monetary losses (e.g. direct economic losses from sectors like agriculture, forestry, environment, human health, etc. were reported to be around US$14.45 billion in China during 2001–2003 and US$128 billion annually in USA) [6, 7]. Biological invasions are usually caused by alien invasive species (but see Valery et al. [8]), which the International Union for Conservation of Nature (IUCN) defines as [9]: (i) 'a species, sub-species or lower taxon occurring outside of its natural range (past or present) and dispersal potential (...)' – i.e. *alien* – that (ii) 'becomes established in natural or semi-natural ecosystems or habitat, is an agent of change, and threatens native biological diversity' – i.e. *invasive*. Invasive species can be plants, animals, or microorganisms and their introduction to invaded ecosystem can be either natural or induced by human activities (e.g. as international trade, travel and tourism).

There has been a constant rise in the number of invasive species successfully getting established in a new habitat [10, 11] and their impacts are anticipated to be significant throughout all ecosystems [12, 13]. Moreover, the global, national and regional spatial patterns of species invasion in future might get worsen due to developments in transportation and international trade [14, 15], increases in human population [15] and changes in climate [16]. Invasive species are of high relevance to both natural and managed ecosystems [17–20]. The risks for our current ecosystems are so substantial that even the Convention on Biological Diversity asks for measures '*to prevent the introduction, control or even eradication of those alien species which threaten ecosystems, habitats or species*' [21].

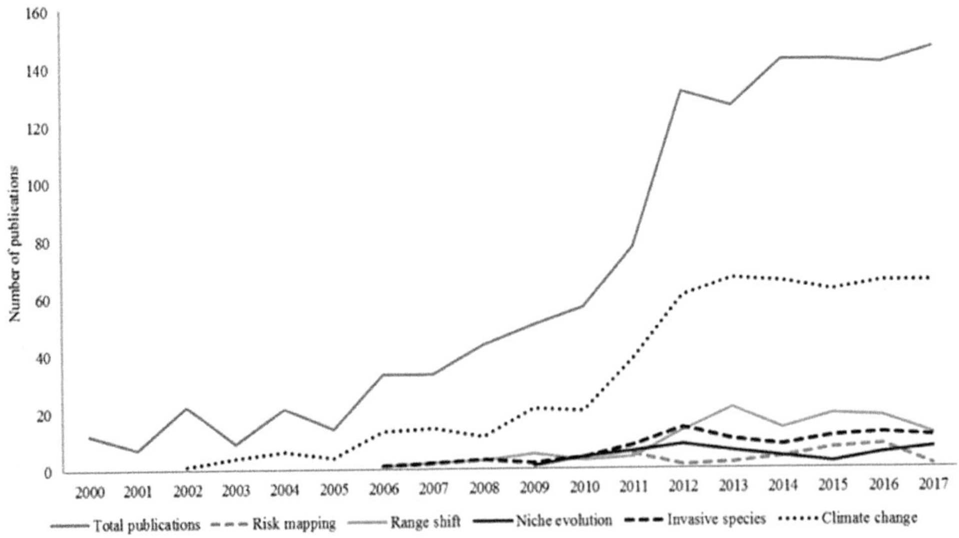

Fig. 13.1. Publications on SDM and their various areas of application.

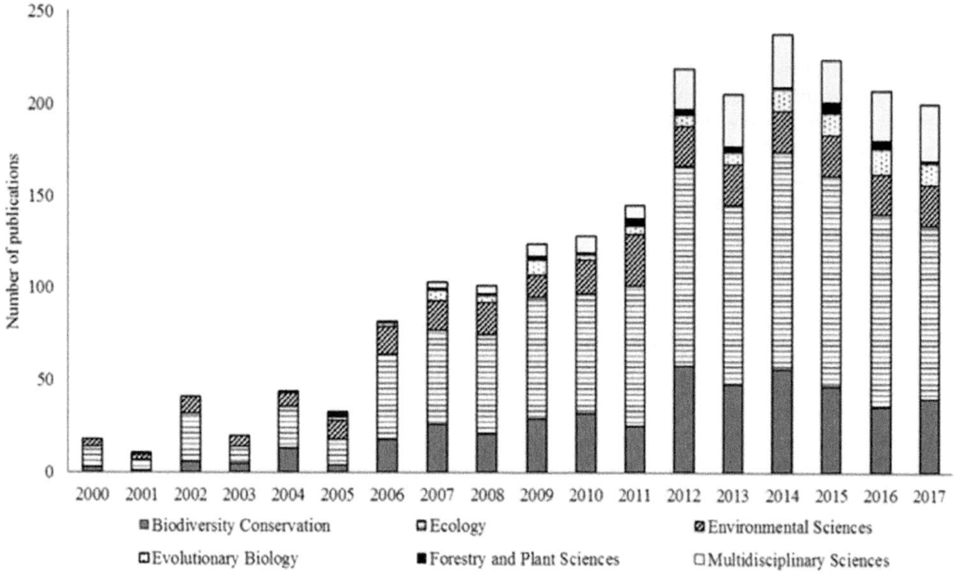

Fig. 13.2. Publications on primary applications of SDM in various fields of ecology and biogeography.

However, introductions of invasive alien species to new ecosystems are both inevitable, and predictable [22]. Early detection and rapid response to incoming aliens are required for a successful response [23]. The sooner invasions can be detected and eradicated, the better as by then species will not have occupied its entire potential range [24]. Early response strategies involve surveying and monitoring of risk areas under threat of invasion to find infestations in their earliest stages of invasion. To identify such areas at risk, that are suitable for the establishment of alien species, climate matching methods show promise as they are based on the classical assumption that species will be able to establish populations in areas outside of their native range that closely match the environmental conditions of their native distributional area [25]. Thus, by matching climatic conditions, researchers can discern meaningful information about alien species' potential establishment areas. These climate matching methods are widely utilized by natural resource managers and agencies, decision makers and nongovernment organizations, who use them for risk analysis and planning of mitigation activities [26]. The use of species distribution models (SDM; cf. Box 13.1) is one such method based on the ecological concept of niches defined as the n-dimensional environmental conditions under which a species can persist in the presence and absence of other species [27]. SDM first became species in Australia [28]. In the 1990s the approach experienced a renaissance caused by simultaneous developments in computer- and statistical sciences [29, 30]. Another 20 years later, within the past decade significant advancements in predictive modelling have occurred with thousands of new publications on SDM, and their applications, in the fields of ecology, conservation, biogeography and evolution [31] (Figure 13.1).

The popularity and rise in the availability of SDM has facilitated their use in invasive species management [29]. However, the robustness of risk maps created using SDM has been questioned due to the way the models are developed and the manner in which output maps are interpreted [52]. Guisan and Thuiller [28] attribute this to the practice of constructing weak ecological baseline assumptions when building an SDM. This might lead to inaccurate assessments of invasion risk, which in turn has the potential to support the selection of sub-optimal response measures, as well as an over-, or under investment in mitigation activities [52].

Box 13.1. Species distribution models

SDM are predictive tools that identify relationships between observed occurrences and environmental predictors by using statistical models or theoretically derived response curves [28, 29]. SDM are also referred to as correlative or statistical models, habitat models, or ecological niche models and are broadly divided into two groups: correlative, and process-based or mechanistic models [32]. These correlative and process-based models are derived from various statistical approaches which include GLM, ordination and classification methods, Bayesian models, locally weighted approaches, environmental envelopes, or combinations of these models [29]. Correlative SDM approaches utilize species occurrence data and associated environmental layers of the study area, to produce maps of probability of occurrence or relative environmental suitability for a species, whereas process-based, or mechanistic niche models, use species functional characteristics and physiological thresholds for model fitting [33]. The correlative SDM make use of existing species occurrence data collected from surveys, online databases or museums/herbaria records for fitting [29]. In practice, most mechanistic models require exhaustive experimental data on species functional traits which becomes a challenge for the target species under study [34]. The correlative approach also differs from the mechanistic from the way it represents species niche. A common understating is that correlative approaches only measures the realized niche, which is a subset of fundamental niche where the species is not absent due to biotic competition. Whereas, the mechanistic approach approximates species fundamental niche [35]. This can be a challenge for correlative approaches as for them the fundamental niche of the species remains unknown. However both of the approaches have their own strengths and weaknesses.

Amongst correlative SDM, several algorithms are available for predicting species potential distributions [36]. MaxEnt is one of the most popular choices for correlative models [37]. MaxEnt is a presence-only-based method which has been successfully applied to model the distributions of invasive species in different parts of the world [e.g. 26, 38–45]. MaxEnt generates a probability estimate of presence (or relative environmental suitability) of a species that varies from 0 (lowest) to 1 (highest). Whereas, CLIMEX is a process-based species distribution modelling tool [46] which generates a climatic suitability index for the species, known as the Eco climatic Index (EI). EI ranges from 0 to 100; values close to 100 represents landscape suitability for the species establishment, whereas 0 represents locations that are unsuitable for the survival of the species. Both of them have been found to be very effective in each of the categories that aim to predict invasive species establishment risk [45, 47–51].

This paper provides an overview of various approaches used to build a robust correlative SDM that incorporate strong ecological baseline assumptions for accurate assessments of the invasion risk, while also reviewing various applications, benefits and challenges of SDM, particularly in the areas of biological invasions. We are consciously excluding process-based models (see Kearney & Porter [33]), as they require a thorough understanding of species physiological responses to environmental factors and are often unavailable for new arriving alien species. We address the following questions:

(1) What are the current applications of SDM in ecology and biological invasions in particular?
(2) What are the benefits and challenges of using SDM to estimate, and spatially project, invasion risk?
(3) What are the important aspects to consider when building a SDM for mapping invasion risk?

Review Text

Current trends in applications of SDM in ecology and invasion biology

SDM are currently in use for a wide range of applications (Table 13.1). From characterizing niche and ecological requirements of a particular species [25, 30, 53, 54], to mapping the potential distribution of plants and animals [16, 38, 45, 55–57], or using output maps in policy making, and to prioritize conservation efforts [58–64]. Williams et al. [65] successfully used

Table 13.1. Examples of SDM applications in the fields of ecology and biogeography

Type of applications	Example reference(s)
Projecting potential impacts of climate change on species distributions	[71–75]
Predicting species invasion	[25, 76–80]
Characterizing the niche and exploring the ecological requirements	[51, 69, 81–83]
Conservation and policy making	[63, 84–89]
Assessing the impacts of land cover change and human footprint on species distributions	[90–95]
Predicting the distribution of rare and endangered species	[65, 66, 84, 96]
Testing an ecological theory	[28, 29, 97–100]
Risk assessment	[49, 52, 78, 101–104]
Assessing disease risk	[81, 105–107]

a suite of SDM as a tool to discover populations of rare plant species with highly specialized habitat needs. Their models included generalized linear models (GLM), artificial neural networks, random forests and MaxEnt. Engler et al. [66] provided an enhanced method for predicting the distribution of rare and endangered species from occurrence and pseudo-absence data by simulating pseudo-absences based on ecological niche factor analysis (ENFA) and concluded that ENFA weighted pseudo-absences could be a possible way to enhance an SDMs explanatory power. Raes et al. [67] assessed botanical richness and endemicity patterns of all species in Flora Malesiana in Borneo using SDM, and concluded that SDM can effectively guide conservation efforts. Esselman and Allan [68] explored SDM applications in a data-limited freshwater setting (inadequate human capacity and technology, lack of investment in research and monitoring) in developing countries of northeastern Mesoamerica to enhance conservation planning and Brotons et al. [69] used data from long-term monitoring programs data to map habitat suitability for 99 bird species. Furthermore, Svenning et al. [70] concluded that SDM can also be used in paleobotany, where it can provide a quantitative ecological perspective, while offering potential for an enhanced contribution of paleobiology to ecology and conservation biology. This compliments with the ability of the SDM to provide predictions of past organisms distributions and assessment of their range determinants. Overall, the applications of SDM have evolved continually over the past decade (Figures 13.1 and 13.2).

Most SDM applications fall in the fields of ecology and biodiversity conservation (Figure 13.1). Within these broad fields SDM are used for diverse applications, primarily related to climate change studies and studying range shifts of species (Figure 13.2). It should however be noted that this involves extrapolating the predictions to novel climates, which often requires extreme cautions (cf. 3.2).

SDM applications related to invasive species and risk mapping are relatively recent but the field is evolving [39, 40, 76, 108–110]. We reviewed current applications of the SDM concerning biological invasions and found that the major applications were around investigating species invasion ecology [26, 41, 42, 79, 109, 111–113], estimate disease risk [81, 105–107], determine possible invasive species range shifts under climate change [71–75] and assess the impacts of land cover change [90, 91] and human footprint [92–95] on invasive species distribution. Although, SDM can be applied in a variety of ways to solve the complex issues related to invasive species, numerous challenges are linked to the process involved in developing SDM and predicting the distributions. In the following sections we review the benefits and challenges associated with the SDM in brief.

Benefits and challenges of using SDM in invasion biology

In regions where invasive species are a significant contributor to the global change in biodiversity and considered one of the key reasons for species extinctions [114], preventive measures assisted by SDM can be of great value. For example, a report prepared for the Canadian Council of Forest Ministers in the year 2009 stated that Canada could have avoided spending US$165 million annually by preventing the introduction and establishment of four damaging a forest invasive alien species: Asian longhorn beetle, emerald ash borer, Sirex wood wasp and sudden oak death disease. Using risk maps to guide surveying and monitoring programs aimed at finding infestations in their earliest stages of invasion appears to be a promising and cost-effective approach [47, 115]. These distribution maps generated from SDM are becoming the favoured guide for resource managers who repeatedly survey for non-native species [78]. Risk maps derived from SDM support decision making for pest management and depict risk probability based on the likelihood of an alien species arrival or establishment [52, 78]. These maps can play a pivotal role in the study of alien species and have the potential to describe where invasive alien species might arrive, establish, spread, or cause harmful impacts [109]. They might also aid in pest management decisions such as international trade regulations, design of surveys and local quarantines [52].

Though SDM has multiple benefits as discussed above in estimating and projecting invasion risk and can be used in a variety of ways to assist decision makers, building a SDM and projecting the distributions of an alien species is not an easy task. Many uncertainties are associated with these projections, particularly when it comes to building a robust SDM for an alien species. Alien species often encounter novel environment settings in their nonnative range and it becomes hard for a SDM to capture the new settings from a native range of the species [48]. The key challenges with respect to using SDM to estimate and spatially project invasive risk have been discussed and highlighted in various reviews. For example, Venette et al. [52] addressed the challenges around unavailability or inadequate information for model construction, choice of model, selection of predictors, calibration and validation of models and lastly interpretation of the outputs. Whereas, Araujo and Guisan [116] identified: clarification of the niche notion, sampling design, parameterization, model selection and predictor contribution and model evaluation as major challenges. We found major challenges of using SDM to estimate and spatially project invasion risk were related to projection (transferability), niche characterization, biotic interactions, species dispersal and uncertainty.

Building SDM for mapping invasion risk

Building SDM for mapping invasion risk is a multifaceted exercise. Below are outlined the required focal to build a correlative SDM [29]:

(1) Collection of species occurrence data.
(2) Assessment of the accuracy and completeness of the occurrence data.
(3) Inspecting the relevance of the environmental variables to be included in the model.
(4) Minimalizing the correlation effect among the predictor variables.
(5) Selecting a suitable algorithm and later fitting the model to train the datasets.
(6) Evaluating the predictive performance on the test data.
(7) Projecting predictions onto geographic space.
(8) Iterating the process to improve the model performance

Each of these steps is associated with a set of challenges related to data and method choice, which can be even more pronounced in the case of invasive species [52]. A number of factors that are involved at each step of SDM development may affect the model predictions, for e.g. biased sampling (step 1), inaccurate data (step 2), autocorrelation among the predictor variables (step 3) etc.

Species observation data, their quality and treatment, appear as a critical aspect to achieve more robust models for accurate predictions of invasion risk. SDM indeed heavily relies on good occurrence data and relevant environmental dataset. Although current availability of high-resolution bio-climatic data on various aspects of environments [26] aids for better SDM, occurrence information for alien species often comes as a challenge due to lack of resources or knowledge. The knowledge on absence information of alien species is also often lacking.

Therefore, different SDM methods have been developed to use either presence-absence (PA) or presence-only (PO) data [69]. In addition, some modelling techniques also use 'pseudo-absence' data for model fitting (e.g. GARP), but these are counted as presence-only methods as the absence of species is not guaranteed regarding these pseudo-absence locations. Yet, selection of pseudo-absence or background sampling is often neglected by modelers, who should try to constrain the pseudo-absence or background sampling to the same spatial extent as presences [117]. The lack of observation data for invasive species demands for more sampling and exchange of data between countries.

SDM are based on the notion of niche conservatism to which invasive species seldom complies. Therefore, issues like transferability, biotic interactions in native and invaded ranges, niche characteristics, species dispersal and uncertainty associated with modelled predictions should get addressed to appropriately represent invasion risk. In the following sections we discuss these aspects and best practices to incorporate them in the SDM.

Invasive species niche characterization

SDM are strongly dependent on the concept of niches in ecology [78] as niche requirements are a key factor in determining suitable areas where an alien species can establish. Distribution models of alien species are generally trained on native distributional areas [25, 71, 118], which often have a higher probability to meet the distributional equilibrium [27]. However, the presence/absence of information from invaded regions may offer additional insights into novel environments and biotic contexts [119]. Some alien species might have different niche requirements in non-native areas than from their native ranges and also can evolve at rapid rates to modify their environmental requirements with a change in the climate i.e. evolutionary niche shift [120], thus it becomes important to check niche conservatism. For example, Fernández et al. [43] studied the ecological niche transferability of an invasive species and concluded that the ecological niche a species holds in their native range is generally a poor predictor of invaded range, despite the fact that niche conservatism has often been assumed when predicting the spread of invasive species [72].

In these cases, SDM will not be able to precisely predict the spread of invasive species, nor the characteristics of the niche for non-native ranges. To limit this risk, Medley [72] used a reciprocal distribution modelling (RDM) approach to investigate niche conservatism for the Asian tiger mosquito (*Aedes albopictus*) and highlighted the importance of the reciprocal models in controlling bi-directional dispersal between native and non-native distributions. RDM is a combination of two models i.e. one created using native occurrences and projected onto invaded regions, and the other using invasive occurrences which are projected back onto the native distribution [43]. If the native model accurately predicts the introduced distribution, and vice versa, the niche has been conserved. It is vital to characterize the niche prior to creating a pest risk map, as it can provide additional novel insights on invasive species ecology and will yield better-informed forecasts of invasive species distributions.

Projecting invasive species distributions

Novel correlative methods (e.g. MaxEnt) [121] outperform more established methods, particularly due to the ability to fit complex functions, including interactions amongst predictors and use of penalty functions to avoid overfitting. The concern is that these correlative models do not perform well when projected to novel environments (transferability) [47]. Reason being their underlying assumption of conservation of niches, which cannot be always true in case of alien species (c.f. 3.1). Since correlative SDM only measures realized niche, the concept of niche conservation has been increasingly doubted because of species environmental variability [35]. Also, the performances of correlative SDM might suffer if not fitted properly while projecting into novel environments.

Another study by Swant et al. [122] raises a similar concern over SDM transferability abilities. It is widely known that correlative models work best in fairly well-sampled regions but much of the efforts are needed when predicting species distributions in unsampled regions (non-native ranges in case of alien species). Peterson et al. [123] chose MaxEnt and GARP to compare model transferability success. They chose three bird species *Caprimulgus vociferus*, *Coccyzus americanus* and *Zenaida macroura* that had fairly broad geographical distributions along with a set of 19 bioclimatic variables in addition to topographical variables. Best subset method was applied in case of GARP while, MaxEnt software was used with default settings except tuning the regularization multiplier value. Authors, found that MaxEnt predictions showed overfitting to the input data and was only transferable at low thresholds. Whereas, GARP had a higher success rate at the prediction with increase commission errors. GARP models constantly showed similarity with species known distributions while MaxEnt models produced an odd pattern in coherence with the input data, interestingly, there were no significant differences between their validation scores. Contrary to it, Swant et al. [122] found higher AUC scores for GARP for their comparative study with MaxEnt. Authors modelled potential distribution of invasive emerald ash borer (*Agrilus planipennis*) in its invaded and native ranges using four ecologically relevant bioclimatic variables. They found that none of the MaxEnt model provided a reliable estimate whereas, GARP when trained in native range performed well. This better performance of GARP leads to the notion that they might deal with spatial bias better than MaxEnt [123]. GARP has been found successful for predicting invasive species distributions. However, it has been criticized for overprediction [39] as it fails to model less important relationships in the data [36].

In an attempt to minimize the prediction errors (false presences and absences) due to poor transferability of SDM and non-equilibrium distribution of alien species, both correlative and mechanistic (cf. Box 1) approaches should be used in cohesion. The predictive performance of MaxEnt has been found to be significantly improved when fitted with outputs from process-based mechanistic models [47]. Authors used MaxEnt and CLIMEX to assess the risk of establishment of western cherry fruit fly in California. They included climatic, topographic and species-specific phenology variables along with human footprint as variables and found that MaxEnt model was improved by including Eco climatic index generated from the CLIMEX models. In recent years, following the same line of development of combining the two approaches (correlative+ mechanistic; cf. Box 1), hybrid models have been developed [124]. These models combine correlative SDM with expert knowledge-driven mechanistic models. The hybrid models overcome limitations of traditional models as they take advantage of both the approaches and yield more reliable predictions. Golding and Purse [125] developed a Bayesian SDM using Gaussian process (GP). This GP model enables the user to incorporate prior ecological knowledge via a prior estimate of a model function [125]. For e.g. effect of moisture limits on a pathogen (sudden oak death: *Phytophthora ramorum*) distributions. This approach can effectively bridge mechanistic and correlative models, such that it retains the information from a mechanistic process while extrapolating the model to novel environments. We recommend critical evaluation of the available modelling options and further refine the model with a hybrid approach to allow more robust estimates of the future distribution of invasive species in novel environments.

Additionally, extreme caution should be taken when projecting alien species distributions, specifically under changing climatic conditions as extrapolation beyond climatic limits in the training data is an unreliable practice [27], since alien species are seldom at equilibrium within their environments. Studies in the past have addressed issues related to extrapolation when projecting SDM into novel environment and suggested linking SDM with landscape, population and physiological models representing processes of change to improve the model extrapolations [43, 126, 127].

Biotic interactions and dispersal constraints

Invasive species are dynamic in nature and can compete with native biological communities such that the complete invasive mechanism is hard to determine or forecast. In the past, scientists have debated the utility of incorporating biotic interactions into SDM [128, 129]. These could include: the presence of competitors/ predators, or the absence of mutualists in SDM. A limited number of studies are available which explicitly include predictors describing biological interactions [52], despite the fact that habitat projections into future climate conditions where biotic interactions may have transformed are likely to result in inaccurate assessments [128].

In a study, Araújo and Luoto [128] used Generalized additive modelling (GAM) to investigate relationships between species and climate; species and host plants; and species and climate + host plants and found that inclusion of biotic interactions significantly increases the explanatory power of the SDM at macro scales. Another similar study carried out by Meier et al. [130] using variance partitioning to estimate the proportion of the variance explained by biotic and abiotic predictors, found that non-inclusion of community composition and other local biotic factors will strongly influence prediction of species distributions. Distribution models of invasive species generally improve after inclusion of information on biotic interactions and the availability of related information on these interactions can advance the SDM projections.

Additionally, the development of SDM for alien species is commonly based on climatic and land use predictor variables. However, the distribution of non-native species is heavily governed by the influence of humans on the landscape [15]. Studies considering the impacts of human footprint on alien species distributions show that human footprint can significantly affect the distribution of alien species [92–95, 131–133]. Thus, the inclusion of related variables explaining human pressure on landscapes becomes imperative when building SDM for alien species. Datasets like 'human footprint' [134], 'Global Human Influence Index (HII)' (http://sedac.ciesin.columbia.edu), 'population density', 'road density', 'urbanization' and 'tourism' available from World resources institute (https://www.wri.org) can be of immense significance [95]. The availability of various datasets explaining human footprint on landscapes provides new opportunities for researchers to model the impact of human influence on the distribution of alien species.

Alien species are mobile in nature and often have traits that facilitate dispersal. However, for most alien species, it is still unknown how much of their complete potential distribution range is represented by observed individuals. This leads to an unknown fundamental niche, although Guisan and Thuiller [28] relate this issue to species competitive and dispersal abilities. So far, most studies have ignored the dispersal limitations of alien species [135, 136], assuming their distribution to be either unlimited, or null [137]. This strategy can work for species with a wide host range and strong dispersal abilities. Despite this, these assumptions can often provide inaccurate information regarding potential distributions and further lead to under- or over-estimation of potential suitable areas. This can also yield greater uncertainties with respect to conservation decisions. Despite these risks, very few studies acknowledge the importance of a species dispersal ability while projecting the distribution across space and time [127, 138, 139].

Also some of the major invasive alien species have been known to be introduced by human mediation [83, 131, 140] and international trade and movement of people are attributed to increase numbers of alien species introductions to novel environments [141]. In their non-native range their movement is either through stratified local diffusion [142] or long distance dispersal associated with human movement [113]. The human footprint variables (for e.g. port and road proximity, roads, navigable rivers, etc.) discussed previously can be directly linked to the vectors and pathways of alien species distributions [94]. However, models are needed to predict distribution based on patterns and density of propagule dispersal addressing both local diffusion and long distance movement. There are several studies that have tried to include information on species dispersal traits to model their distributions [113, 127, 138, 139]. These dispersal traits can then be used to inform spread models such as a modular dispersal framework [143], individual-based spread model [144] and MigClim [136]. Dispersal is a key factor when predicting invasive species distribution, since, generally, all potentially suitable areas cannot be colonizable. Thus we strongly recommend to address dispersal and biotic interactions when building SDM for mapping invasion risk.

Mapping impacts of uncertainty

SDM are complex tools which inevitably include some degree of uncertainty [145]. The uncertainty in SDM predictions and performances results both from incomplete knowledge of the species, and from errors in the specification of the model [146, 147]. Furthermore, many correlative SDM are projected at high resolutions and under climate change scenarios without explicitly addressing uncertainty, making many invasive species risk maps of questionable accuracy [104]. This further affects the decision making and resource allocation measures in conservation planning [148].

In order to lessen the risk of adverse uncertainties in species distribution, SDM should explicitly address levels of uncertainties in their modelled predictions [59]. Various authors have also advocated to depict levels of uncertainties in the modelled predictions [146, 147] but received less consideration. Gould et al. [149] created a tool to provide spatially explicit illustrations of the impact of uncertainty on their modelled projections. Their uncertainty tool uses a Monte Carlo process to produce probabilistic and spatially explicit output. Another approach yet faster is GP models [137] which automatically produces levels of uncertainty in modelled predictions without bootstrapping procedures. The produced prediction uncertainty map measures uncertainty as the variance of the estimated function for the group of predictors [137]. The most recent method to tackle uncertainty in SDM projections is R-based 'mopa' package [150]. The package can handle multi-factor SDM ensemble experiments and can explore various uncertainty factors (e.g. occurrence datasets, pseudo-absence/background data, future climate projections, SDM algorithms, etc.), in addition to assessing contribution of independent factors to the overall uncertainty. It is obvious that conservation and management actions regarding invasive species distributions are linked with huge investments and substantial impacts on ecosystems globally [151]. Therefore, with so much at stake it has become vital to assess the impacts of uncertainty on modelled predictions.

Conclusion

SDM with strong underlying biological assumptions will have better predictive powers and produce invasive species risk maps that will be more likely to forecast precise estimates of invasion risk. Here in this review we have discussed the major aspects to consider when building SDM for mapping invasion risk. We limited ourselves to issues related to model transferability, biotic interactions along with dispersal limitations (local and regional) and uncertainty, which appeared especially relevant in the case of invasive species. However, we do acknowledge other important aspects to consider such as autocorrelation amongst predictors [152], extent and resolution of the study area [153], variable selection, pseudo-absence generation procedures [117, 154] and model evaluation [155]. These discussed challenges need to be addressed to ensure that the predictions match the foreseeable distributional scenario, especially under influence of climate change.

Furthermore, the critical issues and related best practices discussed in this review will aid species distribution modellers in creating more scientifically sound models and ecologically relevant predictions. Future work should be around developing more easy to use hybrid models [124] that are capable of addressing both local diffusion and long distance movement of alien species.

Acknowledgements

The authors wish to thank two anonymous reviewers for valuable feedback on earlier versions of this paper. This work was generously funded by Genome Canada, Genome British Columbia and Genome Quebec within the framework of project bioSAFE (Biosurveillance of Alien Forest Enemies, project number #10106) as part of a Large-Scale Applied Research Project in Natural Resources and the Environment.

References

1. United Nations. The Convention on Biological Diversity [Internet]. Secretariat of the Convention on Biological Diversity; 1992. p. 30. Available from: URL: https://www.cbd.int/convention/ (accessed 11 March 2019).
2. Kettunen M, Genovesi P, Gollasch S, Pagad S, Starfinger U, ten Brink P, *et al.* Technical support to EU strategy on invasive alien species (IAS). Institute for European Environmental Policy (IEEP), Brussels; 2009.
3. Mooney H, Drake J. Biological invasions: a SCOPE program overview [Internet]. Biological Invasions. 1989. p. 491–506.
4. Mooney HA, Hamburg SP, Drake JA. The invasion of plants and animals into California. In: Ecology of Biological Invasions of North America and Hawaii. Springer, New York, NY; 1986. p. 250–72.
5. Drake JA, Mooney HA, Di Castri F, Groves RH, Kruger RH, Rejmanek FJ, *et al.* Biological invasions: A global perspective [Internet]. Scope 1989;37:525.
6. Xu H, Ding H, Li M, Qiang S, Guo J, Han Z, *et al.* The distribution and economic losses of alien species invasion to China. Biological Invasions 2006;8(7):1495–500.
7. Pejchar L, Mooney HA. Invasive species, ecosystem services and human well-being. Trends in Ecology & Evolution 2009;24(9):497–504.
8. Valéry L, Fritz H, Lefeuvre JC, Simberloff D. Ecosystem-level consequences of invasions by native species as a way to investigate relationships between evenness and ecosystem function. Biological Invasions 2009;11(3):609–17.
9. IUCN. Guidelines for the prevention of biodiversity loss caused by alien invasive species. Fifth Meeting of the Conference of the Parties to the Convention on Biological Diversity, Nairobi, Kenya. International Union for Conservation of Nature, Gland, Switzerland; 2000. p. 21.
10. Chornesky EA, Randall JM. The threat of invasive alien species to biological diversity: setting a future course. Annals of the Missouri Botanical Garden 2003;90(1):67–76.
11. Allen EA, Humble LM, Allen E, Humble L. Nonindigenous species introductions: a threat to Canada's forests and forest economy. Canadian Journal of Plant Pathology [Internet] 2002;24(2):103–10.
12. Levine JM, D'Antonio CM. Forecasting biological invasions with increasing international trade. Conservation Biology 2003;17(1):322–6.
13. Musselman LJ. Harmful non-indigenous species in the United States. Economic Botany 1994;48:138–38.
14. Seebens H, Essl F, Dawson W, Fuentes N, Moser D, Pergl J, *et al.* Global trade will accelerate plant invasions in emerging economies under climate change. Global Change Biology 2015;21(11):4128–40.
15. Vitousek PM, Mooney HA, Lubchenco J, Melillo JM. Human domination of earth's ecosystems. Science 1997; 277(5325):494–9. Available from: URL: http://science.sciencemag.org/content/277/5325/494.full
16. Peterson AT, Stewart A, Mohamed KI, Araújo MB. Shifting global invasive potential of European plants with climate change. PLoS One 2008;3(6):1–7.
17. Kenis M, Auger-Rozenberg MA, Roques A, Timms L, Péré C, Cock MJ, *et al.* Ecological effects of invasive alien insects. Biological Invasions 2009;11(1):21–45.
18. Chenje M, Mohamed-katerere J. Invasive alien species. Africa Environ Outlook 2 – Our Environ Our Wealth [Internet]. 2006. p. 331–49. Available from: URL: http://www.unep.org/dewa/africa/docs/en/AEO2_Our_Environ_Our_Wealth.pdf
19. Dodds KJ, Orwig DA. An invasive urban forest pest invades natural environments – Asian long-horned beetle in northeastern US hardwood forests. Canadian Journal of Forest Research [Internet] 2011;41(9):1729–42.
20. Liebhold AM, Halverson JA, Elmes GA. Gypsy moth invasion in North America: a quantitative analysis. Journal of Biogeography 1992;19(5):513–20.
21. Normile D. UN biodiversity summit yields welcome and unexpected progress. Science 2011;331(6016):400.
22. Walther GR, Roques A, Hulme PE, Sykes MT, Pyšek P, Kühn I, *et al.* Alien species in a warmer world: risks and opportunities. Trends in Ecology & Evolution 2009;24(12):686–93.
23. Kaiser BA, Burnett KM. Spatial economic analysis of early detection and rapid response strategies for an invasive species. Resource and Energy Economics 2010;32(4):566–85.
24. Meyerson LA, Mooney HA. Invasive alien species in an era of globalization. Frontiers in Ecology and Environment Ecological Society of America Conference on Ecology an Era Global – Challenges and Opportunities Environmental Science America 2007;5(4):199–208.

25. Peterson AT. Predicting the geography of species' invasions via ecological niche modeling. Quarterly Review of Biology [Internet] 2003;78(4):419–33. Available from: URL: http://www.journals.uchicago.edu/doi/10.1086/378926
26. Stohlgren TJ, Ma P, Kumar S, Rocca M, Morisette JT, Jarnevich CS, *et al.* Ensemble habitat mapping of invasive plant species. Risk Analysis 2010;30(2):224–35.
27. Anderson RP, Lew D, Peterson AT. Evaluating predictive models of species' distributions: criteria for selecting optimal models. Ecological Modelling 2003;162:211–32.
28. Guisan A, Thuiller W. Predicting species distribution: offering more than simple habitat models. Ecology Letters 2005;8(9):993–1009.
29. Elith J, Leathwick JR. Species distribution models: ecological explanation and prediction across space and time. Annual Review in Ecology, Evolution, and Systematics 2009;40:677–97.
30. Peters RH. A critique for ecology. Limnology and Oceanography 1993;38(6):1344–6.
31. Zimmermann NE, Edwards TC, Graham CH, Pearman PB, Svenning JC. New trends in species distribution modelling. Ecography (Cop) 2010;33(6):985–9.
32. Coops N, Waring RH. Combining a generic process-based productivity model and a statistical classification method to predict the presence and absence of tree species in the Pacific Northwest, USA. 2009;(August).
33. Kearney M, Porter W. Mechanistic niche modelling: combining physiological and spatial data to predict species' ranges. Ecology Letters 2009;12(4):334–50.
34. Shabani F, Kumar L, Ahmadi M. A comparison of absolute performance of different correlative and mechanistic species distribution models in an independent area. Ecology and Evolution 2016;6(16):5973–86.
35. Dormann CF, Schymanski SJ, Cabral J, Chuine I, Graham C, Hartig F, *et al.* Correlation and process in species distribution models: bridging a dichotomy. Journal of Biogeography 2012;39(12): 2119–31.
36. Elith J, Graham CH. Do they? How do they? WHY do they differ? On finding reasons for differing performances of species distribution models. 2009;(December 2008).
37. Merow C, Smith MJ, Silander JA. A practical guide to MaxEnt for modeling species' distributions: what it does, and why inputs and settings matter. Ecography (Cop) 2013;36(10):1058–69.
38. Ward DF. Modelling the potential geographic distribution of invasive ant species in New Zealand. Biological Invasions 2007;9(6):723–35.
39. Padalia H, Srivastava V, Kushwaha SP. Modeling potential invasion range of alien invasive species, *Hyptis suaveolens* (L.) Poit. in India: comparison of MaxEnt and GARP. Ecological Informatics 2014;22:36–43.
40. Poulos HM, Chernoff B, Fuller PL, Butman D. Ensemble forecasting of potential habitat for three invasive fishes. Aquatic Invasions 2012;7(1):59–72.
41. Ficetola GF, Thuiller W, Miaud C. Prediction and validation of the potential global distribution of a problematic alien invasive species – The American bullfrog. Diversity and Distributions 2007;13(4):476–85.
42. De Meyer M, Robertson MP, Mansell MW, Ekesi S, Tsuruta K, Mwaiko W, *et al.* Ecological niche and potential geographic distribution of the invasive fruit fly Bactrocera invadens (Diptera, Tephritidae). Bulletin of Entomological Research 2010;100(1):35–48.
43. Fernández M, Hamilton H. Ecological niche transferability using invasive species as a case study. PLoS One 2015;10(3):1–17.
44. Medley KA. Niche shifts during the global invasion of the Asian tiger mosquito, Aedes albopictus Skuse (Culicidae), revealed by reciprocal distribution models. Global Ecology and Biogeography 2010;19(1):122–33.
45. Kumar S, Yee WL, Neven LG. Mapping global potential risk of establishment of *Rhagoletis pomonella* (Diptera: Tephritidae) using MaxEnt and CLIMEX Niche Models. Journal of Economic Entomology 2016;109(5):2043–53.
46. Sutherst RW, Maywald GF. A computerised system for matching climates in ecology. Agriculture Ecosystems & Environment 1985;13(3–4):281–99.
47. Kumar S, Neven LG, Wee LY. Evaluating correlative and mechanistic niche models for assessing the risk of pest establishment. Ecosphere (Washington, DC) 2014;5(7):1–23.
48. Kumar S, Neven LG, Zhu H, Zhang R. Assessing the global risk of establishment of *Cydia pomonella* (Lepidoptera: Tortricidae) using CLIMEX and MaxEnt Niche Models. Journal of Economic Entomology 2015;108(4):1708–19.

49. Elith J. Predicting distributions of invasive species. In: Robinson AP, Walshe T, Burgman MA, Nunn M editors. Invasive Species: Risk Assessment and Management. Cambridge University Press, Cambrigde, UK; 2017. p. 93–129.
50. Webber BL, Yates CJ, Le Maitre DC, Scott JK, Kriticos DJ, Ota N, *et al.* Modelling horses for novel climate courses: insights from projecting potential distributions of native and alien Australian acacias with correlative and mechanistic models. Diversity and Distributions 2011;17(5):978–1000.
51. Kumar S, Neven LG, Yee WL. Assessing the potential for establishment of western cherry fruit fly using ecological niche modeling. Journal of Economic Entomology [Internet] 2014;107(3):1032–44.
52. Venette RC, Kriticos DJ, Magarey RD, Koch FH, Baker RH, Worner SP, *et al.* Pest risk maps for invasive alien species: a roadmap for improvement. Bioscience 2010;60(5):349–62.
53. York P, Evangelista P, Kumar S, Graham J, Flather C, Stohlgren T. A habitat overlap analysis derived from maxent for tamarisk and the south-western willow flycatcher. Frontiers of Earth Science 2011;5(2):120–9.
54. Thum RA, Lennon JT. Comparative ecological niche models predict the invasive spread of variable-leaf milfoil (*Myriophyllum heterophyllum*) and its potential impact on closely related native species. Biological Invasions 2009;12(1):133–43.
55. Peterson AT, Martínez-Meyer E, González-Salazar C, Hall PW. Modeled climate change effects on distributions of Canadian butterfly species. Canadian Journal of Zoology [Internet] 2004;82(6):851–8.
56. Peterson AT, Sánchez-Cordero V, Soberón J, Bartley J, Buddemeier RW, Navarro-Sigüenza AG. Effects of global climate change on geographic distributions of Mexican Cracidae. Ecological Modelling 2001;144(1):21–30.
57. Peterson AT, Williams R, Chen G. Modeled global invasive potential of Asian gypsy moths, Lymantria dispar. Entomologia Experimentalis et Applicata 2007;125(1):39–44.
58. Burgess TI, Scott JK, Mcdougall KL, Stukely MJC, Crane C, Dunstan WA, *et al.* Current and projected global distribution of *Phytophthora cinnamomi*, one of the world's worst plant pathogens. Global Change Biology 2017;23(4):1661–74.
59. Beaumont LJ, Pitman AJ, Poulsen M, Hughes L. Where will species go? Incorporating new advances in climate modelling into projections of species distributions. Global Change Biology 2007;13(7):1368–85.
60. Jeschke JM, Strayer DL. Usefulness of bioclimatic models for studying climate change and invasive species. Annals of the New York Academy of Sciences 2008;1134:1–24.
61. Thomson JR, Fleishman E, Mac Nally R, Dobkin DS. Comparison of predictor sets for species richness and the number of rare species of butterflies and birds. Journal of Biogeography 2007;34(1):90–101.
62. Porfirio LL, Harris RMB, Lefroy EC, Hugh S, Gould SF, Lee G, *et al.* Improving the use of species distribution models in conservation planning and management under climate change. PLoS One 2014;9(11):1–21.
63. Cayuela L, Golicher D, Newton A, Kolb H, de Alburquerque FS, Arets EJMM, *et al.* Species distribution modeling in the tropics: problems, potentialities, and the role of biological data for effective species conservation. Tropical Conservation Science [Internet] 2009;2(3):319–52.
64. del Barrio G, Harrison PA, Berry PM, Butt N, Sanjuan ME, Pearson RG, *et al.* Integrating multiple modelling approaches to predict the potential impacts of climate change on species' distributions in contrasting regions: comparison and implications for policy. Environmental Science & Policy 2006;9(2):129–47.
65. Williams JN, Seo C, Thorne J, Nelson JK, Erwin S, O'Brien JM, *et al.* Using species distribution models to predict new occurrences for rare plants. Diversity and Distributions 2009;15(4):565–76.
66. Engler R, Guisan A, Rechsteiner L. An improved approach for predicting the distribution of rare and endangered species from occurrence and pseudo-absence data. Journal of Applied Ecology 2004;41(2):263–74.
67. Raes N, Roos MC, Slik JWF, Van Loon EE, Ter Steege H. Botanical richness and endemicity patterns of Borneo derived from species distribution models. Ecography (Cop) 2009;32(1):180–92.
68. Esselman PC, Allan JD. Application of species distribution models and conservation planning software to the design of a reserve network for the riverine fishes of northeastern Mesoamerica. Freshwater Biology 2011;56(1):71–88.
69. Brotons L, Thuiller W, Araújo MB, Hirzel AH. Presence-absence versus presence-only modelling methods for predicting bird habitat suitability. Ecography (Cop) [Internet] 2004;27(4):437–48.
70. Svenning JC, Fløjgaard C, Marske KA, Nógues-Bravo D, Normand S. Applications of species distribution modeling to paleobiology. Quaternary Science Reviews [Internet] 2011;30(21–22):2930–47. Available from: URL: http://dx.doi.org/10.1016/j.quascirev.2011.06.012

71. Padalia H, Srivastava V, Kushwaha SP. How climate change might influence the potential distribution of weed, bushmint (*Hyptis suaveolens*)? Environmental Monitoring and Assessment 2015;187(4):210.
72. Medley KA. Niche shifts during the global invasion of the Asian tiger mosquito, *Aedes albopictus* Skuse (Culicidae), revealed by reciprocal distribution models. Global Ecology and Biogeography 2010;19(1):122–33.
73. Vanhanen H, Veteli TO, Päivinen S, Kellomäki S, Niemelä P. Climate change and range shifts in two insect defoliators: gypsy moth and nun moth – A model study. Silva Fennica 2007;41(4):621–38.
74. Dullinger S, Dirnböck T, Grabherr G. Modelling climate change-driven treeline shifts: relative effects of temperature increase, dispersal and invisibility. Journal of Ecology 2004;92(2):241–52.
75. Elith J, Kearney M, Phillips S. The art of modelling range-shifting species. Methods in Ecology and Evolution [Internet] 2010;1(4):330–42.
76. Di Febbraro M, Martinoli A, Russo D, Preatoni D, Bertolino S. Modelling the effects of climate change on the risk of invasion by alien squirrels. Hystrix 2016;27(1):1–8.
77. Williamson M. Explaining and predicting the success of invading species at different stages of invasion. Biological Invasions 2006;8(7):1561–8.
78. Jiménez-Valverde A, Peterson AT, Soberón J, Overton JM, Aragón P, Lobo JM. Use of niche models in invasive species risk assessments. Biological Invasions 2011;13(12):2785–97.
79. Václavík T, Meentemeyer RK. Invasive species distribution modeling (iSDM): Are absence data and dispersal constraints needed to predict actual distributions? Ecological Modelling 2009;220(23):3248–58.
80. Régnière J, Nealis V, Porter K. Climate suitability and management of the gypsy moth invasion into Canada. In: Ecological Impacts of Non-Native Invertebrates and Fungi on Terrestrial Ecosystems. Springer, Dordrecht; 2008. p. 135–48.
81. Meentemeyer RK, Anacker BL, Mark W, Rizzo DM. Early detection of emerging forest disease using dispersal estimation and ecological niche modeling. Ecological Applications 2008;18(2):377–90.
82. Leibold MA. The niche concept revisited: mechanistic models and community context. Ecology 1995;76:1371–82.
83. Lippitt CD, Rogan J, Toledano J, Sangermano F, Eastman JR, Mastro V, *et al.* Incorporating anthropogenic variables into a species distribution model to map gypsy moth risk. Ecological Modelling 2008;210(3):339–50.
84. Breiner FT, Guisan A, Bergamini A, Nobis MP. Overcoming limitations of modelling rare species by using ensembles of small models. Methods in Ecology and Evolution 2015;6(10):1210–8.
85. Morales NS, Fernandez IC, Carrasco BA, Orchard C. Combining niche modelling, land use-change, and genetic information to assess the conservation status of Pouteria splendens populations in Central Chile. bioRxiv [Internet] 2015;2015:026336.
86. Pearson RG. Species' distribution modeling for conservation educators and practitioners. Lessons Conservation 2010;3(3):54–89.
87. Meller L, Cabeza M, Pironon S, Barbet-Massin M, Maiorano L, Georges D, *et al.* Ensemble distribution models in conservation prioritization: from consensus predictions to consensus reserve networks. Diversity and Distributions 2014;20(3):309–21.
88. Daehler CC, Denslow JS, Ansari S, Kuo HC. A risk-assessment system for screening out invasive pest plants from Hawaii and other Pacific Islands. Conservation Biology 2004;18(2):360–8.
89. Kumar S, Stohlgren TJ. Maxent modeling for predicting suitable habitat for threatened and endangered tree *Canacomyrica monticola* in New Caledonia. Journal of Ecology Natural Sciences Vol 1(4) [Internet] 2009;1(4):094–8.
90. Fuller DO, Parenti MS, Hassan AN, Beier JC. Linking land cover and species distribution models to project potential ranges of malaria vectors: an example using *Anopheles arabiensis* in Sudan and Upper Egypt. Malaria Journal [Internet] 2012;11(1):264.
91. Thuiller W, Araújo MB, Lavorel S. Do we need land-cover data to model species distributions in Europe? Journal of Biogeography [Internet] 2004;31(3):353–61. Available from: URL: http://doi.wiley.com/10.1046/j.0305-0270.2003.00991.x
92. Westphal MI, Browne M, Mackinnon K, Noble I. The link between international trade and the global distribution of invasive alien species. Biological Invasions 2008;10(4):391–8.
93. Smolik MG, Dullinger S, Essl F, Kleinbauer I, Leitner M, Peterseil J, *et al.* Integrating species distribution models and interacting particle systems to predict the spread of an invasive alien plant. Journal of Biogeography 2010;37(3):411–22.

94. Gallardo B, Zieritz A, Aldridge DC. The importance of the human footprint in shaping the global distribution of terrestrial, freshwater and marine invaders. PLoS One 2015;10(5): e0125801.
95. Ancillotto L, Strubbe D, Menchetti M, Mori E. An overlooked invader? Ecological niche, invasion success and range dynamics of the Alexandrine parakeet in the invaded range. Biological Invasions 2016;18(2):583–95.
96. Guisan A, Broennimann O, Engler R, Vust M, Yoccoz NG, Lehmann A, *et al.* Using niche-based models to improve the sampling of rare species. Conservation Biology 2006;20(2):501–11.
97. Approach NEW, Ecological TO, On B, Tools NEW, Biodiversity DF, Applied IS, *et al.* Predicting species invasions using ecological niche modeling: new approaches from bioinformatics attack a pressing problem. Bioscience 2001;51(5):363–72.
98. Austin M. Species distribution models and ecological theory: a critical assessment and some possible new approaches. Ecological Modelling 2007;200(1–2):1–9.
99. Soberon J, Peterson AT. Interpretation of models of fundamental ecological niches and species' distributional areas. Biodiversity Informatics [Internet] 2005;2(0):1–10.
100. Hoffman JD, Aguilar-Amuchastegui N, Tyre AJ. Use of simulated data from a process-based habitat model to evaluate methods for predicting species occurrence. Ecography (Cop) 2010;33(4):656–66.
101. McKenney DW, Hopkin AA, Campbell KL, Mackey BG, Foottit R. Opportunities for improved risk assessments of exotic species in Canada using bioclimatic modeling. Environmental Monitoring and Assessment 2003;88(1–3):445–61.
102. Simonson SE, Stohlgren TJ, Tyler L, Gregg WP, Muir R, Garrett LJ. Preliminary assessment of the potential impacts and risks of the invasive cactus moth, *Cactoblastis cactorum* Berg, in the US and Mexico. Final Report to the International Atomic Energy Agency; 2005.
103. Matsuki M, Kay M, Serin J, Floyd R, Scott JK. Potential risk of accidental introduction of *Asian gypsy* moth (Lymantria dispar) to Australasia: effects of climatic conditions and suitability of native plants. Agricultural and Forest Entomology 2001;3(4):305–20.
104. Yemshanov D, Koch FH, Ben-Haim Y, Smith WD. Robustness of risk maps and survey networks to knowledge gaps about a new invasive pest. Risk Analysis 2010;30(2):261–76.
105. Peterson AT, Samy AM. Geographic potential of disease caused by Ebola and Marburg viruses in Africa. Acta Tropica [Internet] 2016;162:114–24. Available from: URL: http://dx.doi.org/10.1016/j.actatropica.2016.06.012
106. Václavík T, Kanaskie A, Hansen EM, Ohmann JL, Meentemeyer RK. Predicting potential and actual distribution of sudden oak death in Oregon: prioritizing landscape contexts for early detection and eradication of disease outbreaks. Forest Ecology and Management 2010;260(6):1026–35.
107. Kluza DA, Vieglais DA, Andreasen JK, Peterson AT. Sudden oak death: geographic risk estimates and predictions of origins. Plant Pathology 2007;56(4):580–7.
108. Srivastava V, Griess VC, Padalia H. Mapping invasion potential using ensemble modelling. A case study on Yushania maling in the Darjeeling Himalayas. Ecological Modelling 2018;385:35–44.
109. Jeschke JM, Strayer DL. Usefulness of bioclimatic models for studying climate change and invasive species. Annals of the New York Academy of Sciences 2008;1134(1):1–24.
110. Roura-Pascual N, Suarez A. The utility of species distribution models to predict the spread of invasive ants (Hymenoptera: Formicidae) and to anticipate changes in their ranges in the face of. Myrmecol News [Internet]. 2008;(August): p. 67–77.
111. Rödder D, Solé M, Böhme W. Predicting the potential distributions of two alien invasive Housegeckos (Gekkonidae: *Hemidactylus frenatus*, *Hemidactylus mabouia*). North-Western Journal of Zoology 2008;4(2):236–46.
112. Stiels D, Schidelko K, Engler JO, van den Elzen R, Rödder D. Predicting the potential distribution of the invasive Common Waxbill *Estrilda astrild* (Passeriformes: Estrildidae). Journal fuer Ornithologie 2011;152(3):769–80.
113. Muirhead JR, Leung B, Van Overdijk C, Kelly DW, Nandakumar K, Marchant KR, *et al.* Modelling local and long-distance dispersal of invasive emerald ash borer *Agrilus planipennis* (Coleoptera) in North America. Diversity and Distributions 2006;12(1):71–9.
114. Holmes TP, Aukema JE, Von Holle B, Liebhold A, Sills E. Economic impacts of invasive species in forests: past, present, and future. Annals of the New York Academy of Sciences 2009;1162:18–38.
115. Brockerhoff EG, Liebhold AM, Richardson B, Suckling DM. Eradication of invasive forest insects: concepts, methods, costs and benefits. New Zealand Journal of Science 2010;40(Suppl.):117–35.
116. Araújo MB, Guisan A. Five (or so) challenges for species distribution modelling. Journal of Biogeography 2006;33(10):1677–88.

117. Liang W, Papeş M, Tran L, Grant J, Washington-Allen R, Stewart S, *et al.* The effect of pseudo-absence selection method on transferability of species distribution models in the context of non-adaptive niche shift. Ecological Modelling 2018;388:1–9.
118. Rodda GH, Jarnevich CS, Reed RN. Challenges in identifying sites climatically matched to the native ranges of animal invaders. PLoS One 2011;6(2).
119. Thuiller W, Richardson DM, Pyšek P, Midgley GF, Hughes GO, Rouget M. Niche-based modelling as a tool for predicting the risk of alien plant invasions at a global scale. Global Change Biology 2005;11(12):2234–50.
120. Sinclair SJ, White MD, Newell GR. How useful are species distribution models for managing biodiverstity under future climates. Ecology and Society 2010;15(1):8.
121. Elith J, Graham CH, Anderson RP, Dudik M, Ferrier S, Guisan A, *et al.* Novel methods improve prediction of species' distributions from occurrence data. Ecography (Cop) 2006;29(2):129–51.
122. Sobek-Swant S, Kluza DA, Cuddington K, Lyons DB. Potential distribution of emerald ash borer: what can we learn from ecological niche models using Maxent and GARP? Forest Ecology and Management [Internet] 2012;281:23–31.
123. Townsend Peterson A, Papeş M, Eaton M. Transferability and model evaluation in ecological niche modeling: A comparison of GARP and Maxent. Ecography (Cop) 2007;30(4):550–60.
124. Gallien L, Münkemüller T, Albert CH, Boulangeat I, Thuiller W. Predicting potential distributions of invasive species: where to go from here? Diversity and Distributions 2010;16(3):331–42.
125. Golding N, Purse BV. Fast and flexible Bayesian species distribution modelling using Gaussian processes. Methods in Ecology and Evolution 2016;7(5):598–608.
126. Aitken SN, Yeaman S, Holliday JA, Wang T, Curtis-Mclane S. Adaptation, migration or extirpation: climate change outcomes for tree populations. Evolutionary Applications 2008;1(1):95–111.
127. Uribe-Rivera DE, Soto-Azat C, Valenzuela-Sánchez A, Bizama G, Simonetti JA, Pliscoff P. Dispersal and extrapolation on the accuracy of temporal predictions from distribution models for the Darwin's frog. Ecological Applications [Internet] 2017;27(5):1633–45. Available from: URL: http://doi.wiley.com/10.1002/eap.1556
128. Araújo MB, Luoto M. The importance of biotic interactions for modelling species distributions under climate change. Global Ecology and Biogeography 2007;16(6):743–53.
129. Wisz MS, Pottier J, Kissling WD, Pellissier L, Lenoir J, Damgaard CF, *et al.* The role of biotic interactions in shaping distributions and realised assemblages of species: implications for species distribution modelling. Biological Reviews 2013;88(1):15–30.
130. Meier ES, Kienast F, Pearman PB, Svenning JC, Thuiller W, Araújo MB, *et al.* Biotic and abiotic variables show little redundancy in explaining tree species distributions. Ecography 2010;33(6):1038–48.
131. Strubbe D, Jackson H, Groombridge J, Matthysen E. Invasion success of a global avian invader is explained by within-taxon niche structure and association with humans in the native range. Diversity and Distributions 2015;21(6):675–85.
132. Randin CF, Dirnböck T, Dullinger S, Zimmermann NE, Zappa M, Guisan A. Are niche-based species distribution models transferable in space? Journal of Biogeography 2006;33(10):1689–703.
133. Masin S, Bonardi A, Padoa-Schioppa E, Bottoni L, Ficetola GF. Risk of invasion by frequently traded freshwater turtles. Biological Invasions 2014;16(1):217–31.
134. Sanderson EW, Jaiteh M, Levy MA, Redford KH, Wannebo AV, Woolmer G. The human footprint and the last of the wild: the human footprint is a global map of human influence on the land surface, which suggests that human beings are stewards of nature, whether we like it or not. BioScience 2002;52(10): 891–904.
135. López-Darias M, Lobo JM, Gouat P. Predicting potential distributions of invasive species: The exotic Barbary ground squirrel in the Canarian archipelago and the west Mediterranean region. Biological Invasions 2008;10(7):1027–40.
136. Engler R, Hordijk W, Guisan A. The MIGCLIM R package – seamless integration of dispersal constraints into projections of species distribution models. Ecography (Cop) 2012;35(10):872–8.
137. Engler R, Randin CF, Vittoz P, Czáka T, Beniston M, Zimmermann NE, *et al.* Predicting future distributions of mountain plants under climate change: does dispersal capacity matter? Ecography (Cop) 2009;32(1):34–45.
138. Bancroft JS, Smith MT. Dispersal and influences on movement for *Anoplophora glabripennis* calculated from individual mark-recapture. Entomologia Experimentalis et Applicata 2005;116(2):83–92.
139. Smith MT, Tobin PC, Bancroft J, Li G, Gao R. Dispersal and spatiotemporal dynamics of Asian longhorned beetle (Coleoptera: Cerambycidae) in China. Environmental Entomology 2004;33(2):435–42.

140. Bartell SM, Nair SK. Establishments risks for invasive species. Risk Analysis 2003;24(4):833–45.
141. Meurisse N, Rassati D, Hurley BP, Brockerhoff EG, Haack RA. Common pathways by which non-native forest insects move internationally and domestically. Journal of Pest Science 2018:1–5.
142. Tobin PC, Cremers KT, Hunt L, Parry D. All quiet on the western front? Using phenological inference to detect the presence of a latent gypsy moth invasion in Northern Minnesota. Biological Invasions 2016;18(12):3561–73.
143. Lustig A, Worner SP, Pitt JPW, Doscher C, Stouffer DB, Senay SD. A modeling framework for the establishment and spread of invasive species in heterogeneous environments. Ecology and Evolution 2017;7(20):8338–48.
144. Adams VM, Petty AM, Douglas MM, Buckley YM, Ferdinands KB, Okazaki T, *et al.* Distribution, demography and dispersal model of spatial spread of invasive plant populations with limited data. Methods in Ecology and Evolution 2015;6(7):782–94.
145. Heikkinen RK, Luoto M, Araujo MB, Virkkala R, Thuiller W, Sykes MT. Methods and uncertainties in bioclimatic envelope modelling under climate change. Progress in Physical Geography [Internet] 2006;30(6):751–77.
146. Beale CM, Lennon JJ. Incorporating uncertainty in predictive species distribution modelling. Philosophical Transactions of the Royal Society B: Biological Sciences [Internet] 2012;367(1586):247–58. Available from: URL: http://rstb.royalsocietypublishing.org/cgi/doi/10.1098/rstb.2011.0178
147. Refsgaard JC, van der Sluijs JP, Højberg AL, Vanrolleghem PA. Uncertainty in the environmental modelling process – A framework and guidance. Environmental Modelling & Software 2007;22(11):1543–56.
148. Barry S, Elith J. Error and uncertainty in habitat models. Journal of Applied Ecology 2006;43:413–23.
149. Gould SF, Beeton NJ, Harris RMB, Hutchinson MF, Lechner AM, Porfirio LL, *et al.* A tool for simulating and communicating uncertainty when modelling species distributions under future climates. Ecology and Evolution 2014;4(24):4798–811.
150. Iturbide M, Bedia J, Gutiérrez JM. Tackling uncertainties of species distribution model projections with package mopa. R Journal 2018;10(1):122–39.
151. Marbuah G, Gren IM, McKie B. Economics of harmful invasive species: a review. Diversity 2014;6(3):500–23.
152. Segurado P, Araujo MB, Kunin WE. Consequences of spatial autocorrelation for niche-based models. Journal of Applied Ecology 2006;43(3):433–44.
153. Anderson RP, Raza A. The effect of the extent of the study region on GIS models of species geographic distributions and estimates of niche evolution: preliminary tests with montane rodents (genus *Nephelomys*) in Venezuela. Journal of Biogeography 2010;37(7):1378–93.
154. Pearce JL, Boyce MS. Modelling distribution and abundance with presence-only data. Journal of Applied Ecology 2006;43(3):405–12.
155. Anderson RP, Lew D, Peterson AT. Evaluating predictive models of species' distributions: criteria for selecting optimal models. Ecological Modelling 2003;162(3):211–32.

14 The threat from *Chromolaena odorata* and options for control

Muhammad Rusdy

Department of Forage Science and Grassland Management, Faculty of Animal Science, Hasanuddin University, Makassar, Indonesia

Abstract

The invasion of natural ecosystems by alien plant species is a serious environmental problem that threatens the productivity, biodiversity and ecosystem functioning. *Chromolaena odorata* is a weed native to tropical America and has become one of the world's worst alien invasive species in the Old World of tropical and subtropical countries. In invaded areas of the Old World, this plant has created many problems both in natural and disturbed ecosystems. To generate information useful for alleviating the problems, literature in Google Scholar, CAB Abstracts, Web of Science, PubMed and Scopus were searched. From those databases, reproduction, distribution and factors contributing to its success as an invader species are presented and discussed. Its threat to agriculture, natural ecosystem and animals are presented. Means of preventing the spread and controlling the growth of the weed are highlighted and discussed.

Keywords: *Chromolaena*, Spread, Distribution, Threat, Management

Review Methodology

Recently published literature in Google Scholar, CAB Abstract, Scopus, PubMed and Web of Science using the key-words of *Chromolaena*, spread, threat, management and control were searched through April to November 2018. We also used the synonym of those words, both singular and plural for searching relevant literature. Of about 200 literature were read, we selected 72 kinds of literature for review. We also observed the growth of *Chromolaena* in the field and asked and the farmers and colleagues for their knowledge concerning the subject.

Introduction

Alien plant invasion is one of the most serious threats to the integrity of the natural and semi-natural ecosystem and biodiversity worldwide [1, 2]. It impacts negatively on agriculture, livelihood, biodiversity and causing economic losses. One of the most invasive weeds in tropical and subtropical areas is *Chromolaena odorata* (L.) R.M King and H. Robinson (hereafter is termed *Chromolaena*). This plant is of a neotropical origin, from the southern regions of USA to northern regions of Argentina. There, it is a pioneer plant in the land that has been overgrazed or otherwise disturbed, and with the advance of growth, this plant is soon replaced by other plant species. However in the Old World, it becomes a transformer species, forming a dense thicket that can smother indigenous vegetation and in most habitats, it is not progressively replaced by

*Corresponding author: muhrusdy79@yahoo.co.id

other plant species [3]. It was introduced to the botanical garden in Calcutta, India in 1845 as an ornamental plant and it has since been spread accidentally or deliberately to other areas of the world. It was introduced to Sri Lanka in 1884 [4] and from there, it spread eastward to China and south into Indonesia [5]. The weed was first reported In Nepal in the 1950s, the Philippines and Papua New Guinea in the 1960s [6] and has spread to Australia in 1994 [7] and most of the Micronesian islands in 2000 [8]. It was probably introduced into Nigeria in 1937 from Sri Lanka [9] and South Africa in the late 1940s from Jamaica [10]. Two biotypes occur in its invasive range i.e. the widespread Asian –West African biotype and another that is restricted in distribution to southern Africa which is more prone to fire-induced mortality [11].

In invaded countries, *Chromolaena* has become a major invasive weed of natural and disturbed ecosystem, from grassland, savanna, forest margin, river bank to abandoned field, disturbed forest, annual/perennial and plantation crops [12]. It can form a dense thicket where it is established that shades out neighbouring other plants. The aggressive nature of the weed in the Old World is mainly caused by lack of a host-specific natural enemy, both arthropods [13] or diseases [14] Several attributes that contribute to its success as invader are huge production and easily dispersal of persistent seeds, rapid growth rates and branching habit, resistance to harsh environmental conditions, its high inflammability and allelopathic traits [12].

The plant can spread through sexual and asexual means but the main spread method is by sexual propagation. Sexual propagation begins when the plant forms flower at 1 year old that increases until the stand reaches 10 years old [15]. The flower is capable of producing large quantities of seed with estimated from 93 000 [16] to 1 600 000 [17] seeds per plant. The light, narrow and hairy seeds are easily dispersed by wind over a short distance and can spread over a long distance by attaching to clothing, fur of animals, machinery etc. but the primary long distance vector responsible for its spread is human activity. In invaded areas, some of these seeds can persist for several years [18].

In the Old World, the arrival and continuous spread of this weed have posed a serious threat for agriculture, natural ecosystem, livelihood and causing economic losses that need to be controlled. Many research results about the threat and control of *Chromolaena* have been reported in the literature. The aims of this paper are to present the threats caused by this weed to the ecosystem and to review the relevant literature on the management practices that can be used to control this weed in natural and agriculture ecosystem.

Threats

Once established, *Chromolaena* grows rapidly. It has extremely fast growth rate (up to 20 mm/day) and often forms a dense thicket of the height of 1.5–2.5 m in open areas and up to 10 m among the trees that can shade out the existing vegetation Table 14.1 above shows some threats of this plant in natural and disturbed ecosystems.

Management

Prevention and control

Prevention

Like other weed management practices, prevention is better and more cost-effective than control. Preventing the spread of *Chromolaena* into new areas will reduce future problems. The spread of seeds mostly occurs due to increasing trade, travel and other accidental introduction. Because the probability of introducing the weed from infested to other non-infested countries is high, a vigilant approach to surveillance and quarantine is required. If prevention is no longer possible, early detection and eradication are important to prevent the spread of the weed.

Controls

1. *Cultural control.* Cultural control using competitive cover plant species can be used to control *Chromolaena*. In The Philippines, planting of *Leucaena leucocephala* has been reported to reduce the weed population in pasture [30], but it is doubtful theweed could be controlled

Table 14.1. Threats from *Chromolaena odorata.*

No.	Threats	Authors
1.	Dense thicket of the weed hinders the movement of machinery, human, livestock and other large mammals.	[19]
2.	Shades out and reducing the growth of forage, preventing the establishment of other plant species, reducing biodiversity conservation and altering the integrity of natural plant community.	[12]
3.	Reduces carrying capacity of grassland for livestock.	[20] and [21]
4.	Can completely close down a ranch	[21]
5.	Reducing the population of banteng (*Bos javanicus*) and deer (*Cervus timorensis*) as caused by the severe reduction of herbage yield.	[22]
6.	Reducing ecotourism potential and game farming areas because of decreasing game population and visibility.	[19]
7.	Threatening the endangered black rhinoceros (*Diceros bicornis*) resulting from reducing its natural feed supply.	[19]
8.	The weed could poison the livestock because of pyrrolizidine and nitrate toxicity.	[23] and [24]
9.	Threatening the continued survival of Nile crocodile in Greater St.Lucia Wetland Park, as a nest in shaded by *Chromolaena* produced a female-biased sex ratio.	[25]
10.	Reduces the area that can be effectively managed by farmers, particularly in small-scale agricultural land.	[26]
11.	Reduces crop yield and increasing the time spent on weeding farms and cost of control.	[26]
12.	The plant harbours grasshopper, *Zonocerus variegatus* that breed in the plant and from there, it moves to cassava, feeding on the leaves and causing yield losses.	[27]
13.	The weed harbours wild pig and snake	[28]
14.	Reducing water run-off catchment as the greater and higher rates of transpiration of this plant compared to indigenous vegetation.	[19]
15.	During the dry season, the plant dries out and may become fuel source that promotes bushfires	[7]
16.	*Chromolaena* Infestation in the early growth stage of oil palm delayed fruiting for 2 years	[29]

successfully if *Leucaena* seeds are oversown in the established *Chromolaena*, as the high growth rates of the *weed*. Planting of legume cover crops such as *Calopogonium muconoides* and *Centrosema pubescens* and grass of *Setaria* sp. immediately after burning could reduce the weed infestation [31]. Komolafe [32] recommended planting of *Pueraria phaseloides*, *Calopogonium muconoides* and *Vigna unguiculata* as cover crops for *Chromolaena* control in tree crop plantation in Nigeria and *Tephrosia purpurea* in Sri Lanka [33]. Wu and Xu [34] reported that burning of the weed followed by planting with signal grass (*Brachiaria decumbens*) effectively suppressed the weed resulting in complete eradication after 3 years. Elephant grass (*Pennisetum purpureum*) might be suitable to be used to control the weed. Observation by the author in Maiwa ranch, Indonesia, showed that after cultivation of the heavily infested areas and planting with elephant grass, the growth of *Chromolaena* was severely reduced, provided the grass is leniently defoliated. The use of cover crops is very beneficial in controlling the weed, as the weed is not tolerant to heavy shade [35].

The slash and burn method of control is commonly practiced in subsistence farming in Asia and Africa. Controlled burning can be the best method of control [21]. Besides killing seeds of the weed on the soil surface, fire was reported to be very effective in eliminating the small and medium-size weed plants [36]. Burning provided a degree of control in northern Queensland, killing a large proportion of the weed of less than 1 m in height with no evidence of seedling recruitment 5 weeks after burning [37]. For long-term control, however, burning should be followed by planting with vegetation cover. Rusdy *et al.* [38] reported that burning of the weed in infested areas without followed by the planting of vegetation cover or mulch application resulted in an infestation of hard seed weed of *Mimosa pudica*. In contrast to small-l or medium-sized plants, burning of large-sized plants can promote the spread of the weed. During the dry season, the dried leaves and stems of this weed are readily burnt, but the stumps and roots remain alive and they rapidly regrow and can cover the area in the succeeding rainy season [39]. The burning of mature plant reduces the standing biomass, but regrowth can occur from remaining stumps, generally more profusely. The stump is the prominent plant part and is the strong sink for nutrients and although the leaves and stems dried up, the stump still has the capability of sprouting [40]. Devendra *et al.* [41] stated that a combination of clearing followed by fire was very effective in reducing densities of *Chromolaena* and much more effective than the single treatment. The clearing followed by fire treatment, however, caused high-intensity canopy fire that severely impacted the native vegetation and changing the system from woody-dominated stage to a grassland-dominated stage. Hauser and Mekoa [42] reported that in a plantain field established secondary forest, biomass productions of *Chromolaena* were larger when biomass of *Chromolaena* and all other plants were burned than when all the biomasses were retained on the soil surface as mulch.

2. *Mechanical control.* Mechanical control has been the predominant method for control of the weed in developing countries. Hand weeding by slashing, uprooting and digging using appropriate tools are the common method of control practiced by the smallholder farmers while the use of tractor-drawn equipment is commonly found in plantation areas.

Slashing reduces the standing weed biomass, but it is not a sustainable solution because the weed can coppice profusely from the remaining stumps. To kill the weed, the remaining stumps and roots must be uprooted or dugout; if not, the plant has to be slashed repeatedly at short time intervals to deplete carbohydrate reserves in the stump and roots. In this case, Are and Folarin [40] recommended slashing of *Chromolaena* four times a year to get effective results in a cocoa plantation. Rusdy *et al* [38] also reported that slashing alone was not effective to control the weed, as much seed remains in the soil and regrowth is common. The most effective means of control for small infestation is to remove the entire plant including roots by uprooting or digging, but follow up clearance is needed because of rapid infestation from the seeds. By establishing an alternative plant cover [43] or mulch application on the bare ground created by the removal of *Chromolaena*, it could increase the efficacy of control substantially. Hand removal of the weed is an effective and environmentally sound control method, especially in the presence of endangered flora or in fragile terrain with small infestations [37]. For larger infestations, the best method is to use machinery such as motorized brush cutter or tractor-drawn mower to remove the top growth of the weed.

The negative effect of uprooting and digging is that they make the soil open to the sun, which promote the exposure of the seeds in the soil to light and with the availability of water, this facilitates seed germination and creates ideal sites for the establishment of the weed. Mechanical control by hand weeding is labour-intensive, slow and costly for large areas and can be effective if the infestation is small and accessible. A combination of other methods such as cultural or chemical control may be more economical and effective. Digging out of the weed, followed by burning and planting with *Brachiaria decumbens* is the most promising method to control the weed in grassland area [38]. Mechanical control followed by cultural or chemical control may be more effective to control the weed than mechanical control only, although this is expensive and not sustainable.

3. *Chemical control. Chromolaena* is most susceptible when herbicides are applied at the

seedling stage or when it is growing vigorously, especially after slashing or shading [44]. Several herbicides applied at the seedling stage or early growth had produced effective results. Soil-applied herbicides such as diuron at the dose of 1.5 kg/ha and atrazine at 2.0 kg/ha were effective for controlling germination and establishment of the weed [45]. Other herbicides such as glyphosate, imazapyr, triclopyr, 2,4-D, fluroxypir, tebuthiuron, picloram, metsulfuron etc. can be used to control the weed. In Nigeria, Ikuenobe and Ayemi [46] recommended glyphosate at 2.4 kg a.i/ha and low rates (0.5 kg a.i/ha or less) of imazapyr. In India, Kumar-Naik *et al.* [47] reported that application of glyphosate at 1.2 to 1.6 a.i/ha and paraquat at 0.4–1.0 kg a.i/ha caused complete drying by 40th and 10th after spraying, respectively. Ikuenobe and Ayemi [46] also reported that triclopyr and hexazinone + diuron were effective in controlling the weed and suppressed it up to 20 weeks after treatment. In South Africa, the use of triclopyr, 2,4,5-T and tebuthiuron were consistently effective, however triclopyr was the most effective [48, 49]. Triclopyr also was more effective than glyphosate and an acceptable level of control could be obtained with a dose of 2.5 dm^3/ha of triclopyr and 3.5–4.5 dm^3/ha of glyphosate [49] and both herbicides were effective when sprayed on normal than slashed *Chromolaena* [36]. Further, Rusdy [50] reported that in grassland area, triclopyr was more suitable to be used to control the weed than glyphosate because triclopyr selectively kill broad-leaves plants such as *Chromolaena* but did not kill forage such as grasses. In Indonesia, Tjitrosemito [22] reported that triclopyr was effective against *Chromolaena*, besides 2,4-D and picloram. Picloram at 1 kg/ha as the recommended rate to control the weed [51].

Besides triclopyr, fluroxypyr is also effective in killing the weed. Brooks *et al.* [52] reported that fluroxypyr and triclopyr-based herbicides controlled 96 and 100% of *Chromolaena*, respectively, metsulfuron-methyl and glyphosate-based herbicides controlled 40 and 57% of the plants respectively, while only 3% of untreated control plants were dead, 12 months after treatment. This also is in line with Vitelly *et al.* [37] that of the foliar herbicides tested, fluroxypyr (70 g/100 L) provided 99% mortality, triclopyr/picloram (103/35/100 L) gave 95% while metsulfuron-methyl (9 g/100 L) killed greater 98% of the weed when applied as basal bark or cut stump. In Thailand, Lueang-a-papong [53] reported that fluroxypyr and triclopyr gave a good control on *Chromolaena*, but fluroxypyr showed a faster and better effect than triclopyr.

The efficacies of triclopyr and glyphosate were different when they are mixed with asulam. A 95% reduction of regrowth of the weed was achieved using a mixture of triclopyr and asulam while 80% reduction of regrowth was achieved using glyphosate alone; however, glyphosate applied alone was more effective than when it was mixed with asulam. A repeat application by the 12 weeks after treatment was necessary to completely kill the weed [54].

Although many herbicides have been reported to give effective results, the recommended herbicides generally failed to give long-term control, as re-infestation of the weed invariably occurs. Also, those herbicides require a repeated application with high dosages, obtaining a long-lasting effect. Therefore, most herbicide-based control is not cost-effective under prevailing economic conditions in developing countries. In addition, it poses some environmental problems. The large biomass yield and extensive root system of *Chromolaena* make it fairly resistant to the herbicidal application [46].

4. *Biological control.* There are many natural enemies of the weed. McFadyen [55] identified more than 200 species of arthropod and two mite species that attack *Chromolaena* in tropical America. Of these, about a quarter is specific to *Chromolaena* [56] and only several of these agents have shown good results in controlling the weed in the Old World countries. Until recently, of the several species tested, the moth *Pareuchaetes psudoinsulata* and gall fly *Cecidochares connexa* were the most widely released and established of agents on *Chromolaena* [7].

Pareuchaetes pseudoinsulata was the first agent deliberately introduced from Trinidad to many other countries. Now, this agent is widespread and locally abundant in many countries. In the early 1980s, It was reported to be established in Sri Lanka [57] and in the mid-1980s, it had established in Sabah, Malaysia but did not cause much damage to the weed [58]. Introduction after 1980s resulted in establishment In Guam and other Pacific islands, Indonesia and Ghana [39]. In Indonesia, it has been established successfully in North Sumatra, but in Java and other Indonesian islands, it has not been reported to establish successfully [59]. In Guam, Rota, Tinian and Saipan, 1 year following the establishment of the moth, the weed was reduced by 100% at three of four sites of *Chromolaena* stand monitored [60]. In Ghana, reduction of competitiveness of *Chromolaena* by *P. pseudoinsulata* positively enhanced forest regeneration and biodiversity [61]. However, Uyi *et al.* [62] reported that in southern Ghana, the damage caused to *Chromolaena* in the field by *Pareuchaetes pseudoinsulata* was insignificant to impact negatively on the weed.

Attempts to introduce and establish *P. pseudoinsulata* in several countries such as Nigeria, South Africa, Ivory Coast, India and Thailand failed [63]. In Nigeria, introduction and release of two agents, *P. pseudoinsulata* and *Apion brunneonigrum* failed to establish, probably because predation and biocontrol efforts were not sustainable [64]. In South Africa, many agents have been released into the field to control the weed. *Pareuchaetes pseudoinsulata* had been mass reared and released into the field but they were disappeared after the first winter and no insects were found subsequently [65]. Another closely related agent, *Pareuchaetes insulata* was released against the weed between 2001 and 2009 but its population level in the fields is low. A closely related agent, *Pareuchaetes aurata aurata* also failed to establish despite several years of releases. The agent *Calycomiza eupatorivora* had also established in South Africa but its impact on *Chromolaena* population was not sufficient to effect adequate suppression of the weed [66]. Although many failures were found, the research on a wide range of insects and pathogen which have potential as a biological control agent against *Chromolaena* in South Africa is still ongoing. Initial releases of stem-boring weevil of *Lixus aemulus* have been made, and host-range testing of the shoot-tip borer *Dichrorampha odorata* indicates that it is suitable for release [67]. A number of possible factors such as predatory pressures, weather pattern insect-induced mechanism by plants [49] may explain why the establishment of *P. pseudoinsulata* is low or erratic.

Another species, *Cecidochares connexa* was imported to Sumatra, Indonesia from Columbia in 1994 [68]. Now, the gall fly has established on most larger Indonesian islands and has significantly reduced the *Chromolaena* population [59], but in South Sulawesi, Indonesia, although it has established, however, the weed is continuously spreading. Introduction of *Cecidochares connexa* from Papua New Guinea to Timor Leste during 2005–2009 also failed to give significant long-lasting results. Control of *Chromolaena* by the gall fly in this country is limited by severe dry season and the widespread use of fire in clearing lands for agriculture [69].

A high degree of efficacy of *Cecidochaeres connexa* has been reported in Papua New Guinea and several Pacific islands. In Papua New Guinea, *Cecidochares connexa* was more effective than *Pareuchaetes pseudoinsulata*. One year after introduction, the gall fly spreads rapidly and exerted control in many parts of the country. *Pareuchates pseudoinsulata*, however, was more limited in its distribution and impact [70]. In Guam and other Micronesian islands, the introduction of the gall fly between 2009 and 2010 resulted in a steady decline in a number of stems, leaves and the height of the weed [71]. The gall fly had also been introduced into Ivory Coast, West Africa In 2003. Post-evaluation conducted 12–13 years later indicates that the gall fly had spread to Ghana and Togo but its density is still low [72].

The advantages of applying biological control are that it is environmentally friendly, agents used are host-specific, self-perpetuating and can be cost-effective in the long term. The disadvantages are slow process, failure to eliminate the weed and it can be fickle.

Conclusion

The huge amount of light and persistent seeds, fast-growing, branching habit of the weed and absence of a natural enemy make *Chromolaena* become a successful invader plant in the Old World. Invasion of this plant poses many threats to agriculture, natural preservation, biodiversity and livelihood. The best management for its spread is prevention, but it is very difficult to apply. Mechanical control is labour-intensive and only suitable in countries where much cheap labour is available. Chemical control also is not cost effective because it fails to give long-lasting control and may be only suitable applied in commercial plantation. Biological control that expected to become the cheapest method of control and resulting in long-lasting effects, its effectiveness is limited in certain areas. Control of this weed is difficult and further research is needed for getting effective, cheap and environmentally safe methods of control.

Acknowledgements

The author thankful to Rector of Hasanuddin University, Indonesia for her readiness to offer some reward to author for the success to publish the article in reputed journal.

References

1. Richardson DM, Cambray CA, Chapman RA, Dean WRJ, Griffith CL, Le Maitre DC, *et al.* Vectors and pathway of biological invasion in South Africa: past, present and future. In: Ruiz G, Carlton J, editors. Invasive Species, Vectors and Management Strategies. Island Press, Washington, D.C.; 2003. p. 294–349.
2. Andreu J, Vila M, Hulme PE. An assessment of stakeholder perception and management of noxious alien plant in Spain Environmental Management. Environment Management 2009;43(6):1244–55.
3. Anonymous. *Chromolaena odorata*. Available from: URL: http://www.arc.agric.za/arc-ppri/Pages/Weeds%20Research/Chromolaena/What-is-Chromolaena.aspx (last accessed 25 June 2018).
4. Grierson AJC. Compositae. In: Dassaanaayake MD, Forsberg FR, editors. A Revised Handbook of the Flora of Ceylon, Volume 1, A.A. Balkema, Rotterdam, The Netherlands; 1980. p. 111–278.
5. McFadyen REC. *Chromolaena* in Asia and the Pacific: spread continues but control prospects improve. In: Zachariades C, Muniappan R, Strathie LW, editors. Proceedings of the Fifth International Workshop on Biological Control and Management of *Chromolaena odorata*. ARC-PPRI, Pretoria, South Africa; 2002. p. 13–18.
6. Henty EE, Pritchard PH. Weeds of New Guinea and Their Control. Botany Bulletin, 7. Department of Forests, Lae, Papua New Guinea; 1973.
7. Muniappan R, Reddy GVP, Lai PY. Distribution and biological control of *Chromolaena odorata*. In: Inderjit S, editor. Invasive Plants: Ecological and Agricultural Aspects. Birkhauser Verlag, Basel, Switzerland; 2005. p. 223–33.
8. Waterhouse BM. Discovery of *Chromolaena odorata* in northern Queensland, Australia. *Chromolaena odorata* Newsletter 1994;9:1–2.
9. Ivens GW. The problems of *Eupatorium odoratum* L. in Nigeria. Pest Articles New Summaries 1974;20:76–82.
10. Zachariades C, von Senger L and Barker NP. Evidence for northern Caribbean origin for the Southern African biotype of *Chromolaena odorata*. In: Day MD, McFadyen RE, editors. Proceedings of the Sixth International Workshop on Biological Control and Management of *Chromolaena odorata*. ACIAR Technical Report 55, ACIAR, Canberra, Australia; 2004. p. 25–7.
11. Zachariades C, Day MD, Muniappan R, Reddy GVP. *Chromolaena odorata* (L.) king and robinson (*Asteraceae*). In: Muniappan R, Reddy GVP, Ramans A, editors. Biological Control of Tropical Weeds Using Arthropods. Cambridge University Press, Cambridge, UK; 2009. p. 130–62.
12. McFadyen REC. *Chromlaena* in Asia and pacific: spread continues but control prospects improve. In: Zachariades C, Muniappan R, Strathie L, editors. Proceedings of the Fifth International Workshop on Biological Control and Management of *Chromolaena odorata*, Durban, October, ARC-PPRI, Pretoria, South Africa; 2000. p. 13–8.
13. Gagne RJ. The *Cecidomiidae* (diptera) associated with *Chromolaena odorata* (L.) King & Robinson (*Compositae*) in the Neotropical Region. Brenesia 1997;12–13:113–31.
14. Barreto RW, Evans HC. The mycobiota of the weed *Chromolaena odorata* in southern Brazil with particular reference to fungal pathogens for biological control. Mycological Research 1994;98(10):1107–16f.

15. Witkowski ETF, Wilson M. Changes in density, biomass, seed production and soil seed banks of the non-native invasive plant, *Chromolaena odorata*, along a 15 year chronosequence. Plant Ecology 2001;152(1):13–27.
16. Weerakoon L. Studies on Biology and Control of *Eupatorium odoratum* (L.) [MSc. thesis]. Vidiodaya Campus, University of Ceylon; 1972.
17. Wilson M. Autecology of invasive alien plant, *Chromolaena odorata* in the Greater St. Lucia Wetland Park [M.Sc. thesis]. University of Witwatersrand, Johannesburg; 1995.
18. Witkowski ETF. Invasion intensity and regeneration potential of non-native invasive plant *Chromolaena odorata* at St. Lucia, South Africa. In: Zachariades C, Muniappan C, Strathie LW, editors. Proceedings of the Fifth International Workshop on Biological Control and Management of *Chromolaena odorata*. Durban, South Africa; 2002, (23–25 October 2000).
19. Goodal J, Zacharias PJK. Managing *Chromolaena odorata* in subtropical grassland in Kwa Zulu-Natal, South Africa. In: Zachariades C, Muniappan R, Strathie LW, editors. Proceedings of the Fifth International Workshop in Biological Control and Management of *Chromolaena odorata*, ARC – PPRI, Durban, South Africa, October 2000, pp. 120–9.
20. Zachariades C, Goodal CM. Distribution, impact and management of *Chromolaena odorata* in Southern Africa. In: Zachariades C, Muniappan R, Strathie LW, editors. Proceedings of the Fifth International Workshop on Biological Control of *Chromolaena odorata*. Durban, South Africa, October; 2002. p. 34–9.
21. Ossom E, Lupupa B, Mhlongo S, Khumalo L. Implication of weed control method on Sandanewze (*Chromolaena odorata*) in Swaziland. World Journal of Agricultural Science 2007;3(6):704–13.
22. Tjitrosemito S. Integrated management of *Chromolaena odorata* emphasizing the classical biological control. Biotropia 1998;11:9–21.
23. Pancho JV, Plucknett DL. *Chromolaena odorata* (L.) King and H. Robinson, A new record of noxious weed in the Philippines. Philippine Journal of Animal Science 1971;8:143–9.
24. Sajise DE, Palls PK, Lales NV, Lales JV. Flowering behavior, pattern of growth and nitrate metabolism of *Chromolaena odorata*. Philippine Weed Science Bulletin 1974;1:17–24.
25. Leslie AJ, Spotila JR. Alien plant threaten Nile crocodile (*Crocodylus niloticus*) breeding in Lake St. Lucia, South Africa. Biological Conservation 2001;98(3):347–55.
26. Luwum P. Control of invasive *Chromolaena odorata*. Thesis submitted to the International Institute for Geo-information Science and Earth. Enschede, The Netherlands; 2002. Available from: URL: https://webapps.itc.utwente.nl/librarywww/papers_2003/msc/nrm/luwum.pdf (last accessed 12 May 2018).
27. Boopre MA. A non-nutritional relationship of *Zonocerus* (*orthopera*) to *Chromolaena* (*asteraceae*) and general implications for weed management. Ecology of *Chromolaena odorata*. Biotrop Special Publication 1991;44:153–61.
28. Day MD, Bofeng I, Nabo I. Successful biological control of *Chromolaena odorata* (*Asteraceae*) by the gall fly *Cecidochares connoxa* (Diptera. *Tephritidae*) in Papua New Guinea. XIII International Symposium on Biological Control of Weeds; 2011. Available from: URL: https://www.invasive.org/proceedings/pdfs/Day%20-%20Bofeng.pdf (last accessed 24 June 2018).
29. Sheldrick RD. The control of Siam weed in Nigeria. Journal of the Nigerian Institute for Oil Palm Research 1968;5(17):7–19.
30. Castilo AC, Moog FA, Pineda C. Introduction of ipil-ipil in 'hagony' infested pastures. Philippine Journal of Animal Industry 1977;32:1–10.
31. Torres DO, Paller EC. The devil weed (*Chromolaena odorata* (L.) R. King and H. Robinson) and its management. SEAWIC Weed Leaflet 1989;4:1–5.
32. Komolafe DA. Weeds and their control in perennial plantation crops in Nigeria. Weeds and their control in the humid and subhumid, tropics. Proceedings of Weeds Conference at the International Institute of Tropical Agriculture, October 1978; 1978. pp. 344–9.
33. Salgado MLM. *Tephrosia purpurea* (Pila) for the control of *Eupatotium* and as green manure in coconut estates. Ceylon Coconut Planters' Review 1972;6(4):160–74.
34. Wu R, Xu X. Cultural control of feijicao (*Chromolaena odorata* (L.) R. King and H. Robinson) by planting signal grass (*Brachiaria decumbens* Stapf) in southern China. J. Biotropia Special Publication 1991;44:83–9.
35. Zhang LL, Wen DZ. Structural and physiological response of two invasive weeds, *Mikania micrantha* and *Chromolaena odorata*, to contrasting light and soil water conditions. Journal of Plant Research 2009;122(1):69–79.

36. Wessels MF. The role of fire and mechanical clearing in the management of *Chromolaena odorata* [MSc. Thesis]. School of Biological and Conservation Sciences, Faculty of Science and Agriculture, University of Kwazulu NatalSouth Africa; 2014.
37. Vitelli JS, Panetta FD, Madigan BA, Van Haaren PEV. Chemical option for the control of Siam Weed (*Chromolaena odorata*) in natural ecosystem. Ecological Management and Restoration 2018;19(1):88–93.
38. Rusdy M, Sjahril R, Riadi M, Budiman. Integration of mechanical and cultural control to manage invasive shrub *Chromolaena odorata* and other weeds under drought conditions in pasture area. Indonesian Journal of Tropical Animal Agriculture 2013;38(1):65–71.
39. Muniappan R, Bamba J. Biological control of *Chromolaena odorata*: successes and failures. In: Spencer NR, editor. Proceedings of the X International Symposium on Biological Control of Weeds 4–14 July 1999, Montana State University, Bozeman, Montana, USA; 2000. p. 81–5.
40. Are LA, Folarin JO. Cultural weed control experiment. Annual Report of Cocoa Research Institute of Nigeria 1970;19:68–89.
41. Devendra R, Mukeshi L, Chavan, Prasad TVR. Growth pattern, sprouting ability and chemical control of *Chromolaena odorata*; 2018. Available from: URL: https:www.academia.edu/7049857/Growth Pattern Sprouting Ability and Chemical control of Chromolaena odorata?auto=download.
42. Hauser S, Mekoa C. Biomass production and nutrient uptake of *Chromolaena odorata* as compared with other weeds in a burned and a mulched secondary forest clearing planted to plantain. Weed Research 2009;49(2):193–260.
43. Gils HV, Delfno J, Rugege D and Janssen L. Efficacy of *Chromolaena odorata* control in a South African conservation forest. South African Journal of Science 2004;100:251–53.
44. Erasmus DJ. A review of mechanical and chemical control of *Chromolaena odorata* in South Africa. Proc. First International Workshop on Biological Control of *Chromolaena*, Bangkok, pp. 34–40. Ag.Exp.Sta. U. Guam, Mangilao; 1988.
45. Abraham CT, Thomas CG, Joseph PA. Herbicides for control of *Chromolaena odorata*. Kerala Agricultural University, Trichur – 2008, 680 654, India.
46. Ikuenobe CE, Ayeni AO. Herbicidal control of *Chromolaena odorata* in oil palm. Weed Research 1998;38:397–404.
47. Kumar-Naik AH, Ramachandra Prasad TV and Basavaraja Naik T. Mortality of Siam weed – *Chromolaena odorata* (L.) R.M. King and H. Robinson. IOSR Journal of Agriculture and Veterinary Science 2016;9(7):11–3.
48. Erasmus DJ, Van Staden J. Screening for candidate herbicides in field trials for chemical control of *Chromolaena odorata*. South African Journal of Plant and Soil 1986;3(2):66–7.
49. Van Staden J. Chemical control of *Chromolaena odorata*: efficacy of triclopyr and glyphosate applied to regrowth. Applied Plant Science 1987;1(1):39–42.
50. Rusdy M. Intergating mechanical and chemical control treatments to manage invasive weed *Chromolaena odorata* (L.) R.M. King and H. Robinson in grassland area. American Journal of Experimental Agriculture 2015;6(3):133–9.
51. Madrid Jr MT. Evaluation of herbicides for the control of *Chromolaena* (l.) King and H. Robinson. Philippine Weed Science Bulletin 1974;1:25–9.
52. Brooks SJ, Gough KL, Campbell SD. Testing the efficacy of low volume herbicide applications on *Chromolaena odorata*. Pakistan Journal of Weed Science Research 2012;18:61–71.
53. Lueang-a-papong. Chemical contro; of siam weed. Available from: URL: http://agris.fao.org/agris-search/do? recordID=TH912035 (last accessed 25 August 2018).
54. Utulu SN. Controlling regrowth of *Chromolaena odorata* (L.) RM King and H. Robinson using herbicide mixtures in young oil palm plantation in Nigeria. In: Prasad UK, Munippan R, Ferrar P, Aeschliman JP, De Foresta H, editors. Proceedings of the Third International Workshop on Biological Control and Management of *Chromolaena odorata*, Abidjan, Cote d'Ivoire, November 1993. Agricultural Experimental Station, University of Guam Publication; 1996. p. 148–56.
55. McFadyen REC. Phytophagous insects recorded from *C. odorata*. *Chromolaena odorata* Newsletter 1988;2:5–23.
56. Waterhouse DF. Biological Control of Weeds: South Asian Prospects. Australian Centre for International Agricultural Research, Canberra; 1994, 302pp.
57. Dharmadhikari PR, Perera PACR, Hassen TMF. The introduction of *Ammalo insulate* for the control of *Eupatorium odoratum* in Sri Lanka. Commonwealth Institute of Biological Control. Technical Bulletin 1977;18:129–35.

58. Crutwell McFadyen RE. *Chromolaena* in Asia and the pacific: spread continues but control prospects improve. In: Zachariades C, Muniappan R, Strathie LW, editors. Proceedings of the Fifth International Workshop on Biological Control and Management of Chromolaena odorata. 23–25 October 2000. APC – PPRI, Durban, South Africa; 2000. p. 13–8.
59. Wilson CG, Eko Budi W. The biological control programme against *Chromolaena odorata* in eastern Indonesia. In: Zacharia-des C, Muniappan R, Strathie LW, editors. Proceedings of the Fifth International Workshop on Biological Control and Management of *Chromolaena odorata*, Durban, South Africa, October 2000, Teeanem, Pietermaritzburg; 2002. p. 34–9.
60. Siebert TF. Biological control of the weed, *Chromolaena odorata* (Asteraceae) by *Pareuchaetes pseudoinsulata* (Lepidoptera: Arctiidae) on Guam and Northern Mariana Islands. Entomophaga 1989;34:531–9.
61. Timbilla JA, Braimah H. Establishment and impact of *Pareuchaetes pseudoinsulata* (*Lepidoptera: Arctiidae*) an exotic predator of the Siam weed, *Chromolaena odorata* (*Asteraceae: Eupatoriae*) in Ghana. In: Spencer NR, editor. Proceedings of the X International Symposium on Biological Control of Weeds, 4–14 July 1999. Montana State University, Bozeman, Montana, USA; 2000. p. 105–11.
62. Uyi UO, Wilson DD and Zachariades C. The impact of *Pareuchaetes pseudoinsulata* (Lepidoptera, Arctiidae) on the growth rate of *Chromolaena odorata* (L.) in southern Ghana. Journal of Applied Science and Environmental Management 2009;13(1):59–65.
63. McFadyen RE. Parasitoids of the arctiid moth *Pareuchaetes pseudoinsulata* (*Lep.: Arctiidae*), an introduced biocontrol agent against the weed *Chromolaena odorata* (Asteraceae), in Asia and Africa. Entomophaga 1997;42(4):467–70.
64. Uyi OO, Ekhator F, Ikuenobe CE, Borokini TI, Aigbokhan EI, Egbon IN, *et al. Chromolaena* invasion in Nigeria: A case for coordinated biological control. Management of Biological Invasion 2014;5(4):377–93.
65. Strathie LWand Zachariades C. Biological control of *Chromolaena odorata* in South Africa: development in research and implementation. In: Zachariades C, Muniappan R, Strathie LW, editors. Proceedings of the Fifth International Workshop on Biological Control and Management of *Chromolaena odorata*, ARC-PPRI, Pretoria, South Africa; 2002. p. 74–9.
66. Uyi OO, Zachariades C, Hill MP. The life history traits of the Arctine moth *Pareuchaetes insulata*, a biological control agent of *Chromolaena odorata* in South Africa. African Entomology 2014;22(3): 611–24.
67. Zachariades C, Strathie LW, Retief E. And Dube N. Progress towards biological control of *Chromolaena odorata* (L.) R.M. King and H.Rob. (Asteraceae) in South Africa. African Entomology 2011;(19):282–302.
68. Desmier de Chenon R, Sipayung A, Sudharto P. A decade of biological control against *Chromolaena odorata* at the Indonesian oil palm, research institute in marihat, pp. 46–52. In: Zachariades C, Muniappan R, Strathie LW, editors. Proc. Fifth International Workshop on Biological Control and Management of *Chromolaena odorata*. 23–25 October 2000. APC – PPRI., Durban, South Africa; 2002. p. 13–8.
69. Day MD, Brito AA, Gutteres ADA, Alves APCC, Paul T, Wilson CG. Biocontrol of *Chromolaena odorata in* timor leste. In: Zachariades C, Strathie LW, Day MD, Muniappan R, editors. Proceedings of the Eight International Workshop on Biological Control of *Chromolaena odorata* and Other *Eupatorieae*, Nairobi, Kenya, 1–2 November 2010. ARC-PPRI, Pretoria; 2013. p. 134–40.
70. Day MD, Bofeng I, Nabo I. Successful biological control of *Chromolaena odorata* (*Asteraceae*) by the gall fly *Cecidochares connexa* (Diptera: Tephritidae) in Papua New Guinea. XIII International Symposium on Biological Control of Weeds; 2011. Available from: URL: http://www.invasiveorg/proceedings/pdfs/Day20-%20Bofeng.pdf (last accessed 18 May 2018).
71. Reddy GVP, Kikuchi RS, Muniappan R. The impact of *Cecidochares connexa* on *Chromolaena odorata* in Guam. In: Zachariades C, Strathie LW, Day MD, Muniappan R, editors. Proceedings of the Eighth International Workshop on Biological Controland Management of *Chromolaena Odorataand* Other Eupatorieae, Nairobi, Kenya, 1–2 November 2010. ARC-PPRI, Pretoria; 2013. p. 128–33.
72. Aigbedion-Atalor PO, Wilson DD, Eziah VY, Paterson ID. The distribution and abundance of the stem-galling fly *Cecidochares connexa* (Macquart) (Diptera: Tephritidae), a biological control agent of *Chromolaena odorata* (L.) (Asteraceae) in Ghana. African Entomology 2018;26(2):471–80.

15 Lymnaeid snails hosts of *Fasciola hepatica* and *Fasciola gigantica* (Trematoda: Digenea): a worldwide review

Antonio A. Vázquez[1,2], Pilar Alda[1,3], Manon Lounnas[1], Emeline Sabourin[1,4], Annia Alba[2], Jean-Pierre Pointier[5] and Sylvie Hurtrez-Boussès[1,6*]

[1] MIVEGEC, University of Montpellier, CNRS, IRD, Montpellier, France. [2] Laboratory of Malacology, Institute of Tropical Medicine Pedro Kourí, Autopista Novia del Mediodía km 6, La Habana, Cuba. [3] Laboratorio de Zoología de Invertebrados I, Departamento de Biología, Bioquímica y Farmacia, Universidad Nacional del Sur, San Juan No. 670, B8000ICN, Bahía Blanca, Buenos Aires, Argentina. [4] Institut de recherche de la Tour du Valat, Arles, France. [5] PSL Research University, USR 3278 CNRS–EPHE, CRIOBE Université de Perpignan, Perpignan, France. [6] Département de Biologie–Ecologie, Faculté des Sciences, Université Montpellier, Montpellier, France

Abstract

Fasciolosis is a snail-borne disease, causing serious public and veterinary health problems worldwide. This disease is produced by infection with *Fasciola hepatica* or *Fasciola gigantica* through the consumption of vegetables or water contaminated with the parasite's metacercarial cysts. Both species of liver flukes are transmitted worldwide by small freshwater snails of the family Lymnaeidae. A global account on the species that are actually or may act as potential hosts of *Fasciola* spp., compiling particular research on their geographical distribution and susceptibility, is needed as a helpful tool in the understanding of fasciolosis transmission, and therefore in the control and prevention programmes of the disease. We have gathered here a comprehensive review of those lymnaeid species that are known to transmit the parasites in the field or that have been experimentally tested. We aim to bring forward the main intermediate hosts by regions in order to facilitate the understanding of worldwide transmission.

Keywords: Lymnaeidae, *Fasciola hepatica*, *Fasciola gigantica*, Fasciolosis, Snail transmission

Review Methodology

We searched the following databases: Scopus, PubMed, Google Scholar, JSTOR, SciELO and SpringerLink. Most common keywords used in database search were: lymnaeid snails, *Fasciola hepatica*, *Fasciola gigantica*, natural prevalence, experimental infection, liver fluke transmission, fasciolosis. Relevant ancient and recent articles related to the snail species reported as hosts of *Fasciola hepatica* and/or *Fasciola gigantica* were included. We used articles written in English, French and Spanish. Up to 160 references were included from all continents. References of local studies that were included in larger scale studies were disregarded to avoid over citing. The names of species referenced in ancient studies are cited here using the latest nomenclature available for each species after recent molecular studies.

*Corresponding author: sylvie.hurtrez@ird.fr

Background

One of the key aspects in the epidemiology of infectious diseases is acquiring a comprehensive view of their potential extension and their means of transmission. It is particularly complicated when a pathogen includes several organisms to complete its life cycle, making it more difficult to manage and prevent. This is especially the case of vector-borne diseases. The understanding of these diseases necessarily involves a knowledge of their chains of transmission and vector species or hosts. Usually, intermediate hosts and vectors are ultimately the target of control programmes because they represent the weakest link in the transmission chain. To accomplish this goal, having an up-to-date understanding of the distribution and actual role of the species involved in transmission is fundamental to prevent and control vector-borne diseases.

Fasciolosis: A Snail-Borne Disease

Fasciolosis is a snail-borne disease causing serious public health problems worldwide, especially in low-income nations [1]. However, it remains as an unsolved problem in both developing and developed countries at veterinary level [2, 3]. Additionally, fasciolosis is considered as one of the most globally spread parasitic diseases, which extends widely across a large variety of habitats in different geographic regions [4]. The disease is mostly caused by the hermaphroditic parasites *Fasciola hepatica* Linnaeus, 1758 and *Fasciola gigantica* Cobbold, 1855 (Platyhelminthes: Trematoda: Digenea). Fasciolosis affects more than 17 million people with nearly 180 million at risk of infection worldwide [5].

However, according to the World Health Organization (WHO), fasciolosis remains a neglected tropical disease [6]. Furthermore, *Fasciola* species are placed in a particular group known as the most 'neglected' among the neglected worms where all the main food-borne trematode infections also stand [7].

Both *Fasciola* species are capable of infecting a very wide spectrum of mammalian definitive hosts, including humans, [8] where parasites reproduce sexually producing thousands of eggs at a daily basis (see Figure 15.1 for a detailed description of the life cycle). Noteworthy, while ruminants are commonly described as the most infected definitive hosts (for which the disease asserts the highest economic losses), several smaller mammals (rodents, lagomorphs, etc.) may have an important role as reservoirs and disseminators of the parasite in the wild [9].

Lymnaeid snails (Gastropoda: Lymnaeidae) act as the intermediate hosts of *F. hepatica* where asexual reproduction of the parasite occurs and a clonal expansion from a single infective larva increases the probability of transmission. The family Lymnaeidae Rafinesque, 1815 is believed to comprise nearly 100 species of small aquatic and amphibious snails [10]. However, about 1200 species have been described worldwide [11]. Lymnaeid snails have a right coiled shell and are known to extend from tropical to temperate regions; some of them occur at very cold latitudes [12–14]. In general, the great ecological plasticity of lymnaeid snails allows them to inhabit a large range of ecosystems which is actually one of the key aspects of the successful extend of fasciolosis.

The family of Lymnaeidae has been the target of several taxonomic revisions in the last two decades due to its importance in the transmission of fasciolosis [15–17]. Vinarski [17] presents the Lymnaeidae divided into two subfamilies, Lymnaeinae (16 genera) and Radicinae (ten genera). Some genera are known to be best suited for a particular species of *Fasciola*, where *F. hepatica* appears to be mainly transmitted by the group of *Galba* whereas *F. gigantica* is commonly related with the *Radix* group [15]. However, different snail habitat preferences (e.g. ponds, ditches, flooded crops), rather than phylogenetic affinity, may better explain the patterns of host specificity by *Fasciola* spp. in different regions [18]. Interestingly, not all lymnaeid species act as intermediate hosts of *Fasciola* spp., and among those that host the parasites, there is a strong variation in the potential of transmission. Indeed, different patterns of susceptibility, host–parasite compatibility and even immunological resistance have been described in this host–parasite system [19–21].

The knowledge of which particular snail species inhabits a risk area for fasciolosis transmission can be used to trigger control campaigns

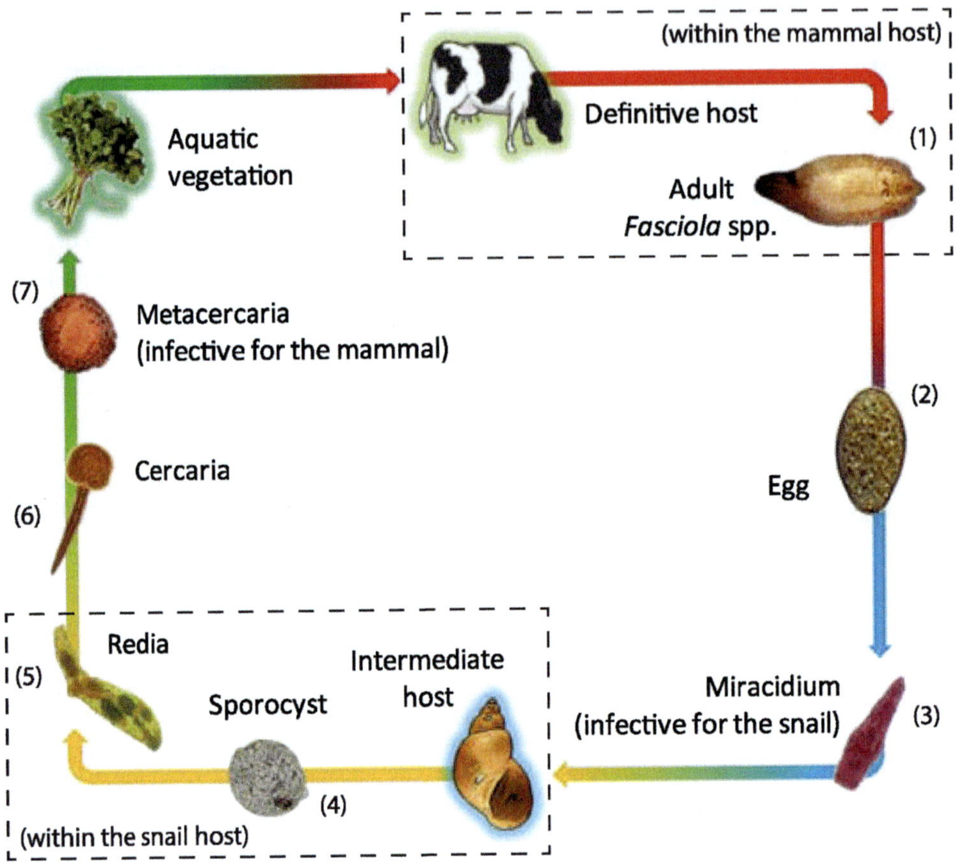

Fig. 15.1. Life cycle of *Fasciola* spp. Adult parasites live in the bile ducts of mammals where sexual reproduction takes place (1). Approximately 40,000 eggs per individual are laid daily and carried out with the host faeces (2). If temperature and humidity conditions are favourable they hatch into miracidia. These larvae must find a susceptible lymnaeid snail in less than 24 h to which they are steered by means of chemotaxis and phototaxis (3). Once in the snail, the miracidium turns into a sporocyst (4) and then produces rediae (5) which, in time, can give place to other generations of daughter rediae. Inside each redia, several cercariae can be produced. These stages are part of the asexual reproduction of the parasite subjected to the compatibility with their host and can last between 30 and 80 days depending on the ecological conditions. When cercariae are mature, they leave the snail, usually producing fatal damages (6). In the water, they swim and encyst forming metacercariae in aquatic or semi-aquatic vegetation or any source of solid substratum (7). Floating metacercariae have also been reported and are a source of infection by contaminated water. Susceptible mammals may close the parasite cycle by eating vegetation with encysted metacercariae. When the parasite reaches the intestine, the juveniles excyst and migrate to the liver where they achieve the adult form. Later, they establish themselves in the bile ducts where sexual maturation takes place.

or mapping risks [22]. The understanding of fasciolosis transmission worldwide requires a global account on the snail species that are vectors or that may act as potential hosts of *Fasciola* spp. The compilation of specific research on their geographical distribution and susceptibility is needed as a helpful tool that may serve in the control and prevention programmes of the disease. In the following sections, we present a review of the global distribution of lymnaeid snails with a known role in *Fasciola* spp. transmission according to the latest systematics published for this family. We will only consider those species which have experimentally proved to be susceptible to infection or that have been described as natural vectors of the parasite in the field (see Table 15.1).

Brief Overview of *Fasciola* spp. Distribution

Fasciola hepatica and *F. gigantica* (known as liver flukes), are the main aetiological agents of fasciolosis worldwide (Figure 15.2). The large diversity of hosts of *F. hepatica* (i.e. many lymnaeid snails and mammal species act as intermediate and definitive hosts, respectively) determine a major role for this species in the global burden of the disease. This species is reported in all continents except Antarctica and it is present in many oceanic islands. Contrastingly, given that *F. gigantica* is usually restricted to the tropical and subtropical region of Asia and Africa, it has a lesser global impact in public and veterinary health [4]. In any case, fasciolosis is a serious problem for public health in some endemic regions such as the Bolivian Altiplano, Peru, Egypt, Iran and southeast Asia where prevalence in the human populations is particularly high [4, 5]. From a veterinary point of view, there are reports of *F. hepatica* infecting over 300 million bovines and nearly 250 million ovine worldwide; together with *F. gigantica* the economic losses ascend up to 3000 million dollars [1, 5]. The localities where fasciolosis reaches the highest prevalence are known for having: (a) high infection rates and intensities in either human and livestock populations, (b) available surface water for compatible lymnaeid snail populations to settle in most of the year and (c) adequate temperature and humidity for the development and survival of each parasite stage in and out of the hosts [5].

Lymnaeid Distribution

The family of Lymnaeidae has a worldwide distribution [11] (Figure 15.3). Although this review focuses only in its role in fasciolosis transmission, it is important to note that several lymnaeid species also serve as hosts of a major number of trematode families, causing diseases in wildlife and domestic fauna as well as humans [68, 69].

Transmission in the Americas

The American lymnaeid species display a vast distribution ranging from north Canada to south of Patagonia as well as through the Caribbean Islands. However, the transmission is not as thoroughly reported in every region.

North American Lymnaeidae

In North America, studies depicting the diversity of Lymnaeidae accumulated in the past but have not been so abundant in the last decades. In the USA, *Galba cubensis* and *Pseudosuccinea columella* were identified as the main vectors of fasciolosis but only after a series of experimental infection trials [29]. However, in Florida (USA), Kaplan *et al.* [70] presented *G. cubensis* as the main host for *F. hepatica*, owing to its broad distribution and to its ability to transmit the parasite throughout the year after presented recovered field prevalence data. Still in the USA, *Galba bulimoides* from Wisconsin was experimentally infected with successful retrieval of metacercariae [46] and *Galba modicella* from Montana was found naturally infected in a goat farm with transmission occurring between spring and summer [33]. Also, in Montana, Knapp *et al.* [45] described the transmission of *F. hepatica* by *Hinkleyia caperata*, which is supposed to occur from the east to the west coasts of North America up north to Alberta [71, 72]. Another related species, *Stagnicola palustris* has been reported infected in eastern Washington [66] with a distribution range from south Alaska to New Mexico [73]. In Canada, although fasciolosis is recognized to occur within dairy cows in Quebec [74], the lymnaeid species implicated are not well studied. However, Hubendick [11] admitted *Galba truncatula*, *S. palustris* and probably *Galba humilis* to reach Canada. In Mexico, *G. humilis* and *G. bulimoides* are known to transmit *F. hepatica* [32] from late summer to early winter mostly in warm and damp sites [75].

In any case, the actual distribution of the lymnaeid species, and particularly those able to transmit *F. hepatica*, in the northern region of North America is poorly known. An ancient

Table 15.1. Summary of the main lymnaeid species host of the liver flukes worldwide with their type localities, main geographic distribution and records of infection with either *Fasciola hepatica* or *Fasciola gigantica* (EI: means that the lymnaeid species has been only experimentally infected)

Lymnaeid species	Type locality	Main geographic range	Liver fluke	Reference
Austropeplea tomentosa (Pfeiffer, 1855)	Auckland (New Zealand)	Australia	*F. hepatica*	[23]
Galba cousini (Jousseaume, 1887)	Chanchu-Yacu (Ecuador)	Ecuador, Colombia, Venezuela	*F. hepatica*	[24–26]
Galba cubensis (Pfeiffer, 1839)	Cuba	Cuba	*F. hepatica*	[27, 28]
		USA		[29]
		Lesser Antilles, Colombia		[30]
		Mexico, Uruguay, Venezuela		[31]
Galba humilis (Say, 1822)	South Carolina (USA)	Mexico	*F. hepatica*	[32]
Galba modicella (Say, 1825)	Owego, New York (USA)	USA	*F. hepatica*	[33]
Galba truncatula (Müller, 1774)	Thangelstedt (Germany)	Argentina, Bolivia, Chile, Peru, Venezuela	*F. hepatica,*	[34–36]
		Africa	*F. gigantica*	[37]
		Ireland, Europe		[38, 39]
Galba viator (d'Orbigny, 1835)	Río Negro (Argentina)	Argentina, Brazil, Peru, Uruguay	*F. hepatica*	[40–42]
Galba viridis (Quoy & Gaimard, 1832)	Guam	Hawaii	*F. hepatica*	[43]
		Micronesia	*F. gigantica*	[44]
		Australia		[23]
Hinkleyia caperata (Say, 1829)	Indiana, USA	North America	*F. hepatica*	[45]
Lymnaea bulimoides (Lea, 1841)	Oregon (USA)	USA	*F. hepatica*	[46]
Lymnaea diaphana (King, 1830)	San Gregorio (Chile)	Argentina, Chile	*F. hepatica*	[47, 48]
Lymnaea stagnalis (Linnaeus, 1758)	Europe	Europe, Asia	*F. hepatica*	[49]
Myxas glutinosa (Müller, 1774)	Fridrichsdal (Denmark)	Urals, Siberia	*F. hepatica* (EI)	[50]
Omphiscola glabra (Müller, 1774)	Fridrichsdal (Denmark)	France	*F. hepatica*	[51]
Pseudosuccinea columella (Say, 1817)	near Philadelphia (USA)	Cuba, USA, Central and South America	*F. hepatica*	[52–54]
		France		[55]
		Africa		[56]
		Australia, New Zealand		[23, 57]
Radix auricularia (Linnaeus, 1758)	Europe	Europe, Southeast Asia	*F. gigantica*	[58]

Table 15.1. *Continued*

Lymnaeid species	Type locality	Main geographic range	Liver fluke	Reference
Radix balthica (Linnaeus, 1758)		Iceland, British Islands, Europe, western Siberia, Asia minor	*F. hepatica*	[12]
Radix gedrosiana (Annadale & Prashad, 1919)	Afghanistan	Afganistan, Iran	*F. gigantica*	[59]
Radix labiata (Rossmaessler, 1835)		Europe, Turkey, Urals	*F. hepatica* (EI)	[12, 60]
Radix lagotis (Schrank, 1803)	Danube river (Germany)	Europe (Danube Basin)		[61]
Radix luteola (Lamark, 1822)	Bengal (India)	Nepal	*F. gigantica*	[16]
Radix natalensis (Krauss, 1848)	Port Natal, Durban (South Africa)	Africa; Saudi Arabia	*F. gigantica*	[62]
Radix ovata (Draparnaud, 1805)	France	France	*F. hepatica*	[63]
Radix peregra (Mülller, 1774)	Fridrichsdal (Denmark)	Central Europe	*F. hepatica* (EI)	[12]
Radix rubiginosa (Michelin, 1831)		Eastern Asia, Malaysia	*F. gigantica*	[64]
Radix swinhoei (Adams, 1866)		South eastern Asia, Philippines	*F. gigantica*	[64]
Stagnicola corvus (Gmelin, 1791)	Thuringia (Germany)	Europe	*F. hepatica*	[65]
Stagnicola fuscus (Pfeiffer, 1821)	Füred, Tihany (Hungary)	Europe	*F. hepatica*	[63, 65]
Stagnicola palustris (Müller, 1774)	Oregon, USA	USA	*F. hepatica*	[66]
		Europe		[67]

species report made by Sterki [76] in eastern North America informed of 13 species of Lymnaeidae including *Lymnaea stagnalis*, *P. columella*, *G. modicella*, *G. humilis*, *Ladislavella elodes* and *Radix auricularia*. However, the systematics of this family in this region should be viewed cautiously due to the lack of molecular studies that would help to clarify the identity of the species described. Given that many lymnaeid species have a large phenotypic variability and are quite difficult to differentiate by shell morphology, the numbers might be overestimated. A more recent list made by Johnson *et al.* [77] on the freshwater snails from Canada and the USA includes 61 species of Lymnaeidae. Among them, 21 correspond to the group of the small-shelled *Galba*. Here again, these numbers are perhaps overestimated due to possible systematic confusions in the absence of molecular studies. In the arctic region, lymnaeid snails are known to occur but are related to the transmission of *Fascioloides magna* parasitizing caribou populations as it has been demonstrated by the observation of parasite stages overwintering in the snails. However, the snail species are not outlined [78]. A recent study carried out by Vinarski *et al.* [79] suggests that the endemic lymnaeid from Greenland, former *Lymnaea vahlii* should be considered as a form of the North American *Ladislavella catascopium* which is known to transmit the liver fluke [15]. In any case, transmission of *F. hepatica* in these high latitudes remains unknown.

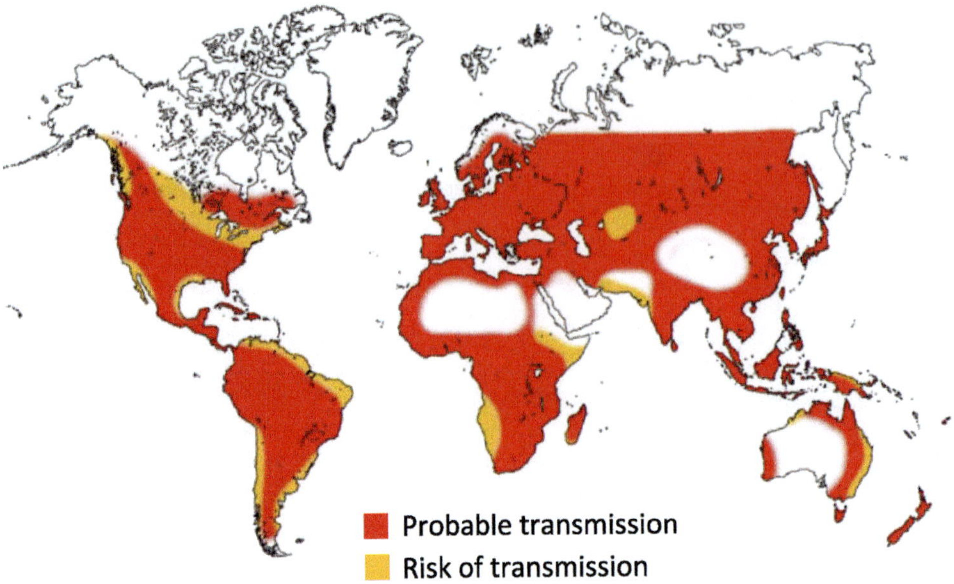

Fig. 15.2. Worldwide transmission of fasciolosis. Map built using data gathered from different reports (Torgerson and Claxton, 1999; WHO, 2013, 2014).

Caribbean Lymnaeidae

In the Caribbean basin, the situation is much less complicated with three species occurring. In Cuba, Vázquez *et al.* [21] infected several populations of *G. cubensis* and *P. columella* with different isolates of *F. hepatica* and reported, according to the levels of susceptibility and compatibility, the former as the main host in the island. *Galba cubensis* is described from Cuba and has an extended distribution in the country while *P. columella* is found only from the west to the central region [80]. Field studies have reported natural prevalences in *G. cubensis* ranging from 0 to 11% [27] and 0 to 34% [28] in localities reported as transmission foci. In Cuba, *P. columella* might play a secondary role in the transmission [21] and has been only recently reported naturally infected by for the first time in the Caribbean region with a prevalence of 1% [52]. Transmission in Cuba is therefore supposed to occur by both species with several human outbreaks and important economic losses due to high cattle infection [3]. Interestingly, at least six natural populations of *P. columella* from Cuba have been found to display an active resistance against *F. hepatica* miracidia infection [20, 21]. Resistant populations can be phenotypically distinguished by a different mantle pigmentation pattern consisting of a number of small well-defined spots in a belt-like shape in the centre [81]. These resistant populations, although frequently rare, are an interesting model to tackle fasciolosis transmission. The mechanisms of resistance are not yet well known.

Both *G. cubensis* and *P. columella* are also described in some Lesser Antilles where transmission of *F. hepatica* had never been reported, like Martinique [82] and Guadeloupe [83]. Both lymnaeid species are also reported in some other Caribbean countries such as the Dominican Republic [84]. In Puerto Rico [85] and Jamaica [86], *G. cubensis* is responsible for *F. hepatica* transmission. Recently, a third species, *Galba schirazensis*, has been reported in the Dominican Republic [87] but it seems that it does not act in fasciolosis transmission. There is a previous report of *Galba viator* from the Dominican Republic [88] but its identification is questionable given the lack of molecular tools at that time. The fact that it is rather the distribution range of *G. cubensis* and that both species are very difficult to separate morphologically, it would probably correspond to the latter.

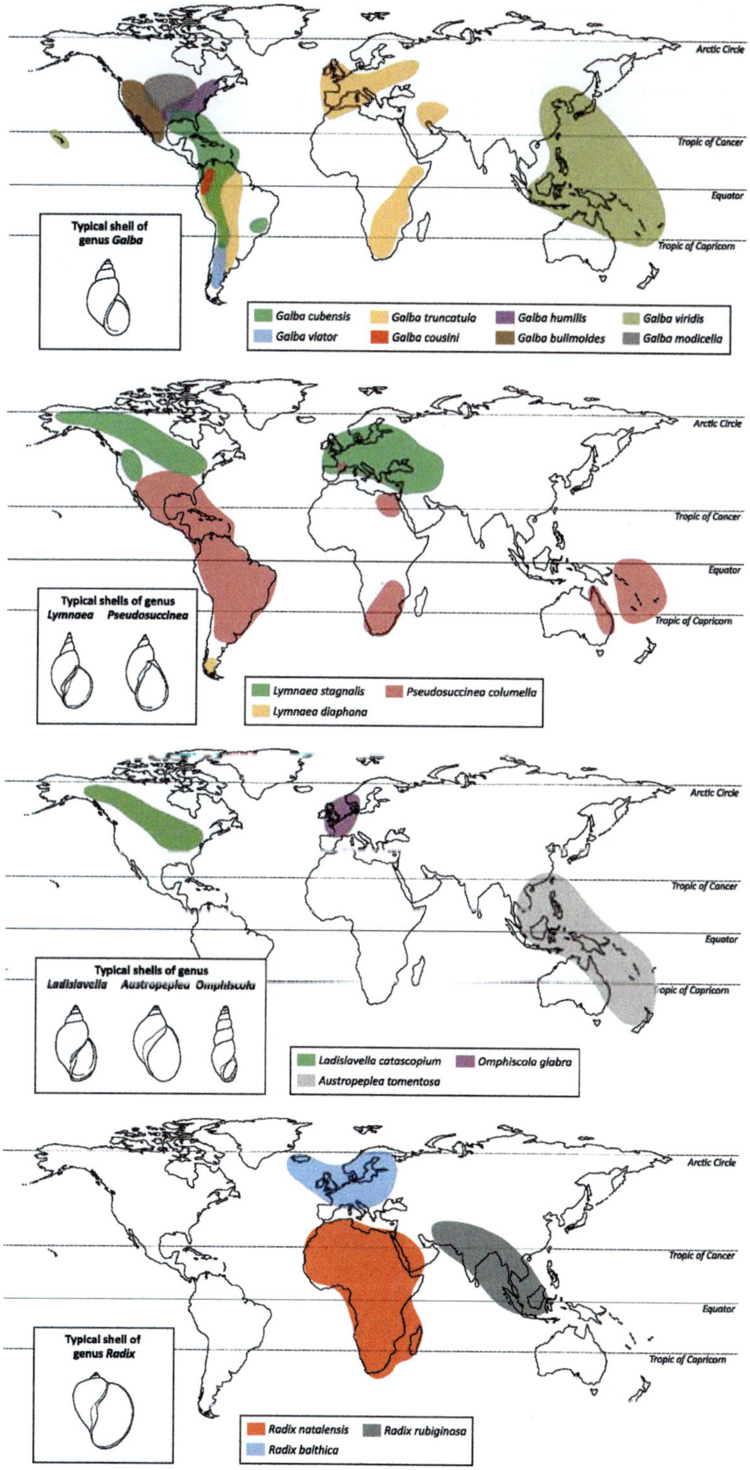

Fig. 15.3. Most common ranges of the main lymnaeid species hosts of *Fasciola* spp.

South American Lymnaeidae

As in North America, the South American scenario becomes again ravelled with a diversity of very morphologically similar species able to transmit the liver fluke. In northern South America, the genus *Galba* is probably mainly responsible for transmission. Several species such as *Galba cousini*, *G. cubensis*, *G. schirazensis* and *G. truncatula* are all reported in this region and most have been found regularly infected with *F. hepatica* [34, 89]. The species *G. cousini* has been found in Venezuela, Colombia and Ecuador [24, 90, 91]. Its role in the transmission in these countries seems restricted to altitudes lying between 2000 and 4000 m [90]. In the last years, *G. schirazensis* was discovered in many types of habitats from Venezuela, Colombia and Ecuador [92] and its role in the transmission of *F. hepatica* remains under discussion. In Venezuela and Colombia, this snail was the only species of lymnaeid found in several active foci, and although not found infected during sampling, it is highly probable that it is responsible for the transmission of *F. hepatica*. In any case, experimental infections carried out with an isolate of *F. hepatica* and a strain of *G. schirazensis* from Colombia showed that seven snails out of 400 harboured several rediae and two contained free cercariae on day 50 post-exposure [19]. In addition, in Ecuador, a 6% natural prevalence was detected in *G. schirazensis* using molecular techniques [93]. However, the actual role of this snail species in a sustained transmission should be confirmed with cercarial shedding and effective metacercarial infectivity. Apart from *Galba* spp., *P. columella* is also reported in Venezuela [34], Colombia [94] and Ecuador [40], and also extends down through Brazil [95, 96] and Argentina [97]. In the Andean countries, particularly the Bolivian Altiplano, *G. truncatula* plays a prominent role in *F. hepatica* transmission [13, 35, 98, 99]. This species may also be a key player in Chile [99] and Argentina [100]. *Galba viator* is also a very important intermediate host of *F. hepatica* in the southern region of Chile and Argentina [41, 101], although current reports of natural prevalence are lacking. According to Kleiman *et al.* [102], this species occurs as far south as beyond parallel 42°S in the Chubut province in the Patagonian region, where a 0.67% natural prevalence of *F. hepatica* was recorded [41]. There were in fact reports of natural infection in *G. viator* and *P. columella* from Argentina as assessed by mean of PCR amplification of the cytochrome *c* oxidase subunit I of *F. hepatica* [103]. Other South American Lymnaeidae are *Lymnaea diaphana* and *Lymnaea rupestris*. The former species is described to serve as a host of *F. hepatica* in the southern part of the Chilean and Argentinean Patagonia [47]. The role of the second species *L. rupestris* in the transmission of *F. hepatica* is not known [104].

Transmission in Eurasia

Transmission in this region has been fairly frequently recorded in several studies except where climatic conditions are probably too harsh for the parasite life cycle, such as in Iceland [105] and probably Siberia. However, several lymnaeid species have been reported from these regions [14, 50, 106] that could eventually act as hosts.

European Lymnaeidae

In Britain and Ireland, there are at least eight lymnaeid species reported by Anderson [107]. However, transmission has been mainly linked to the widespread *G. truncatula* and to the local *Radix balthica* [108]. In continental Europe, several species such as *Radix peregra* have a broad distribution ranging from Nordic countries to southern regions and east to Ukraine. According to Schniebs *et al.* [60], *R. peregra* occurs also in Turkey and the Urals. Although there is no record of this species being naturally infected with *F. hepatica*, its role as well as that of *R. balthica* as incidental hosts in Belgium was demonstrated by Caron *et al.* [12] through series of experimental infections. More recently, the field prevalence of *F. hepatica* was confirmed in *R. balthica* (0.16%) and *G. truncatula* (1.31%) from Belgium and Luxembourg using a multiplex PCR [109]. Another species, *Stagnicola fuscus*, was experimentally tested with *F. hepatica* miracidia originating from France and proved to be partially refractory to infection although the snails have shed cercariae [63]. According to Novobilsky *et al.* [67], *S. fuscus* and *S. palustris* are also potential intermediate hosts of *F. hepatica* in Sweden. Overall, transmission in the field is commonly attributed to *G. truncatula* and *Omphiscola glabra* particularly in France [51, 110, 111]. Recorded prevalences for *G. truncatula* range between 5 and 33% in France and a mean of 11% in Spain [112], and 1.4–1.8% for French *O. glabra* [110]. *Galba truncatula* has been also found naturally infected in several districts of Moscow [113] and up to 85% prevalence was detected in Poland [114]. Although there are no much field data on prevalence on the snail populations, *G. truncatula* is presumed to be a key vector in dairy farms from south Germany and Austria [115] and natural prevalence of 7% is reported from Switzerland through real-time PCR [116]. Recently, some small-scale risk mapping of *F. hepatica* in farms from Belgium has been based on the spatial distribution of *G. truncatula* [22]. From central Europe, several species have been described as potential hosts such as *Catascopia terebra*, *Stagnicola corvus* and *S. fuscus* in the Czech Republic [65]. Several introductions, probably through the aquarium plant trade, have occurred in the past few years and may increase the risk of transmission in some countries. Such introductions include *P. columella* in France [55] and *Austropeplea viridis*, *G. schirazensis* and *G. cubensis* in Spain [92, 117, 118].

Asian Lymnaeidae

There are several species of lymnaeid snails described or reported in northern Asia from the Urals to eastern Russia [14, 50, 106, 119–121]. However, there are no significant reports on *F. hepatica* infection neither in natural nor under experimental conditions. In many other areas of Asia (particularly in the south), fasciolosis is caused by both *F. hepatica* and *F. gigantica* with some regions reporting the existence of hybrid forms of the parasite such as in South Korea and Vietnam [122, 123].

As in many other regions where it has been introduced, *G. truncatula* is held responsible for transmission of *F. hepatica* in some areas in Iran together with *L. stagnalis* [49]. Also, from southwest Asia, transmission of *F. gigantica* is generally linked to the presence of *Radix natalensis* (syn. *Lymnaea arabica*) and *S. palustris* in Saudi Arabia [62, 124], and *Radix gedrosiana* in Iran [59].

Southeast Asia is known to be heavily hit with both species of *Fasciola*. In Thailand, at least three species of lymnaeid snails have been recorded infected with *F. gigantica*: *Radix rubiginosa*, *Radix swinhoei* and *Galba viridis* [64]. Similarly, *G. viridis*, *R. auricularia* and *R. rubiginosa* are the main lymnaeid species recorded in Vietnam with the latter being reported infected in the field [125]. Unfortunately, not all field reports from fasciolosis outbreaks define the snail species responsible for the transmission. In southwest China, for example, after an outbreak of fasciolosis, the transmitting snails in the area were only identified as *Galba* spp. [126]. From Japan, several lymnaeids were experimentally infected using Australian strains of *F. hepatica* by Itagaki *et al.* [127]. Only the species *Austropeplea ollula* and *R. auricularia* resulted infected while *G. truncatula* proved to be refractory.

Transmission in Africa

Fasciolosis in Africa may occur either by infection of *F. hepatica* or *F. gigantica*. One of the most affected regions is the Nile basin in Egypt [128]. In east Africa, both species are also responsible for infections in countries such as Kenya, Uganda, Ethiopia and Sudan [129].

The African Lymnaeidae

One of the lymnaeid species with broader distribution in the continent is *R. natalensis* described from South Africa [56], usually living in steady shallow waters, and considered as a very suitable host for *F. gigantica* [18]. Based on field and experimental studies, this species has been recognized as the main responsible for transmission of *F. gigantica* in Egypt [130]. However, *R. natalensis* is presumed to be also a potential host of *F. hepatica* after a successful experimental infection with French strains of this parasite [131]. It has also been found infected in other countries like Benin [132], Nigeria [133], Ghana [56] and is presumed to be the main transmitter in Ethiopia [134] and Uganda [135]. The distribution range reaches Tanzania and Cameroon but the transmission of either *Fasciola* spp. is not well assessed [136]. In western Africa, *R. natalensis* has been pointed as the main host for *F. gigantica* transmission in Senegal [137]. Outside African mainland, *R. natalensis* is also found in the island of Madagascar [138] where it is the main host of *F. gigantica* [139], and in Reunion Island [140]. The distribution of this species in other African countries includes Somalia and Gambia and is less common along the east and southwestern coasts [56]. According to Brown [56], the similar species *R. peregra* is also found in northwest Africa (Morocco and Algeria) with a single report from Ethiopia.

Moreover, *F. hepatica* transmission in Africa does not follow the same epidemiology than that of *F. gigantica*, with different hosts and distribution. In South Africa, *F. hepatica* is known to be transmitted by *G. truncatula* which inhabits preferably ecosystems of low-flowing water like borders of irrigation channels [141]. This species is also responsible for transmission in Lesotho [142] and Tunisia [143] where horses have been found infected with the parasite [144]. In northeastern Algeria, natural infected *G. truncatula* have been also recorded [145]. Susceptibility to Algerian strains of *F. hepatica* miracidia was experimentally tested showing that the number of obtained metacercariae did not vary regardless the origin of the parasite in that country [146]. Into the west, *G. truncatula* has been also found responsible for transmission of *F. hepatica* infecting sheep in the Middle-Atlas Mountains of Morocco [147]. Studies regarding the compatibility between Moroccan populations of *G. truncatula* and sympatric *F. hepatica* have been done by Belfaiza *et al.* [148] showing that snails may impose a bottleneck on the development of mother rediae which grows more and have a higher production of daughter rediae and cercariae. According to Rondelaud *et al.* [38], a prevalence of 3% has been recorded in field-collected Moroccan *G. truncatula*. In the east, *G. truncatula* is known to maintain *F. hepatica* life cycle in the highlands of Ethiopia [149]. Moreover, *G. truncatula* has been experimentally infected with Egyptian strains of *F. gigantica* with a frequency of 80% of cercariae-shedding snails proving that it may serve as a natural host for this parasite in Egypt [150]. *Galba truncatula* has also been reported from Zaire, Kenya and Tanzania [56].

Another lymnaeid host of *F. hepatica* in Africa is the invasive species *P. columella*. This species has been reported as widely distributed in South Africa [56] but its contribution to fasciolosis transmission in this country is yet to be evaluated [141]. After the discovery of its introduction in South Africa in 1944, it has also been found in Mozambique, Zimbabwe, Zambia and Egypt [56]. However, populations of *P. columella* might be easily mistaken with *R. natalensis* and thus ancient distribution data should be re-evaluated. More recently, Grabner *et al.* [151] molecularly detected the natural infection of Egyptian populations of *P. columella* with several trematodes among which is *F. gigantica*.

Other lymnaeid species occurring in Africa are *L. stagnalis* and *S. palustris*, both recorded from Morocco and Algeria [56]. Transmission of *F. hepatica* by any of these lymnaeids is supposed but not assessed. Finally, it has been recently studied the introduction of *R. rubiginosa* into South Africa which could exacerbate the transmission of *F. gigantica* in this country and whose spread might go unnoticed due to similar shell morphology with the native *R. natalensis* [152].

Transmission in Australia and the pacific islands

Transmission of *F. hepatica* in this region has been usually linked to the native lymnaeid *Austropeplea tomentosa*, particularly in Australia, Tasmania and New Zealand [23, 153]. The role of this species as the main host of *F. hepatica* in Australia was confirmed by Boray [154]. However, in this study, the author also confirms the introduction of three invasive lymnaeid species: *P. columella*, *G. viridis* and *R. auricularia*. Subsequent infection assays were carried out proving that both *P. columella* and *G. viridis* are highly susceptible, whereas *A. tomentosa* remains the most suited host [154]. Later, Molloy and Anderson [23] concluded that although the introduced species may not yet have had an important impact in fasciolosis transmission, the adaptability of *P. columella* to tropical habitats could result in the expansion of *F. hepatica* in the north. Another species was described from Tasmania by Ponder and Waterhouse [153] named as *Kutikina hispida* but transmission capacity has yet to be assessed. Fasciolosis in mainland Australia not only affects bovines but also native marsupial hosts [155]. In the Indonesian archipelago, *R. rubiginosa* has been found infected with *F. gigantica* in irrigated rice fields of West Java [156]. From the Micronesia, particularly the Mariana Islands, the only lymnaeid species reported is *G. viridis* ranked as a native species but no transmission of *Fasciola* spp. has been described so far [44]. *Galba viridis* is also known to occur in Guam, Indonesia, Papua New Guinea and Thailand [157]. However, far northern in the Hawaiian archipelago the species *G. viridis* is a major intermediate host of the liver fluke *F. gigantica* [43]. This species has also been experimentally infected with *F. hepatica* and cercarial shedding was successfully verified [158]. Also, the invasive *P. columella* has been found in Hawaii [157]. In some other Pacific islands, *P. columella* has been reported from Tahiti, Rapa, Tubuai and Rurutu, but its role in the transmission of *F. hepatica* has not been established [159].

Concluding Remarks on Lymnaeidae–*Fasciola* spp. Transmission

As illustrated in this review, although many species may actually serve as intermediate hosts of *Fasciola* spp. in the field, there are a lesser number posing major threats at both regional or world scales. These threats can be due to several factors among which we may highlight: (a) the species is widely distributed and adapted to the circulating strains of liver fluke (shows higher compatibility in terms of surviving infection and redial burden, e.g.), (b) the species is capable of establishing itself in areas of cattle-related human activities (even very anthropized habitats) with ecological preferences that match those of the definitive hosts or (c) the species has a high invasive ability and the lack of genetic diversity of introduced individuals that may facilitate the transmission. Worldwide, the introductions of *G. truncatula* and *P. columella* may have played a key role in the spread of fasciolosis. Both species have a proved susceptibility to *F. hepatica* infection, but more important, a recognized invasive ability allowing both to succeed in new environments [35, 160]. As an attempt to highlight particular species by regions, major threats may be summarized for (a) the Americas: *G. truncatula*, *G. cubensis*, *G. viator* and *P. columella*; (b) Europe: *G. truncatula* and *R. balthica*; (c) Asia (particularly the southern regions): *R. rubiginosa* and *R. auricularia*; (d) Africa: *R. natalensis* and *G. truncatula*; and (e) Oceania: *A. tomentosa* and *G. viridis*.

Although we have reviewed only the main actors in the *Fasciola*–Lymnaeidae interplay by region, we should note that the huge number of described lymnaeid snails in ancient and recent literature may lead to an annoying scenario of who transmit who in many regions. The absence of reliable morphological traits has led to confusion regarding specific identities of lymnaeid populations worldwide. Some species cannot be accurately distinguished because their shell morphology and anatomy are very similar. Even species relatively easy to identify such as *P. columella* or *G. cousini* could have been misidentified in the past due to, for instance, lack of experience in snail identification or the status of the individuals used for the identification (juveniles, worn or broken shells). Thus, studying lymnaeid species often requires an accurate molecular tool. The species identification of fasciolid larval stages found in nature also needs to be always verified, usually by experimental infections or by molecular markers. It is therefore necessary to review in the near future all reported names using the latest molecular tools and data available to reach a more plausible scenario in which, though probably complicated enough, we can picture the world transmission of liver flukes in a more reduced and comprehensible way.

Acknowledgements

We acknowledge our numerous collaborators and students who have contributed to our studies on liver flukes and snails during the last two decades. This work has been financially supported by Ecos-Sud program A16B02, Labex CeMeb (doctoral and ERJ grants to ES, post-doctoral grant to ML), Muse (project REPAS: grant n°170544IA (n°ANR-IDEX-0006) to AAV), Tour du Valat, IPK, IRD, CNRS and University of Montpellier. We are thankful to three anonymous reviewers for their valuable comments that have helped to improve the final version of this paper.

References

1. Robinson MW, Dalton JP. Zoonotic helminth infections with particular emphasis on fasciolosis and other trematodiases. Philosophical Transactions of the Royal Society B 2009;364:2763–76.
2. Rapsch C, Dahinden T, Heinzmann D, Torgenson PR, Braun U, Deplazes P, *et al.* An interactive map to assess the potential spread of *Lymnaea truncatula* and the free-living stages of *Fasciola hepatica* in Switzerland. Veterinary Parasitology 2008;154:242–9.
3. Rojas L, Vázquez A, Domenech I, Robertson LJ. Fascioliasis: can Cuba conquer this emerging parasitosis? Trends in Parasitology 2010;26(1):26–34.
4. Mas-Coma S, Valero MA, Bargues MD. *Fasciola*, lymnaeids and human fascioliasis, with a global overview on disease transmission, epidemiology, evolutionary genetics, molecular epidemiology and control. Advances in Parasitology 2009;69:41–146.
5. Mas-Coma S, Bargues MD, Valero MA. Fascioliasis and other plant-borne trematode zoonoses. International Journal for Parasitology 2005;35:1255–78.
6. WHO. Report of the WHO expert consultation on foodborne trematode infections and taeniasis/cysticercosis. Vientiane: 2009 12–16 October 2009. Report No. WHO/HTM/NTD/PCT/2011.3.
7. WHO. The 'neglected' neglected worms. Geneva: 2007. December Issue 10.
8. Hurtrez-Boussès S, Meunier C, Durand C, Renaud F. Dynamics of host–parasite interactions: the example of population biology of the liver fluke (*Fasciola hepatica*). Microbes and Infection 2001;3:841–9.
9. Ménard A, L'Hostis M, Leray G, Marchandeau S, Pascal M, Roudot N, *et al.* Inventory of wild rodents and lagomorphs as natural hosts of *Fasciola hepatica* on a farm located in a humid area in Loire Atlantique (France). Parasite 2000;7(2):77–82.
10. Strong EE, Gargomini O, Ponder WP, Bouchet P. Global diversity of gastropods (Gastropoda; Mollusca) in freshwater. Hydrobiologia 2008;595:149–66.
11. Hubendick B. Recent Lymnaeidae, their variation, morphology, taxonomy, nomenclature, and distribution. Kungliga Svenska Vetenskapsakademiens Handlingar 1951;4(3):1–223.
12. Caron Y, Lasri S, Losson B. *Fasciola hepatica*: an assessment on the vectorial capacity of *Radix labiata* and *R. balthica* commonly found in Belgium. Veterinary Parasitology 2007;149:95–103.
13. Correa AC, Escobar JS, Noya O, Velázquez LE, González-Ramírez C, Hurtrez-Boussès S, *et al.* Morphological and molecular characterization of Neotropic Lymnaeidae (Gastropoda: Lymnaeoidea), vectors of fasciolosis. Infection, Genetics and Evolution 2011;11:1978–88.
14. Vinarski MV. A new species of stagnicoline snails (Mollusca: Gastropoda: Lymnaeidae) from the extreme North of Western Siberia. Zootaxa 2011;2817:55–8.
15. Bargues MD, Mas-Coma S. Reviewing lymnaeid vectors of fascioliasis by ribosomal DNA sequence analyses. Journal of Helminthology 2005;79:257–67.
16. Correa AC, Escobar JS, Durand P, David P, Jarne P, Pointier JP, *et al.* Bridging gaps in the molecular phylogeny of the Lymnaeidae (Gastropoda: Pulmonata), vectors of fascioliasis. BMC Evolutionary Biology 2010;10:381.
17. Vinarski MV. One, two, or several? How many lymnaeid genera are there? Ruthenica 2013;23(1):41–58.
18. Lotfy WM, Brant SV, DeJong RJ, Le TH, Demiaszkiewicz A, Rajapakse RPVJ, *et al.* Evolutionary origins, diversification, and biogeography of liver flukes (Digenea, Fasciolidae). American Journal of Tropical Medicine and Hygiene 2008;79(2):248–55.
19. Dreyfuss G, Correa AC, Djuikwo-Teukeng FF, Novobilsky A, Höglund J, Pankrác J, *et al.* Differences in the compatibility of infection between the liver flukes *Fascioloides magna* and *Fasciola hepatica* in a Colombian population of the snail *Galba* sp. Journal of Helminthology 2015;89:720–6.
20. Gutiérrez A, Pointier JP, Fraga J, Jobet E, Modat S, Pérez RT, *et al.* *Fasciola hepatica*: identification of molecular markers for resistant and susceptible *Pseudosuccinea columella* snail hosts. Experimental Parasitology 2003;105:211–8.
21. Vázquez AA, Sánchez J, Pointier JP, Théron A, Hurtrez-Boussès S. *Fasciola hepatica* in Cuba: compatibility of different isolates with two intermediate snail hosts, *Galba cubensis* and *Pseudosuccinea columella*. Journal of Helminthology 2014;88(4):434–40.
22. Charlier J, Soenen K, De Roeck E, Hantson W, Ducheyne E, Van Coillie F, *et al.* Longitudinal study on the temporal and micro-spatial distribution of *Galba truncatula* in four farms in Belgium as a base for small-scale risk mapping of *Fasciola hepatica*. Parasites & Vectors 2014;7(1):528.

23. Molloy JB, Anderson GR. The distribution of *Fasciola hepatica* in Queensland, Australia, and the potential impact of introduced snail intermediate hosts. Veterinary Parasitology 2006;137:62–6.
24. Pointier JP, Noya O, Amarista M, Théron A. *Lymnaea cousini* Jousseaume, 1887 (Gastropoda: Lymnaeidae): First record for Venezuala. Mémorias do Instituto Oswaldo Cruz. 2004;99.
25. Brumpt E, Velazquez J, Ucros H, Brumpt L, Mission E. Brumpt et L.-Ch. Brumpt en Colombie et au Venezuela. 1.- Découverte de l'hôte intermédiaire, *Limnaea bogotensis* Pilsbry, de la grande douve, *Fasciola hepatica*, en Colombie. Annales de Parasitologie Humaine et Comparée 1939–1940;17:563–79.
26. Velázquez LE. Synonymy between *Lymnaea bogotensis* Pilsbry, 1935 and *Lymnaea cousini* Jousseaume, 1887 (Gastropoda: Lymnaeidae). Mémorias do Instituto Oswaldo Cruz 2006;101(7):795–9.
27. Vázquez AA, Sánchez J, Alba A, Pointier JP, Hurtrez-Boussès S. Natural prevalence in Cuban populations of the lymnaeid snail *Galba cubensis* infected with the liver fluke *Fasciola hepatica*: small values do matter. Parasitology Research 2015;114:4205–10.
28. Alba A, Vázquez AA, Sánchez J, Fraga J, Hernández HM, Martínez E, *et al.* Assessment of the FasciMol-ELISA in the detection of the trematode *Fasciola hepatica* in field-collected *Galba cubensis*: a novel tool for the malacological survey of fasciolosis transmission. Parasites & Vectors 2016;9:22:1–9.
29. Cruz-Reyes A, Malek EA. Suitability of six lymnaeid snails for infection with *Fasciola hepatica*. Veterinary Parasitology 1987;24(3–4):203–10.
30. Malek EA. Snail hosts of schistosomiasis and other snail-transmitted diseases in Tropical America: a manual. PAHO Scientific Publication. 1985:478.
31. Lutz A. Estudios de zoología y parasitología venezolanas. Caracas: Reimpreso, UCV; 1928.
32. Cruz-Mendoza I, Ibarra F, Quintero MT, Naranjo E, Lecumberri J, Correa D. Seasonal transmission of *Fasciola hepatica* in cattle and *Lymnaea* (*fossaria*) *humilis* snails in central Mexico. Parasitology Research 2005;95:283–6.
33. Rognlie MC, Dimke KL, Potts RS, Knapp SE. Seasonal transmission of *Fasciola hepatica* in Montana, USA, with detection of infected intermediate hosts using a DNA- based assay. Veterinary Parasitology 1996;65:297–305.
34. Pointier JP, Noya O, Alarcón de Noya B, Théron A. Distribution of Lymnaeidae (Mollusca: Pulmonata), intermediate snail hosts of *Fasciola hepatica* in Venezuela. Mémorias do Instituto Oswaldo Cruz 2009;104(5):790–6.
35. Meunier C, Tirard C, Hurtrez-Boussès S, Durand P, Bargues MD, Mas-Coma S, *et al.* Lack of molluscan host diversity and the transmission of an emerging parasitic disease in Bolivia. Molecular Ecology 2001;10:1333–40.
36. Durand P, Pointier JP, Escoubeyrou K, Arenas JA, Yong M, Amarista M, *et al.* Occurrence of a sibling species complex within neotropical lymnaeids, snail intermediate hosts of fascioliasis. Acta Tropica 2002;83:233–40.
37. Dar Y, Vignoles P, Rondelaud D, Dreyfuss G. *Fasciola gigantica*: the growth and larval productivity of redial generations in the snail *Lymnaea truncatula*. Parasitology Research 2002;88:364–7.
38. Rondelaud D, Vignoles P, Dreyfuss G. La Limnée tronquée, un mollusque d'intérêt médical et vétérinaire. Limoges: 2009.
39. Relf V, Good B, Hanrahan JP, McCarthy E, Forbes AB, deWaal T. Temporal studies on *Fasciola hepatica* in *Galba truncatula* in the west of Ireland. Veterinary Parasitology 2011;175:287–92.
40. Paraense WL. Planorbidae, Lymnaeidae and Physidae of Ecuador (Mollusca: basommatophora). Mémorias do Instituto Oswaldo Cruz 2004;99(4):357–62.
41. Kleiman F, Pietrokovsky S, Prepelitchi L, Carbajo A, Wisnivesky-Colli C. Dynamics of *Fasciola hepatica* transmission in the Andean Patagonian valleys, Argentina. Veterinary Parasitology 2007;145:274–86.
42. Carvalho OS, Cardoso PCM, Lira MP, Rumi A, Roche A, Berne E, *et al.* The use of the polymerase chain reaction and restriction fragment length polymorphism technique associated with the classical morphology for characterization of *Lymnaea columella*, *L. viatrix*, and *L. diaphana* (mollusca: Lymnaeidae). Memórias do Instituto Oswaldo Cruz 2004;99(5):503–7.
43. Cowie RH. Catalog and bibliography of the nonindigenous nonmarine snails and slugs of the Hawaiian Islands. Bishop Museum Occasional Papers. 1997;50:66p.
44. Kerr A. Annotated checklist of the aquatic snails of the Mariana Islands, Micronesia. University of Guam Marine Laboratory Technical Report. 2013;147:26p.
45. Knapp SE, Dunkel AM, Han K, Zimmerman LA. Epizootiology of fascioliasis in Montana. Veterinary Parasitology 1992;42:241–6.

46. Foreyt WJ, Todd AC. Experimental infection of Lymnaeid snails in Wisconsin with miracidia of *Fascioloides magna* and *Fasciola hepatica*. The Journal of Parasitology 1978;64(6):1132–4.
47. Bargues MD, Mera y Sierra RL, Artigas P, Mas-Coma S. DNA multigene sequencing of topotypic specimens of the fascioliasis vector *Lymnaea diaphana* and phylogenetic analysis of the genus *Pectinidens* (gastropoda). Mémorias do Instituto Oswaldo Cruz 2012;107(1):111–24.
48. Duffy T, Kleiman F, Pietrokovsky S, Issia L, Schijman AG, Wisnivesky-Colli C. Real-time PCR strategy for rapid discrimination among main lymnaeid species from Argentina. Acta Tropica 2009;109(1):1–4.
49. Yakhchali M, Baran AI, Malekzadeh-Viayeh R. Molecular detection of the infection with *Fasciola hepatica* in field-collected snails of *Galba truncatula* and *Lymnaea stagnalis* from West Azarbaijan, Iran. Archives of Razi Institute 2015;70(3):195–202.
50. Vinarski MV, Grebennikov ME, Shishkoedova OS. Past and present distribution of *Myxas glutinosa* (O.F. Müller, 1774) in the waterbodies of the Urals and Siberia. Journal of Limnology 2013;72(2):336–42.
51. Dreyfuss G, Vignoles P, Rondelaud D. Natural infections of *Omphiscola glabra* (Lymnaeidae) with *Fasciola hepatica* in central France. Parasitology Research 2003;91:458–91.
52. Gutiérrez A, Vázquez AA, Hevia Y, Sánchez J, Correa AC, Hurtrez-Boussès S, *et al.* First report of larval stages of *Fasciola hepatica* in a wild population of *Pseudosuccinea columella* from Cuba and the Caribbean. Journal of Helminthology 2011;85:109–11.
53. Malek EA, Cogswell FB. *Lymnaea* (*Pseudosuccinea*) *columella* in Colombia. The Nautilus 1980;94:112–4.
54. Prepelitchi L, Kleiman F, Pietrokovsky SM, Moriena RA, Racioppi O, Alvarez J, *et al.* First report of *Lymnaea columella* Say, 1817 (Pulmonata: Lymnaeidae) naturally infected with *Fasciola hepatica* (Linnaeus, 1758) (Trematoda: Digenea) in Argentina. Mémorias do Instituto Oswaldo Cruz. 2003.
55. Pointier J, Coustau C, Rondelaud D, Théron A. *Pseudosuccinea columella* (Say 1817) (Gastropoda, Lymnaeidae), snail host of *Fasciola hepatica*: first record for France in the wild. Parasitology Research 2007;101:1389–92.
56. Brown DS. Freshwater Snails of Africa and Their Medical Importance. Second ed. London, UK: Taylor and Francis; 1994.
57. Pullan NB, Whitten LK. Liver fluke *Fasciola hepatica* in New Zealand, Part 1. A spreading parasite in sheep and cattle. New Zealand Veterinary Journal 1972;20:69–72.
58. Al-jibouri M, Al-Mayah S, Hassan H. The factors affecting metacercarial production of *Fasciola gigantica* from *Lymnaea auricularia* snails. Journal of Basrah Researches 2011;37(4E):9–16.
59. Ashrafi K, Massoud J, Holakouei K, Mahmoodi M, Joafshani MA, Valero MA, *et al.* Evidence suggesting that *Fasciola gigantica* might be the most prevalent causal agent of fascioliasis in northern Iran. Iranian Journal of Public Health 2004;33(4):31–7.
60. Schniebs K, Glöer P, Vinarski MV, Hundsdoerfer AK. Intraspecific morphological and genetic variability in the European freshwater snail *Radix labiata* (Rossmaessler, 1835) (Gastropoda: Basommatophora: Lymnaeidae). Contributions to Zoology 2013;82(1):55–68.
61. Bank R. Fauna Europaea project. Systematical and distributional checklist of species-group taxa of continental Mollusca of Europe. version 24. Available from: URL: https:/www.faunaeur.org2011.
62. Lotfy WM, Alsaqabi SM. Climate change and epidemiology of human parasitoses in Saudi Arabia: A review. Journal of Coastal Life Medicine 2016;4(7):580–8.
63. Dreyfuss G, Abrous M, Rondelaud D. The susceptibility of *Lymnaea fuscus* to experimental infection with *Fasciola hepatica*. The Journal of Parasitology 2000;86(1):158–60.
64. Kaset C, Eursitthichai V, Vichasri-Grams S, Viyanant V, Grams R. Rapid identification of lymnaeid snails and their infection with *Fasciola gigantica* in Thailand. Experimental Parasitology 2010;126:482–8.
65. Beran L. A contribution to distribution of genus *Stagnicola* and *Catascopia* (gastropoda: Lymnaeidae) in the Czech Republic. Malacologica Bohemoslovaca 2008;7:70–3.
66. Lang BZ. Snail and mammalian hosts for *Fasciola hepatica* in Eastern Washington. Journal of Parasitology 1977;63:938–9.
67. Novobilsky A, Kašný M, Beran L, Rondelaud D, Höglund J. *Lymnaea palustris* and *Lymnaea fuscus* are potential but uncommon intermediate hosts of *Fasciola hepatica* in Sweden. Parasites & Vectors 2013;6(1):251.
68. Cichy A, Faltýnková A, Zbikowska E. Cercariae (Trematoda, Digenea) in European freshwater snails – a checklist of records from over one hundred years. Folia Malacologica 2011;19(3):165–89.
69. Horák P, Mikeš L, Lichtenbergová L, Skála V, Soldánová M, Brantc SV. Avian schistosomes and outbreaks of cercarial dermatitis. Clinical Microbiology Reviews 2015;28(1):165–90.

70. Kaplan RM, Dame JB, Reddy GR, Courtney CM. The prevalence of *Fasciola hepatica* in its snail intermediate host determined by DNA probe assay. International Journal for Parasitology 1997;27(12):1585–93.
71. Jokinen EH. The Freshwater Snails (Mollusca: Gastropoda) of New York State. The University of the State of New York, New York; 1992.
72. Wu SK. Colorado Freshwater Molluscs. University of Colorado Museum, Colorado; 1989.
73. Leonard AB. A Yarmouthian Molluscan Fauna in the Midcontinent Region of the United States. University of Kansas Paleontological Contributions; 1950. p. 1–48.
74. Bouvry M, Rau MB. Seasonal variations in egg passage of *Fasciola hepatica* in dairy cows in Quebec. Veterinary Parasitology 1986;22:263–73.
75. Cruz-Mendoza I, Figueroa AJ, Correa D, Ramos-Martínez E, Lecumberri-López J, Quiroz-Romero H. Dynamics of *Fasciola hepatica* infection in two species of snails in a rural locality of Mexico. Veterinary Parasitology 2004;121:87–93.
76. Sterki V. Ohio Mollusca: additions and corrections. Ohio Journal of Science 1914;14(5):270–2.
77. Johnson PD, Bogan AE, Brown KM, Burkhead NM, Cordeiro JR, Garner JT, *et al.* Conservation status of freshwater gastropods of Canada and the United States. Fisheries 2013;38(6):246–82.
78. Kutz S, Ducrocq J, Verocai G, Hoar B, Colwell D, Beckmen K, *et al.* Parasites in ungulates of Arctic North America and Greenland: a view of contemporary diversity, ecology, and impact in a world under change. Advances in Parasitology 2012;79:99–252.
79. Vinarski MV, Bolotov I, Schniebs K, Nekhaev IO, Hundsdoerfer AK. Endemics or strangers? The integrative re-appraisal of taxonomy and phylogeny of the Greenland Lymnaeidae (Mollusca: Gastropoda). Comptes Rendus Biologies. 2017.
80. Vázquez AA, Sánchez J, Hevia Y. Distribución y preferencia de hábitats de moluscos hospederos intermediarios de *Fasciola hepatica* en Cuba. Revista Cubana de Medicina Tropical 2009;61(3):248–53.
81. Gutiérrez A, Pointier JP, Yong M, Sánchez J, Théron A. Evidence of phenotypic differences between resistant and susceptible isolates of *Pseudosuccinea columella* (Gastropoda: Lymnaeidae) to *Fasciola hepatica* (Trematoda: Digenea) in Cuba. Parasitology Research 2003;90:129–34.
82. Pointier JP. Invading freshwater snails and biological control in Martinique Island, French West Indies. Mémorias do Instituto Oswaldo Cruz 2001;96 Suppl.: 67–74.
83. Pointier JP, David P. Biological control of *Biomphalaria glabrata*, the intermediate host of schistosomes, by *Marisa cornuarietis* in ponds of Guadeloupe: long-term impact on the local snail fauna and aquatic flora. Biological Control 2004;29:81–9.
84. Gómez J, Vargas M, Malek EA. Moluscos de agua dulce de República Dominicana. Santo Domingo, Rep. Dom.: Instituto de Investigaciones en Bilharzia; 1980.
85. Bendezu P, Frame AD, Frame EL, Bonilla C. Watercress cultivation sites and their relationships to fascioliasis in Puerto Rico. Journal of Agriculture of the University of Puerto Rico 1988;72:405–11.
86. Bundy DAP, Arambulo III PV, Grey CL. La Fasciolosis en Jamaica: aspectos epidemiológicos y económicos de una zoonosis parasitaria transmitida por caracoles. Boletín de la Organización Panamericana de la Salud 1984;96(1):1–20.
87. Bargues MD, Artigas P, Khoubbane M, Flores R, Glöer P, Rojas-García R, *et al. Lymnaea schirazensis*, an overlooked snail distorting fascioliasis data: genotype, phenotype, ecology, worldwide spread, susceptibility, applicability. PLoS ONE 2011;6(9):e24567.
88. Sodeman WA Jr, Rodrick GE, Paraense WL, Vargas M. *Biomphalaria straminea* and other planorbids in the Dominican Republic. Mémorias do Instituto Oswaldo Cruz 1985;80(4):453–6.
89. Alda P, Lounnas M, Vázquez AA, Ayaqui R, Calvopiña M, Celi-Erazo M, *et al.* A new multiplex PCR assay to distinguish among three cryptic *Galba* species, intermediate hosts of *Fasciola hepatica*. Veterinary Parasitology 2018;251:101–5.
90. Uribe N, Becerra WM, Velázquez LE. *Lymnaea cousini*, huésped de *Fasciola hepatica* en el trópico alto andino de Colombia, y sus nuevos haplotipos, confirmados con el marcador mitocondrial del gen de la citocromo oxidasa I. Biomédica 2014;34(4):598–604.
91. Villavicencio A, Carvalho de Vasconcellos M. First report of *Lymnaea cousini* Jousseaume, 1887 naturally infected with *Fasciola hepatica* (Linnaeus, 1758) (Trematoda: Digenea) in Machachi, Ecuador. Mémorias do Instituto Oswaldo Cruz 2005;100(7):735–7.
92. Lounnas M, Correa AC, Alda P, David P, Dubois MP, Calvopiña M, *et al.* Population structure and genetic diversity in the invasive freshwater snail *Galba schirazensis* (Lymnaeidae). Canadian Journal of Zoology 2018;96:425–35.

93. Caron Y, Celi-Erazo M, Hurtrez-Boussès S, Lounnas M, Pointier JP, Saegerman C, *et al.* Is *Galba schirazensis* (Mollusca, Gastropoda) an intermediate host of *Fasciola hepatica* (Trematoda, Digenea) in Ecuador? Parasite 2017;24(24):1–9.
94. Salazar L, Estrada VE, Velázquez LE. Effect of the exposure to *Fasciola hepatica* (Trematoda: Digenea) on life history traits of *Lymnaea cousini* and *Lymnaea columella* (Gastropoda: Lymnaeidae). Experimental Parasitology 2006;114:77–83.
95. Coelho LHL, Lima WS, Guimarães MP. Sympatric and allopatric combinations of *Lymnaea columella* and *Fasciola hepatica* from southern and south-eastern Brazil. Journal of Helminthology 2009;83:285–8.
96. Paraense WL. *Lymnaea columella* in northern Brazil. Mémorias do Instituto Oswaldo Cruz 1983;78:477–82.
97. Moriena RA, Alvarez JD, Pietrokovsky S, Rubel D, Prepelitchi L, Racioppi O, *et al.* Presencia de *Pseudosuccinea columella* naturalmente infestada con *Fasciola hepatica* en Santo Tomé (Corrientes, Argentina). Revista Veterinaria 2008;19(2):147–9.
98. Alarcón EP, Velázquez LE. Descripción morfológica de *Cotylophoron cotylophorum* (Digenea: Paramphistomidae) hallado en bovinos de Rionegro, Antioquia, Colombia. Revista Colombiana de Ciencias Pecuarias 2009;22:168–77.
99. Bargues MD, Artigas P, Khoubbane M, Ortiz P, Naquira C, Mas-Coma S. Molecular characterisation of *Galba truncatula*, *Lymnaea neotropica* and *L. schirazensis* from Cajamarca, Peru and their potential role in transmission of human and animal fascioliasis. Parasites & Vectors 2012;5:174.
100. Bargues MD, Mera y Sierra RL, Gómez HG, Artigas P, Mas-Coma S. Ribosomal DNA ITS–1 sequencing of *Galba truncatula* (Gastropoda, Lymnaeidae) and its potential impact on fascioliasis transmission in Mendoza, Argentina. Animal Biodiversity and Conservation 2006;29(2):191–4.
101. Pointier JP, Cazzaniga N, González-Salas C, Gutiérrez A, Arenas J, Bargues MD, *et al.* Anatomical studies of sibling species within Neotropical lymnaeids, snail intermediate hosts of fascioliasis. Mémorias do Instituto Oswaldo Cruz 2006;101(4):431–5.
102. Kleiman F, Pietrokovsky S, Paraense WL, Wisinivesky-Colli C. Southernmost finding of *Lymnaea viatrix* Orbigny, 1835 (Pulmonata: Lymnaeidae), intermediate host of *Fasciola hepatica* (Linnaeus, 1758) (Trematoda: Digenea), in urban and rural areas of Patagonia, Argentina. Mémorias do Instituto Oswaldo Cruz 2004;99(1):23–4.
103. Cucher MA, Carnevale S, Prepelitchi L, Labbé JH, Wisnivesky-Colli C. PCR diagnosis of *Fasciola hepatica* in field-collected *Lymnaea columella* and *Lymnaea viatrix* snails. Veterinary Parasitology 2006;137:74–82.
104. Paraense WL. *Lymnaea rupestris* sp.N. from southern Brazil (Pulmonata: Lymnaeidae). Mémorias do Instituto Oswaldo Cruz 1982;77(4):437–43.
105. Blair D. Observations and experiments on some larval trematodes of freshwater snails and fish from southern Iceland. Journal of Helminthology 1973;47(4):409–14.
106. Vinarski MV. Three lymnaeid species (Gastropoda: Lymnaeidae) new for the Irtysh River basin (Western Siberia). The Bulletin of the Russian Far East Malacological Society 2008;12:71–8.
107. Anderson R. An annotated list of the non-marine Mollusca of Britain and Ireland. Journal of Conchology 2005;38(6):607.
108. Jones R, Williams H, Dalesman S, Brophy P. Confirmation of *Galba truncatula* as an intermediate host snail for *Calicophoron daubneyi* in Great Britain, with evidence of alternative snail species hosting *Fasciola hepatica*. Parasites & Vectors 2015;8:656.
109. Caron Y, Martens K, Lempereur L, Saegerman C, Losson B. New insight in lymnaeid snails (Mollusca, Gastropoda) as intermediate hosts of *Fasciola hepatica* (Trematoda, Digenea) in Belgium and Luxembourg. Parasites & Vectors 2014;7:66.
110. Dreyfuss G, Vignoles P, Rondelaud D. *Fasciola hepatica*: epidemiological surveillance of natural watercress beds in central France. Parasitology Research 2005;95:278–82.
111. Rondelaud D, Hourdin P, Vignoles P, Dreyfuss G. The contamination of wild watercress with *Fasciola hepatica* in central France depends on the ability of several lymnaeid snails to migrate upstream towards the beds. Parasitology Research 2005;95:305–9.
112. Rondelaud D, Belfaiza M, Vignoles P, Moncef M, Dreyfuss G. Redial generations of *Fasciola hepatica*: a review. Journal of Helminthology 2009;83:245–54.
113. Villavicencio A, Gorochov V, Carvalho de Vasconcellos M. *Lymnaea truncatula* Müller, 1774 (Pulmonata: Lymnaeidae) infected with *Fasciola hepatica* (Linnaeus, 1758) (Trematoda: Digenea), in Moscow districts, Russian Federation. Revista de Patología Tropical 2006;35(1):59–64.

114. Kozac M, Wędrychowicz H. The performance of a PCR assay for field studies on the prevalence of *Fasciola hepatica* infection in *Galba truncatula* intermediate host snails. Veterinary Parasitology 2009;168(1–2):25–30.
115. Knubben-Schweizer G, Torgerson PR. Bovine fasciolosis: control strategies based on the location of *Galba truncatula* habitats on farms. Veterinary Parasitology 2015(208):77–83.
116. Schweizer G, Meli ML, Torgerson PR, Lutz H, Deplazes P, Braun U. Prevalence of *Fasciola hepatica* in the intermediate host *Lymnaea truncatula* detected by real time TaqMan PCR in populations from 70 Swiss farms with cattle husbandry. Veterinary Parasitology 2007;150:164–9.
117. Schniebs K, Glöer P, Quiñonero-Salgado S, Lopez-Soriano J, Hundsdoerfer AK. The first record of *Galba cubensis* (L. Pfeiffer, 1839) (Mollusca: Gastropoda: Lymnaeidae) from open fields of Europe. Folia Malacologica 2018;26(1):3–15.
118. Schniebs K, Glöer P, Vinarski MV, Quiñonero-Salgado S, Lopez-Soriano J, Hundsdoerfer AK. A new alien species in Europe: first record of *Austropeplea viridis* (Quoy & Gaimard, 1833) (Mollusca, Gastropoda, Lymnaeidae) in Spain. Journal of Conchology 2017;42(5):357–70.
119. Vinarski MV, Aksenova OV, Bespalaya YV, Bolotov I, Gofarov MY, Kondakov AV. *Ladislavella tumrokensis*: The first molecular evidence of a Nearctic clade of lymnaeid snails inhabiting Eurasia. Systematics and Biodiversity 2016;14(3):276–87.
120. Vinarski MV, Glöer P. Taxonomic notes on Euro-Siberian freshwater molluscs. 3. *Galba occulta* Jackiewicz, 1959 is a junior synonym of *Lymnaea palustris* var. *terebra* westerlund, 1885*. Mollusca 2008;26(2):175–85.
121. Vinarski MV, Glöer P. Taxonomic notes on Euro-Siberian snails, 4. Re-examination of *Limnaea psilia* Bourguignat 1862, with the description of *Radix parapsilia* n. sp. Archives für Molluskenkunde 2009;138(2):123–36.
122. Agatsuma T, Arakawa Y, Iwagami M, Honzako Y, Cahyaningsih U, Kang S-Y, *et al.* Molecular evidence of natural hybridization between *Fasciola hepatica* and *F. gigantica*. Parasitology International 2000;49:231–8.
123. Le TH, van De N, Agatsuma T, Nguyen TGT, Nguyen QD, McManus DP, *et al.* Human fascioliasis and the presence of hybrid/introgressed forms of *Fasciola hepatica* and *Fasciola gigantica* in Vietnam. International Journal for Parasitology 2008;38(6):725–30.
124. Dajem SMB. Molecular typing of the fresh water snail *Lymnaea arabica*, the possible intermediate host of *Fasciola hepatica*, collected from Saudi Arabia, by RAPD-PCR. Egyptian Academy Journal of Biological Sciences 2012;4(1):173–81.
125. Dung BT, Doanh PN, The DT, Loan HT, Losson B, Caron Y. Morphological and molecular characterization of lymnaeid snails and their potential role in transmission of *Fasciola* spp. in Vietnam. Korean Journal of Parasitology 2013;51(6):657–62.
126. Chen JX, Chen MX, Ai L, Xu XN, Jiao JM, Zhu T, *et al.* An outbreak of human Fascioliasis gigantica in southwest China. PLoS ONE 2013;8(8):e71520.
127. Itagaki T, Fujiwara S, Mashima K, Itagaki H. Experimental infection of *Lymnaea* snails with Australian *Fasciola hepatica*. Japanese Journal of Veterinary Science 1988;50(5):1085–91.
128. Esteban JG, González C, Curtale F, Muñoz-Antoli C, Valero MA, Bargues MD, *et al.* Hyperendemic fascioliasis associated with schistosomiasis villages in the Nile Delta of Egypt. American Journal of Tropical Medicine and Hygiene 2003;69(4):429–37.
129. Malone JB, Gommes R, Hansen J, Yilma JM, Slingenberg J, Snijders F, *et al.* A geographic information system on the potential distribution and abundance of *Fasciola hepatica* and *F. gigantica* in east Africa based on Food and Agriculture Organization databases. Veterinary Parasitology 1998;78:87–101.
130. Dar Y, Rondelaud D, Dreyfuss G. Update of fasciolosis-transmitting snails in Egypt (review and comment). Journal of the Egyptian Society of Parasitology 2005;35(1):477–90.
131. Dar Y, Djuikwo FF, Vignoles P, Dreyfuss G, Rondelaud D. *Radix natalensis* (Gastropoda: Lymnaeidae), a potential intermediate host of *Fasciola hepatica* in Egypt. Parasite 2010;17:251–6.
132. Assogba MN, Dougnon JT, Alkoiret IT, Akpo Y, Boukari M, Youssao AKI. Epidémiologie de la fasciolose bovine au Bénin et essai de lutte biologique. Livestock Research for Rural Development 2011;23(12):248.
133. Adediran OA, Uwalaka EC. Laboratory breeding of *Lymnaea natalensis* (Krauss, 1848), intermediate host of *Fasciola gigantica* (Cobbold, 1856). New York Science Journal 2013;6(8):55–7.

134. Fromsa A, Meharent B, Mekibib B. Major trematode infections of cattle slaughtered at Jimma municipality abattoir and the occurrence of the intermediate hosts in selected water bodies of the zone. Journal of Animal and Veterniary Advances 2011;10(12):1592–7.
135. Howell A, Mugisha L, Davies J, LaCourse EJ, Claridge J, Williams DJL, *et al.* Bovine fasciolosis at increasing altitudes: parasitological and malacological sampling on the slopes of Mount Elgon, Uganda. Parasites & Vectors 2012;5(1):196.
136. Stothard JR, Kristensen TK. Medical and veterinary malacology in Africa. Parasitology Today 2000;16(3):85–6.
137. Vassiliades G. Capacité de résistance à la sécheresse de la limnée (*Limnaea natalensis*), mollusque hôte intermédiaire de *Fasciola gigantica*, au Sénégal. Revue d'Elevage et de Médecine Vétérinaire des Pays Tropicaux 1978;31(1):57–62.
138. Stothard JR, Brémond P, Andriamaro L, Loxton NJ, Sellin B, Sellin E, *et al.* Molecular characterization of the freshwater snail *Lymnaea natalensis* (Gastropoda: Lymnaeidae) on Madagascar with an observation of an unusual polymorphism in ribosomal small subunit genes. Journal of Zoology 2000;252:303–15.
139. Daynes P. La distomatose à Madagascar. Rapport sur des travaux réalisés de *Fasciola gigantica*. Revue d'Elevage et de Médecine Vétérinaire des Pays Tropicaux 1967;20(4):557–62.
140. Barré N, Isautier H, Frandsen F, Mandahl-Barth G. Inventaire des mollusques d'eau douce de la Réunion. Revue d'Elevage et de Médecine Vétérinaires des Pays Tropicaux 1982;35:35–41.
141. Kock KN, Wolmarans CT, Bornman M. Distribution and habitats of the snail *Lymnaea truncatula*, intermediate host of the liver fluke *Fasciola hepatica*, in South Africa. Journal of the South African Veterinary Association 2003;74(4):117–22.
142. Prinsloo JF, Van Eeden JA. Population dynamics of freshwater snails in Lesotho with particular reference to *Lymnaea truncatula*, the intermediate host of *Fasciola hepatica*. Wetenskaplike Bydraes van die PU vir CHO, Reeks B: Natuurwetenskappe 1973;72:1–60.
143. Ghouaidia J, Hammami H. Interaction between the intermediate host of fascioliasis in Tunisia, *Galba truncatula* and a possible competitor, *Melanoides tuberculata* (Muller, 1774): a field study. African Journal of Ecology 2014;52(3):328–33.
144. Amor N, Farjallah S, Said K, Slimane BB. First report of *Fasciola hepatica* in *Equus caballus* host species from Tunisia based on the ribosomal internal transcribed spacer regions. Turkish Journal of Veterinary Animal Science 2011;35(5):319–24.
145. Mekroud A, Benakhla A, Vignoles P, Rondelaud D, Dreyfuss G. Preliminary studies on the prevalences of natural fasciolosis in cattle, sheep, and the host snail (*Galba truncatula*) in north-eastern Algeria. Parasitology Research 2004;92:502–5.
146. Mekroud A, Titi A, Benakhla A, Vignoles P, Rondelaud D. *Fasciola hepatica*: sensibilité des *Galba truncatula* du nord-est algérien à l'infestation expérimentale avec des miracidiums sympatriques. Revue de Médecine Vétérinaire 2006;157(10):494–501.
147. Khallayoune K, Stromberg BE, Dakkak A, Malone JB. Seasonal dynamics of *Fasciola hepatica* burdens in grazing Timahdit sheep in Morocco. International Journal for Parasitology 1991;21(3): 307–14.
148. Belfaiza M, Moncef M, Rondelaud D. *Fasciola hepatica*: an unusual case of adaptation to a Moroccan population of *Galba truncatula*. Parasitology Research 2005;95:374–8.
149. Goll P, Scott J. The interrelationship of *Lymnaea truncatula* and ovine fascioliasis in the Ethiopian central highlands. British Veterinary Journal 1978;134(6):551–5.
150. Dar Y, Rondelaud D, Dreyfuss G. Cercarial shedding from *Galba truncatula* infected with *Fasciola gigantica* of distinct geographic origins. Parasitology Research 2003;89:185–7.
151. Grabner D, Mohamed FAAM, Nachev M, Méabed EMH, Sabry AHA, Sures B. Invasion biology meets parasitology: a case study of parasite spill-back with Egyptian *Fasciola gigantica* in the invasive snail *Pseudosuccinea columella*. PLoS ONE 2014;9(2):e88537.
152. Nadasan DS. The invasive potential of the freshwater snail Radix rubiginosa recently introduced into South Africa [PhD thesis] University KwaZulu-Natal, Durban, South Africa; 2011.
153. Ponder W, Waterhouse J. A new genus and species of Lymnaeidae from the lower Franklin River, South Western Tasmania, Australia. Journal of Molluscan Studies 1997;63:441–68.
154. Boray JC. The potential impact of exotic *Lymnaea* spp. on fascioliasis in Australia. Veterinary Parasitology 1978;4:127–41.
155. Spratt DM, Presidente PJA. Prevalence of *Fasciola hepatica* infection in native mammals in southeastern Australia. Australian Journal of Experimental Biology and Medical Sciences 1981;59(6):713–21.

156. Estuningsih SE, Copeman DB. Trematode larvae in *Lymnaea rubiginosa* and their definitive hosts in irrigated rice fields in West Java. Jurnal Ilmu Ternak dan Veteriner 1996;1(3):200–5.
157. Cowie RH. Invertebrate invasions on Pacific Islands and the replacement of unique native faunas: a synthesis of the land and freshwater snails. Biological Invasions 2001;3:119–36.
158. Lee CG, Cho SH, Lee CY. Metacercarial production of *Lymnaea viridis* experimentally infected with *Fasciola hepatica*. Veterinary Parasitology 1995;58(4):313–8.
159. Pointier JP, Marquet G. Taxonomy and distribution of freshwater molluscs of French Polynesia. The Japanese Journal of Malacology 1990;48:147–60.
160. Lounnas M, Correa AC, Vázquez AA, Dia A, Escobar JS, Nicot A, *et al.* Self-fertilization, long-distance flash invasion and biogeography shape the population structure of *Pseudosuccinea columella* at the worldwide scale. Molecular Ecology 2017;26:887–903.

16 Non-arthropod pests of date palm and their management

H.A.F. El-Shafie[1,2*] and B.M.A. Abdel-Banat[1,2]

[1] Date Palm Center of Research Excellence, King Faisal University, Al-Hofuf, 31982-Al-Hassa, Kingdom of Saudi Arabia. [2] Department of Crop Protection, Faculty of Agriculture, University of Khartoum, 13314-Shambat, Sudan

Abstract

In addition to the arthropod pests, the date palm *Phoenix dactylifera* L. is attacked by numerous non-arthropod pests, which include birds, rodents, bats, nematodes and snails. This review lists 57 species of birds reported in the agro-ecosystem of date palm. Of these, three species are major pests, 35 species are minor pests, 15 species are neutral and four species are birds of prey (raptors). The major bird pest species are the bulbuls, *Pycnonotus* spp., the rose-ringed parakeet *Psittacula krameri* and the house sparrow *Passer domesticus*. Rodents comprise 13 species, of which 12 are rats and mice and one species of porcupine. The major rodents are the roof rat *Rattus rattus*, the Norway rat *Rattus norvegicus*, the house mouse *Mus musculus*, the Nile rat *Arvicanthis niloticus* and the mole rat *Nesokia indica*. The Egyptian fruit bat, *Rousettus aegyptiacus*, is reported to be an important pest of date palm fruits. The review also lists 24 species of nematodes belonging to the following genera: *Meloidogyne*, *Pratylenchus*, *Helicotylenchus* and *Tylenchorhynchus*. Additionally, four species of land snails associated with date palm are listed. The article gives comprehensive information on the major non-arthropod pests of date palm regarding their distribution, biology, behaviour, nature of damage and possible management measures. Non-arthropod pests of minor importance are mentioned as checklists.

Keywords: Non-arthropod pest, Date palm, Pest management, Frugivorous birds, Parasitic nematode, *Phoenix dactylifera*, Rodents

Review Methodology

The authors searched databases such as Google Scholar, Science Direct, Researchgate and others for relevant publications on non-arthropod pests of date palm using the following keywords: vertebrate pests, rodents, frugivorous birds, date palm. Information used in compiling the manuscript was obtained from peer-reviewed articles, books and reports. The authors took photographs documenting damage on date palm by vertebrate pests during the period of writing this manuscript.

Introduction

Non-arthropod pests of date palm include vertebrate (birds, rodents, bats) and invertebrate other than insects (snails and nematodes). The damage to dates and date palms due to birds, rodents, bats, snail and nematode is overlooked and only mentioned briefly at the end of chapters describing insect pests' damage. Vertebrate pests cause substantial damage to date fruits in

*Corresponding author: elshafie62@yahoo.com

the field and storage. For example, birds depredate date fruits while on bunches (sweet *Rutab* and *Tamr* stages), during curing and in storage, particularly the house sparrow, *Passer domesticus* [1, 2]. Rats, on the other hand, can cause up to 30% damage to dates during storage as well as on the tree [3, 4]. Rats can also damage date palm roots, tender shoots and newly emerged inflorescences. Frugivorous bats also make serious damage on date fruits [5]. Many species of nematodes are usually found in the rhizosphere of date palm inflicting serious damage on the root system [6]. Snails can be potential pests of date palm offshoots when humidity is high, where they can inflict serious injury to young offshoots [7]. Data describing quantitative estimates of rodents and birds damage on agricultural crops are very scant and what available are largely based on general reports and qualitative data [8]. The data documenting the damage of non-arthropod pests on date palm are scant and scattered; therefore, this review is written with the aim of consolidating such data in a comprehensive document, which will be available for future utilization on this group of important date palm pests.

Birds

Birds generally feed on dates during the sweet *Khalal*, *Rutab* and *Tamr* maturation stages. The most important depredating bird species on date palm fruit are bulbuls, parakeet, house sparrows, crows and mynas [5, 9]. Bulbuls are by far the most injurious avian pests on date palm in Africa and the Middle East [1, 2]. They are also potential dispersers of noxious weed seeds, besides being vector and reservoir for avian diseases and parasites [10]. Three species of bulbuls are considered serious pests of dates; the common bulbul *Pycnonotus barbatus* in Africa [1, 2], white-cheeked bulbul *Pycnonotus leucogenys* and white-eared bulbul *Pycnonotus leucotis* in Asia [5, 9]. Other bird species that inflict major damage on dates include the rose-ringed parakeet *Psittacula krameri* and the house sparrow, *P. domesticus* [1, 2, 9]. The following paragraphs deal with the major bird pest species.

The White-Eared Bulbul, *P. leucotis* Gould (Passeriformes: Pycnonotidae)

The white-eared bulbul is native to Afghanistan, India, Iraq, Iran, Kuwait, Pakistan and Saudi Arabia. It is also introduced into Bahrain, Oman, Qatar, United Arab Emirates, Israel, Jordan and Syria [11]. It is difficult to differentiate between the males and females of bulbuls due to the absence of a clear sexual dimorphism as the case with some other birds. However, females as a rule are smaller than males and have rather more slender, down curved dark grey bills; they may also have less bright colours [12]. The size of an adult white-eared bulbul is about 18 cm long (from tail to beak) and weighs 30–45 g. Head and throat are black with conspicuous white large patches covering the ears, and hence the common name of the white-eared bulbul (Figure 16.1).

The vent (under tail coverts) is orange-yellow and the tail is dark with white tips, while the rest of body plumage is greyish-brown and the wings are dark. The legs and feet are dark grey and the front from the chest to the belly is light grey. The dark eyes are surrounded by pale bare eye rings. The juvenile bulbuls can be differentiated from the adults by their paler heads, grey ear coverts and pale yellow vents [13]. Bulbuls are usually associated with human habitation and they avoid colonizing open areas without trees because they have short and rounded wings, which are not suitable for long distance flight [14]. Bulbuls make untidy, cup-shaped nests made from fine twigs, dry grass and plant fibres concealed in the branches of many trees. The number of eggs per clutch is usually 2–3 eggs which are creamy-white with purple dots (Figure 16.2).

Bulbuls are monogamous and territorial with males aggressively defending their territory against other males [13]. The female typically builds the nest, lays 2–3 eggs (clutch size) and incubates them. Incubation period is 14–15 days and both parents attend the chicks, which reach the fledgling stage in about 15 days and may leave the nest in a week later. The food of young chicks consists largely of small insects, the proportion of which declines with ageing of hatchlings [13]. The white-eared bulbuls usually

Fig. 16.1. Unsexed adult of the white-eared bulbul, *Pycnonotus leucotis*, perching on a date palm fruit bunch (photo: B.M.A. Abdel-Banat).

Fig. 16.2. Nest of the white-eared bulbul, *Pycnonotus leucotis*, on the midrib of a date palm frond (leaf) (photo: H.A.F. El-Shafie).

feed gregariously in the early hours of the morning (8:00–10:30) and rest at the midday with only a few birds confining their activities to grooming of their feathers and jumping from one tree branch to another without feeding. The second peak of feeding activity is about 2 h before sunset, where so many birds have been observed to feed voraciously on dates. The date fruit can either be consumed completely leaving only the pits attached to the fruit strand or partially eaten (Figure 16.3). Such kind of damage lowers the quality of date fruit and render it unacceptable for human consumption. Bulbuls might cause more damage on dates by repeatedly plucking and dropping the ripe dates as they unsuccessfully attempt to feed on them (Figure 16.3). This usually happens during *tamr* stage when fruits are loosely attached to the bunch strands through their caps.

Rose-Ringed (Ring-Necked) Parakeet, *P. krameri* Scopoli (Psittaciformes: Psittacidae)

The rose-ringed parakeet is native to central Africa and Asia; however, it has been introduced and established in many parts of the world [15]. The parakeet has a medium size of 40 cm and bright green colour. The male can be distinguished from the female by a black collar extending from the throat and below the cheeks that become a rose-coloured ring around the hind neck. The female has a similar colour to the male but lacks the black collar and the rose-coloured ring on the neck. The parakeet has a red, short, round and strongly hooked bill [16]. This species utilizes more than 50 species of trees and shrubs as nesting sites. Parakeets are found near water resources and agricultural fields feeding on fruits, flowers, nectar, grain and cereals depending on their availability [17]. The search for nesting sites starts from May to December and from October to August and copulation takes place in February to May [18].

The rose-ringed parakeet is considered the worst vertebrate pest of agriculture in central Punjab of Pakistan [19]. This bird has been reported to cause considerable damage to date fruits of the cultivar Halawy during July in India. It depredates date fruits throughout the day except for the hot afternoon hours; however, maximum depredation on fruits occurs in the early morning and late afternoon [9]. The parakeet wastes more fruits than it can actually eat through taking a few bites from each fruit before discarding it and seeking another one [19, 20]. The parakeets spend the night in communal roosts in which they gather in the evening and leave these roosting sites in the morning to search for food. The number of parakeets in each roosting site varies considerably; however, it can range from 100 to 10 000 birds [21]. Parakeets prefer trees with a diameter of 0.5–2 m wide and 2–7 m tall from the ground level for nesting. Within such trees, they select large nesting holes with sizes of 5–20 cm for breeding and living [15]. In Europe, it often out-competes native birds for both food and nest sites. Aggressive behaviour of the parakeets including harassment against house sparrow, *P. domesticus*, has

Fig. 16.3. Bird damage on date palm fruits, injury while on the bunch (left) and shedding on the ground (right) (photo: H.A.F. El-Shafie).

been documented [22]. The female incubates the eggs for more than 3 weeks, while she is fed and guarded by her mate. The naked chicks hatch and their eyes open 10 days later: Both parents feed the young for 7–8 weeks, regurgitating partly digested food into their bills. The parakeet reaches sexual maturity in 1–2 years. The clutch size is 3–6 eggs, incubation period is 22–24 days, fledging period is 50–55 days and adult life span is 15–25 years. The clutch sizes of this bird in the northeast England were similar to those in the native range given that food and nesting sites are available [23]. This clearly indicates the reproductive success of the parakeet as invasive species [24]. Waseem *et al.* [15] indicated that the observed variations in clutch size of the rose-ringed parakeet in different parts of the world might be attributed to climate, breeding conditions, nesting availabilities and proportion of predators.

The House Sparrow, *P. domesticus* L. (Passeriformes: Passeridae)

The house sparrow nests in holes, cavities in building and similar places. It has been rarely seen nesting on tree branches. The nest consists of dry grass and fine materials that include feathers, cotton wools, strings and papers, which are used to smoothen the interior of the nest. The female clutch size average 4–5; however, it may be up to 10 eggs. The eggs have creamy to darker colour with some black dots on the chorion. The size of the egg is about 2–2.2 cm long and 1.4–1.6 cm wide [7]. The newly hatched young has bright pink naked skin with closed eyes. Young birds are fed upon small insects by both parent and later in their life on grains [7]. Nhlane [25] gave detailed information on the biology and breeding ecology of the house sparrow. House sparrows damage date fruits through all developmental stages including small green fruits formed just after fruit setting (*Hababouk* stage). It pecks on the fruit in the *Rutab* stage and on the fully ripened dates or *tamr*. The birds spoil the fruit while on the bunch by their droppings. The house sparrow is considered one of the most destructive avian pests of date palm particularly the fruits on bunches, spread on mats for drying and during storage [1, 2]. Guezoul *et al.* [26] evaluated the losses to date fruits caused by the hybrid sparrow (*P. domesticus* × *Passer hispaniolensis*) in the region of Quargla (Algeria). They estimated the losses to about 510 kg/ha or 8.3% of the yield. The house sparrow may occasionally knocks off flowers at the time of date palm blooming and then devours the embryonic fruit [7].

Date Palm as Nesting Site for Birds

The dense cone-shaped canopy of the date palm, together with the unique architecture of the fronds have made the tree more attractive, as nesting site, for numerous bird species (Figures 16.3 and 16.4).

The birds reported in the literature to have nested on date palm include the common myna, the white-eared bulbul, the house sparrow, the rose-ringed parakeet, the monk parakeet, the Indian house sparrow, the laughing turtle dove, the rose-faced lovebird and weaver birds. The laughing turtle dove, *Streptopelia senegalensis*, prefers to nest in the axle of fronds on palms that are more than 4–5 m tall, while it may construct the nest on the midrib of the fronds of date palms <2 m tall (Figure 16.4). Some weaverbirds of the genus *Ploceus* have been reported to hang their nests at the tip of date palm fronds [1]. The monk parakeet usually builds communal nests containing many chambers, which are used for breeding and roosting. Parakeets build their nests around the base of date palm leaf (frond) which eventually merge and form a ring around the trunk [27].

Minor Pestiferous Bird Species Recorded on Date Palm

The bird pest species that inflict mild damage on date palm fruits are listed in Table 16.1.

Other Species of Birds Recorded in the Date Palm Agro-Ecosystem

Several species of birds, other than pestiferous one, have been recorded in the date palm

Fig. 16.4. The laughing turtle dove, *Streptopelia senegalensis* L., nest on the midrib of date palm frond (top, left), nest with two eggs (top, right), nestlings (bottom, left) and juvenile dove (bottom, right) (photo: H.A.F. El-Shafie).

agro-ecosystem (Table 16.2). Many of these birds are insectivorous, and thus play an important role in regulating the harmful insect population as well as other ecological roles.

Biological Control of Bird Pests

Diurnal birds of prey (raptor) including hawks, eagles, kites and vultures may play some roles in controlling the population of rodents and frugivorous birds. The nocturnal owl is by far the most important raptor that has been reported to regulate the population of rodents in the date palm ecosystem. Common buzzard *Buteobuteo*, golden eagles *Aquila chrysaetos*, steppe buzzard *Buteovulpinus* from the family Accipitridae, lesser kestrel and several species from the genus *Falco* of the family Falconidae were also reported to regulate the population of pestiferous vertebrates [28, 34]. The barn owl, *Tyto alba erlangeri*, has been used to control rodents in Israel and Malaysia since 1982. However, the problem with this raptor is the secondary poisoning (when owls feed on rodents or birds poisoned by pesticides). Nesting boxes are erected in areas infested with rodents to lure the owls for establishing themselves in these nests. The use of barn owl is common practice in date palm farms that produce organic dates. This practice of replacing the use of rodenticides with biological control to manage rodents has a great benefit to other raptor birds including kestrels [37]. The use of barn owls as a biological control method could be a more cost-effective alternative, which might also reduce the negative impacts associated with rodenticides [38]. Kan *et al.* [38] used a model that incorporated the spatial patterns of barn owl predation pressure on rodents, and the impact of this predation pressure on nesting choices and agricultural output. They concluded that using barn owls for controlling rodents is profitable.

Bird Management

The development of an effective bird-damage management programme is difficult due to many factors including birds' mobility and their unpredictable changing feeding sites [39]. Other problems in the management of pest bird species on date palm is lack of simple and effective methods of estimating damage upon which management action should be carried out. Additionally, no information on the cost-effectiveness of management measures currently used against bird pests [40]. Lindell *et al.* [41] stressed the significance of the spatial distribution of fruit and non-fruit areas of the farm in bird management. They recommend preparation of bird management plans that take into account the aspect of spatial distribution and emphasized that farmers should invest more in bird management in low-yield years and in block isolated from other blocks of the same crop. Blocks in which trees can avail roosting sites for frugivorous birds should be given more attention in the management plan. Many techniques have been used to manage bird depredations on field crop such as visual repellents, chemical repellents, optical repellents; however, few of these methods are effective [42]. Pre-recorded distress call of parakeet was used as a bio-acoustic technique to manage parakeet in Indian agricultural sectors [43]. As preventive measures, harvested dates should not be left uncovered for long period in the field, to avoid invitation of depredating bulbuls. Spreading of dates on mats for sun drying, which is practiced in some date-growing countries, should be carried out with complete exclusion of birds by using special nets or covers. The use of simple techniques such as netting and nest destruction significantly reduced the damage of Spanish sparrow, *Passer hespaniolensis* on wheat crop [29]. Removal of bird perching places, roosting sites and water resources can help mitigate the depredations of these birds on dates. Pre-harvest fruit bagging is considered one of the most important good agricultural practices used to produce high-quality and healthy fruits [44]. Fruit bagging has been used to help reduce damage in dates due to birds and other injury-causing agents [5, 45], and many types of bags including woven cotton, weave mesh, nylon weave, polyethylene and wax paper were used [9]. The suitability of these types of bags for the protection of dates against birds depend very much on the date palm variety and prevailing climatic conditions. Moreover, this technique has been reported to enhance ripening and fruit quality of some date cultivars [46]. Bunch covers should not affect the ventilation of fruits, which is essential during these stages of fruit development. Early harvesting of dates may be considered one of the most important agro-technical measures to lessen the magnitude of birds' damage on dates.

Rodents (Rats, Mice, Squirrel and Porcupine)

Rats may cause damage to date palm by burrowing underground galleries that negatively affect the irrigation systems as well as the roots of date palm (Figure 16.5).

The roof rat, house mouse and Nile rat can cause severe damage to dates during storage [1]. Losses of dates and coconuts to roof rat (*Rattus rattus*) are thought to be about 30% [8]. Rat damage to offshoots, sucker and mature palms can predispose them to subsequent attack by insect borers, particularly the red palm weevil, *Rhynchophorus ferrugineus* and the dynastic beetle, *Oryctes* spp. Tripathi and Chaudhary [47] reported that dates at ripening stage may suffered squirrel damage of up to 60–80%. Damage to suckers of date palm (*Phoenix dactylifera*) by the Indian crested porcupine, *Hystrix indica*,

Table 16.1. Bird species inflicting minor injury to date palm fruits

Order/family/species	Common name	Reference
Passeriformes, Passeridae		
Passer montanus L.	Tree sparrow	[28]
Passer euchlorus (Bonaparte)	Arabian golden sparrow	[7]
Passer hespaniolensis Temm	Spanish sparrow	[29]
Petronia xanthocollis (Burton)	Yellow-throated sparrow	[5]
Petronia brachydactyla (Bonaparte)	Pale rock sparrow	[5]
Oriolus oriolus L.	The golden oriole	[5]
Passeriformes, Pycnonotidae		
Pycnonotus leucotis Gould	White-eared bulbul	[5]
Pycnonotus barbatus (Desfontaine)	Common bulbul	[1, 2]
Pycnonotus nigricans (Vieilott)	Red-eyed bulbul	[30]
Pycnonotus leucogenys (Gray)	White-cheeked bulbul	[9]
Pycnonotus cafer Deignen	Red-vented bulbul	[10]
Pycnonotus goiavier (Scopoli)	Yellow-vented bulbul	[9]
Pycnonotus xanthopygos (Ehrenberg)	White-spectacled bulbul	[28]
Pycnonotus jocosus L.	Red-whiskered bulbul	[10]
Passeriformes, Sturnidae		
Acridotheres tristis L.	Common myna	[9]
Acridotheres ginginianus (Latham)	Bank myna	[9]
Lamprotornis chloropterus Swainson	Lesser blue-eared starling	[30]
Sturnus vulgaris L.	Common starling	[31]
Passeriformes, Corvidae		
Corvus ruficollis Lesson	Brown-necked raven	[7]
Corvus rhipidurus Hartort	Fan-tailed raven	[7]
Corvus splendens Vieilott	Indian house crow	[9]
Corvus corone cornix L.	Hooded crow	[28]
Corvus macrorhynchos Wagler	Jungle crow	[32]
Pica pica L.	Eurasian magpie	[32]
Passeriformes, Ploceidae		
Quelea quelea L.	Red-billed Quelea	[30]
Passeriformes, Estrildidae		
Amadina erythrocephala L.	Red-headed finch	[30]
Musophagiformes, Musophagidae		
Corythaizoides concolor (Smith, A.)	The grey lourie	[30]
Psittaciformes, Psittacidae		
Myiopsitta monachus Boddaert	Monk parakeet	[33]
Poicephalus rueppellii (Gray, GR.)	Rüpells parrot	[30]
Agapornis roseicollis (Vieilott)	Rosy-faced lovebird	[30]

Table 16.1. *Continued*

Order/family/species	Common name	Reference
Columbiformes, Columbidae		
Streptopelia decipiens (Hartlaub & Finsch)	African mourning dove	[34]
Streptopelia senegalensis L.	Laughing turtle dove	[34]
Streptopelia decaocto (Frivaldszky)	Eurasian collared dove	[34]
Streptopelia turtur L.	Turtle dove	[34]
Columba livia (Gmelin)	Common pigeon	[28]

Table 16.2. Bird species recorded in date palm groves without damaging date fruits

Order/family/species	Common name	Reference
Passeriformes/Muscicapidae		
Cercotrichas podobe Statius Müller	Black scrub robin	[35]
Passeriformes/Leiotrichidae		
Argya squamiceps Cretzschmar	Arabian babbler	[36]
Passeriformes/Laniidae		
Lanius senator L.	Woodchat shrike	[34]
Passeriformes/Motacillidae		
Motacilla alba L.	White wagtail	[34]
Motacilla flava L.	Western yellow wagtail	[34]
Passeriformes/Phylloscopidae		
Phylloscopus collybita (Vieillot)	Common chiffchaff	[34]
Phylloscopus fuscatus (Blyth)	Dusky warbler	[34]
Bucerotiformes/Upupidae		
Upupa epops L.	Common hoopoe	[34]
Coraciformes/Meropidae		
Merops apiaster L.	Bee-eater	[34]
Merops persicus Pallas	Blue-cheeked bee-eater	[34]
Piciformes/Picidae		
Dendrocopos syriacus Ehrenberg	Syrian woodpecker	[28]
Galliformes/Phasianidae		
Alectoris chukar (Gray)	Chukar Partridge	[28]
Alectoris barbara (Reichenow)	Barbary Partridge	[34]
Columbiformes, Columbidae		
Oena capensis L.	Namaqua dove	[1]

[1](El-Shafie, HAF, personal observations).

through uprooting is quite common in Punjab and Balochistan. A small farm near Bhakhar with 500 suckers suffered 100% damage within a month [48]. Lathiya *et al.* [49] studied the faecal pellets of the short-tailed mole rat, *Nesokia indica*, collected from infested date palm orchard in Pakistan. They concluded that dates constituted 40.24% of the mole diet contents, while the date palm stem was 45.9% during the non-fruiting season. This rat is also found

Fig. 16.5. Burrow openings of rodents in a date palm grove, in the root zone of a young palm (left) and under a heap of bunches of refused dates left in the field (right) (photo: H.A.F. El-Shafie).

in India, Bangladesh, Iran, Iraq, Egypt, Syria, North Arabia, Chinese Turkistan and Southern Russian Turkistan [50, 51]. Following are the major rats and mice that inflict serious damage on date palm.

The Mole Rat (Short-Tailed Bandicoot Rat), *N. indica* Gray (Rodentia: Muridae)

This rat is found in Egypt, Afghanistan, Pakistan, Syria and Iraq [8]. It digs long tunnels in sandy soil that can reach 100 meters and about 30 cm below the surface. It feeds on the pith of date palms; it also damages the base of young palms 10–15 years old that eventually lead to fungal infection. This rat is reported to feed on the soft apical meristem of the date palm resulting in the death of the tree [52]. Additionally, the rat depredates date fruits leading to substantial losses. This rat has been reported to inflict economical damage on sugar canes that can reach 50%. It also attacks vegetables and field crops, particularly rice. Due to its fossorial habit, the short-tailed bandicoot rat is recognized as a serious pest that creates many problems with respect to irrigation water management [8]. This species has emerged as a major pest in Nok-Kundi and Dalbandin in Pakistan after a period of 4–5 years of drought. Ahmed *et al.* [53] reported that the addition of 55 dates in the bait of the short-tailed mole rat containing the rodenticides brodifacoum and bromadiolone enhanced the acceptability of the bait by about 69%.

The Roof Rat (Black Rat), *R. rattus* L. (Rodentia: Muridae)

The roof rat is cosmopolitan and commensal usually found in human habitation [8]. It has a medium size of 13–18 cm in length. The tail of this rat is slender and longer than the combined length of head and body. This rat is characterized by having large ears. Colour is grey to jet-black, and the belly is greyish-white or white (Figure 16.6).

The roof rat is a good climber and arboreal in habit. It is omnivorous feeding on different fruits, nuts, sugar cane, vegetables and others. They need to drink about 15–30 mL of water daily, and accordingly, they nest near a water resource. The average weight of the rat is 137 g. The roof rat consumes about 8–12 g of food daily, which amounts to 3–4 kg annually [54]. The male of this rat is polygamous; the female

Fig. 16.6. An adult of the roof rat, *Rattus rattus* (photo: M.M. Al-Ziat).

gives 6–8 litters per year with 4–12 young in each litter, and gestation period is 21–23 days [7]. The roof rat can inflict economic damage on different crops including fruit trees. The damage to dates on the bunch and in storage facilities by the roof rat is well documented [1, 2]. The roof rat becomes active during *rutab* and *tamr* ripening where it climbs date palms and voraciously feed on dates. The roof rat can cause complete removal of date fruits from the bunches, leaving empty strands. It can also inflict severe damage on the inflorescences and immature fruits through all stages of their development [3]. It can also burrow in the trunk of date palm, particularly old ones and makes cavities to nest. It also damages dates during curing and storage. Additionally, this rat can contaminate dates by its urine and droppings. The galleries made by the roof rat in the trunk of date palms facilitate the entry of the devastating red palm weevil, *R. ferrugineus*, and probably other opportunistic insect pests [7].

The Norway Rat (Brown Rat, Cellar Rat), *Rattus norvegicus* (Berkenhout) (Rodentia: Muridae)

This rat measures about 20–25 cm, weighing 283–510 g. The tail of this rat is stout and usually shorter than the length of the head and body together. It burrows in the ground or nest in rubbish piles or garbage dumps [55]. The rat is sexually prolific; a female can give 8–12 offspring every 30 days. Young rats become sexually mature at 3–4 months of age and can reproduce continuously until death. The female may come into heat every 4–5 days and can mate within a day or 2 after giving birth. The average female has 4–6 litters per year [7, 55]. The Norway rat is common in towns and cities where it causes damage to stored food and other properties. It is usually out-competed by indigenous species [55]. The Norway rat consumes 15–25 g of food daily, which corresponds to 6–9 kg annually [54].

The House Mouse, *Mus musculus* L. (Rodentia: Muridae)

The house mouse is cosmopolitan and commensal pest mostly found in connection with domestic premises. It inflicts serious economic losses on stored grains. It also contaminates stocks and damage sacks of stored crops [3, 8]. It has small size with an average weight of 15–20 g. The back of the mouse is grey or brown, while the belly is creamy white. The tail is longer than head and body together. This mouse breeds all the year round, and female gives birth to 6–10 young per litter. The gestation period is 21 days [7]. The species is reported to inflict serious damage on dates either in the field or in storage facilities [1, 2]. According to its body mass, the house mouse consumes 2–3 g of food daily and 0.7–1 kg per year [54].

The Nile Rat (Unstriped Grass Rat), *Arvicanthis niloticus* (E. Geoffroy) (Rodentia: Muridae)

This rat is mostly found in Nile valley and delta in Egypt. It is found in Sudan, Ethiopia, Somalia, Djibouti and Senegal [8]. The Nile rat causes injury to many field crops including grains, vegetable, fruits and sugar cane. Greaves [8] reported that the damage on field crops due to field rodents mainly by the Nile rat was estimated to be US$50–60 million annually. The amount of food eaten by this species of rat is 5–8 g, which is about 2–3 kg per year according to Brooks and Fiedler [54].

Multimammate Rat, *Mastomys natalensis* (A. Smith) (Rodentia: Muridae)

It is distributed in Sudan, Somalia, Djibouti and Mauritania; however, no information is available regarding its damage on field crops. Khairi [1] documented the damage to date fruits by this species in Sudan. The daily and annual food requirement of this rat has been determined by Brooks and Fiedler [54] to be 5–8 g and 2–3 kg, respectively.

Other species of Rodents Recorded in Date Palm Groves

Several other species of rodents reported in the agro-ecosystem of date palm are listed in Table 16.3.

Rat Management

The effectiveness of a rodenticide depends largely on the use of appropriate compound as well as the choice of suitable formulation and application method [8]. The bait of rat commonly used consists of a mixture of zinc phosphide at 30–50 g, 1 kg of millet flour and 3% edible oil. This bait should be placed around infested date palm in bait stations, to avoid poisoning of non-target animals [30]. Sedra [2] reported that poisoned bait used in Morocco against the house mouse *M. musculus* consisted of strychnine sulphate (15 g/kg of food substrate) and Vertox pellets (brodifacoum 0.01% at 100 g/kg substrate). Anticoagulant baits (chlorophacinone and diphacinone) are also available for controlling pocket gophers. Anticoagulant baits are less toxic than strychnine and zinc phosphide, thereby reducing potential mortality from incidental ingestion of these baits by non-target species [57]. The combination baits containing an acute toxicant and anticoagulants gave 90% efficacy against pocket gophers [57]. Addition of denatonium benzoate (DB) in rodenticidal baits maybe an effective tool in reducing accidental ingestion of poison baits by human beings and other non-target species because the compound (DB) is an intensely bitter and non-toxic substance. It is detectable by humans at a concentration of 10 ppb [58]. Bromadiolone is a well-known single-dose second-generation anticoagulant rodenticide worldwide and is highly effective against commensal as well as field rodents. Its mode of action includes inhibition of formation of prothrombin necessary for coagulation of blood, thus causing fatal haemorrhages [47]. Commensal rodents always remain in the proximity of pets, domestic animals and human beings; therefore, their control through toxic rodenticides requires safe bait delivery systems to avoid hazards to non-targets [52, 58].

Bats

Bats are nocturnal flying mammals distributed all over the world. Some bats are insectivorous feeding on a wide range of harmful insects, thus regulating their populations, while other are frugivorous and may cause substantial losses in fruit orchards [7].

The Egyptian Fruit Bat, *Rousettus aegyptiacus* Geoffroy (Chiroptera: Petropodidae)

The Egyptian fruit bat is by far the most serious pest of date palm among mammals. It is reported to have been feeding on dates, in addition to several other crops such as apple,

Table 16.3. Other rodent species reported in the date palm agro-ecosystem

Order/family/species	Common name	Reference
Rodentia, Muridae		
Acomys cahirinus Desmarest	Egyptian spiny mouse	[28]
Tatera indica (Hardwicke)	Indian gerbil	[4]
Tatera robusta (Cretzschmar)	Gerbil	[1]
Gerbillus gerbillus (Olivier)	Lesser Egyptian gerbil	[56]
Gerbillus campestris (Loche)	North African gerbil	[56]
Rodentia, Spalacidae		
Nannospalax ehrenbergi (Nehring)	Palestine mole rat	[28]
Rodentia, Hystricidae		
Hystrix indica Kerr	Indian crested porcupine	[48]

banana, mango, strawberry, pomegranate and peach [59]. The size of the Egyptian fruit bat is about 15 cm, the back is brown and belly is pale brown. Eyes are small and it has prominent strong teeth (Figure 16.7).

This bat uses echo for orientation and occupies orchards, hill caves, dark places and abandoned sites. It lives in large or small communities, where adult bats perch with their legs while the heads are pointing down. The female gives birth to 1–2 offspring only once during the year. Gestation period is 115–120 days; pups are weaned at 6–10 weeks and young ones reach maturity in about 9 months. The life span in captivity may reach 20 years [7]. The main diet of the Egyptian fruit bat was the date fruits as has been reported from Iran, Saudi Arabia and Israel [5, 7, 60].

The feeding signs of Egyptian fruit bat (tooth marks or tiny holes on chewed fruits; red-coloured droppings) are very different from those of birds (the holes are different shapes). Although sometimes we observed birds and bats feeding on fallen fruit beneath the trees, these observations were temporally separated: (fruit bats at night between mid-nights to near sunrise versus bird in the day after sunrise to near sunset) [5]. Bait consisting of pitted dates (dates without the pits) and zinc phosphide (3 g zinc phosphide/100 g pitted dates) can be hanged on or near fruit-bearing palms to control the fruit bats. Bunch covering used for bird management can prevent damage by the fruit bat [7]. Fumigation of bat nesting sites by toxic gases is not recommended because it may also kill beneficial bats that may be living in the same roosting sites.

Nematodes

Date palm is attacked by an array of nematode species (Table 16.4) that cause palm weakening due to their injurious effect on the root system, which eventually hamper the absorption of water and translocation of nutrients.

Nematodes infestation predisposes the date palms to infection by pathogens such as *Fusarium*, *Veritcilum* and *Rhizochtonia*. Nematodes are usually dispersed through movement of contaminated soil, uncleaned equipment and most important the transportation of infested offshoots and palms [7]. The secondary damage by fungi to roots of date palm is aggravated by the infection of root-knot nematodes, which eventually lead to the death of palm roots [66]. The root-knot nematode, *Meloidogyne javanica* was reported to cause mortality percentage of more than 90% to young seedlings of 50 date palm varieties sown in soil heavily infested by this nematode [67]. *Meloidogyne incognita* induced giant cells and gall formation in the roots of susceptible date palm cultivars [62]. Although the list of nematodes reported on date palm seems to contain numerous species, their host–parasite relationship, damage on growth, vigour and yield of date palm remain to be determined [6, 62]. However, it has been reported

Fig. 16.7. The frugivorous bat, *Rousettus* sp., perching on a twig of a tree (photo: M.M. Al-Ziat).

that the nematode *Pratylenchus penetrans* caused puncturing of the epidermal cells and disarrangement of cortical cells of date palm roots, which eventually show large abnormal cavities.

The nematode species *Bursaphelenchus cocophilus*, the causative agent of red ring disease [68], is a serious pest of coconut and oil palms. The South American palm weevil *Rhynchophorus palmarum* vectors this nematode. *B. cocophilus* produces characteristic symptoms of premature senescence of young leaves, irreversible wilt, yield reductions and premature death of palm trees [69]. The red ring nematode has not been reported from Southeast Asia, India and Africa where *R. ferrugineus* prevails [70]. Giblin-Davis [71] stated that *Rhynchophorus* spp. could possibly transmit *B. cocophilus* should it co-exists with these species of weevils. It is pertinent to mention here that the date palm is among the different 17 hosts of red ring nematode, *B. cocophilus* [64, 65].

Nematode Management

Planting of nematode-infested offshoots should be avoided as an effective preventive measure to control nematodes. Date palm plantation should not be intercropped with field and horticultural crops that are susceptible to the nematodes. In

Table 16.4. Nematode species found in the rhizosphere of date palm

Order/family/species	Common name	Reference
Dorylaimida, Longidoridae		
Longidorus africanus Memy	Needle nematode	[6]
Longidorus congoensis Aboul-Eid	Needle nematode	[61]
Xiphinema elongatum Schuurmans Stekhoven &Teunissen	Thread nematode	[61]
Xiphinema americanum Cobb	Thread nematode	[6]
Tylenchida, Heteroderidae		
Meloidogyne javanica (Treub)	Root-knot nematode	[6, 62]
Meloidogyne incognita (Kofoid & white) Chitwood	Root-knot nematode	[6, 62]
Meloidogyne arenaria (Neal) Chitwood	Root-knot nematode	[62]
Tylenchida, Tylenchidae		
Pratylenchus brachyurus (Castillo & Vovlas)	Root-lesion nematode	[6, 62]
Pratylenchus coffeae (Zimmermann) Filipjev & Schuurmans Stekhoven	Root-lesion nematode	[63]
Pratylenchus delattrei Luc.	Root-lesion nematode	[63]
Pratylenchus neglectus (Rensch) Filipjev & Schuurmans Stekhoven	Root-lesion nematode	[63]
Pratylenchus thornei Sher & Allen	Root-lesion nematode	[63]
Pratylenchus penetrans (Cobb) Chitwood & Oteifa	Root-lesion nematode	[61]
Helicotylenchus dihystera (Cobb.)	Spiral nematode	[6]
Tylenchida, Hoplolaimidae		
Helicotylenchus digonicus Perry in Perry, darling & Thorne	Spiral nematode	[6]
Helicotylenchus multicinctus Golden	Spiral nematode	[6]
Helicotylenchus microcephalus Sher	Spiral nematode	[6]
Helicotylenchus indicus Siddiqi	Spiral nematode	[6]
Tylenchida, Criconematidae		
Hemicriconemoides mangiferae Siddiqi	False-sheath nematode	[6]
Tylenchida, Pratylenchidae		
Rotylenchulus anamictus Dasgupta, Raski & Sher	[6]	
Tylenchida, Belonolaimidae		
Tylenchorhynchus nudus Allen	Stunt nematode	[6]
Tylenchorhynchus latus Allen	Stunt nematode	[62]
Tylenchorhynchus goffarti Sturhan	Stunt nematode	[62]
Panagrolaimomopha, Aphelenchoididae		
Bursaphelenchus cocophilus (Cobb) Baujard	Red ring nematode	[64, 65]

order to kill nematodes, soils should be deep ploughed and exposed to the sun or covered with polythene sheets before planting offshoots or adult palms. All tools used in agricultural operations should be clean and nematode-free to avoid spreading of nematodes through contaminated tools. The use of organic manure, weed control and resistant date palm cultivars are effective in the control of parasitic nematodes [62]. For curative treatment, Carbofuran can be used at a rate of 5 g/m^2. The nematicide should be placed around the date palm, mixed thoroughly with soils followed by immediate irrigation. The nematicide application can be repeated after 2–3 months [7].

Land Snail, *Eobania vermiculata* Müller (Stylommatophora: Helicidae)

Recently, the land snails have become important pests of date palm orchards. They inflict serious damage on vegetable, ornamental plants, field crops, fruit trees and date palm offshoots. The population of these pests fluctuates with climatic conditions, being high in more humid areas adjacent to irrigation and drainage canals. The most abundant species of snail on date palm is *E. vermiculata* Müller. The snails are nocturnal and more active during spring and autumn where temperatures and relative humidity are congenial. Their activity is restricted during winter when temperatures are low as well as in extreme temperatures of summer. Large number of snails is killed in summer due to the excessive heat and low humidity [7]. More damage is inflicted on seedlings, offshoots and young palms. Snails usually damage date palm during feeding by rasping the tissues with their tongues (Radula). The snails are found preferably in the leaf axle, which provides more protection and feeding sites (Figure 16.8).

During their movement, snails produce mucus, which contaminates the date fruits, and lower their commercial quality. Other species of snails reported on date palm include the field snail *Monacha obstructa* Pfeiffer, the Carthusian snail *Monacha cartusiana* Pfeiffer and the land snail *Helicopsis vestalis* Pfeiffer [7].

Snail Management

Snails usually thrive well in humid areas with high density of weeds. Thus, they can be managed in date palm groves with cultural practices, which include ploughing and solarization of soil to expose them to adverse conditions and natural enemies. Farm sanitation, proper palm spacing and controlled irrigation and drainage are effective in reducing the population of snails. Mechanically, snails can be controlled by hand picking, particularly when the population is low. Peeled potatoes and citrus fruits can be used as traps for snails, which can eventually be collected and destructed. Tar can be applied at the base of date palm trunk (3 cm-wide band) to prevent the crawling of snail to tender leaves of the date palm. Naturally, rodents, snakes and numerous bird species feed upon snails. In large orchards with high infestation, suitable molluscicides can be used as bait or spraying on infested palms [7].

Fig. 16.8. Snails sheltering at the leaf base of an old date palm (left) and at the base of tender leaves of young date palm (offshoot) (photo: M.M. Al-Ziat).

Conclusions

Non-arthropod pest species can cause substantial economic losses to date palm that may negatively affect the livelihood of farmers if left unmanaged. Parasitic nematodes and land snails should not be overlooked as potential pests of date palm. To sustain a management strategy for non-arthropod pests of date palm that is acceptable economically and socially, the nature and levels of damage caused by this group of pests should be determined. Accordingly, reliable monitoring systems and practical methods of estimating damage are needed. Bird damage on date palm fruits can be effectively minimized by the removal of bird perching places, water resources and roosting sites. Field and storehouses sanitation together with the use of fruit bunch bagging are effective methods in excluding frugivorous birds, rodents and bats. Extension and training of farmers are important components in establishing integrated pest management system for non-arthropod pest species in date palm plantations.

Acknowledgements

The financial support provided by the Date Palm Research Center of Excellence through the research grant number DPRC-1-(2015) is appreciated and sincerely acknowledged.

References

1. Khairi MMA. Date palm status and perspective in Sudan. In: Al-Khayri JM, Jain SM, Johnson DV, editors. Date Palm Genetic Resources and Utilization: Volume 1: Africa and the Americas. Springer Science + Business Media, Dordrecht; 2015. p. 169–191.
2. Sedra MH. Date palm status and perspective in Morocco. In: Al-Khayri JM, Jain SM, Johnson DV, editors. Date Palm Genetic Resources and Utilization: Volume 1: Africa and the Americas. Springer Science + Business Media, Dordrecht; 2015. p. 257–323.
3. Greaves JH. Rodent Pests and Their Control in the Near East. FAO, Plant Production and Protection paper No. 95. Food and Agriculture of the United Nations, Rome, Italy; 1989a. 113 p.
4. Sharma A. Bio-ecology of Indian gerbil, *Tatera indica* (Hardwicke) in relation to soil types [MSc thesis]. Punjab Agricultural University Ludhiana-141004, India; 2016. 74 p.
5. Zohoori H, Kiabi BH, Karami M, Khorasani N, Dareshoori BF, Dakhteh MH. Some observations on fruit used by birds and Egyptian fruit bat *Rousettus aegyptiacus* on Qeshm Island, Southern Iran. Podoces 2007;2(2):156–9.
6. Mani A, Handoo ZA, Livingston S. Plant-parasitic nematodes associated with date palm trees (*Phoenix dactylifera*l.) in the Sultanate of Oman. Nematropica 2005;35:135–43.
7. Al-Ziat MM, Al-Giat SI, Logma HI, Dhafran HA, Al-Abdelsalam KS, editors. Important Diseases and Pests of Date Palm and Their Integrated Control in the Kingdom of Saudi Arabia. FAO; Al-Asr Printing Press, Riyadh, Saudi Arabia; 2002. 369 p. (in Arabic).
8. Greaves JH. Rodent pests of agriculture and their control in the near east region. Arab Journal of Plant Protection 1989b;7:75–82.
9. Swaminathan R, Verma SK. Comparative efficacy of different covers against bird damage and their effects on berry characteristics in date palm. Annals of Arid Zone 2000;39(2):185–9.
10. Islam K, Williams RN. Red-vented bulbul (*Pycnonotus cafer*) and red-whiskered bulbul (*Pycnonotus jocosus*). In: Poole A, Gill F, editors. The Birds of North America, No. 520, The Birds of North America, Inc., Philadelphia, PA; 2000. p. 1–24.
11. BirdLife International. *Pycnonotus leucotis*. The IUCN Red List of Threatened Species 2016: e.T22712687A94342904. Available from: URL: http://dx.doi.org/10.2305/IUCN.UK.2016-3.RLTS.T22712687A94342904.en [cited on 13 January 2018].
12. Beaman M, Madge S. The Handbook of Bird Identification for Europe and Western Palearctic. Publisher Christopher Helm, A & C Black, London; 1998.

13. Srivastava M. Breeding behavior of white-eared bulbul *Pycnonotus leucotis* as observed in a house courtyard at Bikaner, Rajasthan. Our Nature 2010;10:284–285. Available from: URL: https://www.nepjol.info/index.php/ON/article/viewFile/7798/6371 [cited 2018 January 15].
14. Khoury F, Amr Z, Hamidan N, AlHassani I, Mir S, Eid E, *et al.* Some introduced vertebrate species to the Hashemite Kingdom of Jordan. Vertebrate Zoology 2012;62(3):435–51.
15. Waseem M, Ashraf I, Hussain T. Nesting behavior of rose-ringed parakeet (*Psittacula krameri*) in southern Punjab, Pakistan. Science International (Lahore) 2015;27(5):4255–61.
16. Khan HA. Diurnal rhythms of the rose-ringed parakeet (*Psittacula krameri*) in daylight hours in its communal roost. Online Journal of Biological Sciences 2002;2(8):551–3.
17. Ali S, Ripley S. Handbook of Birds of India and Pakistan. Oxford University Press, Bombay, UK; 1968.
18. Sarwar M, Beg MA, Khan AA, Shahwar D. Breeding behavior and reproduction in rose-ringed parakeet. Pakistan Journal of Zoology 1989;21:131–8.
19. Ahmed S, Khan HA, Javed M, UR-Rehman K. An estimation of rose-ringed parakeet (*Psittacula krameri*) depredations on citrus, guava, and mango in orchard fruit farm. International Journal of Agriculture and Biology 2012;14(1):149–52.
20. Ali MH, Rao BHL, Rao MA, Ali PS. Bird damage in maize. Journal of Bombay Natural History Society 1982;79:201–4.
21. Khan HA. Effect of tree species and size on the availability of nest cavities of rose-ringed parakeet (*Psittacula krameri*) in central Punjab, Pakistan. Pakistan Veterinary Journal 1999;19(3):145–8.
22. Covas L, Senar JC, Roqué L, Quesada J. Records of fatal attacks by rose-ringed parakeets *Psittacula krameri* on native avifauna. Revista Catalanad, Ornitologia 2017;33:45–9.
23. Lambert MS, Massei G, Bell J, Berry L, Haigh C, Cowan DP. Reproductive success of rose-ringed parakeets *Psittacula krameri* in a captive UK population. Pest Management Science 2009;65:1215–18.
24. Butler CJ, Cresswell W, Gosler A, Perrings C. The breeding biology of rose-ringed parakeets *Psittacula krameri* in England during a period of rapid population expansion. Bird Study 2013;60:527–32.
25. Nhlane MED. The breeding biology of the house sparrow *Passer domesticus* at Blantyre, Malawi. Ostrich: Journal of African Ornithology 2017;71(1):80–2.
26. Guezoul O, Sekour M, Souttou K, Doumandij S. Estimated damage caused by the hybrid sparrow *Passer domesticus* × *Passer hispaniolensis* on dates (*Phoenix dactylifera*) in two palm groves in Quargla. Lebanese Scientific Journal 2010;11:3–9.
27. Postigo J-L, Shwartz A, Strubbe D, Mun`oz A-R. Unrelenting spread of the alien monk parakeet *Myiopsitta monachus* in Israel. Is it time to sound the alarm? Pest Management Science 2016;73:349–53.
28. Radwan ES. The current status of the date palm tree (Phoenix dactylifera L.) and its uses In the Gaza strip, Palestine [MSc thesis]. The Islamic University-GazaPalestine; 2017. 158 p.
29. Belkacem AA, Sekour M, Doumandji S. Effectiveness of mesh netting and nest destruction in protection of crops against attack by Spanish sparrow *Passer hespaniolensis*. African Journal of Agricultural Research 2012;7(32):4575–80.
30. Zaid A, de Wet PF. Pollination and bunch management. In: Zaid A, editor. Date Palm Cultivation. Food and Agriculture Organization (FAO), Plant production and protection paper No. 156, Rev.1. FAO, Rome, Italy; 2002. p. 145–75.
31. Royall Jr WC. Starlings do damage to crops in Arizona. Progressive Agriculture in Arizona 1962;14(3):14–16. (http://hdl.handle.net/10150/299177).
32. Abu-Anaia AR. Protection of dates from bird pests. The Blessed Tree 2011;3(2):72–6. (In Arabic)
33. Tinajero R, Rodríguez Estrella R. Monk parakeet (*Myiopsitta monachus*), a new species breeding in the southern part of Baja California. Acta Zoológica Mexicana (n. s.) 2015;31(2):190–7.
34. Guezoul O, Chenchouni H, Sekour M, Ababsa L, Souttou K, Doumandji S. An avifaunal survey of mesic manmade ecosystems 'Oases' in Algerian hot-hyper arid lands. Saudi Journal of Biological Sciences 2013;20:37–43.
35. Collar N. Black Scrub-robin (*Cercotrichas podobe*). In: del Hoyo J, Elliott A, Sargatal J, Christie DA, de Juana E, editors. Handbook of the Birds of the World Alive. Lynx Edicions, Barcelona, Spain; 2018. Available from: URL: https://www.hbw.com/node/58482 [cited 2018 April 16].
36. Collar N, Robson C. Arabian babbler (*Argya squamiceps*). In: del Hoyo J, Elliott A, Sargatal J, Christie DA, de Juana E, editors. Handbook of the Birds of the World Alive. Lynx Edicions, Barcelona, Spain; 2018. Available from: URL: https://www.hbw.com/node/59551 [cited 2018 April 7].

37. Charter M. Using barn owls (*Tyto alba erlangeri*) for biological control in Israel. 2018. Available from: URL: http://www.owls.org/ewExternalFiles/barn_owls_in_israel.pdf [cited 2018 April 8].
38. Kan I, Motro Y, Horvitz N, Kimhi A, Leshem Y, Om-Tov Y, *et al.* Agricultural rodent control using barn owls: Is it profitable? American Journal of Agricultural Economics 2013;96 (3):733–52.
39. Linz GE, Homan HJ, Werner SJ, Hagy HM, Bleier WJ. Assessment of bird-management strategies to protect sunflowers. Bioscience 2011;61:960–70.
40. Tracey J, Bomford M, Hart Q, Saunders G, Sinclair R. Managing Bird Damage to Fruit and Other Horticultural Crops. Bureau of Rural Sciences, Canberra, Goanna Print, Canberra, Australia; 2007.
41. Linddell CA, Steensma KMM, Curtis PD, Boulanger JR, Carroll JE, Burrows C, *et al.* Proportions of bird damage in tree fruits are higher in low-fruit-abundance contexts. Crop Protection 2016;90:40–8.
42. Kale M, Balfors B, Mörtberg U, Bhattacharya P, Chakane S. Damage to the agricultural yield due to birds, present repelling techniques and its impacts: an insight from the Indian perspective. International Journal of Applied Agricultural Research 2011;6(3):223–36.
43. Suubramanya S. Maize. Bird damage and management. Journal of Biosciences 1982;19(4):369–80.
44. Sharma RR, Reddy SVR, Jhalegar MJ. Pre-harvest fruit bagging: a useful approach for plant protection and improved post-harvest fruit quality – a review. The Journal of Horticultural Science and Biotechnology 2014;89(2):101–13.
45. Harhash MM, AL-Obeed RS. Effect of bunch bagging color on yield and fruit quality of date palm. American–Eurasian Journal of Agricultural and Environmental Science 2010;7:312–19.
46. Awad MA. Increasing the rate of ripening of date palm fruit (*Phoenix dactylifera* L.) cv. 'Helali' by pre-harvest and postharvest treatments. Postharvest Biology and Technology 2007;43:121–7.
47. Tripathi RS, Chaudhary V. Principles and practices of rodent pest management. In: Prasad D, Singh A, editors. Advances in Plant Protection Sciences. Akansha Publishing House, New Delhi, India; 2005. p. 245–59.
48. Hafeez S, Ashfaq M, Khan GS, Khan ZH. Damage inflicted by the Indian crested porcupine, *Hystrix indica*, on forestry and agricultural systems in Punjab, Pakistan. Journal of Asian and African Studies 2011;47(2):168–75.
49. Lathiya SB, Ahmed SM, Pervez A, Khadijah E. Food habits of mole rat (*Nesokia* sp.) in date-palm orchards of district Chinghai Baluchistan, Pakistan. Sarhad Journal of Agriculture 2010;26(4):595–8.
50. Poche RM, Sultan P, Evans SJ, Haque ME, Karim MA, Saddique MA. *Nesokia indica* from Bangladesh. Mammalia 1982;64:547–9.
51. Roberts TJ. Field Guide to the Small Mammals of Pakistan. Oxford University Press; 2005. 280 p.
52. Abdul-Soad AA, Mahdi SM, Markhand GS. Date palm status and perspective in Pakistan. In: Al-Khayri JM, Jain SM, Johnson DV, editors. Date Palm Genetic Resources and Utilization: Volume 2: Asia and Europe. Springer Science + Business Media, Dordrecht; 2015. p. 153–205.
53. Ahmed SM, Pervez A, Khan F, Zafar H, Sahar N, Khan J. Enhancement of poison bait acceptance through taste additives in short-tailed mole rat, *Nesokia indica* (Gray) infesting date palm orchards in Nok-Kundi and Mashkale in province Balochistan, Pakistan. Journal of Phytopathology and Pest Management 2017;4(2):13–21.
54. Brooks J, Fiedler L. Vertebrate pests: damage on stored foods. In: Mejia D, Lewis B, editors. Vertebrate Pests: Post-Harvest Operations. Food and Agriculture Organization of the United Nations (FAO), US Department of Agriculture (USDA), Washington, DC; 1999. p. 1–26.
55. Nelson S. Rat-feeding injury to plant in Hawaii. 2012. Available from: URL: https://www.ctahr.hawaii.edu/oc/freepubs/pdf/MP-7.pdf [cited 2018 April 8].
56. Soliman S, Mohallal EME. A survey of the mammalian fauna of Siwa oasis, Egypt. Egyptian Journal of Zoology 2014;61(56):171–85. doi: 10.12816/0005514.
57. Witmer GW, Baldwin RA, Moulton RS. Identifying possible alternative rodenticide baits to replace strychnine baits for pocket gophers in California. Crop Protection 2017;92:203–6.
58. Tripathi RS, Chaudhary V. Bio-efficacy of denatonium benzoate added formulation of bromadiolone against commensal rodents. Crop Protection 2016;80:132–7.
59. Madkour G. *Rousettus aegyptiacus* (Megachiroptera) as a fruit eating bat in A.R. Egypt. Agricultural Research Review 1977;55:167–72.
60. Korine C, Izhaki I, Arad Z. Is the Egyptian fruit-bat *Rousettus aegyptiacus* a pest in Israel? An analysis of the bat's diet and implications for its conservation. Biological Conservation 1999;88(3):301–6.
61. Lamberti F, Greco N, Zaouchi H. A nematological survey of date palms and other major crops in Algeria. FAO Plant Protection Bulletin 1975;23:156–60.

62. Youssef MMA. Nematodes associated with date palm and their control measures. Archives of Phytopathology and Plant Protection 2013;47(4):425–8.
63. Mani A, Al Hinai MS, Handoo ZA. Occurrence, population density, and distribution of root-lesion nematodes, *Pratylenchus* spp. in the Sultanate of Oman. Nematropica 1997;27:209–19.
64. Schuilling M, van Dither JBM. Red ring disease in the Paricatubaoil palm estate, Para, Brazil. A case study. Zeitschrift für Angewandte Entomologie 1981;91:154–69.
65. Griffith R, Giblin-Davis RM, Koshy PK, Sosamma VK. Nematode parasites of coconut and other palms. In: Luc M, Sikora RA, Bridge J, editors. Plant Parasitic Nematodes in Subtropical and Tropical Agriculture. CAB International, Oxon, UK; 2005. p. 493–527.
66. Carpenter JB, Elmer HS. Pests and Diseases of the Date Palm. Handbook No. 527 United States Department of Agriculture, Washington, DC; 1978.
67. Carpenter JB. Root-knot nematode damage to date palm seedlings in relation to germination and stage of development. Date Growers Institute Report 1964;41:10–14.
68. Oehlschlager AC, McDonald RS, Chinchilla CM, Patschke SN. Influence of a pheromone-based mass-trapping system on the distribution of *Rhynchophorus palmarum* (Coleoptera: Curculionidae) in oil palm. Environmental Entomology 1995;24:1005–12.
69. Oehlschlager AC, Chinchilla C, Castillo G, Gonzalez L. Control of red ring disease by mass trapping of *Rhynchophorus palmarum* (Coleoptera: Curculionidae). Florida Entomologist 2002;85:507–13.
70. Murphy ST, Briscoe BR. The red palm weevil as an alien invasive: biology and the prospects for biological control as a component of IPM. Biocontrol News Information 1999;20:35–46.
71. Giblin-Davis RM. Interactions of nematodes with insects. In: Khan MW, editor. Nematode Interactions. Springer, Dordrecht; 1993. p. 302–44.

17 The use of environmental DNA as an early warning tool in the detection of new freshwater invasive non-native species

Rosetta C. Blackman*, Bernd Hänfling and Lori Lawson-Handley
School of Environmental Sciences, University of Hull, Cottingham Road, Hull, HU6 7RX, UK

Abstract

Successful removal and eradication of invasive non-native species (INNS) relies on early detection, ideally with pathway screening, to prevent any spread or outbreak in to a new region. The collection of DNA from environmental samples, known as environmental DNA or eDNA, is a revolutionary technique allowing the screening of target species without their prior collection or observation. This technique has been referred to as a 'game changer' for biodiversity monitoring, and the potential for INNS detection is particularly appealing. Current freshwater monitoring methods are heavily dependent on specimens being in sufficient number to be collected, thus making early detection at the beginning of an invasion, when numbers are likely to be low, challenging. However, the DNA of new INNS may still be prevalent in the environment, even at low densities, making detection possible. Here we review the recent developments in the application of eDNA for the detection of INNS, and assess the promise and pitfalls for its more widespread application.

Keywords: Invasive non-native species, Monitoring, Environment DNA, Early warning, Detection

Review Methodology

Sources were identified by regular examination of current and previously published literature and by examination of citation lists in relevant articles.

Introduction

The early detection of a new invasive non-native species (INNS) has long been recognized as a crucial step in preventing the establishment of a sustainable population. In order to make eradication or containment a possibility, potential pathways and high-risk sites must be monitored effectively to detect an invasion at the earliest possible stage [1–6]. A systematic examination of potential threats can be an important first step to focus preventative efforts. Commonly referred to as 'horizon scanning', potential INNS threats are identified by using a risk assessment-based approach. In a recent UK study, priority INNS were ranked in terms of risk of arrival, establishment and impact on native ecosystems; the quagga mussel, *Dreissena rostriformis bugensis*, was the highest scoring [5]. The quagga mussel arrived in the UK shortly after

*Corresponding author: rosiecblackman@gmail.com

the publication of the horizon scanning study, but was only detected by traditional monitoring methods once it had become established. This emphasizes the need to develop monitoring tools with a greater sensitivity, in order to monitor potential pathways and high-risk sites, and allow detection of an invasion at the earliest possible stage and make eradication or containment a possibility.

Legislative requirements often promote routine monitoring for the prevention of new INNS arrivals. For example, in Europe the most significant of these is the recently adopted European Union regulation 'on the prevention and management of the introduction and spread of invasive alien species' (EU Regulation 1143/2014). This regulation imposes restrictions on INNS of Union concern, and requires Member States to undertake research, monitoring and surveillance of INNS. Surveillance systems are required to include both targeted and general surveys and should adopt the monitoring systems already in place for other directives such as the Water Framework Directive (WFD 2000/60/EC).

The WFD is another substantial piece of legislation aimed to improve and enhance the ecological status of water-bodies in Europe by 2027. As part of its delivery, monitoring is carried out to assess progress, including biological elements (macroinvertebrates, fish and phytobenthos) and high-impact INNS. The current methods used to monitor aquatic environments, such as trapping, netting and electrofishing, rely heavily on the capture of the target specimen and identification by trained staff, and although not specifically aimed at early detection of INNS, these methods are currently the most likely mechanisms to pick up new INNS once they have become established. However, these methods are often inefficient for detecting small or juvenile specimens, and species which are problematic to identify, or at low densities. For example, the standard macroinvertebrate 3 min kick net sample (with 1 min hand search) method will only recover approximately 62% of families and 50% of species at a given site [7]. Although increased sampling time does correlate with the number of families/species collected, this is often not possible due to mounting cost and time constraints. Sample collection can also be inconsistent; with variation between samples, river habitat and collection, being noted [7]. When a species is in low abundance, it is likely to be missed by physical collection methods [5, 8], and even if collected, morphological identification of these new and often cryptic INNS is highly dependent on the experience and skills of the analyst [9].

Recent and rapid developments in the use of molecular tools, such as environmental DNA (eDNA), are providing new opportunities for early detection and pathway screening of INNS and other species of interest (see Table 16.1 and references [6, 10]). The term eDNA refers to the DNA from cells or tissue shed by organisms into the environment from skin, faeces, urine, saliva or decaying material [6, 11–13]. For eDNA samples taken in aquatic systems, water is typically filtered to capture the eDNA, or eDNA is precipitated with ethanol. There are two main downstream applications for the prepared eDNA samples: the detection of a single species via 'targeted' detection methods such as quantitative polymerase chain reaction (PCR); or the detection of whole communities or groups of taxa via high-throughput sequencing (HTS) approaches such as metabarcoding (see Figure 17.1). Both methods offer opportunities for non-invasive sampling and potentially allow greater taxonomic resolution of cryptic species and greater sensitivity for the detection of rare or elusive species compared with conventional methods.

Due to increased trade and transport, INNS introductions have now reached an all-time high [65]. This increased pressure, coupled with the shortcomings of current methods, highlights the clear need for more effective detection methods for INNS. Development of molecular tools may provide a much needed alternative for monitoring pathways and detecting species before establishment, with additional benefits including identification of cryptic species, monitoring spread, heightened biosecurity and understanding impact. In this review, we investigate the opportunities for early detection of INNS with eDNA and ask whether it will be the answer to future INNS monitoring needs. We also evaluate the relative pros and cons of the targeted and whole community approaches and suggest how these tools can best be applied for the detection of INNS.

Table 17.1. Examples of eDNA case studies to detect invasive or non-native species

References	Environment/sample type	Taxon	Method	Target
[14]	Natural water bodies	Mosquito (*Aedes koreicus*)	qPCR (Probe)	I
[15]	Ponds	African clawed frog (*Xenopus laevis*)	qPCR (Probe)	I
[16]	Experimental ponds	African jewelfish (*Hemichromis letourneuxi*)	qPCR (Probe)	I
[17, 18]	Ponds	American bullfrog (*Rana catesbeiana = Lithobates catesbeianus*)	Standard PCR	I
[19]	River	American wedge clam (*Rangia cuneata*)	Standard PCR	I
[20, 21]	Pond water and sediments	Amphibian chytrid fungus (*Batrachochytrium dendrobatidis*)	qPCR (Dye and Probe)	P
[14]	Natural water bodies	Asian bush mosquito (*Aedes j. japonicus*)	qPCR (Probe)	I
[14]	Natural water bodies	Asian tiger mosquito (*Aedes albopictus*)	qPCR (Probe)	I
[3[1], 22[2], 23[3]]	Chicago area waterway (large river and canal complex)[1,2]; bait shop tanks[3]	Bighead carp (*Hypophthalmichthys nobilis*)	Standard PCR	I
[52]	River	Big head goby *(Ponticola kessleri)*	Standard PCR	I
[22]	River	Black carp (*Mylpharyngodon piceus*)	Standard PCR	I
[24[5], 25[5]]	Pond/lake; stream; mesocosm	Bluegill sunfish (*Lepomis macrochirus*)	qPCR[4,5] (Probe); ddPCR[5]	I
[26]	Large wetlands	Boa constrictor (*Boa constrictor*)	qPCR (Probe)	I
[27, 28]	Aquaria and ponds	Brazilian waterweed (*Egeria densa*)	qPCR (Probe)	I
[30[6], 31[7]]	Headwater streams	Brook trout (*Salvelinus fontinalis*)	Standard PCR[6]; qPCR (Probe)[6,7]	I
[26[8], 32[9]]	Large wetlands	Burmese python (*Python molurus bivittatus*)	qPCR (Probe)[8]; Standard PCR[9]	I
[3]	Ships ballast	Chinese mitten crab (*Eriocheir sinensis*)	Standard PCR on Microfluidic chip	I
[33]	Rivers	Chinese grand Salamander (*Andrias davidianus*)	qPCR (Probe)	
[34[10]; 35[11]]	Ponds, mesocosm and river	Common carp (*Cyprinus carpio*)	qPCR (Probe)[10]; qPCR (Dye)[11]	I
[37, 38]	Large lakes	Crayfish plague (*Aphanomyces astaci*)	qPCR (Probe)	P
[39]	Cucumber root, soil and scarid flies	Cucumber wilt fungus (*Fusarium oxysporum* f. sp. *cucumerinum*)	qPCR (Dye); pseudo-nested PCR.	P
[40]	Lakes and rivers	Dreissenid mussels	qPCR (Probe)	I
[23]	Bait shop tanks	Eurasian rudd (*Scardinius erythrophthalmus*)	Standard PCR	I
[27]	10 g of leaf in 1 L water	European frogbit (*Hydrocharis morsus-ranae*)	Standard PCR	I

Continued

Table 17.1. Continued

References	Environment/sample type	Taxon	Method	Target
[27]	10 g of leaf in 1 L water	Fanwort (*Cabomba caroliniana*)	Standard PCR	I
[41]	Aquaria	Fathead minnow (*Pimephales promelas*)	Standard PCR	I
[9]	Rivers	Freshwater shrimp (*Gammarus fossarum*)	Metabarcoding	NN
[27]	10 g of leaf in 1 L water	Giant salvinia (*Salvinia molesta*)	Standard PCR	I
[42]	Ships ballast	Golden (*Limnoperna fortuneii*)	Standard PCR on Microfluidic chip	I
[22[12], 23[13]]	Rivers[12]; Bait shop tanks[13]	Goldfish (*Carassius auratus*)	Standard PCR	I
[22[14], 44[15]]	Large experimental tanks[14]; Rivers[14]; Ponds[15]	Grass carp (*Ctenopharyngodon idella*)	Standard PCR[14]; qPCR[15] and ddPCR[15]	I
[26]	Large wetlands	Green anaconda (*Eunectes murinus*)	qPCR (Probe)	I
[52]	River	Monkey goby *(Neogobius fluviatilis)*	Standard PCR	I
[45]	Lakes	Narrow-clawed crayfish (*Astacus leptodactylus)*	qPCR (Probe)	I
[46]	Container/River	New Zealand Mud Snail (*Potamopyrgus antipordarum*)	qPCR (Probe)	I
[47]	River basin	Nile tilapia (*Oreochromis niloticus*)	Standard PCR	I
[47]	River	North African catfish (*Clarias gariepinus*)	Standard PCR	I
[26]	Large wetlands	North African python (*Python sebae*)	qPCR (Probe)	I
[48]	Simulated ballast water	Northern Pacific seastar (*Asterias amurensis*)	Standard PCR	I
[51]	Aquaria	Northern Pike (*Esox Lucius*)	qPCR (Probe)	NN
[49]	Lakes and rivers	Northern River Crangonyctid (*Crangonyx pseudogracilis*)	Standard PCR	NN
[50]	River	Northern snakeshead (*Channa argus*)	Metabarcoding; ddPCR	I
[27]	10 g of leaf in 1 L water	Parrot feather (*Myriophyllum aquaticum*)	Standard PCR	I
[47]	River	Prussian Carp (*Carassius gibelio*)	Standard PCR	I
[41]	Aquaria and ponds	Pumpkinseed (*Lepomis gibbosus*)	Standard PCR	I
[42[16]; 75[17]]	Ships ballast[16]; Rivers[17]	Quagga mussel (*Dreissena rostriformis bugensis*)	Standard PCR on Microfluidic chip[16]; standard PCR[17] and qPCR (Dye)[17]	I
[52]	River	Racer goby *(Babka gymnotrachelus)*	Standard PCR	I
[54[10]; 55[19]]	Ponds[18,19]; Aquaria[19]	Red swamp crayfish (*Procambarus clarkii*)	qPCR (Probe)	I
[56]	Aquaria and pond	Red-eared slider (*Trachemys scripta*)	Standard PCR	I

Continued

Table 17.1. Continued

References	Environment/sample type	Taxon	Method	Target
[23[20], 43[21], 52[22]]	Bait shop tanks[20]; Tanks[21]; River[22]	Round goby (*Neogobius melanostomus*)	Standard PCR[20,21,22]; qPCR (Probe)[21] and ddPCR[21]	I
[57]	Lakes	Rusty crayfish (*Orconectes rusticus*)	qPCR (Dye)	I
[59]	Pond	Siberian sturgeon (*Acipenser baerii*)	Standard PCR	I
[3[23], 22[24], 23[25]]	Chicago area waterway (large river and canal complex)[23,24]; bait shop tanks[25]	Silver carp (*Hypophthalmichthys molitrix*)	Standard PCR	I
[45[26], 58[27], 60[28]]	Lakes[26,27]; mesocosm[28]	Signal crayfish (*Pacifastacus leniusculus*)	qPCR (Probe)[26,28]; qPCR (Dye)[27]	I
[41]	Aquaria	Sunbleak (*Leucaspius delineatus*)	Standard PCR	I
[47[29], 61[30]]	River basin[29]; Aquaria and ponds[30]	Topmouth gudgeon (*Pseudoasbora parva*)	Standard PCR	I
[23]	Bait shop tanks	Tubenose goby (*Proterorhinus semilunaris*)	Standard PCR	I
[62]	Large rivers and estuary	Wakasagi smelt (*Hypomesus nipponensis*)	qPCR (Probe)	I
[27]	10 g of leaf in 1 L water	Water hyacinth (*Eichhornia crassipes*)	Standard PCR	I
[27]	10 g of leaf in 1 L water	Water lettuce (*Pistia stratiotes*)	Standard PCR	I
[27]	10 g of leaf in 1 L water	Water soldier (*Stratiotes aloides*)	Standard PCR	I
[29]	Aquaria and ponds	Water thyme (*Hydrilla verticillata*)	qPCR (Probe)	I
[52]	River	Western tubenose goby (*Proterorhinus semilunaris*)	Standard PCR	I
[63]	Bait shop tanks	White perch (*Morine americana*)	Metabarcoding	I
[26]	Large wetlands	Yellow anaconda (*Eunectes notaeus*)	qPCR (Probe)	I
[27]	10 g of leaf in 1 L water	Yellow floating heart (*Nymphoides peltata*)	Standard PCR	I
[64]	Container	Zebra fish (*Danio rerio*)	Standard PCR	I
[53]	Lakes and rivers	Zebra mussel (*Dreissena polymorpha*)	Standard PCR	I

Target: I, invasive; NN, non-native; P, pathogen.
The superscript numbers in references column relates to the environment/sample type and individual species referenced in that row.

Will eDNA Revolutionize Detection of Invasive Non-native Species?

All organisms shed DNA into their environment, but the nature of the DNA, the absolute amount and rate at which it is shed vary greatly between species [66]. Our ability to detect eDNA depends on both the ecology and physiology of the species, as well as the environment in which they live [66]. The first eDNA studies focused on describing historical fauna and flora from ancient sediments, where, given the right conditions,

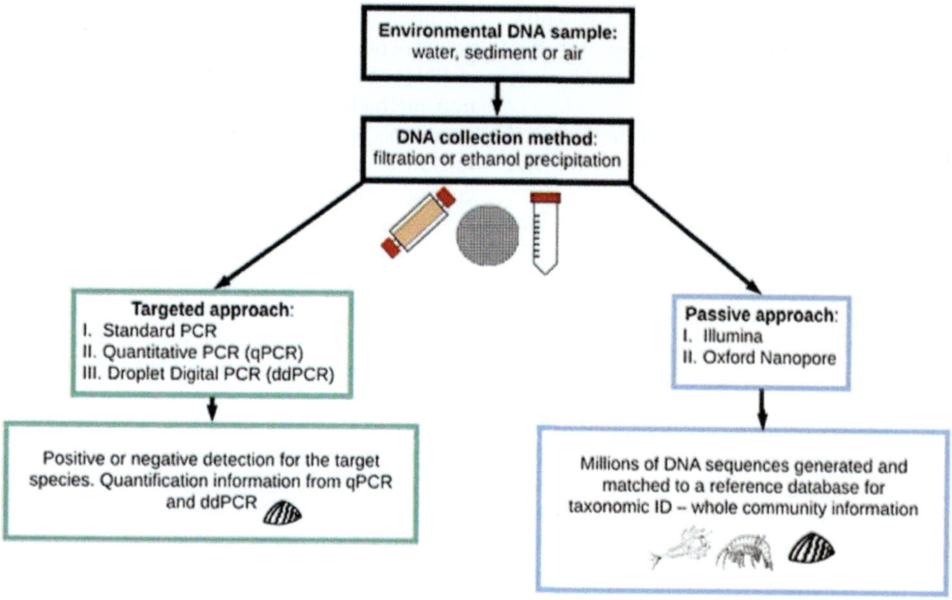

Fig. 17.1. Environmental DNA workflow. Detailing the two downstream methods for eDNA analysis: species-specific detection and community-based detection.

DNA can be preserved for tens of thousands of years [67]. In aquatic systems, eDNA is likely to be a combination of whole cells and degraded fragments and its persistence within the system can vary significantly, dependent on the abiotic and biotic properties of the environment, as we discuss in *Factors influencing detection of INNS using eDNA* (also see [66]). Several reviews have already explored in detail the use of eDNA as a tool for biodiversity assessment; detection of INNS has featured in a number of general (e.g. [6, 10, 12, 66, 68, 69]) and specific reviews [68] and a highly influential report has also been published, which focused on INNS [70], but the field has moved forward at a considerable rate since their publication. It is therefore timely to review recent advances and evaluate the promise and pitfalls of eDNA for the detection and monitoring of INNS.

Targeting the Invaders: the Species-Specific Approach

The most commonly used method for INNS detection using eDNA is the species-specific approach, which uses species-specific primers during PCR to amplify the target DNA present in the sample [71]. There are three types of PCR, which can be used for species-specific detection. Firstly, standard PCR (st-PCR) relies on using an agarose gel to visualize the positive or negative detection in the form of the correct size band. Secondly, quantitative real-time PCR (qPCR) which uses either dye-based detection or probe-based technology to show amplification of target DNA in real time. Finally, droplet digital (ddPCR) is a relatively new method based on microfluidics. During ddPCR, samples are randomly partitioned into several thousand individual droplets, which are then amplified by PCR, essentially allowing individual droplets of sample to be screened for target DNA [43]. Both qPCR and ddPCR allow quantification of the target DNA within the sample and have greater sensitivity over st-PCR [43, 72]; however, st-PCR requires less specialist equipment, reagents and training than qPCR, does not require laborious construction of standard curves, and is generally cheaper. One study estimated that reagents and consumables for st-PCR are 1.5–2-fold lower than for qPCR, while st-PCR machines are 4× cheaper than those for qPCR [41]. By using st-PCR, results of amplification can only be confirmed using electrophoresis and running the PCR product on a gel to determine the presence of the target DNA; however, this is a potential source of contamination as samples are often run in parallel and PCR product could easily spread to neighbouring wells.

All three methods rely heavily on the specificity of the primers, which involves extensive testing, development and validation via *in silico*, *in vitro* and *in situ* [71]. The great majority of eDNA studies have used the targeted approach; with over 100 species-specific assays now available (see [6] and Table 16.1). For example, the first application of eDNA to monitor INNS was by Ficetola *et al.* [18] who demonstrated the successful detection of the American bull frog (*Rana catesbeiana*=*Lithobates catesbeianus*), using st-PCR in natural wetlands in France. Their study showed detection at low densities of both tadpoles in mesocosm (as low as 0.3 individuals per litre) and adults in natural ponds, and crucially had no false-positive detection. This study highlighted the future application of eDNA methods to detect invasive, rare and elusive species because of its sensitivity at low densities. Subsequently, a comparison demonstrated eDNA also had a greater sensitivity to detect this species over traditional sampling (diurnal and nocturnal surveys) 38 of the 49 surveyed using eDNA detected the target species compared with seven of the 49 using traditional methods [17]. Since the original American bull frog study, others have been able to successfully detect a range of species using this approach in both lentic and lotic systems (see Table 16.1).

Perhaps the other most well-studied example, in terms of the detection of an INNS using eDNA, is that of the Asian carp in North America. Jerde *et al.* [3] developed and tested species-specific primers designed for st-PCR for both the silver, *Hypophthalmichthys molitrix* and bighead, *H. nobilis* carp. An extensive eDNA sampling campaign showed that the invasion front of these species had been previously underestimated by conventional sampling, to the extent that both species had been picked up upstream of an electric barrier constructed to prevent the spread of the species into the Great Lakes. This study, alongside others by this group [3, 22] consistently highlight the effective use of eDNA to detect invasive non-native fish in the Great Lakes and surrounding waterways. The 'rapid evolution' of eDNA technology was stressed by Mahon *et al.* [22] who reiterated the need for quantitative information and biodiversity assessment using eDNA. Subsequently, Turner *et al.* [73] published new eDNA capture and extraction methods, alongside a *Hypophthalmichthys* genus-specific qPCR assay, allowing heightened sensitivity compared with the original st-PCR developed in 2010. Probability of detection was 94.8% using qPCR compared with 4.2% using st-PCR, demonstrating the considerable increased sensitivity achieved with qPCR. This fast and continuous development of eDNA tools by this group highlights the pace in which this field is moving and the possibilities for further application of eDNA as a routine tool for other species.

With a species-specific approach, one of the main hurdles is primer specificity. Several studies have demonstrated that primer specificity is a key for successful detection and the prevention of false positives, with the risk of amplification of non-target species [19, 27, 30, 71]. This is particularly important when trying to detect invasive or rare species, which may be in low abundances and have closely related native or common species within the same waterbody. Other issues will influence the detection of target taxa, Mächler *et al.* [49] were unable to detect *Tinodes waeneri* using high specificity primers in the field, despite successful detection using kick-net sampling, stating the biology of the species as a possible cause for this. In contrast, this study also showed that when using an 'unspecific primer pair' for the detection of *Gammarus pulex*, they were still able to positively match 88.8% of the sequences produced to the target species. This may indicate that there is an optimal specificity required when designing species-specific primers, and this is likely to vary between target organisms.

Development of species-specific primers can be a lengthy process, requiring substantial investment in time and development, which must include stringent validation, and in some cases may not prove successful in field application (as with the case of *T. waeneri* [49]). A key consideration is whether to employ standard or quantitative approaches. Quantitative approaches are often favoured due to greater sensitivity, but in many cases, the sensitivity of st-PCR may be adequate. Studies that have evaluated the detection rate of st-PCR have provided quite encouraging results; Jerde *et al.* [3] for example, determined that silver carp (*H. molitrix*) DNA could be detected by st-PCR from just seven copies of the target DNA per PCR reaction. Similarly, limits of detection were quite impressive for st-PCR assays (in the region of 4.6×10^{-4} to 3×10^{-2} ng/ul) for four invasive fish species (*Lepomis gibbosus, Leucaspius delineatus, Pimephales promelas* and *Pseudorasbora parva* [41]). For many applications, this level of sensitivity will most likely be sufficient, but the detection rates of st-PCR can, if required, be potentially improved by performing a second round of 'nested' or 'pseudo-nested' PCR. In nested PCR, a second pair of species-specific oligonucleotide primers are designed that are internal to the original primer pair. In pseudo-nested PCR, one internal primer is used together with one of the original primers. PCR product from the first reaction is used as template DNA for the nested or pseudo-nested reaction. There is evidence that this second PCR step can improve detection rates [39, 74]; there is a downside to the increased sensitivity of nested and pseudo-nested PCRs: the use of PCR product as template DNA in a second reaction presents great challenges for avoiding contamination, which is a very real concern in eDNA studies.

Species detection

Few studies have directly compared the performance of st-PCR and qPCR for species detection; however, in a recent study on invasive zebra and quagga mussels (*Dreissena polymorpha* and *D. rostriformis bugensis*, respectively) efficacy of st- and qPCR detection, methods were compared from laboratory, lake and river samples and st-PCR was found to be not only cheaper and simpler, but also more robust and less prone to false positives or negatives than qPCR [75]. To our knowledge, only one study has so far compared st-PCR, qPCR and ddPCR for species detection from environmental samples. The three methods were compared directly for their ability to detect round goby (*Neogobius melanostomus*) in controlled mesocosm experiments [43]. Water samples were collected from tanks with goby density varying from low, to medium to high, over eight time points ranging from 30 min to 24 h after introduction of the fish [43]. Round goby DNA was detected in all samples using st-PCR, qPCR or ddPCR, demonstrating that the methods are comparable for presence–absence detection. More direct comparisons between methods are needed to determine whether this is more widely seen.

In summary, even st-PCR assays, which are less sensitive than qPCR or ddPCR, have been repeatedly shown to outperform established survey methods in terms of detection sensitivity (e.g. [18]). Although the most sensitive methods are often desirable for monitoring purposes, there is often a trade-off between high sensitivity and false positives, and for some applications, st-PCR, which is cheaper, less technically challenging and in some cases less prone to false positives [74], can be more suitable than qPCR or ddPCR. qPCR is preferable for detecting species that are rare or in environments where eDNA concentration is low but may be more sensitive than st-PCR to PCR inhibitors that may be present in environmental samples, which could lead to false negatives. Probe-based qPCR approaches have the benefit of higher specificity than dye-based qPCR, but from the limited data available, limits of detection are similar.

Abundance estimation

A key difference between st-PCR and quantitative PCR approaches is the ability of the latter to estimate the concentration of DNA in a sample, and in principle the abundance of target species. Whether eDNA methods can go beyond describing simple presence–absence and reliably be used for estimating abundance or biomass is one of the key questions currently occupying the eDNA research community. qPCR is obviously quantitative by definition, and several studies have reported a positive but not necessarily linear relationship between biomass or organism density and eDNA concentration. For example, Thomsen *et al.* [76] found a significant correlation between the average number of DNA molecules (estimated by probe-based qPCR) and the estimated population size (density) of great crested newts (*Triturus cristatus*) or spadefoot toads (*Pelobates fuscus*) in natural ponds. In mesocosm studies of the same species, eDNA concentration was strongly related to the number of individuals in the tank, and the length of time the animals had been in the mesocosm [76]. The concentration of eDNA dropped dramatically once the animals were removed from the tanks, and was no longer detected by day 14 post-removal [76]. Similarly, Takahara *et al.* [77] found that concentration of eDNA (again via probe-based qPCR) was positively correlated with common carp biomass in both aquaria and experimental ponds. Goldberg *et al.* [46] also reported a significant relationship between population density of invasive New Zealand mudsnails (*Potamapyrgus antipodarum*) and eDNA concentration in aquaria. Applying eDNA for quantitative estimates is obviously likely to be more challenging in larger wetlands and in running rather than standing water bodies. Encouragingly, eDNA concentration was found to be positively associated with density, biomass and proportion of transects occupied for Rocky Mountain tailed frogs (*Ascaphus montanus*) and Idaho giant salamanders (*Dicamptodon aterrimus*) in natural streams [78]. In this example, proportion of transects occupied was the best predictor of eDNA concentration, followed by density and biomass. In a recent pioneering use of the approach, Bylemans *et al.* [79] successfully used qPCR to monitor changes in nuclear and mitochondrial eDNA concentrations over time and thus to monitor spawning activity of the endangered Macquarie perch (*Macquaria australasica*) in Australia.

Results from other studies demonstrate that relationships between organism abundance and eDNA concentration are not always confirmed or straightforward. No clear correlation was found between eDNA concentration and the number of great crested newts (*T. cristatus*) in ponds [13] or hellbenders (*Cryptobranchus alleganiensis*) in lotic environments [80]. In the hellbender study, this could be because the field surveys underestimate the true numbers, or that the eDNA estimate was influenced by individuals upstream of the sample site [80]. Abundance estimates are likely influenced by a multitude of properties of the organism and environment. Klymus *et al.* [81] performed controlled experiments in aquaria with bighead and silver carp and found that the amount of eDNA increased linearly with fish biomass but that quantification of eDNA samples was highly variable even when sampling the same individual under controlled conditions. In this case, feeding behaviour (but not water temperature) altered the rate of eDNA production. Feeding increases metabolism leading to higher faecal and eDNA production. Environmental conditions such as temperature, water chemistry and exposure to UV could also influence eDNA abundance estimates. For example, in one experiment, fish released more eDNA in warm water than in cold water and eDNA concentration better reflected fish abundance/biomass at higher temperature [82]. Spatial and temporal variability in eDNA concentration could have a large impact on our ability to make quantitative inferences with eDNA. Sampling in proximity to an organism, or at a time of day when the organism is more active, could result in higher abundance estimates than sampling further away or when the organism is at rest. However, Pilliod *et al.* [78] found no effect on time of day, distance from target animals or sample location within a stream on the concentration of tailed frog or giant salamander eDNA. Herder *et al.* [70] reviewed the power of many eDNA studies for abundance estimation and suggested that varying rates of eDNA persistence together with dilution, inhibition and other factors were responsible for this poor performance, but suggested that when sampling is performed in similar water types, at the same time of the year, and using the same methods and personnel, it is likely that relative differences in densities can be detected using target approaches.

The majority of studies that have attempted to estimate abundance have used a qPCR (particularly probe-based) approach, but ddPCR may have advantages for abundance estimation. In the direct comparison of ddPCR and probebased qPCR for detecting round goby, *Neogobius melanostomus* mentioned above, variability in DNA concentration between sample replicates was lower with ddPCR than with qPCR, suggesting quantification with the former method is more consistent [43]. Overall, there was no statistical difference between platforms in terms of eDNA concentration estimates; however, 9/24 point estimates differed from an expected 1:1 ratio and there was a tendency for these differences to be greater at low fish densities, indicating the platforms may differ in their results at low target DNA concentrations [43]. Other work has suggested that ddPCR may be better suited for the detection of low-concentration DNA since each ddPCR result is based on essentially thousands of reactions as opposed to a single qPCR reaction, and it is not reliant on calibration curves, which can be inaccurate at low concentrations [83]. ddPCR and qPCR have been rigorously compared in mesocosm experiments for estimation of common carp abundance and biomass [84]. A significant, positive correlation was found between eDNA concentration estimated from both methods, but as with the round goby study, variance between replicates was lower for ddPCR than for qPCR [84]. There was also a significant correlation between carp abundance (or biomass) and eDNA concentration estimated via both methods, but the correlation was slightly better for ddPCR, leading the authors to conclude that ddPCR is more efficient than qPCR for accurately estimating fish abundance/biomass, particularly at low concentrations of eDNA [84].

Finally, combining quantitative information from qPCR or ddPCR with site occupancy (i.e. the number of sites in which a species is detected) information is likely to substantially improve abundance estimates [85]. Although this measure quantifies spatial distribution rather than abundance, the two measures are positively correlated [86] and so site occupancy can be considered as a proxy for abundance. The use of occupancy models, which quantify the proportion of an area occupied by a species and account for temporal as well as spatial variation, further improves such quantification [87].

In summary, targeted approaches are powerful, sensitive, have the ability to provide quantitative information and have been routinely demonstrated to outperform established surveys for detection. The main drawback is that they by definition focus on single species. In the case of INNS monitoring, this can be a considerable disadvantage non-target INNS present in the sample may be missed. Non-targeted or whole community analyses, sometimes referred to as 'passive detection' methods, therefore offer considerably greater promise for assessments of general biodiversity, as discussed below.

Passive Detection: Whole Community Approach

The second method for screening eDNA samples utilizes HTS technology to identify whole communities of taxa by generating millions of DNA sequences. The most commonly used current method is metabarcoding, which uses PCR to amplify target DNA; however, in this case, the primers target a *group* of taxa (i.e. fish, invertebrates, mammals, etc.) rather than a specific species. Metabarcoding libraries are run on an HTS platform such as an Illumina MiSeq or HiSeq, allowing a large number of samples to be simultaneously sequenced in a single run and produces large datasets of sequences, which can then be compared with a reference database of sequences bioinformatically. Other technologies, such as PCR-free approaches and Oxford Nanopore's MinION platform, are starting to be explored for eDNA work, but as the focus so far has been on metabarcoding, we discuss this approach in detail here and return to alternative technologies in '*Future molecular developments*'.

Species detection

Metabarcoding is a powerful tool for bioassessment and has been applied to lentic, lotic and marine systems to generate community-level data (see Table 16.1). Several studies have detected a greater number of taxa with this method compared with traditional sampling [88, 89]. Metabarcoding is particularly appealing as a passive surveillance tool for both routine and high-risk pathway monitoring, and it is widely thought, as costs fall, this would be a suitable tool to use for detecting new and unexpected INNS, but so far the number of studies that have used metabarcoding for INNS detection are limited. Mahon and co-workers first demonstrated the effectiveness of metabarcoding for monitoring high-risk pathways – in this case the bait trade – water samples collected from bait shops in the North American Great Lakes area [23, 63]. Brown *et al.* [90] recently carried out an extensive metabarcoding survey of zooplankton from 16 major Canadian ports, with the specific objective of INNS detection. Twenty-four non-native species were identified, of which 11 were previously undetected.

Abundance estimation

The number of sequence reads generated during metabarcoding could be considered a proxy for abundance, but both in principle and in reality this relationship is complex. There are several opportunities for bias during the molecular and bioinformatics pipelines; such as preferential amplification during PCR (primer bias), and completeness of taxonomic reference databases, which can favour detection of some species over others. In spite of these challenges, several studies have demonstrated a positive correlation between sequence reads and species biomass or rank abundance, although the relationship is not necessarily a simple linear one (e.g. [89, 91–94]). A metabarcoding approach could be combined with site occupancy to provide an additional estimate of species abundance. This combined approach was taken by Hänfling *et al.* [89] in a study of the fish community of Lake Windermere, England. In this example, sequence read counts and site occupancy both correlated with rank abundance (assessed using long term data from gill-net and hydroacoustics surveys) but the latter correlation was stronger. This study along with other indicates that site occupancy could be a more appropriate surrogate for relative abundance than the number of sequence reads [87, 89, 93, 94]. A considerable amount of work is ongoing to try to improve abundance estimation both using PCR based and PCR-free approaches, but at present, a combination of site occupancy information and read count seem to offer a reasonable approximation. Metabarcoding could also be combined with qPCR or ddPCR if additional quantitative information is needed for a small number of species. Further work is needed to test the methods in more challenging habitats such as large rivers and marine environments.

One particular benefit of metabarcoding over targeted methods, is the additional information obtained for the invaded community. Metabarcoding and other whole community approaches allow studies to go much further than INNS detection, for example in investigating impact of INNS on species richness or interactions. Furthermore as demonstrated by Hänfling *et al.* [89], eDNA metabarcoding data can be used to describe the ecological status of a water body, which could also lead to a greater understanding of community change and dynamics over time if an INNS becomes established in a new region [50]. As highlighted in part above, as this method is relatively new, there are a number of elements still under development. As with the species-specific approach, metabarcoding relies on primer specificity to all taxa within a group and some primers show a bias, amplifying certain taxa in preference to others [92]. Bioinformatic data analysis can also be highly complex and relies heavily on a reference database of DNA sequences to match the data produced to known species sequence records [50, 63]. Effort is being taken globally to add to existing reference databases such as the Barcode of Life (BOLD http://www.boldsystems.org/) and to link DNA barcode data with species record databases (e.g. as in the UK National Biodiversity Network (NBN) Atlas https://nbnatlas.org/), however obtaining funds for database construction remains a significant challenge for this work.

Future Molecular Developments

In the studies discussed above, researchers have aimed to amplify the eDNA using PCR, as the concentration of DNA from environmental samples is thought to be low. However, PCR-based amplification relies heavily on the specificity of the primers being developed (both target and metabarcoding). Amplification of non-target DNA in a targeted approach or primer bias (preferential amplification for selected taxa) in a metabarcoding approach will have major implications for the detection of INNS through false positives and negatives (see *Avoiding false positives and negatives*, below) and has been highlighted by a number of studies [50, 92, 95]. Therefore, a number of recent developments aim to use PCR-free detection of target species. Metagenomic or PCR-free sequencing involves the random fragmentation of all DNA in a sample. Metagenomics sequencing has been applied to invertebrate community samples (e.g. [96–98]) and for the detection of trophic interactions [96–98] but has yet to be applied to eDNA in its strictest sense. Metagenomics for eDNA could be challenging due to the high proportion of non-target DNA found in the sample. Methods such as enriching or capturing

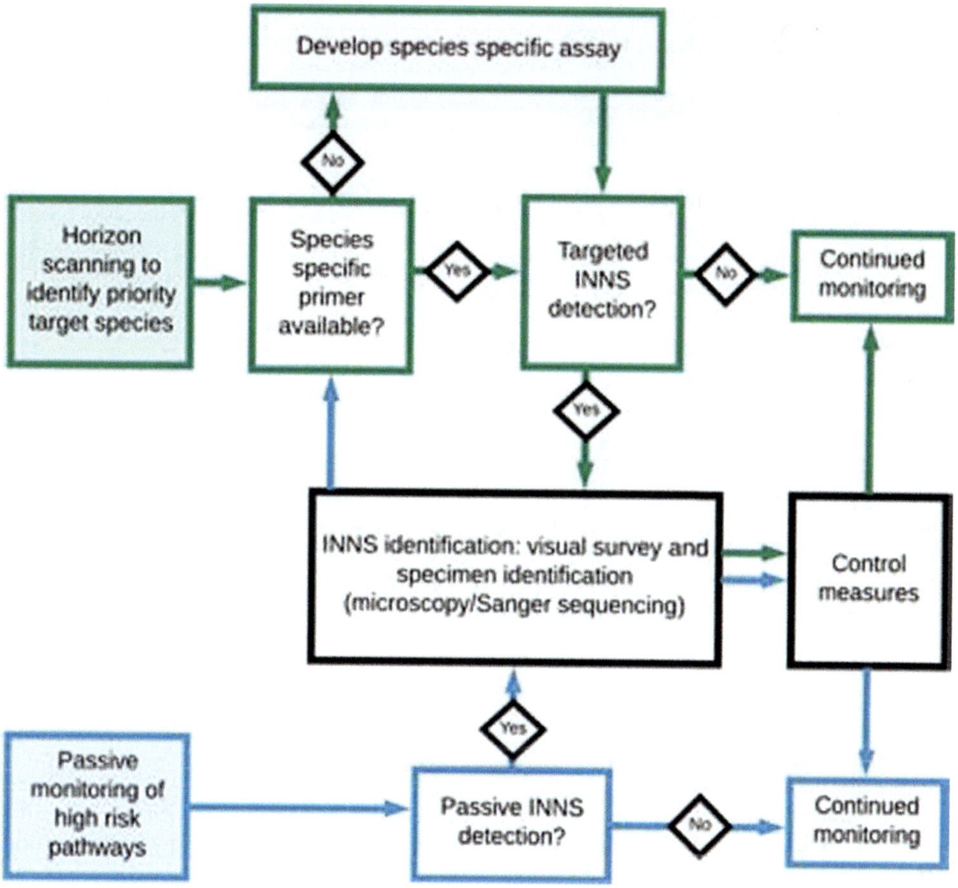

Fig. 17.2. Proposed INNS monitoring strategy during eDNA method development. (a) (blue) Surveillance of INNS using metabarcoding approaches, enabling systematic screening at priority vectors (ballast water, reservoirs, etc.) for those INNS not highlighted by horizon scanning exercises. (b) (green) Targeted approach enabling screening of priority vectors using developed species-specific primers and highlighting those species which require primers being developed.

the target DNA may improve the detection rates, but so far this technique remains challenging. For example, in an investigation of arthropod trophic interactions, enriched shotgun metagenomics detected only 15–239 reads out of two million that could be assigned to species other than the predator [98]. Nanopore technologies, such as the Oxford Nanopore MinION (https://nanoporetech.com/products/minion), have the potential to increase the power of such approaches through significantly higher throughput and read length, but dealing with the volume of information from taxonomically diverse communities (potentially all taxa without enrichment) will pose new and considerable bioinformatics challenges. Although not yet applied for targeted monitoring of INNS, ON technologies have already proven rapid and effective for identifying pathogen outbreaks in the field [99, 100]. A similar approach could be employed for early detection of priority INNS, with the main advantages of this technology being the portability (it is suitable for use in field applications even where Internet connection is limited) and the opportunity for rapid results (within 24 h, [101]).

Targeted or Metabarcoding Approaches: Which Road to Take?

Choice of approach will clearly depend on the objectives of the study or monitoring programme. If a single species is under surveillance, a metabarcoding approach would be overkill, and metabarcoding may be less sensitive than quantitative PCR approaches for single species detection. So far, few studies have carried out detailed comparisons to determine which approach is most suitable. However, Simmons *et al.* [50] compared targeted detection using both st-PCR and ddPCR to metabarcoding for the detection of Asian Carp in the Muskingum River Watershed in Ohio. ddPCR detected bighead carp at ten sites compared with no detection with st-PCR or metabarcoding. However, the metabarcoding analysis highlighted a previously unrecorded INNS, the northern snakehead, *Channa argus*, which was subsequently confirmed using ddPCR. The study demonstrates that although metabarcoding may be less sensitive than ddPCR (or qPCR) for detection, it has great potential for passive surveillance and if used to complement targeted approaches will greatly improve INNS detection and management.

A combination of both targeted and passive approaches would be the most powerful strategy, allowing immediate surveillance of water bodies utilising targeted primers for high priority known or anticipated species, alongside a metabarcoding approach for the detection of previously unrecorded or unexpected species. In Figure 17.2, we suggest an amalgamated strategy for the detection of INNS with eDNA, summarising the steps needed for initial detection, verification and control.

Factors Influencing Detection of INNS with eDNA

Since the study by Ficetola *et al.* [18], a number of elements have been highlighted which will influence the detection of a species or communities by eDNA, including sample volume and processing. Rees *et al.* [12] carried out a detailed review of current sampling and laboratory processes concluding there was no single suitable methodology, as the volume of water and sampling strategy are highly dependent on the system being sampled (see [12] for further discussion). This study, along with several others, has highlighted our need to understand how eDNA behaves within the system being sampled [12, 18, 66, 102]. For clarity, we use the definitions by Goldberg *et al.* [103] to describe the three main factors influencing successful species detection from eDNA: production, degradation and transport.

DNA production

Little is currently known about the rates of DNA shedding by individual species. Several studies have focused on the detection of fish and amphibians as it is thought these groups readily shed DNA into their environment, however any variability of DNA production by different taxa, which is likely linked to biomass, will influence the ability to detect them [3, 10, 31, 54, 78, 81, 102]. Although a number of macroinvertebrates have been successfully detected using both targeted and metabarcoding techniques [9, 49, 72, 104, 105]. And a study by Tréguier *et al.* [54] demonstrated the difficulty of detecting the invasive red swamp crayfish, *Procambarus clarkii*, when in low abundances, suggesting that low amounts of extracellular DNA in the water is a major limitation for crayfish detection compared with fish and amphibians. This has important implications when trying to detect INNS, particularly at the early stages of invasion.

Degradation

Factors influencing the persistence of eDNA within an environment include temperature, microbial community and UV; however, full understanding of all the potential influences on eDNA as it moves through an aquatic system is lacking and further research is needed (see [66] for further discussion). Detection success is highly variable between systems, with degradation rates typically faster in marine than freshwater habitats. In general, studies have shown that eDNA can remain detectable for between 7 and 21 days [12, 66, 103, 106].

Transport

Detection of eDNA also strongly depends on flow rate, with DNA in lentic systems clearly persisting for longer than in lentic systems. Identifying how the target DNA is transported through differing systems is essential for developing appropriate sampling strategies and methods [31, 107]. The availability of eDNA within lentic systems may reflect processes such as diffusion and circulation of water and so eDNA is readily available to samplers, whilst in lotic systems eDNA availability is likely to be influenced by hydrological dynamics [31, 107, 108]. These studies also highlight the influence of changing substrate type; as eDNA flows through the system, it will be subject to retention in the substrate, due to slower flows, and then resuspension. However, Shogren *et al.* showed stochastic variation in the retention of DNA by these differing substrates sizes with no clear consensus [107] and so further research is required.

Despite eDNA availability being influenced by the above processes, studies have also shown that eDNA is able to persist for long stretches in lotic systems. Deiner and Altermatt [104] showed in a study of *Daphnia longispina* and *Unio tumidus* that eDNA of both species was picked up more than 9 km downstream. A similar study by Jane *et al.* [31] using fish concluded that detection was successful up to 239.5 m downstream. Both suggesting that although factors affecting the persistence of eDNA are in force, detection is still possible. A paper by Deiner *et al.* [105] summarized the effects of flow and river dynamics nicely as acting as a 'conveyor belt' for biodiversity information, both terrestrial and aquatic, and indicates that monitoring within a lotic system could detect catchment-level biodiversity including INNS. This approach could enable a wider picture of community diversity to be collected before targeting more specific locations for INNS.

Avoiding False Positives and Negatives

The application of eDNA for biodiversity monitoring has been rapid and its applications are likely to extend dramatically as fundamental questions such as sample method, approach (targeted versus metabarcoding) and downstream analysis are overcome. As with most new technologies, in the initial development of a tool, there has been little standardization in terms of sample and laboratory protocols, and for newly detected INNS, molecular or morphological verification steps should ideally be employed, as outlined in Figure 17.2 and in Blackman *et al.* [9].

With regards to INNS detection, the need to minimize false positives and negatives must remain a priority in future studies [102]. To ensure against incorrect results, these tools must be developed with 'gold standard' protocols in mind [71], with the inclusion of blanks at all stages of processing (field sampling, filtering/precipitation, extraction and laboratory processing). False-positive detection is a particular issue, when typically tissue samples or standards of the target species are included in a PCR plate reaction. One method to overcome this would be the use of synthetic oligonucleotides as described by Wilson *et al.* [109], with the addition of readily detectable inserts (with both st-PCR and qPCR), this enables the synthetic DNA to be separated from actual target DNA in a reaction and would prevent false-positive results; however, these would need to be designed and developed on a species-specific basis.

Inhibition, as mentioned previously, is also a key concern when ensuring against false-negative detection and is highly dependent on seasonal change [31]. Typically, inhibition such as this is solved by dilutions of the sample or the addition of BSA; however, dilution of samples will result in lower copy number to be detected [31]. As part of eDNA great crested newt detection, synthetic DNA is also used as a control in this case for the detection of inhibition as each sample is spiked during preservation and so possible inhibition or DNA deterioration can be detected at extraction and the PCR stage [13].

As discussed previously, there are several factors affecting the persistence of eDNAwithin an aquatic environment and sensitivity levels vary depending on the molecular approach being used [43, 50]. Negative detections could result from either of these factors rather than actual absence of the target species. Both biological and temporal sample replicates are needed to provide the greatest confidence in detections and avoid false results. To allow for continued method development and robust protocols, consortiums such as DNAqua-net [110] and the US and UK DNA working groups aim to address and progress many of the issues highlighted above. This form of collaboration allows researchers from multiple disciplines to further demonstrate and advance our ability to deliver successful biodiversity monitoring using eDNA.

Conclusion

Significant progress has been made in this field since the first use of eDNA to detect an INNS in a natural system <10 years ago [18]; including the development of a large number of species-specific assays and progress in community-based analysis. The development of species-specific primers may require a great deal of investment in terms of primer design, optimization and validation, but the need for these tools is prevalent, and in the case of high priority INNS such as those identified via 'horizon scanning', which may be the most suitable method of early detection. The endorsement of HTS approaches in recent studies has also given invasion biologists future opportunities for passive surveillance, understanding impact on community structure, biosecurity testing and identification of cryptic species.

Some fundamental elements still remain largely unresolved and widely unregulated in terms of sampling strategy or analysis; however, this is not due to a lack of research but rather the huge variation in target INNS, systems being studied and detailed species knowledge. There is a strong need to develop appropriate validation processes and consistent field and laboratory standards, but without stifling progress as the techniques are progressing so rapidly. Perhaps most importantly, if we are to use eDNA for the detection of new INNS, targeted or community based, we need to have an understanding of both the molecular methods and the species which we are trying to detect. Whether we target a fish species in a lake or a mussel in a river, to use this method effectively, ecological understanding must play a key role in developing new tools.

Acknowledgements

The authors would like to thank Tim Jones, Andrew Sheard, Helena Sinclair and Dr Helen Roy, for their helpful suggestions and the three anonymous reviewers whose comments strengthened and improved the manuscript.

References

1. Hulme PE. Beyond control: wider implications for the management of biological invasions. Journal of Applied Ecology 2006;43(5):835–47.
2. Davis MA. Invasion Biology. Oxford University Press, Oxford, UK; 2009.
3. Jerde CL, Mahon AR, Chadderton WL, Lodge DM. 'Sight-unseen' detection of rare aquatic species using environmental DNA. Conservation Letters 2011;4(2):150–7.
4. Lodge DM, Turner CR, Jerde CL, Barnes MA, Chadderton L, Egan SP, *et al.* Conservation in a cup of water: estimating biodiversity and population abundance from environmental DNA. Molecular Ecology 2012;21(11):2555–8.
5. Roy HE, Peyton J, Aldridge DC, Bantock T, Blackburn TM, Britton R, *et al.* Horizon scanning for invasive alien species with the potential to threaten biodiversity in Great Britain. Global Change Biology 2014;20(12):3859–71.
6. Lawson Handley L. How will the 'molecular revolution' contribute to biological recording? Biological Journal of Linnean Society of London 2015;115(3):750–66.
7. Furse MT, Wright JF, Armitage PD, Moss D. An appraisal of pond-net samples for biological monitoring of lotic macro-invertebrates. Water Research 1981;15(6):679–89.
8. Trebitz AS, Hoffman JC, Darling JA, Pilgrim EM, Kelly JR, Brown EA, *et al.* Early detection monitoring for aquatic non-indigenous species: optimizing surveillance, incorporating advanced technologies,

and identifying research needs. Journal of Environmental Economics and Management 2017;202(Pt 1):299–310.
9. Blackman R, Constable D, Hahn C, Sheard A, Durkota J, Hänfling B, *et al.* Detection of a new non-native freshwater species by DNA metabarcoding of environmental samples – first record of *Gammarus fossarum* in the UK. Aquatic Invasions 2017;12(2):177–89.
10. Thomsen PF, Willerslev E. Environmental DNA – an emerging tool in conservation for monitoring past and present biodiversity. Biological Conservation 2015;183:4–18.
11. Taberlet P, Coissac E, Pompanon F, Brochmann C, Willerslev E. Towards next-generation biodiversity assessment using DNA metabarcoding. Molecular Ecology 2012;21(8):2045–50.
12. Rees HC, Maddison BC, Middleditch DJ, Patmore JRM, Gough KC. REVIEW: the detection of aquatic animal species using environmental DNA – a review of eDNA as a survey tool in ecology. Journal of Applied Ecology 2014;51(5):1450–9.
13. Biggs J, Ewald N, Valentini A, Gaboriaud C, Dejean T, Griffiths RA, *et al.* Using eDNA to develop a national citizen science-based monitoring programme for the great crested newt (*Triturus cristatus*). Biological Conservation 2015;183:19–28.
14. Schneider J, Valentini A, Dejean T, Montarsi F, Taberlet P, Glaizot O, *et al.* Detection of invasive mosquito vectors using environmental DNA (eDNA) from water samples. PLoS ONE 2016;11(9):e0162493.
15. Secondi J, Dejean T, Valentini A, Audebaud B, Miaud C. Detection of a global aquatic invasive amphibian, *Xenopus laevis*, using environmental DNA. Amphibia-reptilia 2016;37(1):131–6.
16. Moyer GR, Díaz-Ferguson E, Hill JE, Shea C. Assessing environmental DNA detection in controlled lentic systems. PLoS ONE 2014;9(7):e103767.
17. Dejean T, Valentini A, Miquel C, Taberlet P, Bellemain E, Miaud C. Improved detection of an alien invasive species through environmental DNA barcoding: the example of the American bullfrog *Lithobates catesbeianus*. Journal of Applied Ecology 2012;49(4):953–9.
18. Ficetola GF, Miaud C, Pompanon F, Taberlet P. Species detection using environmental DNA from water samples. Biology Letters 2008;4(4):423–5.
19. Ardura A, Zaiko A, Martinez JL, Samulioviene A, Semenova A, Garcia-Vazquez E. eDNA and specific primers for early detection of invasive species – a case study on the bivalve *Rangia cuneata*, currently spreading in Europe. Marine Environmental Research 2015;112(Pt B):48–55.
20. Kirshtein JD, Anderson CW, Wood JS, Longcore JE, Voytek MA. Quantitative PCR detection of *Batrachochytrium dendrobatidis* DNA from sediments and water. Diseases of Aquatic Organisms 2007;77:11–5.
21. Hyman OJ, Collins JP. Evaluation of a filtration-based method for detecting *Batrachochytrium dendrobatidis* in natural bodies of water. Diseases of Aquatic Organisms 2012;97(3):185–95.
22. Mahon AR, Jerde CL, Galaska M, Bergner JL, Lindsay Chadderton W, Lodge DM, *et al.* Validation of eDNA surveillance sensitivity for detection of Asian carps in controlled and field experiments. PLoS ONE 2013;8(3):e58316.
23. Nathan LR, Jerde CL, Budny ML, Mahon AR. The use of environmental DNA in invasive species surveillance of the Great Lakes commercial bait trade. Conservation Biology 2015;29(2):430–9.
24. Takahara T, Minamoto T, Doi H. Using environmental DNA to estimate the distribution of an invasive fish species in ponds. PLoS ONE 2013;8(2):e56584.
25. Doi H, Takahara T, Minamoto T, Matsuhashi S, Uchii K, Yamanaka H. Droplet digital polymerase chain reaction (PCR) outperforms real-time PCR in the detection of environmental DNA from an invasive fish species. Environmental Science & Technology 2015;49(9):5601–8.
26. Hunter ME, Oyler-McCance SJ, Dorazio RM, Fike JA, Smith BJ, Hunter CT, *et al.* Environmental DNA (eDNA) sampling improves occurrence and detection estimates of invasive Burmese pythons. PLoS ONE 2015;10(4):e0121655.
27. Scriver M, Marinich A, Wilson C, Freeland J. Development of species-specific environmental DNA (eDNA) markers for invasive aquatic plants. Aquatic Botany 2015/4;122:27–31.
28. Fujiwara A, Matsuhashi S, Doi H, Yamamoto S, Minamoto T. Use of environmental DNA to survey the distribution of an invasive submerged plant in ponds. Freshw Science 2016;35(2):748–54.
29. Matsuhashi S, Doi H, Fujiwara A, Watanabe S, Minamoto T. Evaluation of the environmental DNA method for estimating distribution and biomass of submerged aquatic plants. PLoS ONE 2016;11(6):e0156217.
30. Wilcox TM, McKelvey KS, Young MK, Jane SF, Lowe WH, Whiteley AR, *et al.* Robust detection of rare species using environmental DNA: the importance of primer specificity. PLoS ONE 2013;8(3): e59520.

31. Jane SF, Wilcox TM, McKelvey KS, Young MK, Schwartz MK, Lowe WH, *et al.* Distance, flow and PCR inhibition: eDNA dynamics in two headwater streams. Molecular Ecology Resources 2015;15(1):216–27.
32. Piaggio AJ, Engeman RM, Hopken MW, Humphrey JS, Keacher KL, Bruce WE, *et al.* Detecting an elusive invasive species: a diagnostic PCR to detect Burmese python in Florida waters and an assessment of persistence of environmental DNA. Molecular Ecology Resources 2014;14(2):374–80.
33. Fukumoto S, Ushimaru A, Minamoto T. A basin-scale application of environmental DNA assessment for rare endemic species and closely related exotic species in rivers: a case study of giant salamanders in Japan. Journal of Applied Ecology 2015;52(2):358–65.
34. Eichmiller JJ, Bajer PG, Sorensen PW. The relationship between the distribution of common carp and their environmental DNA in a small lake. PLoS ONE 2014;9(11): e112611.
35. Turner CR, Barnes MA, Xu CCY, Jones SE, Jerde CL, Lodge DM. Particle size distribution and optimal capture of aqueous macrobial eDNA. Methods in Ecology and Evolution 2014;5(7):676–84.
36. Uchii K, Doi H, Yamanaka H, Minamoto T. Distinct seasonal migration patterns of Japanese native and non-native genotypes of common carp estimated by environmental DNA. Ecology and Evolution 2017;7(20):8515–22.
37. Vrålstad T, Knutsen AK, Tengs T, Holst-Jensen A. A quantitative TaqMan MGB real-time polymerase chain reaction based assay for detection of the causative agent of crayfish plague *Aphanomyces astaci*. Veterinary Microbiology 2009;137(1–2):146–55.
38. Strand DA, Jussila J, Johnsen SI, Viljamaa-Dirks S, Edsman L, Wiik-Nielsen J, *et al.* Detection of crayfish plague spores in large freshwater systems. Journal of Applied Ecology 2014;51(2):544–53.
39. Scarlett K, Tesoriero L, Daniel R, Guest D. Detection and quantification of *Fusarium oxysporum f. sp cucumerinum* in environmental samples using a specific quantitative PCR assay. European Journal of Plant Pathology 2013;137(2):315–24.
40. Peñarrubia L, Alcaraz C, de Vaate AB, Sanz N, Pla C, Vidal O, *et al.* Validated methodology for quantifying infestation levels of dreissenid mussels in environmental DNA (eDNA) samples. Scientific Reports 2016;6:39067.
41. Davison PI, Créach V, Liang W-J, Andreou D, Britton JR, Copp GH. Laboratory and field validation of a simple method for detecting four species of non-native freshwater fish using eDNA. Journal of Fish Biology 2016;89(3):1782–93.
42. Mahon AR, Barnes MA, Senapati S, Feder JL, Darling JA, Chang H-C, *et al.* Molecular detection of invasive species in heterogeneous mixtures using a microfluidic carbon nanotube platform. PLoS ONE 2011;6(2):e17280.
43. Nathan LM, Simmons M,Wegleitner BJ, Jerde CL, Mahon AR. Quantifying environmental DNA signals for aquatic invasive species across multiple detection platforms. Environmental Science & Technology 2014;48(21):12800–6.
44. Hunter ME, Dorazio RM, Butterfield JSS, Meigs-Friend G, Nico LG, Ferrante JA. Detection limits of quantitative and digital PCR assays and their influence in presence-absence surveys of environmental DNA. Molecular Ecology Resources 2017;17(2):221–9.
45. Agersnap S, Larsen WB, Knudsen SW, Strand D, Thomsen PF, Hesselsøe M, *et al.* Monitoring of noble, signal and narrow-clawed crayfish using environmental DNA from freshwater samples. PLoS ONE 2017;12(6):e0179261.
46. Goldberg CS, Sepulveda A, Ray A, Baumgardt J, Waits LP. Environmental DNA as a new method for early detection of New Zealand mudsnails (*Potamopyrgus antipodarum*). Freshwater Science 2013;32(3):792–800.
47. Keskin E. Detection of invasive freshwater fish species using environmental DNA survey. Biochemical Systematics and Ecology 2014;56:68–74.
48. Deagle BE, Bax N, Hewitt CL, Patil JG. Development and evaluation of a PCR-based test for detection of Asterias (Echinodermata: Asteroidea) larvae in Australian plankton samples from ballast water. Marine and Freshwater Research 2003;54(6):709–19.
49. Mächler E, Deiner K, Steinmann P, Altermatt F. Utility of environmental DNA for monitoring rare and indicator macroinvertebrate species. Freshwater Science 2014;33(4):1174–83.
50. Simmons M, Tucker A, Chadderton WL, Jerde CL, Mahon AR. Active and passive environmental DNA surveillance of aquatic invasive species. Canadian Journal of Fisheries and Aquatic Sciences 2015;73(1):76–83.
51. Olsen JB, Lewis CJ, Massengill RL, Dunker KJ, Wenburg JK. An evaluation of target specificity and sensitivity of three qPCR assays for detecting environmental DNA from Northern Pike (*Esox lucius*). Conservation Genetics Resources 2015;7(3):615–7.

52. Adrian-Kalchhauser I, Burkhardt-Holm P. An eDNA assay to monitor a globally invasive fish species from flowing freshwater. PLoS ONE 2016;11(1):e0147558.
53. Lance RF, Carr MR. Detecting eDNA of Invasive Dreissenid Mussels: Report on Capital Investment Project. ANSRP Technical Notes Collection. ERDC/TN ANSRP-12-2. U.S. Army Engineer Research and Development Center, Vicksburg, MS, USA; 2012. Available from: URL: http://el.erdc.usace.army.mil/publications.cfm?Topic=technote&Code=ansrp.
54. Tréguier A, Paillisson J-M, Dejean T, Valentini A, Schlaepfer MA, Roussel J-M. Environmental DNA surveillance for invertebrate species: advantages and technical limitations to detect invasive crayfish *Procambarus clarkii* in freshwater ponds. Journal of Applied Ecology 2014;51(4):871–9.
55. Cai W, Ma Z, Yang C, Wang L, Wang W, Zhao G, *et al.* Using eDNA to detect the distribution and density of invasive crayfish in the Honghe-Hani rice terrace world heritage site. PLoS ONE 2017;12(5):e0177724.
56. Davy CM, Kidd AG, Wilson CC. Development and validation of environmental DNA (eDNA) markers for detection of freshwater turtles. PLoS ONE 2015;10(7):e0130965.
57. Dougherty MM, Larson ER, Renshaw MA, Gantz CA, Egan SP, Erickson DM, *et al.* Environmental DNA (eDNA) detects the invasive rusty crayfish *Orconectes rusticus* at low abundances. Journal of Applied Ecology 2016;53(3):722–32.
58. Larson ER, Renshaw MA, Gantz CA, Umek J, Chandra S, Lodge DM, *et al.* Environmental DNA (eDNA) detects the invasive crayfishes *Orconectes rusticus* and *Pacifastacus leniusculus* in large lakes of North America. Hydrobiologia 2017;800(1):173–85.
59. Dejean T, Valentini A, Duparc A, Pellier-Cuit S, Pompanon F, Taberlet P, *et al.* Persistence of environmental DNA in freshwater ecosystems. PLoS ONE 2011;6(8):e23398.
60. Dunn N, Priestley V, Herraiz A, Arnold R, Savolainen V. Behavior and season affect crayfish detection and density inference using environmental DNA. Ecology and Evolution 2017;7(19):7777–85.
61. Davison PI, Copp GH, Créach V, Vilizzi L, Britton JR. Application of environmental DNA analysis to inform invasive fish eradication operations. Naturwissenschaften 2017;104(3–4):35.
62. Baerwald MR, Schreier BM, Schumer G, May B. Detection of threatened delta smelt in the gut contents of the invasive Mississippi silverside in the San Francisco estuary using TaqMan assays. Transactions of the American Fisheries Society 2012;141(6):1600–7.
63. Mahon AR, Nathan LR, Jerde CL. Meta-genomic surveillance of invasive species in the bait trade. Conservation Genetics Resources 2014;6:563–7.
64. Collins RA, Armstrong KF, Holyoake AJ, Keeling S. Something in the water: biosecurity monitoring of ornamental fish imports using environmental DNA. Biological Invasions 2013;15(6):1209–15.
65. Hulme PE. Trade, transport and trouble: managing invasive species pathways in an era of globalization. Journal of Applied Ecology 2009;46(1):10–8
66. Barnes MA, Turner CR. The ecology of environmental DNA and implications for conservation genetics. Conservation Genetics 2016;17(1):1–17.
67. Willerslev E, Hansen AJ, Gilbert MTP, Shapiro B. Diverse plant and animal genetic records from Holocene and pleistocene sediments. Science 2003;300(5620):791–5.
68. Darling JA, Mahon AR. From molecules to management: adopting DNA-based methods for monitoring biological invasions in aquatic environments. Environmental Research 2011;111(7):978–88.
69. Valentini A, Taberlet P, Miaud C, Civade R, Herder J, Thomsen PF, *et al.* Next-generation monitoring of aquatic biodiversity using environmental DNA metabarcoding. Molecular Ecology 2016;25(4):929–42.
70. Herder JE, Valentini A, Bellemain E, Dejean T, van Delft JJCW, Thomsen PF, *et al.* Environmental DNA: A review of the possible applications for the detection of (invasive) species. Stichting RAVON, Nijmegen. Report 2013-104.
71. Goldberg CS, Turner CR, Deiner K, Klymus KE, Thomsen PF, Murphy MA, *et al.* Critical considerations for the application of environmental DNA methods to detect aquatic species. Methods in Ecology and Evolution [Internet]. 2016. Available from: URL: http://dx.doi.org/10.1111/2041-210X.12595.
72. Thomsen PF, Kielgast J, Iversen LL, Møller PR, Rasmussen M, Willerslev E. Detection of a diverse marine fish fauna using environmental DNA from seawater samples. PLoS ONE 2012;7(8):e41732.
73. Turner CR, Miller DJ, Coyne KJ, Corush J. Improved methods for capture, extraction, and quantitative assay of environmental DNA from Asian bigheaded carp (*Hypophthalmichthys* spp.). PLoS ONE 2014;9(12):e114329.

74. Stoeckle BC, Kuehn R, Geist J. Environmental DNA as a monitoring tool for the endangered freshwater pearl mussel (*Margaritifera margaritifera* L.): a substitute for classical monitoring approaches? Aquatic Conservation 2016;26(6):1120–9.
75. De Ventura L, Kopp K, Seppälä K, Jokela J. Tracing the quagga mussel invasion along the Rhine river system using eDNA markers: early detection and surveillance of invasive zebra and quagga mussels. Management of Biological Invasions 2017;8(1):101–12.
76. Thomsen PF, Kielgast J, Iversen LL, Wiuf C, Rasmussen M, Gilbert MTP, *et al.* Monitoring endangered freshwater biodiversity using environmental DNA. Molecular Ecology 2012;21(11):2565–73.
77. Takahara T, Minamoto T, Yamanaka H, Doi H, Kawabata Z. Estimation of fish biomass using environmental DNA. PLoS ONE 2012;7(4):e35868.
78. Pilliod DS, Goldberg CS, Arkle RS, Waits LP. Estimating occupancy and abundance of stream amphibians using environmental DNA from filtered water samples. Canadian Journal of Fisheries and Aquatic Sciences 2013;70(8):1123–30.
79. Bylemans J, Furlan EM, Hardy CM, McGuffie P, Lintermans M, Gleeson DM. An environmental DNA (eDNA) based method for monitoring spawning activity: a case study using the endangered Macquarie perch (*Macquaria australasica*). Methods in Ecology and Evolution 2017;8(5):646–55.
80. Spear SF, Groves JD, Williams LA, Waits LP. Using environmental DNA methods to improve detectability in a hellbender (*Cryptobranchus alleganiensis*) monitoring program. Biological Conservation 2015;183:38–45.
81. Klymus KE, Richter CA, Chapman DC, Paukert C. Quantification of eDNA shedding rates from invasive bighead carp *Hypophthalmichthys nobilis* and silver carp *Hypophthalmichthys molitrix*. Biological Conservation 2015;183:77–84.
82. Lacoursiere-Roussel A, Rosabal M, Bernatchez L. Estimating fish abundance and biomass from eDNA concentrations: variability among capture methods and environmental conditions. Molecular Ecology Resources 2016;16(6):1401–14.
83. Mahon AR, Rohly A, Budny M, Elgin E, Jerde C, Chadderton W, *et al.* Environmental DNA Monitoring and Surveillance: Standard Operation Procedures. Report to the United States Army Corps of Engineers, Environmental Laboratories, Cooperative Environmental Studies Unit, Vicksburg, Mississippi; 2010.
84. Doi H, Uchii K, Takahara T, Matsuhashi S, Yamanaka H, Minamoto T. Use of droplet digital PCR for estimation of fish abundance and biomass in environmental DNA surveys. PLoS ONE 2015;10(3):e0122763.
85. MacKenzie DI, Nichols JD, Lachman GB, Droege S, Andrew Royle J, Langtimm CA. Estimating site occupancy rates when detection probabilities are less than one. Ecology 2002;83(8):2248–55.
86. MacKenzie DI, Nichols JD. Occupancy as a surrogate for abundance estimation. Animal Biodiversity and Conservation 2004;27(1):461–7.
87. Schmidt BR, Kéry M, Ursenbacher S, Hyman OJ, Collins JP. Site occupancy models in the analysis of environmental DNA presence/absence surveys: a case study of an emerging amphibian pathogen. Methods in Ecology and Evolution 2013;4(7):646–53.
88. Port JA, O'Donnell JL, Romero-Maraccini OC, Leary PR, Litvin SY, Nickols KJ, *et al.* Assessing vertebrate biodiversity in a kelp forest ecosystem using environmental DNA. Molecular Ecology 2016;25(2):527–41.
89. Hänfling B, Lawson Handley L, Read DS, Hahn C, Li J, Nichols P, *et al.* Environmental DNA metabarcoding of lake fish communities reflects long-term data from established survey methods. Molecular Ecology 2016;25(13):3101–19.
90. Brown EA, Chain FJJ, Zhan A, MacIsaac HJ, Cristescu ME. Early detection of aquatic invaders using metabarcoding reveals a high number of non-indigenous species in Canadian ports. Diversity and Distributions 2016;22(10):1045–59.
91. Evans NT, Olds BP, Turner CR, Renshaw MA, Li Y, Jerde CL, *et al.* Quantification of mesocosm fish and amphibian species diversity via eDNA metabarcoding. Molecular Ecology Resources 2016;16(1):29–41.
92. Elbrecht V, Leese F. Can DNA-based ecosystem assessments quantify species abundance? Testing primer bias and biomass–sequence relationships with an innovative metabarcoding protocol. PLoS ONE 2015;10(7):e0130324.
93. Stoeckle MY, Soboleva L, Charlop-Powers Z. Aquatic environmental DNA detects seasonal fish abundance and habitat preference in an urban estuary. PLoS ONE 2017;12(4):e0175186.

94. Elbrecht V, Vamos EE, Meissner K, Aroviita J, Leese F. Assessing strengths and weaknesses of DNA metabarcoding-based macroinvertebrate identification for routine stream monitoring. Methods in Ecology and Evolution 2017;8(10):1265–75.
95. Shokralla S, Gibson J, King I, Baird D, Janzen D, HallwachsW, *et al.* Environmental DNA Barcode Sequence Capture: Targeted, PCR-free Sequence Capture for Biodiversity Analysis from Bulk Environmental Samples. bioRxiv 2016;087437.
96. Zhou X, Li Y, Liu S, Yang Q, Su X, Zhou L, *et al.* Ultra-deep sequencing enables high-fidelity recovery of biodiversity for bulk arthropod samples without PCR amplification. Gigascience 2013;2(1):4.
97. Paula DP, Linard B, Andow DA, Sujii ER, Pires CSS, Vogler AP. Detection and decay rates of prey and prey symbionts in the gut of a predator through metagenomics. Molecular Ecology Resources 2015;15(4):880–92.
98. Paula DP, Linard B, Crampton-Platt A, Srivathsan A, Timmermans MJTN, Sujii ER, *et al.* Uncovering trophic interactions in arthropod predators through DNA shotgun-sequencing of gut contents. PLoS ONE 2016;11(9): e0161841.
99. Greninger AL, Naccache SN, Federman S, Yu G, Mbala P, Bres V, *et al.* Rapid metagenomic identification of viral pathogens in clinical samples by real-time nanopore sequencing analysis. Genome Medicine 2015;7(1):99.
100. Quick J, Loman NJ, Duraffour S, Simpson JT, Severi E, Cowley L, *et al.* Real-time, portable genome sequencing for Ebola surveillance. Nature 2016;530(7589):228–32.
101. Votintseva AA, Bradley P, Pankhurst L, Del Ojo Elias C, Loose M, Nilgiriwala K, *et al.* Same-day diagnostic and surveillance data for tuberculosis via whole-genome sequencing of direct respiratory samples. Journal of Clinical Microbiology 2017;55(5):1285–98.
102. Roussel J-M, Paillisson J-M, Tréguier A, Petit E. The downside of eDNA as a survey tool in water bodies. Journal of Applied Ecology 2015;52(4):823–6.
103. Goldberg CS, Strickler KM, Pilliod DS. Moving environmental DNA methods from concept to practice for monitoring aquatic macroorganisms. Biological Conservation 2015;183:1–3.
104. Deiner K, Altermatt F. Transport distance of invertebrate environmental DNA in a natural river. PLoS ONE 2014;9(2):e88786.
105. Deiner K, Fronhofer EA, Mächler E, Walser J-C, Altermatt F. Environmental DNA reveals that rivers are conveyer belts of biodiversity information. Nature Communications 2016;7:12544.
106. Strickler KM, Fremier AK, Goldberg CS. Quantifying effects of UV-B, temperature, and pH on eDNA degradation in aquatic microcosms. Biological Conservation 2015;183:85–92.
107. Shogren AJ, Tank JL, Andruszkiewicz E, Olds B, Mahon AR, Jerde CL, *et al.* Controls on eDNA movement in streams: transport, retention, and resuspension. Scientific Reports 2017;7(1):5065.
108. Jerde CL, Olds BP, Shogren AJ, Andruszkiewicz EA, Mahon AR, Bolster D, *et al.* Influence of stream bottom substrate on retention and transport of vertebrate environmental DNA. Environmental Science & Technology 2016;50(16):8770–9.
109. Wilson CC, Wozney KM, Smith CM. Recognizing false positives: synthetic oligonucleotide controls for environmental DNA surveillance. Methods in Ecology and Evolution 2016;7(1):23–9.
110. Leese F, Altermatt F, Bouchez A, Ekrem T, Hering D, Meissner K, *et al.* DNAqua-Net: developing new genetic tools for bioassessment and monitoring of aquatic ecosystems in Europe. Riogrande Odontologia 2016;2:e11321.

18 Weed biological control in the Greater Mekong Subregion: status and opportunities for the future

Michael D. Day[1]*, Shicai Shen[2]*, Gaofeng Xu[2], Fudou Zhang[2] and Rachel L. Winston[3]

[1] Department of Agriculture and Fisheries, Ecosciences Precinct, GPO Box 267, Brisbane, Queensland 4001, Australia. [2] Agricultural Environment and Resource Research Institute, Yunnan Academy of Agricultural Sciences, Kunming Yunnan 650205, China. [3] MIA Consulting, 316 N, Hansen Ave., Shelley, ID 83274, USA

Abstract

The status of weed biological control in the six Greater Mekong Subregion (GMS) countries was assessed, and opportunities to develop biological control in the region are proposed. Twenty biological control agents targeting nine weed species have been introduced into GMS countries, with 13 agents establishing on seven weed species. Another nine biological control agents have spread unintentionally into the region. These agents are having a slight to high impact on their target weeds. However, the number of weeds in the region that have been targeted for biological control is only a fraction of the number of weeds (45) in the region that have been targeted for biological control elsewhere. This presents a tremendous opportunity to expand weed biological control in the region and reduce the dependency on herbicides while increasing productivity. There are numerous highly effective biological control agents present elsewhere that could be introduced to assist with the management of the region's most important weeds, such as *Pistia stratiotes*, *Salvinia molesta*, *Chromolaena odorata* and *Mikania micrantha*. Along with the introduction of biological control agents, there is also a scope to increase the capacity and training in biological control within GMS countries to develop effective weed control and management.

Keywords: Host specific, *Chromolaena odorata*, *Mikania micrantha*, *Salvinia molesta*

Review Methodology

Information on the status of weed biological was obtained from the on-line catalogue. Information on the list of alien invasive weeds which are present in each of the six GMS countries was obtained from various web-based databases (key words: country, invasive plant) and publications. We supplemented this with local knowledge in each country not present in any database or publication. We then applied various criteria (biological control targets, non-target impacts, medium to high impacts) to finalise a list of weeds that were targets for biological control in at least one country and that had biological control agents that were low-risk and causing medium to high impacts against the target weed.

*Corresponding author: michael.day@daf.qld.gov.au, shenshicai2011@aliyun.com

Introduction

The Greater Mekong Subregion (GMS) is a natural economic area bound by the Mekong River Basin, covering 2.6 million square kilometres with a combined population of approximately 326 million people [1, 2]. The GMS countries are Cambodia, the People's Republic of China (specifically Yunnan Province and Guangxi Zhuang Autonomous Region), Lao People's Democratic Republic (Lao PDR), Myanmar, Thailand and Vietnam. Originating in China's Qinghai Province, the Mekong River (known in China as Lancang Jiang and Thailand as Mae Nam Khong), flows through Yunnan Province, China and forms the border between Myanmar and Lao PDR and most of the border between Lao PDR and Thailand. It then flows through Cambodia and southern Vietnam into a rich delta, before emptying into the South China Sea. The total length of the river is 4880 km, including 2161 km in China (1237 km in Yunnan Province) [2, 3].

The GMS is one of the most abundant areas of global biodiversity. The region is rich in water resources and fertile arable land. It is the main production and export area for rice worldwide; Thailand and Vietnam's rice exports have been the largest and the second largest in the world, respectively, for many years [4], and the rubber production base of GMS is the largest in the world. In addition, the production of tropical fruit and aquatic products is also of significant economic importance in the region [5].

The six GMS countries belong to the developing countries of the world, and agriculture is the main economic pillar, accounting for a high proportion of their GDP, especially Myanmar, Lao PDR and Cambodia [4]. Most poor and malnourished people that live in rural areas of the region are dependent on subsistence and small-scale agriculture. Thus, sustainable agricultural production is highly important to food security in GMS countries.

Due to complex geographical and ecological conditions, the GMS is seriously afflicted with pests and invasive alien species, and the risk is increasing through cross-border trade and travel [6–8]. The threats posed by pests and invasive alien species are far beyond the management ability of any one country in the GMS and require effective regional cooperation. Currently, a great number of agricultural pests and invasive alien species, especially the rice planthopper, rice blast and noxious weeds, have become severe, resulting in huge economic losses and even threatening sustainable development of agriculture and livelihoods. In China alone, economic losses due to invasive alien species exceeds US$14 billion per year [9]. Therefore, there is an urgent need to cope with this situation through GMS plant protection workers taking joint action immediately [8].

In order to improve sustainable agricultural development and strengthen prevention and control of agricultural pests in GMS, the Plant Protection Working Group in GMS (PPWG-GMS) was established in August 2011 (under approval of the Third Steering Committee Meeting of the Exchange and Cooperation Consortium for Agricultural Science and Technology in GMS). The delegates represented the Cambodian Agricultural Research and Development Institute, Yunnan Academy of Agricultural Sciences, National Agriculture and Forestry Research Institute of Lao PDR, Myanmar Academy of Agricultural, Forestry, Livestock, and Fishery Sciences, Department of Agriculture, Ministry of Agriculture and Cooperatives of Thailand and Vietnam Academy of Agricultural Sciences. Much has been accomplished by membership institutions and experts of the PPWG-GMS since its founding, including multilateral cooperation and technology exchange, joint field surveys, technology training and green control demonstrations for rice and vegetable pests [10].

The costs of managing agricultural pests in the GMS, particularly invasive alien plants, can be extremely high. In China, the manual removal of *Eichhornia crassipes* (Mart.) Solms (Pontederiaceae) (water hyacinth) between 1991 and 2001 cost over US$12 million per year, but it was not effective [11]. Unfortunately, figures to control weeds in most countries in Asia are unavailable. However, elsewhere, the cost of weed control also runs into the millions of dollars. The cost to control *E. crassipes* on Lake Victoria in East Central Africa is over US$8.3 million per annum (p.a.), while Egypt spends approximately US$7 million p.a. to control aquatic weeds [12].

Considering that thousands of hectares have been invaded by these and other weeds in Asia, it is unlikely that these and many other plant species will be effectively controlled by mechanical and chemical means alone. Consequently, there is an urgent need to find economically sustainable solutions to existing and emerging weed problems in GMS countries. Biological control is a low-risk, self-sustainable alternative used to manage many weeds in many countries [13–15]. Indeed, China, Myanmar, Thailand and Vietnam have already deliberately introduced biological control agents to control some of their worst weeds. Numerous other weed biological control agents have also spread unintentionally into the region [15].

Training workshops and field surveys of invasive alien plants have recently been conducted in some GMS countries through cooperation between Yunnan Academy of Agricultural Science, China, the Centre for Agriculture and Bioscience International and the Queensland Department of Agriculture and Fisheries, Australia [16]. However, the biological control programme of invasive alien plants in GMS countries could be strengthened further due to the lack of specialist researchers in weed biological control in the region. This paper reviews the current status of classical biological control efforts against introduced weeds in the six GMS countries and identifies low-risk and effective biological control agents that could be introduced into GMS to help manage some of the worst weeds in the region.

Materials and Methods

The status of weed biological control in the six GMS countries was determined by extracting information from Winston *et al.* [15]. To this dataset, we added information from reports and communications acquired since its publication, resulting in the most current status of weed biological control activities in the region [16, 17].

A list of weeds present in each of the six GMS countries that are biological control targets elsewhere was obtained by first extracting from various databases and lists of alien plant species in each GMS country [18–20]. We supplemented this with local knowledge in each country, adding alien plant species present in a country but not present in any database.

From this compilation, we extracted all plant species that have been the target for weed biological control in at least one country worldwide (using [15]). We then removed all plant species that are considered native to Southeast Asia, leaving a list of plant species exotic to the region and that are targeted for weed biological control in at least one other country.

The plant list was reduced further by removing species for which there were only biological control agents under list 2 in Winston *et al.* [15]. These are biological control agents considered native to a country, yet employed to help manage weed populations within that country. We also removed plant species for which the biological control agents were only under list 3 of Winston *et al.* [15]. These are biological control agents that have been unintentionally introduced or have spread naturally into a country. Biological control agents in both these groups have never been subjected to host specificity testing, so they have not been proven suitable for release into a new country and should not be recommended for additional introductions. Biological control agents found in these two lists but also in list 1 (classical biological control agents intentionally released elsewhere in the world) were not removed.

We next removed plant species for which the only biological control agents available had reports of significant impacts against non-target species. We defined significant impacts as the biological control agents could cause damage and maintain populations on a desirable non-target species. Plant species with biological control agents that were reported to cause only occasional minor damage due to spillover effects were not removed [15].

Finally, the list was further reduced by removing plant species for which the only biological control agents available are ineffective. This resulted in a list of weed species for which there was at least one biological control agent that was considered low risk (i.e. host specificity testing had been conducted and specificity confirmed) and that was causing medium to high impacts on their target weed in at least one country elsewhere.

Tables of biological control agents that could be introduced into the six GMS countries (if target weeds were deemed sufficiently problematic) were prepared based on whether the biological agents were already present in Asia, present in other countries outside Asia and that have been recently released but impacts on their target weeds have not yet been determined.

There was no attempt to prioritize which weed species in each country should be targeted, as weed lists were based on their occurrence in a particular country and not whether they were causing impacts. The prioritization of weeds is, therefore, best left to the relevant authorities in each country. However, we have identified the most effective biological control agents available for each plant species, should biological control be considered for a particular weed.

Results

At least two weed biological control agents have established in each of the six GMS countries (Table 18.1). However, only four of the six countries have deliberately introduced weed biological control agents. Thailand has been the country most active in weed biological control, deliberately introducing 14 biological control agents against six weed species. China has also been very active, deliberately introducing nine biological control agents against five weed species.

Twenty biological control agents have now been deliberately introduced into the GMS region to control nine weed species. Of these, 13 biological control agents have established in at least one country. Nine biological control agents have been either unintentionally introduced or have spread naturally into the region from neighbouring countries (Table 18.2). A tenth species, *Orthezia insignis* Browne (Hemiptera: Ortheziidae), which is often found on *Lantana camara* L. (Verbenaceae), is also present in the region. However, this species attacks several non-target plants and is not considered a bona fide biological control agent; it is not recommended for introduction or re-distribution.

Of the biological control agents that have established in the region, the flea beetle *Agasicles hygrophila* Selman & Vogt (Coleoptera: Chrysomelidae) for the control of *Alternanthera philoxeroides* (Mart.) Griseb. (Amaranthaceae) and *Ophraella communa* LeSage (Coleoptera: Chrysomelidae) for the control of *Ambrosia artemisiifolia* L. (Asteraceae) are probably the most successful agents. Two beetles, *Neochetina bruchi* Hustache (Coleoptera: Erirhinidae) and *N. eichhorniae* Warner, introduced to control *E. crassipes*, cause variable degrees of control, depending on the location and types of infestations [15] (Table 18.2).

Most of the biological control agents established in the GMS region that are causing

Table 18.1. The number of weed species targeted for biocontrol and the number of biocontrol agents that have been deliberately introduced (intentional) and biocontrol agents that were not deliberately introduced but have been reported (unintentional) in the Greater Mekong Subregion countries

	Intentional introductions			Unintentional introductions		Total introductions	
Country	No. of weed species	No. of agents released	No. of agents established	No. of weed species	No. of agents established	No. of weed species	No. of agents established
Cambodia	0	0	0	2	2	2	2
China	5	9	6	7	10	9	16
Laos	0	0	0	2	3	2	3
Myanmar	2	3	3	3	5	4	6[1]
Thailand	6	14	9	3	6	8	15
Vietnam	3	7	4	3	4	5	8

[1]Two *Acanthoscelides* spp. had spread from Thailand and were later deliberately released to supplement field populations.

medium to high impacts against their target weeds are not present in all the countries with the weed. GMS countries where these biological control agents could be introduced are listed in Table 18.2.

Some biological control agents present in the region cause no or only slight damage to their target weeds, while others have not been assessed for their weed control efficacy. Biological control agents that are ineffective or have not been assessed have not been recommended for introduction to other countries (Table 18.2). However, those agents not assessed could be considered for introduction into other countries if their impacts were ever shown to be moderate to high.

There are 45 alien weed species present in the GMS region that are targets for biological control in at least one country worldwide (Table 18.3). Many of these weed species are quite widespread and are found in most GMS countries; nine species occur in all six countries. However, 14 species are reported to occur in only one of the GMS countries.

Ten biological control agents found elsewhere in Asia, and causing medium to high impacts on their target weeds, could be introduced into the GMS region where their respective target weed occurs (Table 18.4). Some of the more widely established and effective agents include *Cyrtobagous salviniae* Calder & Sands (Coleoptera: Erirhinidae) for the control of *Salvinia molesta* D.S. Mitch. (Salviniaceae) and *Cecidochares connexa* Macquart (Diptera: Tephritidae) for the control of *Chromolaena odorata*. Both of these agents are already present in four other countries in Asia.

Over 40 biological control agents causing medium to high impacts on 32 weed species are present outside of Asia (Table 18.5). These agents could be introduced into GMS countries if target weed infestations warrant it. Of these agents, *Neohydronomus affinis* Hustache (Coleoptera: Curculionidae) for the control of *Pistia stratiotes* L. (Araceae) is the most effective and widespread agent, being utilized in 19 countries. *Cactoblastis cactorum* Berg (Lepidoptera: Pyralidae) for the control of *Opuntia* spp. (Cactaceae) is also effective and widespread, and is also utilized in 19 countries.

Numerous biological control agents have only recently been released and/or their impacts have not yet been assessed (Table 18.6). These biological control agents have been thoroughly tested and deemed suitable for release, although releases have been made in only a few countries to date. These agents could be considered for introduction into the GMS, subject to field evaluations in countries where they have been released.

Discussion

Biological control of weeds has a long history in Asia, commencing in 1836 when the cochineal *Dactylopius confusus* Cockerell (Hemiptera: Dactylopiidae) was intentionally distributed within India to control *Opuntia monacantha* (Willd.) Haw. (Cactaceae). Since then, 42 biological control agents have been introduced into 15 countries in Asia to control 19 weed species [15]. Within the GMS, biological control agents have been deliberately released into only four countries. However, biological control agents can move, and all six GMS countries now contain at least two biological control agents. These agents are causing slight to high impacts on their target weeds.

Despite some success in weed biological control in the GMS, only one biological control agent has been deliberately released in the region in the last 20 years. The reasons for this lack of recent activity are many [21–23]. For some countries, there is no clear process or regulatory pathway to facilitate the introduction of biological control agents [23]. Such is the case in Cambodia and Lao PDR where biological agents have never been deliberately introduced. Other countries have processes in place, but they are reluctant to approve the release of a biological control agent. There is a lingering perception that either the biological control agent would not control the weed in all regions and therefore it is not useful [23], or that biological control is risky and that biological control agents could evolve or mutate to attack economically important or other desirable plant species [21].

There remains a lack of understanding of insect–plant interactions and host specificity testing, and what is involved when assessing potential biological control agents. Weed biological control has been practised since 1902, resulting in nearly 500 biological control agents being deliberately released on nearly 200 weed species in over 90 countries [15]. The vast majority of biological control agents were thoroughly tested against many plant species (sometimes against up to 100 plant species) prior to their release, and they have been established in the field for many years, even decades, without any reports of non-target impacts.

The two beetles *N. bruchi* and *N. eichhorniae* to control *E. crassipes* have been deliberately introduced into over 35 countries, while *Teleonemia scrupulosa* Stål (Hemiptera: Tingidae) and *Uroplata girardi* Pic (Coleoptera: Chrysomelidae), both biological control agents for *L. camara*, have each been introduced deliberately into over 25 countries; no non-target impacts have been reported for all four agents [15]. Yet, the introduction of these and other biological control agents is met with resistance due to perceived risks to the environment and agriculture.

A major hurdle is that for some countries, the impact of weeds on agriculture, the environment and human health has not been fully documented. The lack of understanding in the importance of weeds results in a reduced need to implement biological control or other management strategies. Where the importance of weeds has been documented, there is often a lack of capacity for weed biological control programmes, in terms of infrastructure and resources (qualified people and funding) [22]. The introduction of biological control agents invariably involves importing agents into a quarantine facility. For many countries, this is not practical or feasible, and they must rely on testing conducted elsewhere and on funding from donor organizations [24].

As some countries do not have the capacity or processes in place to implement biological control, or have strategic plans to manage weeds, farmers and land managers are often left controlling weeds in an *ad hoc* fashion. In some cases, invasive species are utilized for pig food (*E. crassipes*) [25], firewood (*L. camara*) [13] or medicinal purposes (*Mikania micrantha*) [26]. Invasive species are typically inferior to other species in terms of these uses, and their combined detriments often exceed their minimal benefits. Still, once invasive species are being utilized, they can be perceived as useful and therefore should not be controlled, exacerbating the reluctance to implement biological control [23, 26, 27]. With weeds becoming an ever-increasing problem affecting agriculture, food security and livelihoods, and with conventional control such as the use of herbicides being unsustainable and costly, there are tremendous opportunities to establish or expand weed biological control programmes in GMS countries. The benefits of weed biological control are enormous, not just economically but also in terms of human and environmental health through the reduction in the use of herbicides, which can poison users, crops and waterways [14].

The benefit:cost ratios of weed biological control for various countries range from 11:1 in Zimbabwe to over 4000:1 in South Africa [28]. In Australia, where the average benefit:cost ratio for all weed biological control projects is approximately 23:1, the benefit of the cactus biocontrol programme alone is over 300:1. Much of this is due to *Opuntia stricta* (Haw.) Haw. (Cactaceae) (prickly pear) being successfully controlled by *C. cactorum* for over 85 years [29]. In Sri Lanka, the benefits of biological control of *S. molesta* in Sri Lanka was estimated at 53:1 in terms of production and over 1600:1 in terms of labour costs [30]. Doeleman [30] highlights the high financial benefits from the biological control of *S. molesta* and how biological control opens up new prospects for other weeds where chemical control is not feasible.

In the GMS region, the number of alien invasive weeds that have been deliberately targeted for biological control (nine) is only a fraction of the total number of weeds in the region that have been targeted for weed biological control elsewhere (45). This demonstrates the huge potential for expanding weed biological control programmes in the region to manage some of the region's worst weeds. Even within the region, established biological control agents causing medium to high impacts on their target weed in at least one country could be moved into another GMS country where they are not already present.

Utilizing low-risk and effective agents, particularly from within the same region, is highly successful [14, 24]. First, the agents are climatically adapted and likely to establish. Second, these agents have been thoroughly tested and have often been present in the field for many

Table 18.2. Status of weed biocontrol agents deliberately released (intentional) and/or spread of their own accord (unintentional) into the Greater Mekong Subregion countries and the potential GMS countries in which they could be introduced

Weed family	Weed species	Biocontrol agent family	Biocontrol agent species	Guild	KHM	CHN	LAO	MMR	THA	VNM
Amaranthaceae	*Alternanthera philoxeroides* (Mart.) Griseb.	Chrysomelidae	*Agasicles hygrophila* Selman & Vogt	Leaf-feeding		I E H	*	*	I E H	*
Asteraceae	*Ageratina adenophora* (Spreng.) R. M. King & H. Rob.	Mycosphaerellaceae	*Passalora ageratinae* Crous & A.R. Wood	Pathogen		U E ?	~		~	~
		Tephritidae	*Procecidochares utilis* Stone	Stem-galling		I E S	*		I E S	*
	Ambrosia artemisiifolia L.	Chrysomelidae	*Ophraella communa* LeSage	Leaf-feeding		U E H				
			Zygogramma suturalis Fabricius	Leaf-feeding		I F				
		Tortricidae	*Epiblema strenuana* Walker	Stem-boring		I E M				
	Chromolaena odorata (L.) R. M. King & H. Rob.	Erebidae	*Pareuchaetes pseudoinsulata* Rego Barros	Leaf-feeding	~	~	~	~	I F	I F
		Eriophyidae	*Acalitus adoratus* Keifer	Leaf-galling	~	U E S	U E S	U E S	U E S	U E S
		Tephritidae	*Cecidochares connexa* Macquart	Stem-galling					I F	
	Mikania micrantha Kunth	Pucciniaceae	*Puccinia spegazzinii* De Toni	Pathogen		I F				
	Parthenium hysterophorus L.	Pucciniaceae	*Puccinia abrupta* Dietel & Holw. var. partheniicola (H.S. Jacks.) Parmelee	Pathogen		U E ?		*	*	*
		Tortricidae	*Epiblema strenuana* Walker1	Stem-boring		*		*	*	*

Fabaceae	*Leucaena leucocephala* (Lam.) de Wit	Chrysomelidae	*Acanthoscelides macrophthalmus* Schaeffer	Seed-feeding	*	U E S	*	*	U E ?	U E ?
	Mimosa pigra L.	Brentidae	*Coelocephalapion aculeatum* Fall	Flower-feeding					I F	
		Chrysomelidae	*Acanthoscelides puniceus* Johnson	Seed-feeding	~	~	U E ?	I E ?	I E S	I E N
			Acanthoscelides quadridentatus Schaeffer	Seed-feeding	~	~	U E ?	I E ?	I E S	I E N
			Chlamisus mimosae Karren	Leaf-feeding	~	~	~	~	I ?	I ?
		Sesiidae	*Carmenta mimosa* Eichlin & Passoa	Stem-boring	*	*	*	*	I F	I E M
Lamiaceae	*Clerodendrum chinense* (Osbeck) Mabb.	Chrysomelidae	*Phyllocharis undulata* (L.)	Leaf-feeding					I E ?	
Pontederiaceae	*Eichhornia crassipes* (Mart.) Solms	Crambidae	*Niphograpta albiguttalis* Warren	Petiole-boring	*	*	*	*	I E ?	*
			Xubida infusella Walker	Crown-boring	~	~	~	~	I E ?	~
		Erirhinidae	*Neochetina bruchi* Hustache	Crown-boring	*	I E V	*	*	I E H	I E ?
			Neochetina eichhorniae Warner	Crown-boring	U E ?	I E V	*	I E ?	I E H	I ?
		Miridae	Eccritotarsus catarinensis Carvalho	Leaf-feeding		I F				
Verbenaceae	*Lantana camara* L. (sens. lat.)	Agromyzidae	*Calycomyza lantanae* Frick	Leaf-mining	U E ?	U E S	~	~	U E S	U E S

Continued

Table 18.2. Continued

Weed family	Weed species	Biocontrol agent family	Biocontrol agent species	Guild	KHM	CHN	LAO	MMR	THA	VNM
			Ophiomyia lantanae Froggatt	Seed-mining	~	I E S	~	U E S	U E S	U E S
		Erebidae	*Hypena laceratalis* Walker	Leaf-feeding	~	U E S	~	~	~	~
		Pterophoridae	*Lantanophaga pusillidactyla* Walker	Flower-feeding	~	I E S	~	U E S	U E S	~
		Tingidae	*Teleonemia scrupulosa* Stål	Sap-sucking	*	*	*	*	U E S	*
		Tortricidae	*Crocidosema lantana* Busck	Flower-feeding	~	U E S	~	~	~	~

Countries: KHM, Cambodia; CHN, China; LAO, Laos; MMR, Myanmar; THA, Thailand; VNM, Vietnam. Status: I, intentionally introduced, U, unintentionally introduced; E, established; F, failed to establish; ?, unknown. Impact: H, high; M, moderate; N, none, S, slight; V, variable; ?, unknown.
*Potential countries where low-risk and effective biocontrol agents could be introduced (based on weed occurrence in each country, not weed density); ~weed is present in the country but the biocontrol agent is not recommended for introduction due to only causing minimal impacts.
[1]*Epiblema strenuana* has established on *Ambrosia artemisiifolia* but can be released on *Parthenium hysterophorus*.

Table 18.3. List of alien invasive weeds present in the Greater Mekong Subregion countries that have been deliberately targeted for biocontrol in at least one country worldwide and for which there are low-risk and effective biocontrol agents causing moderate to high impacts to their target weed elsewhere

Family	Species	Common name	Weed origin	KHM	CHN	LAO	MMR	THA	VNM
Amaranthaceae	*Alternanthera philoxeroides*	Alligator weed	South America		*	*	*	*	*
Apocynaceae	*Cryptostegia grandiflora* R. Br.	Rubber vine	Madagascar					*	
Araceae	*Pistia stratiotes* L.	Water lettuce	tropical America	*	*	*	*	*	*
Asteraceae	*Ageratina adenophora*	Crofton weed	Mexico		*	*		*	*
	Ambrosia artemisiifolia	Common ragweed	North America		*				
	Chromolaena odorata	Chromolaena	Tropical and subtropical America	*	*	*	*	*	*
	Elephantopus mollis Kunth	Tobacco weed	Central America						*
	Mikania micrantha	Mile-a-minute	Central and South America	*	*	*	*	*	*
	Parthenium hysterophorus	Parthenium weed	North, Central, South America		*		*	*	*
	Xanthium strumarium L.	Noogoora burr	North, Central, South America		*			*	*
Azollaceae	*Azolla filiculoides* Lam.	Water fern	North, Central, South America	*	*		*		
Basellaceae	*Anredera cordifolia* (Ten.) Steenis	Madeira vine	South America		*				
Bignoniaceae	*Dolichandra unguis-cati* (L.) L. G. Lohmann	Cat's claw creeper	Tropical America		*				
	Spathodea campanulata Beauv.	African tulip tree	Africa		*	*		*	
	Tecoma stans (L.) Juss. ex Kunth var. stans	Yellow bells	Tropical America					*	
Cactaceae	*Cereus jamacaru* DC.	Queen of the night	South America						*

Continued

Table 18.3. Continued

Family	Species	Common name	Weed origin	KHM	CHN	LAO	MMR	THA	VNM
	Opuntia elatior Mill.	Prickly pear	Caribbean, Central America					*	
	Opuntia ficus-indica (L.) Mill.	Indian fig	Mexico	*	*				*
	Opuntia monacantha (Willd.) Haw.	Drooping prickly pear	Argentina, Brazil, Paraguay, Uruguay		*			*	*
	Opuntia stricta (Haw.) Haw.	Common prickly pear	North and South America, Caribbean		*				*
Cucurbitaceae	*Coccinia grandis* (L.) Voigt	Ivy gourd	East Africa	*	*			*	*
Fabaceae	*Acacia cyclops* A. Cunn. ex G. Don	Red eye	Australia					*	
	Acacia dealbata Link	Silver wattle	Australia		*				
	Acacia decurrens (Wendl.) Willd.	Green wattle	Australia						*
	Acacia longifolia (Andrews) Willd.	Sydney golden wattle	Australia				*		*
	Acacia mearnsii De Wild.	Black wattle	Australia		*				*
	Acacia melanoxylon R. Br.	Australian blackwood	Australia					*	
	Acacia saligna (Labill.) H. L. Wendl.	Port Jackson willow	Australia						*
	Leucaena leucocephala	Leucaena	Mexico, Central America	*	*	*	*	*	*
	Mimosa diplotricha C. Wright	Sensitive plant	Tropical America	*	*	*	*	*	*
	Mimosa pigra	Giant sensitive plant	Tropical America	*	*	*	*	*	*

Continued

Table 18.3. Continued

Family	Species	Common name	Weed origin	KHM	CHN	LAO	MMR	THA	VNM
	Parkinsonia aculeata L.	Parkinsonia	Tropical and subtropical America	*		*		*	*
	Prosopis juliflora (Sw.) DC.	Mexican thorn	Colombia, Ecuador, Mexico, Peru, Venezuela	*			*	*	*
	Vachellia nilotica (L.) P.J.H.Hurter & Mabb	Prickly acacia	Indian subcontinent				*		*
Haloragaceae	*Myriophyllum aquaticum* (Vell.) Verdc.	Parrot's feather	South America	*	*			*	*
Malvaceae	*Sida acuta* Burm. f.	Spinyhead sida	Tropical America	*	*	*	*	*	*
	Sida rhombifolia L.	Paddy's lucerne, common sida	Tropical America	*	*			*	*
Melastomataceae	*Clidemia hirta* (L.) D. Don	Koster's curse	Tropical Central and South America					*	*
Myrtaceae	*Melaleuca quinquenervia* (Cav.) S. T. Blake	Broad-leafed paperbark	Australia		*			*	*
Poaceae	*Arundo donax* L.	Giant reed	Mediterranean Europe, Asia		*		*	*	
Pontederiaceae	*Eichhornia crassipes*	Water hyacinth	South America	*	*	*	*	*	*
Salviniaceae	*Salvinia molesta* D.S. Mitch.	Salvinia	Brazil		*	*		*	
Solanaceae	*Solanum sisymbriifolium* Lam.	Wild tobacco	South America		*				
	Solanum viarum Dunal	Tropical soda apple	South America		*		*	*	*
Verbenaceae	*Lantana camara*	Lantana	Central and South America	*	*	*	*	*	*

Countries: KHM, Cambodia; CHN, China; LAO, Laos; MMR, Myanmar; THA, Thailand; VNM, Vietnam.
*Indicates in which country the weed is present.

Table 18.4. Weed biocontrol agents established elsewhere in Asia that are causing moderate to high impacts against the target weed and that could be released into the Greater Mekong Subregion countries where the target weed occurs

Weed family	Weed species	Biocontrol agent family	Biocontrol agent species	Guild	No. of countries agent present	Asian countries agent present	Average impact	KHM	CHN	LAO	MMR	THA	VNM
Asteraceae	*Ambrosia artemisiifolia*	Chrysomelidae	*Zygogramma bicolorata* Pallister	Leaf-feeding	6	India, Nepal, Pakistan	Medium		*				
	Chromolaena odorata	Tephritidae	*Cecidochares connexa*	Stem-galling	14	India, Indonesia, Philippines, Timor Leste	Medium to high	*	*	*	*	*	*
	Mikania micrantha	Pucciniaceae	*Puccinia spegazzinii*	pathogen	6	Taiwan	High	*	*	*	*	*	*
	Parthenium hysterophorus	Chrysomelidae	*Zygogramma bicolorata*	Leaf-feeding	6	India, Nepal, Pakistan	Variable to medium		*		*	*	*
	Xanthium strumarium	Pucciniaceae	*Puccinia xanthii* Schwein.	Pathogen	4	Sri Lanka, Timor Leste	Variable to medium		*			*	*
Cactaceae	*Opuntia elatior*	Dactylopiidae	Dactylopius opuntiae Cockerell	Sap-sucking	2	India, Indonesia	High					*	
	Opuntia ficus-indica	Dactylopiidae	*Dactylopius opuntiae*	Sap-sucking	5	Israel	High	*	*				*
	Opuntia monacantha	Dactylopiidae	Dactylopius ceylonicus Green	Sap-sucking	5	India, Sri Lanka	High		*			*	*
	Opuntia stricta	Dactylopiidae	*Dactylopius opuntiae*	Sap-sucking	6	India, Sri Lanka	High		*				*
Fabaceae	*Mimosa diplotricha*	Psyllidae	Heteropsylla spinulosa Muddiman, Hodkinson & Hollis	Sap-sucking	14	Timor Leste	High	*	*	*	*	*	*

Salviniaceae	*Salvinia molesta*	Erirhinidae	Cyrtobagous salviniae Calder & Sands	Leaf-feeding	21	India, Indonesia, Malaysia, Sri Lanka	High		*	*		*	
Verbenaceae	*Lantana camara*	Chrysomelidae	*Octotoma scabripennis* Guérin-Méneville	Leaf-mining	7	India	Medium	*	*	*	*	*	*
			Uroplata girardi Pic	Leaf-mining	24	India, Philippines	Medium	*	*	*	*	*	*

Countries: KHM, Cambodia; CHN, China; LAO, Laos; MMR, Myanmar; THA, Thailand; VNM, Vietnam.
*Indicates in which country the weed is present.

Table 18.5. Biological control agents that are not present in Asia but are causing moderate to high impacts elsewhere and could be introduced into the Greater Mekong Subregion countries where the target weed occurs

Weed family	Weed species	Biocontrol agent family	Biocontrol agent species	Guild	Total countries agent is present	Average impacts elsewhere	KHM	CHN	LAO	MMR	THA	VNM
Amaranthaceae	*Alternanthera philoxeroides*	Pyralidae	*Arcola malloi* Pastrana	Leaf-feeding	3	Variable to high		*	*	*	*	*
Apocynaceae	*Cryptostegia grandiflora*	Chaconiaceae	*Maravalia cryptostegiae* (Cummins) Ono	Pathogen	2	Variable					*	
Araceae	*Pistia stratiotes*	Curculionidae	*Neohydronomus affinis* Hustache	Leaf-feeding	19	High	*	*	*	*	*	*
Asteraceae	*Elephantopus mollis*	Tephritidae	*Tetraeuaresta obscuriventris* Loew	Stem-galling	5	Slight to medium						*
	Parthenium hysterophorus	Curculionidae	*Listronotus setosipennis* Hustache	Flower-feeding	2	Variable to high		*		*	*	*
		Curculionidae	*Smicronyx lutulentus* Dietz	Seed-feeding	2	Variable		*		*	*	*
Azollaceae	*Azolla filiculoides*	Erirhinidae	*Stenopelmus rufinasus* Gyllenhal	Leaf-feeding	17	High	*	*		*		
Bignoniaceae	*Dolichandra unguis-cati*	Tingidae	*Carvalhotingis visenda* Drake & Hambleton	Sap-sucking	2	Medium		*				
Cactaceae	*Cereus jamacaru*	Pseudococcidae	*Hypogeococcus festerianus* Lizer y Trelles	Sap-sucking	1	High						*
		Cerambycidae	*Nealcidion cereicola* Fisher	Cladode-boring	1	High						*
	Opuntia ficus-indica	Pyralidae	*Cactoblastis cactorum* Berg	Cladode-boring	19	Medium to high	*	*				*

	Opuntia monacantha	Pyralidae	*Cactoblastis cactorum*	Cladode-boring	19	Medium to high		*			*	*
	Opuntia stricta	Pyralidae	*Cactoblastis cactorum*	Cladode-boring	19	Medium to high		*				*
Cucurbitaceae	*Coccinia grandis*	Curculionidae	*Acythopeus cocciniae* O'Brien & Pakaluk	Leaf-mining	3	Medium to high	*	*			*	*
		Sesiidae	*Melittia oedipus* Oberthür	Stem-boring	3	Medium to high	*	*			*	*
Fabaceae	*Acacia cyclops*	Cecidomyiidae	*Dasineura dielsi* Rübsaamen	Flower-galling	1	Variable to high					*	
		Curculionidae	*Melanterius servulus* Pascoe	Seed-feeding	1	Variable to high					*	
	Acacia dealbata	Curculionidae	*Melanterius maculatus* Lea	Seed-feeding	1	Medium		*				
	Acacia decurrens	Curculionidae	*Melanterius maculatus*	Seed-feeding	1	Medium						*
	Acacia longifolia	Curculionidae	*Melanterius ventralis* Lea	Seed-feeding	1	High				*		*
		Hymenoptera	*Trichilogaster acaciaelongifoliae* Froggatt	Flower-galling	2	High				*		*
	Acacia mearnsii	Cecidomyiidae	*Dasineura rubiformis* Kolesik	Flower-galling	2	Variable to high		*				*
	Acacia melanoxylon	Curculionidae	*Melanterius acaciae* Lea	Seed-feeding	1	High					*	

Continued

Table 18.5. Continued

Weed family	Weed species	Biocontrol agent family	Biocontrol agent species	Guild	Total countries agent is present	Average impacts elsewhere	KHM	CHN	LAO	MMR	THA	VNM
	Acacia saligna	Curculionidae	*Melanterius compactus* Lea	Seed-feeding	1	High						*
		Pileolariaceae	*Uromycladium tepperianum* (Sacc.) McAlpine	Pathogen	1	High						*
	Mimosa pigra	Curculionidae	*Chalcodermus serripes* Fåhraeus	Seed-feeding	1	High	*	*	*	*	*	*
		Geometridae	*Macaria pallidata* Warren	Leaf-feeding	1	Variable to high	*	*	*	*	*	*
		Chrysomelidae	*Nesaecrepida infuscata* Schaeffer	Leaf-feeding	1	High	*	*	*	*	*	*
		Gracillariidae	*Neurostrota gunniella* Busck	Leaf-feeding	1	High	*	*	*	*	*	*
	Prosopis juliflora	Gelechiidae	*Evippe* sp. #1	Leaf-feeding	1	Variable	*			*	*	*
	Vachellia (syn. Acacia) nilotica	Geometridae	*Chiasmia assimilis* Warren	Leaf-feeding	1	Variable to high				*		*
Haloragaceae	*Myriophyllum aquaticum*	Chrysomelidae	*Lysathia* sp.	Leaf-feeding	2	High	*	*			*	*
Malvaceae	*Sida acuta*	Chrysomelidae	*Calligrapha pantherina* Stål	Leaf-feeding	5	High	*	*	*	*	*	*
	Sida rhombifolia		*Calligrapha pantherina*	Leaf-feeding	5	High	*	*			*	*
Melastomataceae	*Clidemia hirta*	Phlaeothripidae	*Liothrips urichi* Karny	Sap-sucking	4	Variable to high					*	*
Myrtaceae	*Melaleuca quinquenervia*	Psyllidae	*Boreioglycaspis melaleucae* Moore	Sap-sucking	2	Medium to high		*			*	*

		Cecidomyiidae	*Lophodiplosis trifida* Gagné	Leaf-galling	1	High		*			*	*
		Curculionidae	*Oxyops vitiosa* Pascoe	Leaf-feeding	2	Slight to high		*			*	*
		Pucciniaceae	*Puccinia psidii* G. Winter	Pathogen	1	High		*			*	*
Poaceae	*Arundo donax*	Eurytomidae	*Tetramesa romana* Walker	Stem-boring	2	High		*		*	*	
Pontederiaceae	*Eichhornia crassipes*	Mycosphaerellaceae	*Cercospora piaropi* Tharp	Pathogen	3	Slight to high	*	*	*	*	*	*
		Miridae	*Eccritotarsus catarinensis*	Sap-sucking	3	Variable to high	*	*	*	*	*	*
Solanaceae	*Solanum sisymbriifolium* Lam.	Chrysomelidae	*Gratiana spadicea* Klug	Leaf-feeding	1	Medium		*			*	*
	Solanum viarum Dunal		*Gratiana boliviana* Spaeth	Leaf-feeding	1	High		*		*	*	*
Verbenaceae	*Lantana camara*	Eriophyidae	*Aceria lantanae* Cook	Flower-galling	5	Variable to high	*	*	*	*	*	*
		Miridae	*Falconia intermedia* Distant	Sap-sucking	2	Medium to high	*	*	*	*	*	*
		Agromyzidae	*Ophiomyia camarae* Spencer	Leaf-mining	11	Variable to high	*	*	*	*	*	*

Countries: KHM, Cambodia; CHN, China; LAO, Laos; MMR, Myanmar; THA, Thailand; VNM, Vietnam.
*Indicates in which country the weed is present

Table 18.6. Biological control agents that are not present in Asia but have been recently released elsewhere, and are showing early promise in controlling their target weed, and could be introduced into the Greater Mekong Subregion countries where the target weed occurs

Weed family	Weed scientific name	Agent family	Agent scientific name	Guild	Total countries agent released	KHM	CHN	LAO	MMR	THA	VNM
Asteraraceae	*Ageratina adenophora*	Pucciniosiraceae	*Baeodromus eupatorii* (Arthur) Arthur	Pathogen	1		*	*		*	*
Basellaceae	*Anredera cordifolia*	Chrysomelidae	*Plectonycha correntina* Lacordaire	Leaf-feeding	1		*				
Bignoniaceae	*Dolichandra unguis-cati*	Buprestidae	*Hedwigiella jureceki* Obenberger	Leaf-mining	2		*				
	Spathodea campanulata	Eriophyidae	*Colomerus spathodeae* Carmona	Leaf-galling	1		*	*		*	
	Tecoma stans var. stans	Coccinellidae	*Mada polluta* Mulsant	Leaf-feeding	1					*	
		Agromyzidae	*Pseudonapomyza* sp.	Leaf-mining	1					*	
Fabaceae	*Parkinsonia aculeata*	Geometridae	*Eueupithecia cisplatensis* Prout	Leaf-feeding	1	*		*		*	*
			Eueupithecia vollonoides Hausmann	Leaf-feeding	1	*		*		*	*
Poaceae	*Arundo donax*	Diaspididae	*Rhizaspidiotus donacis* Leonardi	Sap-sucking	2		*		*	*	
Pontederiaceae	*Eichhornia crassipes*	Delphacidae	*Megamelus scutellaris* Berg	Sap-sucking	2	*	*	*	*	*	*
Verbenaceae	*Lantana camara*	Uropyxidaceae	*Prospodium tuberculatum* (Spegazzini) Arthur	Pathogen	2	*	*	*	*	*	*

Countries: KHM, Cambodia; CHN, China; LAO, Laos; MMR, Myanmar; THA, Thailand; VNM, Vietnam.
*Indicates in which country the weed is present.

years. Therefore, any non-target impacts would have been detected by now. This provides greater confidence that the agents are host specific and may circumvent the need for additional host specificity testing. For instance, *A. hygrophila* could be introduced into Lao PDR, Myanmar and Vietnam from China and/or Thailand where the agent has established and is causing high impacts on its target weed, *A. philoxeroides*. Similarly, the two beetles *N. bruchi* and *N. eichhorniae*, introduced to control *E. crassipes*, could be introduced into several GMS countries where they are not already present [15].

For biological control agents found outside the region, i.e. Australia, the Pacific or South Africa, it might be necessary to test a few additional plant species prior to importation, depending on what species have been tested previously. Agents tested prior to their release in one country may not have been tested against locally important related plant species found in other countries where the agents are being considered for release. For instance, *N. bruchi*, a biological control agent for *E. crassipes*, was tested on over 260 plant species in total in ten countries [25], prior to being deliberately released in over 35 countries [15]. The rust pathogen *Puccinia spegazzinii* de Toni (Pucciniaceae) was tested against a combined 287 species in five countries, prior to being imported into numerous countries for the control of *Mikania micrantha* Kunth (Asteraceae) [26]. Both agents have been present in the field for many years and have not been observed causing non-target impacts.

Prior to releasing any biological control agent, it may be prudent to conduct field surveys to determine if the particular species is already present in the country. Biological control agents can and do disperse into new areas, as many weed infestations are contiguous across province and country borders. *Calycomyza lantanae* Frick (Diptera: Agromyzidae), a biological control agent for *L. camara*, was released into only three countries worldwide (Australia, Fiji and South Africa), but it is now found in another 26 countries, including 12 countries in Asia [13, 15]. Twenty-five biological control agents targeting 14 weed species have spread naturally into at least one of 25 countries in Asia [15].

Because the six GMS countries have not released any biological control agents recently, or in some cases ever, it would be advantageous to investigate ways to highlight the importance of invasive plants and promote weed biological control in the region. Workshops could be held for scientists and regulators on the occurrence, distribution and socio-economic impacts of invasive weeds, which would raise the awareness and importance of invasive species in the region. Workshops and training programmes could be conducted to cover all aspects of weed biological control, including potential benefits and risks, as well as the processes involved in implementing biological control programmes. Field visits should be considered, which would give participants the chance to view weed infestations, assess their impacts on agriculture and the environment, and the opportunity to view biological control agents in action, whether they were deliberately introduced or spread naturally into the country. Host specificity testing experiments could also be conducted to demonstrate how biological control agents are evaluated prior to their release in the field. Numerous projects involving invasive species and biological control have incorporated citizen science components, videos and other information packages. Many of these are freely available on various websites (e.g. http://www.cabi.org).

Ultimately, it is up to individual countries to take the initiative to manage invasive species. This paper has outlined opportunities to decrease the impacts of invasive weeds in GMS countries through weed biological control. By incorporating a complete package involving all aspects of weed biological control, weed management in the GMS countries can be improved, and the impacts of weeds on productivity and the environment, as well as the use of herbicides can be decreased.

Acknowledgements

This research was supported by Yunnan Provincial Key Fund Program (2010CC002), Middle-aged and Young Academic Leader Training Foundation of Yunnan Province (2014HB039 and 2018HB054), Applied Basic Research Foundation of Yunnan Province (2017FB049) and International Science & Technology Cooperation Program of Yunnan Provincial Science and Technology Department (2014IA009).

References

1. Asian Development Bank (ADB). (2004) The Greater Mekong Subregion: Beyond Borders: Regional Cooperation Strategy and Program (2004–2008). Available from: URL: https://www.adb.org/sites/default/files/institutional-document/32613/files/gms-rcsp.pdf. Accessed March 2004.
2. Zhang JL, Zhao M. Survey on agricultural corporation in Greater Mekong Subregion. World Agriculture 2011;6:15–21.
3. Zhang Y, He DM, Lu Y, Feng Y, Reznick J. The influence of large dams building on resettlement in the Upper Mekong river. Journal of Geographical Sciences 2013;23:947–857.
4. Li XD, Wang SW. Agricultural trade and corporation in Greater Mekong Subregion. World Agriculture 2009;11:56–9.
5. Wang Y, Liu DM, Chen SB. Study on the agricultural products trade competitiveness of main GMS countries. China Soft Science 2010;46:88–96.
6. Shen SC, Zhang FD, Xu GF, Li TL, Wu D. Occurrence and infestation of invasive weed in crop field in Yunnan. Southwest China Journal of Agricultural Sciences 2012;25:554–61.
7. Yu YH, Gao XY, Zeng XR, Long XZ, Wei DW, Qin JL, *et al.* Current status of invasive alien pests in Guangxi Province and Vietnam. Journal of Biosafety 2016;25:171–80.
8. Shen SC, Xu GF, Li DY, Clements DR, Jin GM, Yin XX, *et al.* Occurrence and damage of invasive alien plants in Dehong Prefecture, western of Yunnan Province. Acta Ecologica Sinica 2017;37:195–200.
9. Xu H, Ding H, Li M, Qiang S, Guo J, Han Z, *et al.* The distribution and economic losses of alien species invasion to China. Biological Invasions 2006;8:1495–1500.
10. Xie Q. Review and prospect of agricultural cooperation between Yunnan and GMS. Yunnan Agriculture 2013;4:10–13.
11. Ding J, Wang R, Fu W, Zhang G. Water hyacinth in China: Its distribution, problems and control status. In: Julien MH, Hill MP, Center TD, Ding J, editors. Proceedings of the Second Meeting of the Global Working Group for the Biological and Integrated Control of Water Hyacinth. ACIAR, Canberra; 2001. p. 29–32.
12. Kasulo V. The economics of invasive species: costs, incentives, risk analysis and the role of donors. In: Lyons EE, Miller SE, editors. Invasive Species in Eastern Africa: Proceedings of a workshop Held at ICIPE, July 5–6, 1999. ICIPE Science Press, Nairobi. 2000. p. 42–7.
13. Day MD, Wiley CJ, Playford J, Zalucki MP. Lantana: current management status and future prospects. ACIAR Monograph No. 102. ACIAR, Canberra, Australia. 2003. p. 128.
14. Julien MH, Scott JK, Orapa W, Paynter Q. History, opportunities and challenges for biological control in Australia, New Zealand and the Pacific islands. Crop Protection 2007;26:255–65.
15. Winston RL, Schwarzlander M, Hinz HL, Day MD, Cock MJW, Julien MH. Biological Control of Weeds: A World Catalogue of Agents and Their Target Weeds. USAD Forest Service, Forest Health Technology Enterprise Team, Morgantown, West Virginia, FHTET-2014-04, 2014. 1–838.
16. Shen SC, Day MD, Xu GF, Li DY, Jin GM, Yin XX, *et al.* The current status of biological control of weeds in southern China and future options. Acta Ecologica Sinica 2018;38:157–64.
17. Winston RL, Schwarzländer M, Hinz HL, Day MD, Julien MH, editors. (2018) Biological Control of Weeds: AWorld Catalogue of Agents and Their Target Weeds, Online Edition. USDA Forest Service, Forest Health Technology Enterprise Team, Morgantown, West Virginia, USA. Hosted by The University of Georgia – Center for Invasive Species and Ecosystem Health. Available from: URL: https://www.ibiocontrol.org/catalog/. Accessed 16 May 2018.
18. European and Mediterranean Plant Protection Organization (EPPO). (2017) EPPO Global Database. Available from: URL: https://gd.eppo.int/taxon/. Accessed September 2017.

19. GRIIS. (2017) Global Register of Introduced and Invasive Species. Available from: URL: http://www.griis.org/. Accessed September 2017.
20. Invasive Species Specialist Group (ISSG). (2017) Global Invasive Species Database. Available from: URL: http://www.iucngisd.org/gisd/. Accessed September 2017.
21. McFadyen REC. Biological control of weeds. Annual Review of Entomology 1998;43:369–93.
22. Fowler SV. Trivial and political reasons for the failure of classical biological control of weeds: a personal view. In: Spencer NR, editor. Proceedings of the X International Symposium for the Biological Control of Weeds. USDA Agricultural Research Service, Bozeman. 2000. p. 169–72.
23. Ghosheh HZ. Constraints in implementing biological weed control: a review. Weed Biology and Management 2005;5:83–92.
24. Dovey L, Orapa W, Randall S. The need to build biological control capacity in the Pacific. In: Cullen JM, Briese DT, Kriticos DJ, Lonsdale WM, Morin L, Scott JK, editors. Proceedings of the XI International Symposium on Biological Control of Weeds. CSIRO Entomology, Canberra; 2004. p. 36–41.
25. Julien MH, Griffiths MW, Wright AD. Biological control of water hyacinth: the weevil *Neochetina bruchi* and *N. eichhorniae*: biologies, host ranges, and rearing, releasing and monitoring techniques for biological control of *Eichhornia crassipes*. ACIAR Monograph No. 60. 1999. p. 87.
26. Day MD, Clements DR, Gile C, Senaratne KADW, Shen S, Weston LA, *et al.* Biology and impacts of Pacific islands invasive species: *Mikania micrantha* Kunth (Asteraceae). Pacific Science 2016;70:257–85.
27. Bhagwat SA, Breman E, Thekaekara T, Thornton TF, Willis KJ. A battle lost? Report on two centuries of invasion and management of *Lantana camara* L. in Australia, India and South Africa. PLoS ONE 2012;7(3):e32407.
28. van Wilgen BW, de Lange WJ. The costs and benefits of biological control of invasive alien plants in South Africa. African Entomology 2011;19:504–14.
29. Page AR, Lacey KL. Economic impact assessment of Australian weed biological control. CRC for Australian Weed Management Technical Series 2006. No. 10. 151 pp.
30. Doeleman JA. Biological control of *Salvinia molesta* in Sri Lanka: an assessment of costs and benefits. ACIAR Technical Reports. 1989. No. 12: l4 pp.

19 The natural enemies of privets (*Ligustrum*: Oleaceae): a literature review, with particular reference to biological control

Richard H. Shaw*, Matthew J.W. Cock and Harry C. Evans
CABI, Bakeham Lane, Egham, Surrey TW20 9TY, UK

Abstract

The arthropod and fungal natural enemies of privets (*Ligustrum* spp., Oleaceae) are listed, based on the data from taxonomic and nomenclatural websites, printed and electronic literature searches, including CAB Abstracts. Initially, the lists were compiled to identify and assess those natural enemies occurring on *Ligustrum robustum* in its indigenous Asian range as part of a classical biological control programme for the island of La Réunion, where this privet species has become a problematic invasive alien weed. However, because other species of *Ligustrum* are also posing invasive problems in other parts of the world, wherever they have been introduced, the lists have been expanded to cover all species in the genus. These records of natural enemies of *Ligustrum*, together with the distribution data, should enable those involved with the issues of invasive privets to assess the potential of classical biological control as an option for their management. As a further aid, an overview of those natural enemies considered to have the highest potential as classical biological control agents for *L. robustum* is included.

Keywords: Classical biological control, Fungi, Insects, Invasive alien weeds, Invertebrates, La Réunion, Sri Lanka, Biocontrol, *Ligustrum robustum* subsp. *walkeri*

Review Methodology

Natural enemy records were compiled separately for invertebrates (mainly insects) and fungi. For invertebrates, records were located from CAB Abstracts (records featuring '*Ligustrum*' and 'pest'), and various references documenting food plants of specific insect groups and/or authoritative accounts of particular groups (e.g. [1–9]), publications arising from the assessment of the scope for biological control of other species of *Ligustrum* [10, 11], and Internet searches. The survey of insect natural enemies of *Ligustrum sinense* Lour. in China by Zhang *et al.* [11] is included in our summary, but only those species which were confirmed to feed on *L. sinense*, as opposed to just being found on this plant. Data on the wider host range and distribution were extracted from similar sources and the Crop Protection Compendium [12].

For fungi, Petrak's and Saccardo and Trotter's lists were consulted [13, 14], and updated with a search of the database of the US National Fungus Collections [15], and those of the official nomenclatural repositories for the kingdom Fungi: *Index Fungorum* (www.indexfungorum.org) and MycoBank (www.mycobank.org/), as well as herbaria in the UK and Sri Lanka.

*Corresponding author: r.shaw@cabi.org

The records are compiled in separate tables for arthropods and fungi. The references provided in these tables are not comprehensive, most were selected to represent the more accessible or relevant publications. Nomenclature and classification have been updated: plants based on The Plant List [16], arthropods from a variety of hard copy and Internet sources including those mentioned above and Rider [17], Beccaloni *et al.* [18], Gilligan *et al.* [19], Nuss *et al.* [20], and Poole [21] and fungi according to the *Dictionary of the Fungi* [22], *Index Fungorum* and MycoBank.

Introduction

The present review was generated originally as a background search to assess the natural enemies occurring on *Ligustrum robustum* subsp. *walkeri* (Decne.) P.S. Green (Oleaceae) in its Asian centre of origin, as part of a classical biological control programme for the island of La Réunion, where this privet has become a problematic invasive alien species [23, 24]. Subsequently, the search was expanded to include all species of the genus *Ligustrum*, with particular emphasis on those that are invasive aliens in other parts of the world. We feel that this unpublished research and the data generated need to be placed in the public domain as a guide for ecologists, biocontrol practitioners and stakeholders who may be involved in future decision-making on the options available for the management of invasive alien privets.

Species of the genus *Ligustrum* – with a predominantly Eurasian origin [25] – have been moved around the world as ornamental plants, especially for hedging purposes, since the mid 1800s [26]. However, several of these privet species have since become invasive in their exotic locations and, in particular, are now posing a threat to native ecosystems from the Mascarenes to Australia and New Zealand in the Old World, as well as to those in the New World, notably in Argentina, Brazil and the USA [27–36]. In addition, several of these invasive *Ligustrum* species are also classified as poisonous plants, and have been implicated in allergies such as hay fever and asthma [37, 38].

Classical biological control offers an economic, environmentally benign and long-term solution for the management of invasive alien weeds. Successful programmes have been conducted against many problematic exotic weeds, including woody species, using both arthropod and fungal natural enemies [39–41], but, thus far, never against species of the genus *Ligustrum*.

The first step in any programme involves literature and database searches for natural enemies: typically, for those arthropods and fungi that have been recorded from the centre of origin or diversity of the target plant species, based on the premise that these may be coevolved natural enemies and, therefore, with a high degree of specificity to their plant host. This is followed by surveys in the plant's natural range to collect and identify natural enemies and then to assess their potential as classical biological control agents, factoring in field observations of host impact and specificity, as well as on preliminary evaluations of their compatibility with screening procedures, such as ease of handling and breeding or culturing issues. On this basis, many of the natural enemies are eliminated and a short list is drawn up of potential agents to undergo the protracted process of screening to determine their specificity and efficacy. The natural enemies of *L. robustum* (Roxb.) Blume were assessed using these parameters and further studies were concentrated on several potential insect agents [24, 42], and the leaf-feeding uraniid *Epiplema albida* Hampson was considered safe for introduction to La Réunion [43, 44].

Ligustrum robustum and its Phylogeny

The genus *Ligustrum* is placed in the family Oleaceae of the Lamiales, which includes 24 other families of which Acanthaceae, Bignoniaceae, Orobanchaceae, Plantaginaceae, Scrophulariaceae and Verbenaceae are amongst the better known [45]. *Ligustrum* comprises around 40 species and is indigenous throughout most of the temperate and tropical Old World except Africa and the coldest regions [25], and introduced elsewhere. Recent molecular studies have revealed that *Ligustrum* and *Syringa* (lilacs) are closely related and are treated as the subtribe Ligustrinae within the tribe Oleeae [46].

Ligustrum robustum subsp. *walkeri* was originally described as a valid species, but has also been treated as a variety of *L. robustum*, and is currently accepted as a subspecies [16, 47, 48]. The indigenous distribution of *L. robustum* is restricted to India, Sri Lanka and Bangladesh to Vietnam. Within this range, subspecies *robustum* was described from Bangladesh (Silhet) and is found from northeast India to Thailand and Vietnam [48]. Subspecies *walkeri* was thought to occur in southern India ('Kerala and Tamil Nadu in the Western Ghats, especially the Nilgiri and Palni Hills, at altitudes up to 1500 m') and Sri Lanka ('Central, Sabaragamuwa and Uva Provinces, in montane regions at altitudes between 450 and 2000 m') [25, 47], but Green [48, 49] now treats the southern Indian populations as a separate species, *Ligustrum perrottetii* A.DC. *Ligustrum robustum* is no longer considered to occur in China and the taxon *L. robustum* subsp. *chinense* from southern China [50] is now considered to be a synonym of *Ligustrum expansum* Rehder [51].

In the Mascarene Islands, *L. robustum* subsp. *walkeri* appeared first in Mauritius where it was introduced around the turn of the nineteenth century [52, 53]. It now forms dense, impenetrable thickets, and its presence has been correlated with the inability of native vegetation to re-establish following disturbance [27, 54]. *Ligustrum robustum* subsp. *walkeri* arrived in La Réunion in 1969, becoming invasive in the forests of the island where it is still spreading [55].

Initially, precise information about the geographic origin of the introduced *L. robustum* in the Mascarene Islands was lacking, in part because *Ligustrum* is a taxonomically difficult genus. Molecular techniques were utilized alongside traditional techniques and historical research to elucidate the taxonomy and the exact area of origin of the introduced *Ligustrum* [56, 57]. Dried leaf samples were collected during natural enemy surveys and native material of *L. robustum* subsp. *walkeri* from Sri Lanka, *L. robustum* subsp. *robustum* from north eastern India (Assam and Meghalaya) and the closely related *L. perrottetii* from southern India (the Western Ghats) were compared with introduced material from La Réunion and Mauritius using chloroplast RFLP markers and RAPDs. Sri Lankan and introduced material was monomorphic for the same chloroplast DNA haplotype that was absent from south and north-east Indian *Ligustrum*. This material was also clearly distinguished from Indian *Ligustrum* by RAPDs, which also indicated that *L. robustum* subsp. *walkeri* in Sri Lanka is more similar to the southern Indian *L. perrottetii* than to *L. robustum* subsp. *robustum* from north-east India. This would seem to indicate that either *L. perrottetii* should be included as a subspecies of *L. robustum*, or that subsp. *walkeri* should be treated as a separate species; however, thus far, no taxonomic change has been made. Milne and Abbott [57] concluded that the material introduced and established on Mauritius and La Réunion is derived from the Sri Lankan *L. robustum* subsp. *walkeri*. As a result, the search for natural enemies and the studies of potential biological control agents were concentrated in Sri Lanka [24, 56, 58].

Methods

The present literature review was originally made in 1996 and was based on the literature only up until that date. More recent publications were consulted in preparing this list for publication, including some based on our subsequent research [42–44, 59, 60].

Very few natural enemy species were recorded from *L. robustum* itself in the literature survey (apart from our work). Records from other *Ligustrum* spp. were also compiled, to give insight into (1) which types of natural enemy might also be expected to occur on *L. robustum*, and (2) the patterns of natural enemy specificity documented within *Ligustrum* and related genera.

Hobbyist and insect display breeders of large Lepidoptera such as silk moths (Saturniidae) use privet as a substitute food plant for many species in captivity, as it remains in leaf throughout the winter. This seems to account for all records of this family feeding on *Ligustrum* sp (p). (original sources and [61]), apart from some species feeding on introduced privets in the Americas.

The degree of host specificity required for a weed biological control agent depends on the target area, and its indigenous and economic plants. In the case of La Réunion, there are no indigenous *Ligustrum* spp. or *Syringa* spp. *Noronhia broomeana* Horne ex Oliv. is the only endemic species of Oleaceae and is a key test plant species. Other key test plants include *Noronhia emarginata* (Lam.) Thouars (a Madagascan native naturalized in Mauritius and La Réunion), *Olea europaea* subsp. *africana* (Mill.) P.S. Green and *Olea lancea* Lam. (Mascarene endemics), although subsp. *africana* may no longer be recognized, and *O. lancea* may prove to be a synonym of the widespread *Olea capensis* L. [43]. In addition, olives (*Olea* spp.) are cultivated and ornamentals grown there include *Ligustrum ovalifolium* Hassk. and *Jasminum* spp. Hence natural enemy species only recorded from *Ligustrum* spp. or genera only recorded from Oleaceae are likely to reflect oligophagous or monophagous species worth investigating for host specificity (we treat monophagous as restricted to *Ligustrum* spp., oligophagous as restricted to Oleaceae and polyphagous as having a wider host beyond Oleaceae).

Results

The results are presented in two tables: Table 19.1 for invertebrates and Table 19.2 for fungi. Since the recent dawn of the molecular era, mycology has undergone a radical change and fungal systematics has become a dynamic and ever changing discipline [62]. Most significantly, it has now become possible to confirm, or to link for the first time, asexual (anamorph) and sexual (teleomorph) stages using DNA sequencing. Previously, because of the morphological and phenological differences between asexual and sexual reproduction, and the rarity in or absence from the life cycle of one of these morphs, separate names were assigned to these stages – the dual system of fungal nomenclature – even when the associations were suspected or known: proven by obtaining asexual cultures from the sexual spores in the Ascomycota; or by plant inoculation in the biotrophic (non-culturable) rust fungi of the Basidiomycota. However, recent changes to Article 59 of the International Code of Nomenclature – which originally allowed for the naming of both asexual and sexual states of the fungi with unique Latin binomials – now ensures that the scientific name should be based on the principle of priority in the 'one fungus-one name policy' [62, 63]. This is reflected in Table 19.2 where, in some cases, both the old and new genus or species names are included when name priority has yet to be determined. In fact, disparity still exists between the various fungal databases concerning nomenclature and systematics. In addition, for some of the species listed – especially, those recorded during the nineteenth century – the taxonomy has not been 'tested' and the validity of many of the names needs to be verified. For this reason, dubious taxonomic records, as well as those records of generalist saprophytic fungi with a wide 'host' (substrate) range, have been omitted from the list (Table 19.2).

Discussion

When this literature survey of the natural enemies of *Ligustrum* spp. was first compiled in 1996 (R.H. Shaw unpublished), 73 invertebrates and 105 fungi were listed. Here, we list 213 invertebrates and 109 fungi, an increase of 192 and 4%, respectively. We suggest this difference in invertebrate numbers is due to a combination of newly published knowledge, greater accessibility of older information via the Internet, and for some groups such as Lepidoptera, a more comprehensive literature search. The difference for fungi is net of these factors and a more rigorous exclusion of species that are unlikely to be primary pathogens. These differences are a clear indication of the need for caution when making quantitative and qualitative comparisons between natural enemy lists for biological control targets compiled from the literature in different ways and at different dates.

Fungi

In the case of the fungal natural enemies of *L. robustum*, the majority of fungi collected from the centre of origin – mainly comprising necrotrophic species of the Ascomycota (see Table 19.2) – had a restricted occurrence and a limited impact on the host *in situ*, and/or proved to be difficult to manipulate in the initial host-range screening. This was the problem with *Thedgonia ligustrina* (Boerema) B. Sutton, which was identified from literature searches as the fungal species with the most potential, in terms of host impact and specificity (see Table 19.2, Figure 19.1). The latter illustrates that this pathogen can be highly damaging under natural conditions on indigenous *Ligustrum vulgare* L. in the UK, but this was never seen in the field in Asia on *L. robustum* [24]. Thus, for the long-term screening studies, none of the fungal pathogens were considered to warrant inclusion and, as detailed above, only insects were evaluated further. However, several years after the biological control programme had closed down, without the release of any agents, a disease outbreak on *L. robustum* subsp. *walkeri* in La Réunion was reported, causing severe defoliation at a number of sites on the island, with no apparent symptoms on the hedges of *L. ovalifolium* and *L. sinense* in a nearby village (C. Lavergne, pers. com., 25/09/2003). Specimens subsequently received at CABI were identified as *T. ligustrina*, with severe leaf damage and abundant sporulation. How the fungus reached the island remains a mystery, as is the absence of infection on other *Ligustrum* species, since records show that it has a wide host range within the genus (Table 19.2). This would suggest that host-specific pathotypes of the pathogen may occur. Unfortunately, further reports on the impact of the fungus on the invasive *L. robustum* subsp. *walkeri*, and its host range on ornamental privets, are not available. Phylogenetically, this is an extremely interesting fungus because, although its morphology superficially resembles the cercosporoid fungi – leading it to be placed in the Mycosphaerellaceae of the Dothideales (now Capnodiales [237], see Table 19.2) – DNA sequence analysis has now revealed that it belongs to a completely unrelated and unknown family (*incertae sedis*) within the Heliotales [233]. From the field evidence in the UK and La Réunion, it is also a highly damaging pathogen and, therefore, should be a priority agent for classical biological control of invasive alien privets.

A further analysis of Table 19.2 shows that there is an interesting complex of rust species associated with the genus, especially in eastern Asia. Disappointingly, and strangely, no rusts have been reported in the literature, or were recorded during the surveys, on *L. robustum* in western Asia. Conversely, in both China and Japan, *Puccinia klugkistiana* (Dietel) Jing X. Ji & Kakish has been reported from some of the other problematic invasive alien privets (*L. lucidum* W.T. Aiton, *L. ovalifolium*, *L. sinense*), where it can cause significant host damage [162]. Traditionally, rusts have been the 'fungal pathogens of choice' for classical biological control programmes because of their obligatory biotrophic life style – and hence rigid host specificity – as well as their often debilitating host impact and long-distance dispersal efficiency [40]. This rust species, therefore, warrants inclusion in any future classical biological control programme against invasive privets, such as *L. lucidum* (glossy privet) and *L. sinense* (Chinese privet).

Invertebrates

The invertebrate natural enemies of *Ligustrum* spp. found in our literature review can be divided into polyphagous (more than one family), oligophagous (restricted to Oleaceae) and monophagous (restricted to *Ligustrum* sp (p).) (Table 19.3), although the latter two categories may reflect lack of knowledge of other hosts rather than reality. For the purposes of classical biological control, the first category is discounted, the second category gives an indication of natural enemy genera which may contain monophagous species, while the third category is most likely to contain species of potential biological control interest. Surveys should take the latter two groups into particular consideration. Species and genera found on surveys that have not previously been recorded from *Ligustrum* will need evaluation, based initially on literature and field observations.

Based on the results of this literature survey, we conclude that surveys for natural

Table 19.1. Phytophagous invertebrate natural enemies recorded from *Ligustrum* spp. in the literature

Classification species	Host plant (s)	Feeding style	Location	References
ACARI Eriophyiidae				
Aculus ligustri (Keifer) (=*Phyllocoptes ligustri*)	*Ligustrum* sp(p).	Leaf cell sap sucker	W Palaearctic, USA	[8]
Aceria ligustri (Keifer) (=*Eriophyes ligustri*)	*Ligustrum* sp(p).	Bud and leaf cell sap sucker	SE Europe, California	[8]
Tenuipaplpidae				
Brevipalpus obovatus (Donnadieu)	Highly polyphagous including *Ligustrum* sp(p).	Sap sucker	Cosmopolitan	[8]
COLEOPTERA Chrysomelidae				
Argopistes tsekooni Chen	Oleaceae including *Ligustrum* spp.	Leaf miner and feeder	China	[11, 64–66]
Argopistes biplagiatus Motschulsky	Oleaceae including *Fraxinus*, *Osmanthus*, *Ligustrum*, *Syringa* sp(p).	Larval leaf miner and adult leaf feeder	Japan	[67]
Argopistes coccinelliformis Csiki	Oleaceae including *Osmanthus*, *Ligustrum*, Syringa sp(p).	Larval leaf miner and adult leaf feeder	Japan	[68]
Dermorhytis lewisi Jacoby	*Ligustrum robustum*	Larval root feeder and adult leaf edge feeder	Sri Lanka	[59, 69]
Dermorhytis ornatissima Baly	*L. robustum*	Larval root feeder and adult leaf edge feeder	Sri Lanka	[59, 69]
Hyphasis spp. (2)	*L. robustum*	Adult leaf feeder, larva unknown	Sri Lanka	[59]
Longitarsus bimaculatus? Baly	Oleaceae, primarily *Fraxinus*, but also *Ligustrum* sp(p).	Root (larvae) and leaf (adults) feeder	China	[11, 70]
Curculionidae				
Deporaus tigris Sawada	*Ligustrum* spp.	Leaf roller	Japan	[71]
Euwallacea fornicatus (Eichoff)	Polyphagous pest of trees and bushes	Stem borer	Widespread invasive pest in tropics	[12, 59]
Lepropus immunis (Walker)	Polyphagous on trees of at least four families	Adult leaf feeder	Sri Lanka	[59]
Lignyodes enucleator (Panzer)	Oleaceae including *Fraxinus*, *Ligustrum*, *Syringa* sp(p).	Presumed leaf feeder	Turkey and Poland	[72]
Ochyronmera ligustri Warner	*Ligustrum*, *Syringa* sp(p).	Leaf feeder	E Asia, Introduced in the USA	[73, 74]

Continued

Table 19.1. Continued

Classification species	Host plant (s)	Feeding style	Location	References
Otiorhynchus aurifer Boheman	Polyphagous including *Ligustrum* sp(p).	Larval root feeder, adult leaf feeder	Europe	[75]
Otiorhynchus crataegi Germar	Polyphagous including *Ligustrum* sp(p).	Larval root feeder, adult leaf feeder	Europe	[76, 77]
Otiorhynchus meridionalis Gyllenhal	Polyphagous including *Ligustrum* sp(p).	Larval root feeder, adult leaf feeder	Europe, introduced in N America	[78]
Otiorhynchus rotundatus Siebold	Polyphagous including *Ligustrum* sp(p).	Leaf feeder	Poland	[79, 80]
Otiorhynchus smreczynskii Cmoluch	Polyphagous including *Ligustrum* sp(p).	Larval root feeder, adult leaf feeder	E and N Europe	[81]
Pseudocneorhinus bifasciatus (Roelefs)	Numerous ornamentals including *Ligustrum* sp(p).	Root feeder	Japan, introduced in the USA	[82]
Xylosandrus arquatus (Sampson)	Pest of camphor (*Cinnamomum camphora*, Lauraceae); *L. robustum*	Stem borer	Sri Lanka	[59]
Meloidae				
Lytta vesicatoria (Linnaeus)	Polyphagous including *Ligustrum* sp(p).	Leaf feeder	West Palaearctic	[83]
Platypodidae				
Megaplatypus ruciat (Chapuis)	Polyphagous including *Ligustrum* sp(p).	Stem borer	South America, introduced in Italy.	[84]
DIPTERA Cecidomyiidae				
Asphondylia sphaera Monzen	*Ligustrum* spp and short-term alternate hosts of other families	Gall former	Japan	[85]
Placochela nigripes (Löw) (=*P. ligustri*)	*Sambucus*, *Ligustrum* sp(p).	Gall former	Europe	[86]
HEMIPTERA Aphididae				
Aphis crinosa (Paik)	*Ligustrum* spp.; also *on Lonicera*, *Viburnum* spp.	Sap sucker	Japan and Korea	[1]
Aphis gossypii Glover	Highly polyphagous including *Ligustrum* sp(p).	Sap sucker	Cosmopolitan	[1]
Aphis spiraecola Patch	Highly polyphagous including *Ligustrum* sp(p).	Sap sucker	Cosmopolitan	[1]
Aulacorthum ibotum Essig & Kuwana	*Ligustrum* spp.	Sap sucker	Japan and Korea	[1]
Ceratopemphigus zehntneri Schouteden	*L. robustum* subsp. *rucia* and presumably another species in China	Gall inducer and sap sucker	Sri Lanka, China	[1, 60]

Continued

Table 19.1. Continued

Classification species	Host plant (s)	Feeding style	Location	References
Macrosiphum euphorbiae (Thomas)	Highly polyphagous including *Ligustrum* sp(p).	Sap sucker	Cosmopolitan	[1]
Prociphilus bumeliae (Schrank)	Alternating between Oleaceae and *Abies*	Sap sucker	Europe	[1]
Prociphilus ligustrifoliae (Tseng & Tao)[1]	*Ligustrum* spp. And an unknown alternate host	Sap sucker	China, Japan	[1, 87]
Prociphilus oriens Mordvilko	Polyphagous including *Ligustrum* sp(p).; alternating on *Abies*	Sap sucker	India to Japan	[1]
Prociphilus osmanthae Essig & Kuwana	*Osmanthus* and maybe *Ligustrum* spp. as well as an unknown alternate host	Sap sucker	Japan, maybe NE India	[1]
Myzus ascalonicus Doncaster	Highly polyphagous including *Ligustrum* sp(p).	Sap sucker	Cosmopolitan	[1]
Myzus ligustri (Mosley)	*Ligustrum* spp.	Sap sucker	Europe, introduced in Nearctic	[1, 77]
Myzus persicae (Sulzer)	Highly polyphagous including *Ligustrum* sp(p)., alternating on *Prunus*	Sap sucker	Cosmopolitan	[1]
Aphrophoridae				
Aphrophota sp.	*Ligustrum* sp(p).	Sap sucker	China	[11]
Clovia bipunctata (Kirby)	Polyphagous including *Ligustrum* sp(p).	Sap sucker	China	[11]
Cicadellidae				
Docotettix cornutus Ribaut	Pomegranate, olive and privet	Sap sucker	Turkey	[88, 89]
Coccidae				
Ericerus pela Chavannes	*Fraxinus* and *Ligustrum* spp.	Sap sucker	China	[90]
Parlatoria oleae (Colvée)	Primarily on *Olea* sp(p), but polyphagous including on *Ligustrum* sp(p).	Sap sucker	USA, Europe and Middle East	[12, 91]
Pseudaulacaspis pentagona (Targioni)	Polyphagous pest including *Ligustrum* sp(p).	Sap sucker	Cosmopolitan	[11, 92]
Coreidae				
Cletus punctiger (Dallas)	Polyphagous pest including *Ligustrum* sp(p).	Sap sucker	China	[11, 12]
Homoeocerus sp.	*Ligustrum* sp(p).	Sap sucker	China	[11]

Continued

Table 19.1. Continued

Classification species	Host plant (s)	Feeding style	Location	References
Riptortus pedestris (Fabricius)	Polyphagous pest including *Ligustrum* sp(p).	Sap sucker	Sri Lanka to China	[11, 12]
Flatidae				
Geisha sp.	*Ligustrum* sp(p).	Sap sucker	China	[11]
Lawana sp.	*Ligustrum* sp(p).	Sap sucker	China	[11]
Metcalfa pruinosa (Say)	Polyphagous including *Ligustrum* sp(p).	Sap sucker	N Americs, introduced in S Europe, Korea	[93]
Salurnis marginella (Guérin-Méneville)	Polyphagous including *Ligustrum* sp(p).	Sap sucker	China and SE Asia	[11]
Miridae				
Adelphocoris fasiaticollis Reuter	Polyphagous including *Ligustrum* sp(p).	Sap sucker	China	[11]
Pentatomidae				
Acrosternum impicticorne (Stål)	Soyabean pest also recorded from *Ligustrum lucidum*	Sap sucker	Brazil	[94, 95]
Antiteuchus tripterus (Stål)	Cocoa pest also recorded from *L. lucidum*	Sap sucker	Neotropical	[95]
Antestiopsis ruciate (Fabricius) (=*A. picturata* (Distant))	Coffee pest also recorded from flowers of *L. robustum*	Sap sucker (on flowers)	India, Sri Lanka	[17, 59, 96]
Carbula humerigera (Uhler) (=*C. obtusangula* Reuter)	Polyphagous including *Ligustrum* sp(p).	Sap sucker	China	[11]
Erthesina fullo (Thunberg)	Polyphagous including *Ligustrum* sp(p).	Sap sucker	China to SE Asia	[11, 17]
Eysarcoris sp.	*Ligustrum* sp(p).	Sap sucker	China	[11]
Loxa deducta Walker	*L. lucidum*, but must also be found on indigenous hosts	Sap sucker	Brazil	[95]
Menida sp.[2]	*Ligustrum* sp(p).	Sap sucker	China	[11]
Nezara viridula (Linnaeus)	Polyphagous pest including *Ligustrum* sp(p).	Sap sucker	Pantropical	[11, 95, 97]
Pentatoma sp.	*Ligustrum* sp(p).	Sap sucker	China	[11]
Pallantia macunaima Grazia	*L. lucidum*, but must also be found on indigenous hosts	Sap sucker	Brazil	[95]
Palomena angulosa Motschulsky	Polyphagous pest including *Ligustrum* sp(p).	Sap sucker	Japan to SE Asia	[11, 12]

Continued

Table 19.1. Continued

Classification species	Host plant (s)	Feeding style	Location	References
Ricaniidae				
Pochazia sp.	*Ligustrum* sp(p).	Sap sucker	China	[11]
Scutelleridae				
Hyperoncus[3] *lateritius* Westwood	Polyphagous including *Ligustrum* sp(p).	Sap sucker	China	[11]
Tingidae				
Leptoypha hospital Drake & Poor	*Ligustrum* sp(p).	Cell sap sucker	China to W Malaysia	[98, 99]
Omoplax majorcarinae Guilbert	*Ligustrum* sp(p).	Cell sap sucker	Bonin Islands (Japan)	[100]
HYMENOPTERA				
Tenthredinidae				
Macrophya pilotheca Wei & Ma	*Ligustrum* spp.	Leaf feeder	China	[101]
Macrophya punctumalbum (Linnaeus)	*Ligustrum*, *Fraxinus* and *Syringa*	Leaf feeder	Europe, Canada, USA and Russia	[102, 103]
LEPIDOPTERA				
Brahmaeidae[4]				
Brahmaea certhia (Fabricius)	Oleaceae: *L. lucidum*, *Osmanthus fragrans*	Leaf feeder	China, SE Russia and Korean Peninsula	[4, 10]
Brahmaea hearseyi White	Oleaceae: *Ligustrum* spp.	Leaf feeder	NE Himalyas to Philippines	[4, 10]
Brahmaea porphyrio Chu and Wang	Oleaceae including *Ligustrum*, *Olea*, *Osmanthus*, *Syringa* sp(p).	Leaf feeder	China	[10][5]
Brahmaea wallichii (Linnaeus)	Oleaceae: *Ligustrum* spp. including *L. robustum*, *Osmanthus* spp.	Leaf feeder	India to China, Taiwan and Japan	[10, 104, 105]
Cossidae				
Zeuzera pyrina (Linnaeus)	Polyphagous pest of fruit and other trees	Wood borer	Widespread globally	[12, 59]
Crambidae				
Anania coronata (Hufnagel)	Mainly on *Sambucus*, but on other species including *Ligustrum* sp(p).	Leaf feeder	Holarctic	[5, 106]
Cydalima perspectalis (Walker)	Polyphagous including *Ligustrum sinense*	Leaf feeder	East Palaearctic, India; introduced in West Palaearctic where it is a pest of box (*Buxus*)	[11, 107]

Continued

Table 19.1. Continued

Classification species	Host plant (s)	Feeding style	Location	References
Palpita celsalis (Walker) (=*P. annulata* (Fabricius))	*Ligustrum robustum*, *Chionanthus zeylandica* (Oleaceae)	Leaf feeder	Sri Lanka to China	[59, 108]
Palpita gracilalis (Hulst)[6]	*Ligustrum* sp(p).	Leaf feeder	Nearctic	[5, 9]
Palpita inusitata (Butler)	*L. sinense*	Leaf feeder	China, Japan	[11]
Palpita nigropunctalis (Bremer)[7]	*Ligustrum* spp.; also *Fraxinus* sp(p).	Flowers, fruit and leaves	E Asia	[10, 11, 109]
Palpita quadristigmalis (Guenée)	*Ligustrum*, *Olea*, *Osmanthus* sp(p).	Leaf feeder	Nearctic, southern S America	[3, 9, 110]
Palpita vitrealis (Rossi) (=*Palpita unionalis*)	Mostly Oleaceae, especially *Olea*, *Jasminum* sp(p)., but also *Ligustrum* sp(p).	Leaf feeder	Old World tropics	[12, 111]
Pycnarmon cribrata (Fabricius)	Polyphagous including *L. sinense*	Leaf feeder	S and SE Asia	[11]
Erebidae				
Amata sp.	*Ligustrum* sp(p). and other genera	Leaf feeder	China	[10]
Aloa lactinea (Cramer)	Polyphagous including *L. sinense*	Leaf feeder	India to Japan to Philippines, Sundaland	[9, 11]
Arna bipunctapex (Hampson)	Polyphagous including *Ligustrum* sp(p).	Leaf feeder	China	[4]
Callindra similis (Moore)	*Ligustrum* sp(p). and other genera	Leaf feeder	NE India to China	[10]
Halysidota ruscheweghi Dyar	Polyphagous including *Ligustrum* sp(p).	Leaf feeder	Southern S America	[3]
Ischyja manlia (Cramer)	Polyphagous including *Ligustrum* sp(p).	Leaf feeder	India to Japan and Australia	[9, 10]
Lemyra phasma (Leech)	Polyphagous including *Ligustrum* sp(p).	Leaf feeder	NE India to China	[9, 10]
Orgyia leucostigma (Smith & Abbott)	Polyphagous on trees and shrubs including *Ligustrum* sp(p).	Leaf feeder	E USA	[7, 9, 12]
Pangrapta cana (Leech)	*L. sinense*	Leaf feeder	China	[11]
Pangrapta grisangula (Hampson)	*L. robustum*	Leaf feeder	South India, Sri Lanka	[4, 59]
Pangrapta obscurata (Butler)	*Ligustrum* spp.	Leaf feeder	China, Japan	[10]
Pangrapta perturbans (Walker) (=*P. trimantesalis* (Walker))	*Ligustrum* spp.	Leaf feeder	Bangladesh to Japan	[8]

Continued

Table 19.1. Continued

Classification species	Host plant (s)	Feeding style	Location	References
Spilarctia subcarnea (Walker)	Polyphagous including *L. sinense*	Leaf feeder	E Palaearctic	[11]
Eupterotidae				
Apona ligustri Mell	*Ligustrum japonicum*	Leaf feeder	China	[4]
Ganisa cyanogrisea Mell	*L. japonicum*, *L. lucidum*	Leaf feeder	China	[4]
Palirisa cervina Moore	*L. japonicum*	Leaf feeder	SE Asia	[4]
Gelechiidae				
Parastega hemisigna Clarke	*L. lucidum*	Leaf feeder	Argentina	[3, 112, 128]
Geometridae				
Abraxas grossularia (Linnaeus)	Polyphagous, a minor pest of fruit bushes, but not typically on *Ligustrum*	Leaf feeder	Europe and North America.	[4, 9, 113]
Acasis viretata (Hübner)	Polyphagous including *Ligustrum* spp.	Flowers, leaves and buds	Eurasia	[10, 113]
Aethalura intertexta (Walker) (=*anticaria* Walker)	Polyphagous on deciduous trees and shrubs	Leaf feeder	S Canada and E USA	[7]
Agaraeus parva (Hedeman)	Polyphagous including *Ligustrum* sp(p).	Leaf feeder	N India to SE Russia	[10]
Alsophila aescularia (Denis & Schiffermüller)	Polyphagous on deciduous trees	Leaf feeder	Temperate Eurasia	[6, 9]
Angerona prunaria (Linnaeus)	Polyphagous on deciduous trees and shrubs	Leaf feeder	Temperate Eurasia	[6]
Apeira syringaria (Linnaeus)	Caprifoliaceae and Oleaceae including *Ligustrum* sp(p).	Leaf feeder	Europe	[6]
Ascotis selenaria (Denis & Schiffermüller)	Polyphagous including *Ligustrum* sp(p).	Leaf feeder	S Europe to Japan, Africa	[9, 10, 114]
Biston marginata Shiraki	Polyphagous including *Ligustrum* sp(p).	Leaf feeder	China, Japan, Vietnam	[9, 10]
Chloroclystis v-ata Haworth	Polyphagous including *Ligustrum* sp(p).	Leaf feeder	Palaearctic	[6, 9]
Comostola rubripunctata (Warren)	*Ligustrum tsusimense*	Leaf feeder	Japan	[6]
Crocallis elinguaria (Linnaeus)	Polyphagous on deciduous trees and shrubs	Leaf feeder	Europe	[6]
Ectropis crepuscularia (Guenée)	Polyphagous on deciduous trees and shrubs	Leaf feeder	Holarctic	[7, 9, 113]

Continued

Table 19.1. Continued

Classification species	Host plant (s)	Feeding style	Location	References
Garaeus specularis Moore	Polyphagous including *Ligustrum* sp(p).	Leaf feeder	India to Japan	[6, 115]
Hemistola dijuncta (Walker)	*Ligustrum tschonoskii*	Leaf feeder	Japan	[6]
Lycia hirtaria (Clerk)	Polyphagous on deciduous trees and shrubs	Leaf feeder	Europe	[6, 9]
Menophra abruptaria (Thunberg)	Oleaceae including *Ligustrum*, *Syringa* sp(p)	Leaf feeder	Europe	[6]
Naxa seriaria (Motschulsky)	*Ligustrum* sp(p). and other genera	Leaf feeder	SE Russia, China	[6, 10, 11, 115]
Ourapteryx sambucaria (Linnaeus)	Polyphagous on deciduous trees and shrubs	Leaf feeder	Europe	[6]
Paleacrita vernata (Peck)	Polyphagous on deciduous trees and shrubs including *Ligustrum* sp(p).	Leaf feeder	E Nearctic	[7, 9]
Percnia luridaria (Leech)	*L. sinense*	Leaf feeder	China	[11]
Perizoma impromissata (Walker)	Polyphagous including *Ligustrum* sp(p).	Leaf feeder	Neotropical	[3, 6]
Peribatodes rhomboidaria (Denis & Schiffermüller)	Polyphagous on deciduous trees	Leaf feeder	Europe	[6, 9]
Phthonosema invenustaria (Leech)	Polyphagous including *Ligustrum* sp(p).	Leaf feeder	Far E Asia	[6]
Problepsis deliaria Guenée	Oleaceae including *L. robustum*	Leaf feeder	India, Sri Lanka	[59]
Problepsis superans (Butler)	*Ligustrum* sp(p).	Leaf feeder	Far East Asia	[10, 116]
Sabulodes aegrotata (Guenée) (=*caberata* auct.)	Polyphagous including *Ligustrum* sp(p).	Leaf feeder	W coastal Nearctic	[7, 9]
Scopula caricaria (Reutti)	Asteraceae in Europe; Zhang *et al.*'s [11] record from *L. sinense* seems unlikely	Leaf feeder	Palaearctic	[11, 117]
Trichopteryx polycommata (Denis et Schiffermüller)	Caprifoliaceae and Oleaceae including *Ligustrum* sp(p).	Leaf feeder	Temperate Eurasia	[6, 10]
Gracillariidae				
Caloptilia cuculipennella (Hübner)	Oleaceae: *Fraxinus*, *Ligustrum*, *Syringa* and *Jasminum* sp(p).	Leaf miner	Palaearctic	[9, 106, 118]

Continued

Table 19.1. Continued

Classification species	Host plant (s)	Feeding style	Location	References
Caloptilia fraxinella (Ely)	Oleaceae: mainly *Fraxinus* spp. but also *Ligustrum* sp(p).	Leaf miner	Nearctic	[5]
Caloptilia japonica (Kumata)	*Ligustrum* spp.	Leaf miner	Japan	[118]
Caloptilia syringella (Fabricius)	Oleaceae: *Fraxinus*, *Ligustrum*, *Syringa* and *Jasminum* sp(p).	Leaf miner	Europe, introduced Nearctic	[9, 106]
Metriochroa fraxinella Kumata	Oleaceae, including *Fraxinus*, *Ligustrum* s(p).	Leaf miner	Japan	[119, 120]
Phyllocnistis citrella Stainton	Pest of citrus (Rutaceae)	Leaf miner	Widespread and invasive in tropics	[12, 59]
Phyllocnistis sp.	*L. japonicum*	Leaf miner	Japan	[121]
Hepialidae				
Endoclita variegata (Walker)	Polyphagous including *Ligustrum* spp.	Wood borer	Sri Lanka to Japan, Borneo and Java	[12]
Endoclita excrescens (Butler)	Polyphagous including *L. sinense*	Stem borer	Far E Palaearctic	[9, 11]
Lasiocampidae				
Artace obumbrata Köhler	*Ligustrum* sp. (introduced)	Gregarious leaf feeder	Argentina	[3, 112]
Titya proxima (Burmeister)	Polyphagous including *Ligustrum* spp.	Leaf feeder	Neotropical	[3, 112]
Limacodidae				
Stroter intermissa (Walker)	*Ligustrum* sp(p). and probably polyphagous	Leaf feeder	Africa	[5]
Lycaenidae				
Artopoetes pryeri (Murray)	Oleaceae including *Ligustrum* spp. and *Syringa* spp.	Presumed leaf feeder	East Asia	[4, 10]
Celastrina argiolus (Linnaeus)	Polyphagous including *Ligustrum* sp(p).	Flower and fruit feeder	Holarctic	[122]
Laeosopis roboris (Esper)	*Fraxinus*; *Ligustrum* sp(p) is recorded as a laboratory host.	Leaf feeder	SW Europe	[9, 123]
Noctuidae				
Cerma cerintha (Treitschke)	Mostly Rosaceae, but also *Ligustrum* spp.	Leaf feeder	E Nearctic	[7]
Craniophora ligustri (Denis and Schiffermüller)	Mainly Oleaceae *Fraxinus*, *Ligustrum*, *Syringa* sp(p). but other deciduous trees recorded	Leaf feeder	Temperate Eurasia	[5, 10, 113]

Continued

Table 19.1. Continued

Classification species	Host plant (s)	Feeding style	Location	References
Cranniophora malesiae Holloway	Oleaceae: *Olea*, *Ligustrum* and *Osmanthus*	Leaf feeder	NE Himalaya to New Guinea	[124]
Euplexia lucipara (Linnaeus)	Polyphagous including *Ligustrum* sp(p).	Leaf feeder	Palaearctic	[10, 125][9]
Harrisimemna trisignata (Walker)	Polyphagous including *Ligustrum* sp(p).	Leaf feeder	Nearctic	[5]
Polyphaenis sericata (Esper)	Polyphagous including *Ligustrum* sp(p).	Leaf feeder	S Europe to WAsia	[5]
Nymphalidae				
Parathyma ranga (Moore)	Oleaceae including *Chionanthus*, *Fraxinus*, *Ligustrum*, *Olea*, *Osmanthus* sp(p).	Leaf feeder	India to SE Asia	[4, 10]
Oceophoridae				
Gonionota bourquiniella (Köhler)	*Ligustrum* and *Solanum* sp(p).		Brazil to Argentina	[3]
Papilionidae				
Papilio multicaudata W.F. Kirby	Polyphagous on deciduous trees and shrubs including *Ligustrum* sp(p).	Leaf feeder	W Nearctic	[122]
Papilio rutulus Lucas	Polyphagous on deciduous trees and shrubs including *Ligustrum* sp(p).	Leaf feeder	California	[126]
Psychidae				
Chalioides kondonis Kondo	*Ligustrum* sp(p). and other genera	Leaf feeder	China, Japan	[9, 10]
Eumeta minuscula (Butler)	Polyphagous pest of tea	Leaf feeder	E Asia	[9, 10, 12]
Eumeta variegata Snellen	Polyphagous pest	Leaf feeder	SE Asia	[9–12]
Oiketicus townsendi Townsend	Polyphagous including *Ligustrum* sp(p).	Leaf feeder	E Nearctic	[127]
Pyralidae				
Endotricha theonalis (Walker)	*L. sinense*	Presumed leaf feeder	China, Japan	[11]
Laodamia eulepidella (Hampson) (=*Phycita eulepidella*)	Oleaceae including *Ligustrum robustum*	Flower bud feeder	South and south-east Asia, Australia	[20, 59]
Saturniidae				
Automeris coresus (Boisduval)	Polyphagous including *Ligustrum* sp(p).	Leaf feeder	Neotropical	[3, 128, 129]
Automeris naranja Schaus	Polyphagous including *Ligustrum* sp(p).	Leaf feeder	Neotropical	[3, 130]

Continued

Table 19.1. Continued

Classification species	Host plant (s)	Feeding style	Location	References
Eacles imperialis (Drury)	Polyphagous including *Ligustrum* sp(p).	Leaf feeder	Neotropical	[3, 129]
Eupackardia calleta (Westwood)	Typically on *Leucophyllum* (Scrophulariaceae), but also on Oleaceae including *Ligustrum* sp(p).	Leaf feeder	Central America to Southern USA	[131]
Hylesia nigricans (Berg)	Polyphagous including *Ligustrum* sp(p).	Leaf feeder	Neotropical	[3, 129]
Rothschildia jacobaeae (Walker)	Polyphagous including *Ligustrum* sp(p).	Leaf feeder	Neotropical	[3, 129, 132]
Sesiidae				
Podosesia syringae (Harris)	Oleaceae, including *Fraxinus*, *Syringa*, *Ligustrum*, *Olea*, *Chionanthus* sp(p).	Stem borer	Nearctic	[133, 134]
Sphingidae				
Acherontia atropos (Linnaeus)	Mainly Solanaceae but polyphagous including *Ligustrum* sp(p).	Leaf feeder	Europe and Africa	[9, 135]
Acherontia lachesis (Fabricius)	Polyphagous	Leaf feeder	S and SE Asia to New Guinea	[9, 10, 135]
Acherontia styx Westwood	Polyphagous, including *Ligustrum* spp. as a minor food plant	Leaf feeder	Arabia to SE Asia	[4, 9, 10, 135]
Ceratomia undulosa (Walker)	Oleaceae including *Fraxinus*, *Ligustrum*, *Chionanthus* sp(p).	Leaf feeder	Eastern N America	[136]
Dolbina elegans A. Bang Haas	Oleaceae including *Fraxinus* and *Ligustrum* sp(p).	Leaf feeder	Bulgaria to Iran	[135]
Dolbina exacta Staudinger	Oleaceae including *Fraxinus*, *Ligustrum*, *Syringa* sp(p).	Leaf feeder	Far E Asia	[137]
Dolbina inexacta (Walker)	Oleaceae: *Chionathus*, *Fraxinus*, *Ligustrum* (including *L. robustum*), *Olea*, *Osmanthus*, *Syringa*	Leaf feeder	Pakistan to SE Asia	[4, 137]
Dolbinia tancrei Staudinger	Oleaceae including *Fraxinus*, *Ligustrum*, *Olea*, *Osmanthus* sp(p).	Leaf feeder	Far E Asia	[10, 137]
Kentrochrysalis consimilis Rothschild & Jordan	*Ligustrum obtusifolium*	Leaf feeder	Far E. Asia	[137]

Continued

Table 19.1. Continued

Classification species	Host plant (s)	Feeding style	Location	References
Kentrochrysalis streckeri (Staudinger)	Oleaceae including *Fraxinus*, *Ligustrum*, *Syringa* sp(p).	Leaf feeder	E Asia	[4, 10, 137]
Manduca rustica (Fabricius)	Polyphagous including *Ligustrum* spp.	Leaf feeder	Neotropical	[3, 129]
Psilogramma increta (Walker)	Polyphagous including *Ligustrum* spp.	Leaf feeder	S and E Asia	[9, 10, 137]
Psilogramma menephron (Cramer)	Polyphagous including *L. robustum* and other *Ligustrum* spp.	Leaf feeder	Indo-Australian	[4, 9–11, 138]
Sphinx chersis (Hübner)	Mainly Oleaceae including *Ligustrum* spp., but also Rosaceae, Salicaceae	Leaf feeder	Eastern N America	[136]
Sphinx kalmiae J.E. Smith	Mainly Oleaceae *Fraxinus* spp. but also including *Ligustrum* spp.	Leaf feeder	Eastern N America	[136]
Sphinx ligustri Linnaeus	Polyphagous including *Ligustrum* spp.	Leaf feeder	Temperate Europe to Japan	[10, 135]
Tortricidae				
Ancylis apicella (Denis & Schiffermüller)	Mainly on Rhamnaceae, but polyphagous including *Ligustrum* sp(p).	Leaf tier and feeder	Europe to Russia	[2, 139]
Archips compitalis Razowski	Polyphagous including *Ligustrum* sp(p).	Leaf roller; also predator of *E. pela*	China	[140]
Archips podanus (Scopoli)	Polyphagous pest on trees and shrubs, including *Ligustrum* sp(p).	Flower, bud and leaf feeder	Europe	[12, 141]
Archips rosana (Linnaeus)	Polyphagous on trees and shrubs, including *Ligustrum* sp(p).	Flower, bud and leaf feeder	Europe, introduced Nearctic	[9, 12, 142]
Argyrotaenia fletcheriella (Köhler)	Polyphagous including *Ligustrum* sp(p).		Southern S America	[3]
Bonagota salubricola (Meyrick) and/or *B. cranaodes* (Meyrick)	Apple pest, polyphagous including *Ligustrum* sp(p).	Flower, fruit and leaf feeder	South America	[3, 143, 144]
Choristoneura longicellanus (Walsingham)	Polyphagous on trees and shrubs, including *Ligustrum* sp(p).		Far E. Asia	[2, 9]
Clarkeulia bourquini (Clarke)	Polyphagous including *Ligustrum* sp(p).	Flower, fruit and leaf feeder	Brazil to Argentina	[2, 3, 143, 145]
Cryptoptila immersana (Walker)	Polyphagous including *Ligustrum* sp(p).		Australia	[2, 9]

Continued

Table 19.1. Continued

Classification species	Host plant (s)	Feeding style	Location	References
Ditula angustiorana (Haworth)	Polyphagous on trees and shrubs, including *Ligustrum* sp(p).	Flower, fruit and leaf feeder	Europe	[2]
Epiphyas postvittana (Walker)	Polyphagous pest, including *Ligustrum* sp(p).	Fruit and leaf feeder	Australia; introduced Pacific, UK	[4, 9, 12, 146]
Eupoecilia ambiguella (Hübner)	Polyphagous on trees and shrubs	Fruit feeder	Europe and Russia	[9, 146]
Exapate congelatella (Clerck)	Polyphagous on trees and shrubs, including *Ligustrum* sp(p).	Leaf tier and feeder	Europe to China	[9, 146]
Clepsis consimilana (Hübner)	Polyphagous including *Ligustrum* sp(p).	Leaf tier and feeder, mainly on old or dead leaves	Palaearctic; introduced USA and Madagascar	[146, 147]
Homona magnanima Diakonoff	Polyphagous pest, especially on tea	Leaf tier and feeder	E Asia	[9, 10, 12]
Lobesia botrana (Denis and Schiffermüller)	Polyphagous pest, including *Ligustrum vulgare*	Flower and fruit feeder	Palaearctic	[9, 12, 139]
Lozotaenia forsterana (Fabricius)	Polyphagous including *Ligustrum* sp(p).	Leaf tier and feeder	Europe to Siberia	[9, 10, 146]
Pandemis heparana (Denis & Schiffermüller)	Polyphagous including *Ligustrum* sp(p).	Leaf tier and feeder	Palaearctic	[2, 9, 148]
Minutargyrotoza calvicaput (Walsingham)	*L. tschonoskii*	Leaf tier and feeder	Japan	[148]
Pseudargyrotoza conwagana (Fabricius)	Oleaceae, *Fraxinus*, *Ligustrum*, *Syringa* sp(p). and Berberaceae, *Berberis* sp(p).	Flower and fruit feeder	Europe to China	[10, 146]
Rhopobota naevana (Hübner)	Polyphagous including *Ligustrum* sp(p).	Leaf tier and feeder	Palaearctic, South Asia	[2, 59, 149]
Syricoris lacunana (Denis & Schiffermueller)	Polyphagous including *Ligustrum* sp(p).	Leaf, flower tier and feeder	Palaearctic	[139]
Uraniidae				
Epiplema albida (Hampson)	*Ligustrum robustum*	Leaf feeder	Southern India, Sri Lanka	[43, 44, 59]
Yponomeutidae				
Prays oleae (Bernhard)	Olive pest also known from *Ligustrum* sp(p).	Leaf and fruit feeder	Mediterranean	[12]
Zelleria japonicella Moriuti	*L. tschonoskii*, *Syringa persica*		Japan	[150]

Continued

Table 19.1. Continued

Classification species	Host plant (s)	Feeding style	Location	References
ORTHOPTERA Tettigoniidae				
Leptophyes punctatissima (Bosc)	Polyphagous including *Ligustrum* sp(p).	Leaf feeder	Europe	[151, 152]
THYSANOPTERA Thripidae				
Dendrothrips ornatus (Jablonowski)	Polyphagous including *Ligustrum* sp(p).	Sap sucker	Europe, introduced in N America and China	[153, 154]
NEMATODA Heteroderidae				
Cactodera cacti Filipjev & Schuurmans Stekhoven	Mainly Cactaceae, but recorded from *Ligustrum* sp(p).	Root feeder	Cosmopolitan	[12, 155]
Heterodera cyperi Golden, Rau & Cobb	Mainly grasses and sedges, but recorded from *Ligustrum* sp(p).	Root feeder	Pantropical	[156]

[1]This may be a synonym of *P. osmanthae*.
[2]A genus in need of revision [17].
[3]*Hyperoncus* Stål, 1871 appears to be an unavailable homonym of *Hyperoncus* Hesse, 1867 (see [17], http://www.marinespecies.org/aphia.php?p=taxdetails&id=849211).
[4]Zhang *et al.* [11] record *Brahmaea ledereri* Rogenhofer as a leaf feeder on *L. sinense* in China, but this species is restricted to Turkey [157].
[5]Also http://jiuzhai.com/index.php/research/animal/3157-brahmaea-porphyrio-chu-et-wang.html – how to cite?.
[6]This is an unavailable homonym, but there is no replacement name available yet.
[7]Many misidentifications in literature: http://globiz.pyraloidea.org/.
[8]http://www.jpmoth.org/~dmoth/80_Noctuidae/09Catocalinae/4343_Pangrapta_perturbans/Pangrapta%20perturbans.htm.
[9]This record from India may be a misidentification.

enemies of *L. robustum* subsp. *walkeri* in its indigenous range may find the types of oligophagous or monophagous natural enemies shown in Table 19.4. However, not all these types of natural enemy were found in Sri Lanka, and other potentially oligophagous or monophagous species were found that are not included in the literature-based list, reflecting that their food plants and biology have not previously been documented. We note that *E. albida* did not appear in the literature survey, but it is the only species for which a full host range study was completed [43]. In contrast, all the other species that were the subject of preliminary host range testing were already recognized from the literature survey (Table 19.4).

The records in Table 19.1 show that oligophagous species that feed on *Ligustrum* spp. and other Oleaceae have frequently been recorded on *Syringa* spp. as would be expected from these to genera being placed together as the Ligustrinae [46], as well as on *Fraxinus* and *Osmanthus* spp. and less often on *Olea*, *Jasminum* and *Chionanthus* spp. All these genera except *Jasminum* are placed in the subfamily Oleeae: *Fraxinus* in the tribe Fraxinini is closest phylogenetically to Ligustrini, followed by *Chionanthus*, *Olea* and *Osmanthus* in the tribe Oleinae, while *Jasminum* in the subfamily Jasmineae is the

Table 19.2. List of fungi recorded from the genus *Ligustrum*

Classification[1] species	Host	Location[2]	Reference[3]
Basidiomycota			
Pucciniales			
Aecidium ligustri F. Strauss[4]	*Ligustrum vulgare*	Germany	[158]
Aecidium ligustricola Cummins	*Ligustrum* sp.	China	[159]
Caeoma ligustri (Rabenh.) G. Winter	*L. vulgare*	Germany	[158]
Puccinia abchazica Tranzschel & Erem.	*L. arafa*	Russia	[160]
Puccinia (*Aecidium*) *klugkistiana* (Dietel) Jing X. Ji & Kakish.[5]	*L. ibota*, *L. japonicum*, *L. lucidum*, *L. obtusifolium*, *L. ovalifolium*, *L. quithonil*, *L. sinense*	China, Japan	[159, 161, 162]
Puccinia obtusata G.H. Otth ex E. Fisch.	*L. obtusifolium*	Japan	[163]
Uredo amami-ohshimaensis Morim.	*L. ovalifolium*	Japan	[164]
Cantharellales			
Ceratobasidium sp.[6]	*L. robustum s.l.*, subsp. *walkeri*	India, Sri Lanka	[24]
Ascomycota			
Xylariales			
Anthostoma bavaricum Petr.	*L. vulgare*	Germany	[165]
Anthostomella cavarica (Petr.) P.M.D. Martin	*L. vulgare*	Austria	[166]
Dendrophoma ligustrina Kirschst.	*L. vulgare*	Germany	[167]
Eutypa ligustri Richon	*L. vulgare*	France	[168]
Seiridium ceratosporum (De Not.) Nag Raj	*L. robustum* subsp. *walkeri*	Sri Lanka	[169]
Pleosporales			
Ascochyta ligustri Sacc. & Speg.	*L. vulgare*	Armenia, Italy	[170]
Ascochyta ligustrina Pass.	*L. vulgare*	France	[171]
Coniothyrium ligustri Brunaud	*L. ovalifolium*, *L. japonicum*	France, Greece	[172]
Corynespora cassiicola (Berk. & M.A. Curtis) C.T. Wei	*L. sinense*	USA	[173, 174]
Corynespora ligustri Y.L. Guo	*L. lucidum*,	China	[175]
Cucurbitaria ligustri Fabre	*L. lucidum*, *L. vulgare*	France, Ukraine	[176]
Epicoccum ligustri Henn.	*L. vulgare*	Brazil	[177]
Helminthosporium ligustri Meng Zhang & T.Y. Zhang	*L. quihoui*	China	[178]
Hendersonia ligustricola (Girz.) Merezhko	*L. vulgare*	Ukraine	[179]

Continued

Table 19.2. Continued

Classification[1] species	Host	Location[2]	Reference[3]
Leptosphaeria sp.[6]	*L. robustum* subsp. *walkeri*	Sri Lanka	[24]
Massarina ligustri (G.H. Otth) Sacc.	*L. communis*	Switzerland	[172]
Phaeodothis ligustri Sawada	*L. nokoensis*	Taiwan	[180]
Phoma lavitskii Zerova	*L. vulgare*	Russia	[181]
Phoma ligustrina Thüm.	*L. japonica*, *L. ibota*	France, USA	[182]
Phoma ligustri N.F. Buchw.	*L. ovalifolium*	Faröes (Denmark)	[183]
Phoma sp.	*L. robustum* subsp. *walkeri*	Sri Lanka	[24]
Pleospora ligustri Politis	*L. japonicum*	Greece	[184]
Rechingierella (*Zopfia*) *boudieri* (Arnaud) D. Hawksw.[5]	*L. vulgare*	France	[185]
Capnodiales			
Asterina (*Shivomyces*) *ligustricola* Hosag. & Kamar[5]	*L. travancoricum*	India	[186]
Cercospora (*Cercosporina*) *adusta* Heald & Wolf[5]	*L. amurense*, *L. californicum*, *L. ibolium*, *L. japonicum*, *L. lucidum*, *L. ovalifolium*	China, Korea, USA	[187, 188]
Cercospora ligustri Roum.	*L. japonicum*, *L. obtusifolium*, *L. vulgare*	France, China, USA	[171]
Cercospora ligustricola T.L. Tai	*L. lucidum*, *L. sinense*	China	[188]
Mycosphaerella sp. 1[6]	*L. robustum* subsp. *walkeri*	La Réunion	[24]
Mycosphaerella sp. 3[6]	*L. robustum* subsp. *walkeri*	Sri Lanka	[24]
[5]*Passalora* (*Cercospora*) *oleacearum* (Chidd.) U. Braun	*L. japonicum*, *L. nilgherrense*, *L. robustum s.l.*	China, India	[24, 189, 190][6]
Pseudocercospora ligustri Deighton	*L. japonicum*, *L. lucidum*, *L. obtusifolium*, *L. ovalifolium*, *L. sinense*	Korea, USA	[191]
Ramularia ligustrina Maubl.	*Ligustrum* sp.	France	[192]
Ramularia sp.[6]	*L. robustum* subsp. *walkeri*	Sri Lanka	[24]
Septoria japonica Thüm.	*L. japonicum*, *L. lucidum*	China, Italy	[182]
Septoria (*Mycosphaerella*) *ligustri* (Desm.) J.J. Kickx[5]	*L. vulgare*	Armenia, France, UK	[182, 193]
Septoria ligustrina Sacc.	*L. vulgare*	Austria	[194]
Sphaerulina sp.[6]	*L. robustum* subsp. *walkeri*	Sri Lanka	[24]
Magnaporthales			
Clasterosporium ligustri O.J. Rich.	*L. vulgare*	France	[195]
Glomerellales			
Colletotrichum aotearoa B. Weir & P.R. Johnst.	*L. lucidum*	New Zealand	[196]
Colletotrichum acutatum J.H. Simmonds[6]	*L. robustum* subsp. *walkeri*	Sri Lanka	[24]

Continued

Table 19.2. Continued

Classification[1] species	Host	Location[2]	Reference[3]
Colletotrichum ligustri Lobik	*L. vulgare*	Russia	[197]
Glomerella (*Colletotrichum gloeosporioides*) *cingulata* (Stoneman) Spauld. & H. Schrenk[5]	*L. lucidum*, *L. japonicum*, *L. ovalifolium*, *L. sinense*, *L. vulgare*	China, Hungary, Ukraine, USA	[198, 199]
Diaporthales			
Cytospora teheranica Esfand.	*L. vulgare*	Iran	[200]
Diaporthe ligustri Allesch.	*L. vulgare*	Denmark	[171]
Diaporthe ligustrina Petr.	*L. vulgare*	Czech Republic	[201]
Diplodina ligustri Delacr.	*L. vulgare*	France	[172]
Gnomonia cingulata Beck	*L. vulgare*	Austria	[202]
Leucostoma (*Valsa*) *auerswaldi* f. *ligustrina* (Nitschke) Höhn.[5]	*L. vulgare*	Germany	[203]
Mycocryptospora (*Cryptospora*) *anthostomoides* (Rehm) J. Reid & C. Booth[5]	*L. vulgare*	Germany	[204]
Phomopsis brachyceras Grove	*L. vulgare*	UK	[205]
Phomopsis ligustri-vulgaris Petr.	*L. vulgare*	Czech Republic, USA	[206]
Phomopsis spp. 1 and 2[6]	*L. robustum* subsp. *walkeri*	Sri Lanka	[24]
Schiffnerula (*Questieriella*) *pulchra* (Sacc.) Petr.[5]	*L. vulgare*	Italy	[207–209]
Valsa cypri (Tul.) Tul. & C. Tul.	*L. vulgare*	France, Germany, Italy	[207]
Microthyriales			
Cyclotheca kamatii T.S. Viswan. & Ananthan.	*L. nilgherrense*, *L. robustum s.l.*	India	[24, 210][6]
Botryosphaeriales			
Asteromella (*Phyllosticta*) *ligulariicola* (Melnik) Vanev & Aa[5,7]	*L. obtusifolium*	Japan	[211]
Camarosporium ligustri Berl. & Destrée	*L. vulgare*	Belgium	[212]
Diplodia ligustri Westend.	*L. vulgare*, *Ligustrum* sp.	Belgium, China, Italy	[182]
Diplodia ligustricola Girz.	*L. vulgare*	Russia	[213]
Fusicoccum ligustri Died.	*L. vulgare*	Germany	[214]
Guignardia sp.[6]	*L. robustum* subsp. *walkeri*	Sri Lanka	[24]
Macrophyllosticta ligustri Sousa da Câmara	*L. vulgare*	Portugal	[165]
Microdiplodia mamma Allesch.	*L. vulgare*	Germany	[215]
Microdiplodia ligustri Hollós	*L. vulgare*	Hungary	[197]

Continued

Table 19.2. Continued

Classification[1] species	Host	Location[2]	Reference[3]
Microdiplodia ligustricola Girz.	*L. vulgare*	Russia	[216]
Phyllosticta ibotae Nann.	*L. ibota*	Italy	[202]
Phyllosticta ligustri Sacc.	*L. japonicum, L. obtusifolium, L. quihoui, L. vulgare*	Armenia, China, Italy, Korea, Portugal, Ukraine	[170]
Phyllosticta ligustrina Sacc. & Speg.	*L. obtusifolium, L. vulgare*	Italy, Japan	[170]
Phyllosticta ovalifolii Brunaud	*L. ovalifolium*	France, USA	[195, 217]
Phyllosticta thuemenii Tassi	*L. japonicum*	Ukraine	[218]
Phyllosticta sp.[6]	*L. robustum* subsp. *walkeri*	Sri Lanka	[24]
Sphaeropsis ligustri N.F. Buchw.	*L. ovalifolium*	Faeröes (Denmark)	[183]
Zukaliopsis gardiniae Sawada	*Ligustrum* sp.	Taiwan	[219]
Ostropales			
Schizoxylon (Stictis) ligustri (Schwein.) Sherwood[5]	*Ligustrum* sp.	USA	[220, 221]
Patellariales			
Lecanidion clavisporum (Berk. & Broome) Sacc. & D. Sacc.	*Ligustrum* sp.	UK	[215]
Meliolales			
Irenopsis ligustri M.S. Patil & Maham.	*L. nilgherrense*	India	[222]
Meliola ligustri Hosag.	*L. robustum s.l.*	India	[223]
Erysiphales			
[5]*Erysiphe (Microsphaera) katumotoi* (U. Braun) U. Braun & S. Takam.[5]	*L. obtusifolium, L. ovalifolium*	Japan	[224]
Erysiphe (Microsphaera) ligustri (Homma) U. Braun & S. Takam.[5]	*L. obtusifolium, L. ovalifolium, L. tschonoskii*	Japan, Korea	[224]
Hypocreales			
Tubercularia ligustri Cooke	*Ligustrum* sp.	UK	[195]
Pezizales			
Camptomeris sp.[6]	*L. robustum s.l.*	India	[24]
Coleophoma sp.[6]	*L. robustum s.l.*	India	[24]
Pithya arethusa Velen.	*L. vulgare*	Czech Republic	[165]
Solicorynespora ligustri Jian Ma & X.G. Zhang	*L. quihoui*	China	[225]
Dothideales			
Dothidasteroma sp.[6]	*L. robustum s.l.*	India	[24]

Continued

Table 19.2. Continued

Classification[1] species	Host	Location[2]	Reference[3]
Pleurostromella ligustrina Petr.	*L. vulgare*	Czech Republic	[226]
Protothyrium sp.[6]	*L. robustum s.l.*	India	[24]
Chaetosphaeriales			
Zignoella ligustrina Pass.	*L. vulgare*	Italy	[172]
Helotiales			
Bispora antennata (*monilioides*) (Pers.) E.W. Mason[5]	*Ligustrum* sp.	Italy	[171, 227]
Durella ligustrina Velen.	*L. vulgare*	Czech Republic	[165]
Gloeosporium ligustri Henn.	*L. vulgare*	Brazil	[177]
Gloeosporium ligustrinum Sacc.	*L. perottetii*	France	[182]
Gloeosporium virgulatum G.F. Atk.	*L. vulgare*	USA	[172]
Hymenoscyphus (*Helotium*) *ligustri* (Velen.) Svrček[5]	*L. vulgare*	Czech Republic	[181, 228]
Myxofusicoccum ligustrinum Petr.[6]	*L. vulgare*	Czech Republic	[201]
Myxosporium cingulatum Beck	*L. vulgare*	Austria	[197]
Sclerotium ligustri Săvul. & Sandu	*L. vulgare*	Romania	[229]
Thedgonia ligustrina (Boerema) B. Sutton	*L. japonicum*, *L. lucidum*, *L. obtusifolium*, *L. ovalifolium*, *L. robustum* subsp. *walkeri*, *L. vulgare*	Austria, Bulgaria, France, Germany, India, Iran, Japan, The Netherlands, Poland, Romania, South Korea, Sri Lanka, La Réunion, UK	[24, 230–234][6]
Thedgonia sp. 1[6]	*L. robustum* subsp. *walkeri*	Sri Lanka	[24]
Thedgonia sp. 2[6]	*L. robustum s.l.*	India	[24]
Tympanis ligustri Tul.	*L. japonicum*, *L. vulgare*	France, Germany, UK	[235]

[1]Classification according to Kirk *et al.* [22].
[2]Country records, old names updated to modern geography.
[3]Oldest, most accessible reference cited first, and, when appropriate, followed by references updating the taxonomic status and/or new host records: in the Saccardo and Petrak references, the page number of the protologue is given for each species; similarly, for books and monographs, the page number for the protologue is given after the title.
[4]Authors of fungal names standardized and abbreviated according to Kirk and Ansell [236]; *Index Fungorum* (www.indexfungorum.org).
[5]Correct genus name, according to Kirk *et al.* [22]: in some cases, amended following the one fungus-one name policy [62, 63]; original genus or alternative name (asexual or sexual stage) in parentheses.
[6]Unpublished records from CABI surveys in India and Sri Lanka, 1997–2000, host from India listed as *L. robustum sensu lato* (*s.l.*); specimens identified to genus by CABI Bioscience Biosystematics and Molecular Biology Programme, species level not determined but most considered to be undescribed taxa.
[7]Taxonomic affiliation (Order) not confirmed – *incertae sedis* – genus assigned to most appropriate Order.

Fig. 19.1. *Ligustrum vulgare* in natural chalk grassland habitat, North Downs, Surrey, UK. (A) Defoliated canopies of common privet along chalk ridge; (B) leaf spots of *Thedgonia ligustrina*, causing heavy damage and leading to premature leaf fall; (C–D) detail of *Thedgonia* leaf spots with characteristic tan-orange, central areas of sporulation.

Table 19.3. Breakdown of number of species of invertebrate natural enemy recorded to feed on *Ligustrum* sp(p). by family and specificity (summarized from Table 19.1)

	Monophagous (restricted to *Ligustrum* sp(p).)	Oligophagous (restricted to Oleaceae)	Polyphagous (not restricted to Oleaceae)	Not clear[1]
Acari	2		1	
Coleoptera	2	7	11	1
Diptera		1		1
Hemiptera	4	1	25	12
Hymenoptera	1	1		
Lepidoptera	20	30	88	1
Orthoptera			1	
Thysanoptera			1	
Nematoda			2	
Total	29	40	131	15

[1]Includes species expected to have alternate hosts, or only identified to genus.

Table 19.4. Comparison of the likely monophagous invertebrate groups identified from the literature survey and the filed surveys

Order: family	Likely oligophagous species or guilds based on literature review	Actual or probably oligophagous species found in surveys
Acari: Eriophyiidae	Gall-forming mites	None
Coleoptera: Chrysomelidae	Larval leaf miners such as *Argopistes* spp.	*Argopistes coccinelliformis*
Coleoptera: Chrysomelidae	None	Leaf feeders, two *Hyphasis* spp.
Coleoptera: Chrysomelidae	Larval root feeders such as *Longitarsus* spp.	Two *Dermorhytis* spp.
Diptera: Cecidomyiidae	Gall formers such as *Asphondylia* spp.	None
Hemiptera: Tingidae	Lace bug cell sap suckers such as *Leptoypha* spp.	None
Lepidoptera: Crambidae	Leaf rollers and feeders such as *Palpita* spp.	*Palpita celsalis*
Lepidoptera: Erebidae	Leaf feeders such as *Pangrapta* spp.	*Pangrapta grisangula*
Lepidoptera: Geometridae	Leaf feeders such as *Cosmostola*, *Hemistola*, *Problepsis* spp.	*Problepsis deliaria*
Lepidoptera: Gracillariidae	Leaf miners such as *Caloptilia* spp.	None
Lepidoptera: Uraniidae	None	Leaf feeder, *Epiplema albida*

least closely related. Apart from *L. robustum* subsp. *walkeri*, one species of *Olea*, two species of *Chionanthus* and five species of *Jasminum* are indigenous in Sri Lanka [48]. In addition, cultivated varieties of *Jasminum* are commonly planted. Including these genera in the field surveys for natural enemies will give a quick indication as to whether a particular species is likely to be monophagous on *Ligustrum*.

Conclusions

Very many natural enemies of *Ligustrum* species exist in the native range and amongst these are likely to be potential biological control agents worthy of further study. It is hoped that the contents of this natural enemy review will assist in adding options to the future management of these troublesome and widely introduced weeds.

Acknowledgements

This literature survey was originally funded by La Région Réunion and facilitated by the Office National des Forêts. It was updated, expanded and prepared for publication with support from CABI. CABI is an international intergovernmental organization and we gratefully acknowledge the core financial support from our member countries (and lead agencies) including the UK (Department for International Development), China (Chinese Ministry of Agriculture), Australia (Australian Centre for International Agricultural Research), Canada (Agriculture and Agri-Food Canada), The Netherlands (Directorate-General for International Cooperation) and Switzerland (Swiss Agency for Development and Cooperation). See https://www.cabi.org/about-cabi/who-we-work-with/key-donors/ for details.

References

1. Blackman R, Eastop V. 2016. Aphids on the World's crops. An online identification and information guide. Available from: URL: http://www.aphidsonworldsplants.info/index.htm (last accessed May 2016).
2. Brown JW, Robinson G, Powell JA. 2008. Food plant database of the leafrollers of the world (Lepidoptera: Tortricidae) (Version 1.0). Available from: URL: http://www.tortricid.net/foodplants.asp.
3. Pastrana JA. Los Lepidópteros Argentinos sus Plantas Hospedoras y otros Substratos Alimenticios. South American Biological Control Laboratory USDA-ARS and Sociedad Entomológica Argentina, Tucuman, Argentina; 2004. p. xvi + 334.
4. Robinson GS, Ackery PR, Kitching IJ, Beccaloni GW, Hernández LM. Hostplants of the Moth and Butterfly Caterpillars of the Oriental Region. The Natural History Museum, London; 2001. p. 744.
5. Robinson GS, Ackery PR, Kitching IJ, Beccaloni GW, Hernández LM. HOSTS – A Database of the World's Lepidopteran Hostplants. Natural History Museum, London; 2010. Available from: URL: http://www.nhm.ac.uk/hosts.
6. Scoble MJ (editor). Geometrid Moths of the World: A Catalogue. CSIRO Publishing, Collingwood, Victoria, Australia; 1999, 2 vols. p. 1200.
7. Tietz HM. An Index to the Described Life Histories, Early Stages and Hosts of the Macrolepidoptera of the Continental United States and Canada. Allyn Museum of Entomology, Srasotoa, FL; 1972, 2 vols. p. iv + 1041.
8. Vacante V. The Handbook of Mites of Economic Plants: Identification, Bio-Ecology and Control. CABI, Wallingford, UK; 2016. p. xvii + 872.
9. Zhang BC. Index of Economically Important Lepidoptera. CABI, Wallingford, UK; 1994. p. 599.
10. Zheng H, Wu Y, Ding J, Binion D, Fu W, Reardon R. Invasive Plants of Asian Origin Established in the United States and Their Natural Enemies, Volume 1. 2nd ed. U.S. Department of Agriculture Forest Health Technology Enterprise Team, FHTET 2004–05, Morgantown, WV; 2006. p. vii + 147.
11. Zhang YZ, Hanula JL, Sun JH. Survey for potential insect biological control agents of *Ligustrum sinense* (Scrophulariales: Oleaceae) in China. Florida Entomologist 2008a;91(3):372–82.
12. CABI. 2018. Crop Protection Compendium.
13. Petrak F. List of New Species and Varieties of Fungi, New Combinations and New Names Published 1936–1939. Commonwealth Mycological Institute, Kew, UK; 1950. p. 117.
14. Saccardo PA, Trotter A. Sylloge Fungorum Omnium Hucusque Cognitarum. Johnson Reprint Corporation, New York; 1972. p. 766.
15. Farr DF, Rossman AY. 2017. Fungal Databases. US National Fungus Collections: ARS-USDA. Retrieved 25/07/2017. Available from: URL: https://ntars-grin.gov/fungaldatabases/.
16. The Plant List. 2013. The Plant List. Version 1.1. Available from: URL: http://www.theplantlist.org/ (accessed 10 April 2016).
17. Rider D. 2018. Pentatomoidea Home page. Available from: URL: https://www.ndsu.edu/pubweb/~rider/Pentatomoidea/index.htm (last accessed 9 January 2018).
18. Beccaloni G, Scoble M, Kitching I, Simonsen T, Robinson G, Pitkin B, *et al.* (editors) (2003) The Global Lepidoptera Names Index (LepIndex). Available from: URL: http://www.nhm.ac.uk/entomology/lepindex (accessed May 2016).
19. Gilligan TM, Baixeras J, Brown JW, Tuck KR. 2014. T@RTS: Online World Catalogue of the Tortricidae (Ver. 3.0). Available from: URL: http://www.tortricid.net/catalogue.asp.
20. Nuss M, Landry B, Mally R, Vegliante F, Tränkner A, Bauer F, *et al.* 2018. Global Information System on Pyraloidea. Available from: URL: http://www.pyraloidea.org (accessed January 2018).
21. Poole RW. Noctuidae. Lepidopterorum Catalogus. (New Series, edited by J.B. Heppner, Fasc. 118, 3 vols). E.J. Brill, Leiden, The Netherlands; 1989. p. 1314.
22. Kirk PM, Cannon PF, Minter DW, Stalpers JA. Dictionary of the Fungi. 10th ed. CAB International, Wallingford, UK; 2008.
23. Shaw RH. The Biological Control of *Ligustrum robustum* ssp. *walkeri* on the Island of La Réunion. Literature review (Project XB 1140). Unpublished report. CABI Bioscience, Ascot, UK; 1999. p. 30.
24. Shaw R, Djeddour D, Gassmann A (editors). 2001. The Biological Control of *Ligustrum robustum* subsp. *walkeri* on the Island of *Reunión*. Unpublished report. CABI Bioscience, Ascot, UK, 37 pp. + appendices.

25. Green PS. Oleaceae. In: Dassanayake MD, Clayton WD (editors). A Revised Handbook to the Flora of Ceylon, Volume 6. Amerind, New Delhi, India; 1988. p. 254–71.
26. Coates AM. Garden Shrubs and Their Histories. E.P. Dutton, New York; 1965.
27. Lorence D-H, Sussman RW. Exotic species invasion into Mauritius wet forest remnants. Journal of Tropical Ecology 1986;2:147–62.
28. Macdonald IAW, Thebaud C, Strahm WA, Strasberg D. Effects of alien plant invasions on native vegetation remnants on La Réunion (Mascarene Islands, Indian Ocean). Environmental Conservation 1991;18:51–61.
29. Cronk QCB, Fuller JL. Plant Invaders: The Threat to Natural Ecosystems. Chapman & Hall, London; 1995.
30. Swarbrick JT, Timmins SM, Bullen KM. The biology of Australian weeds. 36. *Ligustrum lucidum* Aiton and *Ligustrum sinense* Lour. Plant Protection Quarterly 1999;14(4):122–30.
31. Clout MN, Lowe S. Invasive species and environmental changes in New Zealand. In: Mooney HA, Hobbs RJ, editors. Invasive Species in a Changing World. Island Press, Washington, DC; 2000. p. 369–83.
32. Ehrenfeld JG. Exotic invasive species in urban wetlands: environmental correlates and implications for wetland management. Journal of Applied Ecology 2008;45:1160–9.
33. Hoyos LE, Gavier-Pizarro GI, Kuemmerle T, Bucher EH, Radeloff VC, Tecco PA. Invasion of glossy privet (*Ligustrum lucidum*) and native forest loss in the Sierras Chicas of Córdoba, Argentina. Biological Invasions 2010;12:3261–75.
34. Greene BT, Blossey B. Lost in the weeds: *Ligustrum sinense* reduces native plant growth and survival. Biological Invasions 2011;14:139–50.
35. Hanula JL, Horn S. Removing an invasive shrub (Chinese privet) increases native bee diversity and abundance in riparian forests of the southeastern United States. Insect Conservation and Diversity 2011;4:275–83.
36. Oliveira TWG, Milani JEF, Blum CT. Phenological behaviour of the invasive species *Ligustrum lucidum* in an urban forest fragment in Curitiba, Parana State, Brazil. Floresta 2014;46:371–8.
37. Connor HE. The Poisonous Plants in New Zealand. DSIR, Wellington, New Zealand; 1997.
38. Cariñanos P, Alcázar P, Galán C, Domínguez E. Privet pollen (*Ligustrum* sp.) as potential cause of pollinosis in the city of Cordoba, south-west Spain. Allergy 2002;57:92–7.
39. McFadyen REC. Biological control of weeds. Annual Review of Entomology 1998;43:369–93.
40. Evans HC. Biological control of weeds with fungi. In: Kempken F, editor. The Mycota XI, Agricultural Applications. 2nd ed. Springer Verlag, Berlin; 2013. p. 145–72.
41. Winston RL, Schwarzländer M, Hinz HL, Day MD, Cock MJW, Julien MH (editors). Biological Control of Weeds: AWorld Catalogue of Agents and Their Target Weeds. 5th ed. USDA Forest Service, Forest Health Technology Enterprise Team, Morgantown, West Virginia, USA; 2014. FHTET-2014–04. p. 838.
42. Sakalasooriya MB, Wijayagunasekara HNP, Marambe B, Shaw R. Screening of potential biocontrol agents for *Ligustrum robustum* subsp. *walkeri* in Sri Lanka for possible release in the island of La Réunion. Tropical Agricultural Research 2001;13:353–63.
43. Shaw RH, Cock MJW. Host specificity of *Epiplema albida*, a potential biological control agent for Sri Lankan privet in the Mascarene Islands. Insects 2017;8(77):11. doi: 10.3390/insects8030077.
44. Cock MJW, Shaw RH. The biology of *Epiplema albida* (Hampson) (Lepidoptera, Uraniidae, Epipleminae) on Sri Lankan privet, *Ligustrum robustum* subsp. *walkeri* (Oleaceae). Tropical Lepidoptera Research 2017;27(2):96–100.
45. Stevens PF. 2016. Angiosperm Phylogeny Website. 2001 onwards; version 12, July 2012 [and more or less continuously updated since]. Available from: URL: http://www.mobot.org/MOBOT/research/APweb/ (accessed 1 January 2016).
46. Wallander E, Albert VA. Phylogeny and classification of Oleaceae based on *rps16* and *trnL-F* sequence data. American Journal of Botany 2000;87:1827–41.
47. Green PS. *Ligustrum robustum* subsp. *walkeri* (Oleaceae). Kew Bulletin 1985;40:130.
48. Green PS. Synopsis of the Oleaceae from the Indian Sub-Continent. Kew Bulletin 2003;58(2):257–95.
49. Green PS. *Ligustrum* (Oleaceae) in Southern India. Kew Bulletin 1990;45:693–6.
50. Green PS. Taxonomic notes relating to *Ligustrum* (Oleaceae). Kew Bulletin 1995;50:379–86.
51. Qin XK. A new system of *Ligustrum* (Oleaceae). Acta Botanica Yunnanica 2009;31(2):97–116. [In Chinese with English abstract].

52. Vaughan RE, Wiehe PO. Studies on the vegetation of Mauritius I. A preliminary survey of the plant communities. Journal of Ecology 1937;25:314–43.
53. Rouillard G, Guého J. Le Jardin Botanique de Curepipe. Curepipe Municipality, Mauritius; 1990. p. 49.
54. Lorence D-H, Sussman RW. Diversity, density, and invasion in a Mauritian wet forest. Monographs of the Systematics of the Missouri Botanical Garden 1988;25:187–204.
55. Lavergne C, Rameau JC, Figier J. The invasive woody weed *Ligustrum robustum* subsp. *walkeri* threatens native forests on La Réunion. Biological Invasions 1999;1(4):377–92.
56. Shaw RH, Milne RI. The use of molecular techniques in the classical biological control programme against an invasive *Ligustrum* species in La Réunion. [Abstract]. In: Spencer NR, editor. Proceedings of the X International Symposium on Biological Control of Weeds, July 4–14. Montana State University, Bozeman, Montana, USA; 1999. p. 303.
57. Milne RI, Abbott RJ. Geographic origin and taxonomic status of the invasive privet, *Ligustrum robustum* (Oleaceae), in the Mascarene Islands, determined by chloroplast DNA and RAPDs. Heredity 2004;92:78–87.
58. Gaskin JF, Bon M-C, Cock MJW, Cristofaro M, De Biase A, De Clerck-Floate R, *et al.* Applying molecular-based approaches to classical biological control of weeds. Biological Control 2011;58:1–21.
59. Sakalasooriya MB, Wijayagunasekara HNP, Marambe B, Shaw R. Identification of potential biological control agents for *Ligustrum robustum* ssp. *walkeri*. Tropical Agricultural Research 2000;12:205–16.
60. Cock MJW, Shaw RH, Blackman RL. On the biology of *Ceratopemphigus zehntneri* Schouteden (Hemiptera: Pemphigidae), a gall forming aphid on *Ligustrum robustum* ssp. *walkeri* (Oleaceae), in Sri Lanka. Zootaxa 2010;2614:46–52.
61. Lemaire C. The Saturniidae of America. Les Saturniidae Américains. Hemileucinae. 3 vols. Goecke & Evers, Keltern, Germany; 2002. Part A 1–688, Part B 689–1388, Part C plates 1–126, ES1–ES14.
62. Hawksworth DL. A new Dawn for the naming of fungi. MycoKeys 2011;1:7–20.
63. Gams W. Recent changes in fungal nomenclature and their impact on naming of microfungi. In: Li D-W, editor. Biology of Microfungi. Springer International Publishing, Switzerland; 2016. p. 7–23.
64. Zhang YZ, Hanula JL, Sun JH. Host specificity of *Argopistes tsekooni* (Coleoptera: Chrysomelidae), a potential biological control agent of Chinese privet. Journal of Economic Entomology 2008b;101(6):1146–51.
65. Yu L, Qian X, Li J. Biological characteristics and control of *Argopistes tsekooni*. Chinese Bulletin of Entomology 2009;46(6):906–10. [In Chinese with English summary].
66. Zhang YZ, Sun JH, Hanula JL. Biology and life history of *Argopistes tsekooni* (Coleoptera: Chrysomelidae) in China, a promising biological control agent of Chinese privet. Annals of the Entomological Society of America 2009;102(3):508–16.
67. Inoue TS. Host plant preference of fleabeetle *Argopistes biplagiatus* Motschulsky adults (Coleoptera: Chrysomelidae). Applied Entomology and Zoology 1991;26(3):403–6.
68. Inoue T, Shinkaji N. Studies on the life history of the fleabeetle, *Argopistes coccinelliformis* Csiki (Coleoptera: Chrysomelidae). II. Developmental ability on various host trees. Japanese Journal of Applied Entomology and Zoology 1989;33(4):223–30. [In Japanese with English summary].
69. Jacoby M. The Fauna of British India, Including Ceylon and Burma. Coleoptera. Chrysomelidae, Volume I. Taylor & Francis, London; 1908. p. xx + 534, 2 plates.
70. Pei YH, Li HM, Kong F, Li YM, Kang P, Sun XG. The morphology and behavior of *Longitarsus bimaculatus*. Chinese Bulletin of Entomology 2008;45(3):401–4.
71. Sawada Y. A systematic study of the family Rhynchitidae of Japan (Coleoptera, Curculionidae). Humans and Nature 1993;2(1):1–93.
72. Cmoluch Z, Letowski J, Smardzewska Z. A contribution to the knowledge of the weevils (Coleoptera, Curculionidae) of Poland. II. Annales Universitatis Mariae Curie Sklodowska, C Biologia 1978;33:405–9.
73. Baker JR. Insect and Related Pests of Flowers and Foliage Plants: Some Important, Common, and Potential Pests in the Southeastern United States. North Carolina Cooperative Extension Service, North Carolina, USA; 1994. p. v + 106.
74. Cuda JP, Zeller MC. First record of *Ochyromera ligustri* (Coleoptera: Curculionidae) from Chinese privet in Florida. Florida Entomologist 1998;81(4):582–4.
75. Heijerman T, Drost MBP. *Otiorhynchus aurifer*, een Zuid-Europese snuitkever ingeburgerd in Nederland (Coleoptera: Curculionidae). [*Otiorhynchus aurifer*, a South European weevil established

in the Netherlands (Coleoptera: Curculionidae).]. Entomologische Berichten 2000;60(5):84–8. [In Dutch with English summary].

76. Heikertinger F. *Otiorrhynchus crataegi* Germ, und *mastix* Ol., zwei Zierstrauchschädlinge derWiener Gärten. [*Otiorrhynchus crataegi* Germ. and *O. mastix* Ol., two pests of ornamental shrubs in the gardens of Vienna]. Verhandlungen der Zoologisch-botanischen Gesellschaft in Wien 1924;73:119–28. [In German].
77. Alford DV. Plant Pests. A Natural History of Pests of Farms and Gardens. Collins, London; 2011. p. x + 500.
78. Heijerman T, Hellingman S. *Otiorhynchus meridionalis*, een nieuwe invasieve snuitkeversoort voor de fauna van Nederland (Coleoptera: Curculionidae). [*Otiorhynchus meridionalis*, an invasive weevil species new to the fauna of The Netherlands (Coleoptera: Curculionidae).]. Entomologische Berichten 2009;69(3):95–100. [In Dutch with English summary].
79. Lengerken H. Lebensweise und Entwicklung des Fliederschäd-lings *Otiorrhynchus rotundalus*, Siebold. [The life-history and development of the lilac pest, *Otiorrhynchus rotundalus*, Siebold.]. Zeitschrift fur Angewandte Entomologie 1919;5(1):67–83. [In German].
80. Petryszak B. Notes on *Otiorhynchus rotundatus* Siebold, 1837 (Col., Curculionidae). Przeglad Zoologiczny 1979;23(4):327–9. [In Polish with English summary].
81. Balalaikins M, Bukejs A. *Otiorhynchus smreczynskii* (Coleoptera: Curculionidae) – a new to Estonia and Lithuania weevil species with notes on its occurrence and bionomy in the Eastern Baltic region. Acta Zoologica Lituanica 2011;21(4):263–7.
82. Buchanan LL. The Japanese weevil, *Pseudocneorhinus bifasciatus* Roelofs, in America (*P. setosus* of American records, not Roelofs) (Coleoptera, Curculionidae). Bulletin of the Brooklyn Entomological Society 1946;41(4):143.
83. Mondera A, Priore V, Saracino V, Monaco F. *Lytta vesicatoria L.*: serious deflorations in young olive-groves of Medio Ionio Cosentino. Informatore Fitopatologico 1995;45(11):35–7.
84. Giménez RA, Etiennot AE. Host range of *Platypus mutatus* (Chapuis, 1865) (Coleoptera: Platypodidae). Entomotropica 2003;18(2):89–94.
85. Uechi N, Yukawa J. Host range and life history of *Asphondylia sphaera* (Diptera: Cecidomyiidae): use of short-term alternate hosts. Annals of the Entomological Society of America 2006;99(6): 1165–71.
86. Tokuda M, Harris KM, Yukawa J. Morphological features and molecular phylogeny of *Placochela* Rübsaamen (Diptera: Cecidomyiidae) with implications for taxonomy and host specificity. Entomological Science 2005;8(4):419–27.
87. Zheng FK. The bionomics of *Prociphilus ligustrifoliae* (Tseng et Tao) (Homoptera: Aphididae) and control measures. Insect Knowledge Kunchong Zhishi 1981;18(3):110–2.
88. Lodos N, Kalkandelen A. Preliminary list of Auchenorrhyncha with notes on distribution and importance of species in Turkey. XX. Family Cicadellidae: Deltocephalinae; Fieberiellini, Stirelini and Tetartostylini. Turkiye Bitki Koruma Dergisi 1986;10(1):25–32.
89. Başpinar H, Yıldırım EM, Xing JC. Determination and population fluctuations of Cicadellidae (Hemiptera: Cicadomorpha) species in pomegranate orchards in Aydin Province, Turkey. Türkiye Entomoloji Dergisi 2013;37(1):3–11.
90. Cen MJ, Ji HX. A preliminary report on *Ericerus pela* Chavannes in Guangxi. Insect Knowledge 1988;25:230–2. [In Chinese].
91. McConnell HS. *Parlatoria oleae* Colv., a pest of privet in Maryland. Journal of Economic Entomology 1930;23(1):142–4.
92. Bobb ML, Weidhaas JA, Ponton LF. White peach scale: life history and control studies. Journal of Economic Entomology 1973;66:1290–2.
93. Grozea I, Gogan A, Vlad M, Virteiu AM, Stef R, Carabet A, *et al.* A new problem for western Romania: *Metcalfa pruinosa* Emiptera: Flatidae. Bulletin of University of Agricultural Sciences and Veterinary Medicine Cluj-Napoca. Horticulture 2015;72(1):74–80.
94. Grazia J, del Vecchio MC, Hildebrand R. Estudo das ninfas de pentatomídeos (Heteroptera) que vivem sobre soja (Glycine max (L.) Merrill): IV – *Acrosternum impicticorne* (Stål, 1872). [Study of the nymphs of pentatomids (Heteroptera) that live on soyabean (Glycine max (L.) Merrill): IV – *Acrosternum impicticorne* (Stål, 1872).] Anais da Sociedade Entomológica do Brasil 1982;11(2):261–8.
95. Panizzi AR, Grazia J. Stink bugs (Heteroptera, Pentatomidae) and an unique host plant in the Brazilian subtropics. Iheringia, Série Zoologia 2001;90:21–35.

96. Irulandi S, Kumar PKV, Sreedharan K. Observations on the incidence and biology of the pentatomid bug, *Antestiopsis cruciata* (Fabricius) (Hemiptera: Pentatomidae) on coffee. Journal of Coffee Research 2003;31(1):95–8.
97. Schaefer CW, Panizzi AR. Heteroptera of Economic Importance. CRC Press, Boca Raton, FL, USA; 2000. p. 856.
98. Zhang Y, Hanula JL, Horn S, Braman SK, Sun JH. Biology of *Leptoypha hospita* (Hemiptera: Tingidae), a potential biological control agent of Chinese privet. Annals of the Entomological Society of America 2011;104(6):1327–33.
99. Zhang YZ, Hanula JL, O'Brien J, Horn S, Braman K, Sun JH. Evaluation of the impacts of herbivory by lace bugs on Chinese privet (*Ligustrum sinense*) survival and physiology. Biological Control 2013;64(3):299–304.
100. Guilbert E. Western Pacific Tingidae (Heteroptera): new species and new records. Memoirs of the Queensland Museum 2001;46(2):543–54.
101. She DS, Feng FJ. Study on biological properties of *Macrophya pilotheca*. Journal of Zhejiang Forestry Science and Technology 2010;30(6):59–61. [In Chinese with English summary].
102. Hoebeke ER, Johnson WT. A European privet sawfly, *Macrophya punctumalbum* (L.): North American distribution, host plants, seasonal history and descriptions of the immature stages (Hymenoptera: Tenthredinidae). Proceedings of the Entomological Society of Washington 1985;87(1): 25–33.
103. Piekarska-Boniecka H, Kadłubowski W, Siatkowski I. A study of bionomy of the privet sawfly (*Macrophya punctumalbum* (L.)) (Hymenoptera, Tenthredinidae) – a pest of park plants. Acta Scientiarum Polonorum – Hortorum Cultus 2008;7(1):3–11.
104. Sevastopulo DG. On the food-plants of Indian Bombyces (Heterocera). Journal of the Bombay Natural History Society 1940;41:817–27.
105. Konno K, Okada S, Hirayama C. Selective secretion of free glycine, a neutralizer against a plant defense chemical, in the digestive juice of the privet moth larvae. Journal of Insect Physiology 2001;47(12):1451–7.
106. Emmet AM (editor). A Field Guide to the Smaller British Lepidoptera. British Entomological and Natural History Society, London; 1979. p. 271.
107. Mally R, Nuss M. Phylogeny and nomenclature of the box tree moth, *Cydalima perspectalis* (Walker, 1859) comb. n., which was recently introduced into Europe (Lepidoptera: Pyraloidea: Crambidae: Spilomelinae). European Journal of Entomology 2010;107:393–400.
108. Lin ZG, Lei YL, Guo LZ. Bionomics of a new pest, *Palpita annulata* on *Ligustrum quihoui*. Chinese Bulletin of Entomology 2005;42(1):74–7. [In Chinese with English summary].
109. Gotoh T, Hagino Y, Doke K. Life cycle of the lilac pyralid *Palpita nigropunctalis* (Bremer) (Lepidoptera: Crambidae) on five oleaceous tree species. Journal of Asia-Pacific Entomology 2011;14:195–200.
110. Holland WJ. The Moth Book: A Guide to the Moths of North America. Doubleday; 1903. p. xxiv + 479.
111. Vassilaina-Alexopoulou P, Santorini AP. Some data on the biology of *Palpita unionalis* Hubner (Lepidoptera: Pyralidae), under laboratory conditions. Annales de l'Institut Phytopathologique Benaki 1973;10(4):320–6.
112. Hayward KJ. Datos para el estudio de la ontogenia de lepidópteros argentines. Miscelánea Instituto Migual Lillo, Universidad Nacional de Tucumán 1969;31:1–142.
113. Porter J. The Colour Identification Guide to Caterpillars of the British Isles. Viking, London; 1997. p. xii + 275.
114. Wysoki M. A bibliography of the giant looper, Boarmia (*Ascotis*) *selenaria* Schiffermuller, 1775 (Lepidoptera: Geometridae) for the years 1913–1981. Phytoparasitica 1982;10(1):65–70.
115. Holloway JD. The moths of Borneo: family Geometridae, subfamily Ennominae. Malayan Nature Journal 1993;47:1–309.
116. Hu LX, Chen Y, He ZS, Zou ZW, Xia B. Demographic analysis of the fitness of *Problepsis superans* (Lepidoptera: Geometridae) feeding on three *Ligustrum* (Lamiales: Oleaceae) species. Journal of Economic Entomology 2014;107(3):1045–54.
117. Robineau R. (coordinator) Guide des Papillons Nocturnes de France. Delachaux & Niestlé, Paris; 2007. p. 288.
118. Kumata T. A taxonomic revision of the *Gracillaria* group occurring in Japan (Lepidoptera: Gracillariidae). Insect Matsumurana, new series 1982;26:1–186.

119. Kumata T. Japanese species of the subfamily Oecophyllembiinae Réal et Balachowsky (Lepidoptera: Gracillariidae), with descriptions of a new genus and eight new species. Insecta Matsumurana, new series 1998;54:77–131.
120. Kobayashi S, Huang GH, Nakamura A, Hirowatari T. Four new species of Gracillariidae (Lepidoptera) from China and Japan, and description of the pupal morphology of the genera *Corythoxestis*, *Eumetriochroa*, *Guttigera*, and *Metriochroa*. Zootaxa 2013;3619(2):101–29.
121. Ayabe Y, Minoura T, Hijii N. Plasticity in resource use by the leafminer moth *Phyllocnistis* sp. in response to variations in host plant resources over space and time. Journal of Forest Research 2015;20(1):213–21.
122. Scott JA. The Butterflies of North America: A Natural History and Field Guide. Stanford University Press, Stanford, CA; 1986. p. xiii + 583.
123. Verhulst G. The rearing of *Laeosopis roboris* Esper, (1793). SHILAP Revista De Lepidopterologia 1986;14(54):43–6.
124. Holloway JD. The moths of Borneo: family Noctuidae, trifine subfamilies: Noctuinae etc. Malayan Nature Journal 1989;42:57–226.
125. Sevastopulo DG. On the food-plants of Indian Agaristidae and Noctuidae (Heterocera). Journal of the Bombay Natural History Society 1941;42(1–2):421–30.
126. Graves SD, Shapiro AM. Exotics as host plants of the California butterfly fauna. Biological Conservation 2003;110:413–33.
127. Davis DR. Bagworm moths of the western Hemisphere. Bulletin of the United States National Museum 1964;244:233.
128. Bourquin F. Mariposas Argentinas. The author, Buenos Aires, Argentina; 1945. p. 209.
129. Rizzo HF. Catálogo de lepidópteros hallados en la Facultad de Agronomia y Veterinaria de Buenos Aires. Universidad de Buenos Aires Facultad de Agronomia y Veterinaria, Publicación Interna 1971;2:1–35.
130. Schreiter R. Notas entomo-biológicas y otras. Preparados para su publicación por Kenneth J. Hayward. Acta Zoologica Lilloana 1943;1:7–44.
131. Passoa VA, Passoa S. Distribution and hostplant records for *Eupackardia calleta* from southeastern Texas with notes on mandibular morphology of Attacini (Saturniidae). Journal of the Lepidopterists' Society 1999;53(4):133–7.
132. Rizzo HF. Redescriptión y biologia de *Rothschildia jacobaeae* (Walker) (Lepidoptera, Saturniidae). Revista de la Socieded Entomológia Argentina 1965;28:17–32.
133. Appleby JE. Observations on the life history and control of the lilac borer. Journal of Economic Entomology 1973;66:248–9.
134. Brown LN, Mizell III RF. The clearwing borers of Florida (Lepidoptera: Sesiidae). Tropical Lepidoptera 1993;4(4):1–21.
135. Pittaway AR. 2016. Sphingidae of the Western Palaearctic (including Europe, North Africa, the Middle East, western Siberia and western Central Asia). Available from: URL: http://tpittaway.tripod.com/sphinx/list.htm (last accessed 11 April 2016).
136. Tuttle JP. The Hawk Moths of North America. Wedge Entomological Research Foundation, Washington, DC; 2007. p. xviii + 253.
137. Pittaway AR, Kitching IJ. 2016. Sphingidae of the Eastern Palaearctic (including Siberia, the Russian Far East, Mongolia, China, Taiwan, the Korean Peninsula and Japan). Available from: URL: http://tpittaway.tripod.com/china/china.htm (accessed 11 April 2016).
138. Holloway JD. The Moths of Borneo: Superfamily Bombycoidea: Families Lasiocampidae. Eupterotidae, Bombycidae, Brahmaeidae, Saturniidae, Sphingidae. Southdene Sdn. Bhd., Kuala Lumpur, Malaysia; 1983. p. 199.
139. Bradley JD, Tremewan WG, Smith A. British Tortricoid Moths. Tortricidae: Olethreutinae. The Ray Society, London; 1979. p. viii + 336.
140. Li CS. A study of the privet leafroller (*Archips compitalis* Razowski) (Lepidoptera: Tortricidae), a newly recorded predator of the Chinese wax-insect (*Ericerus pela* (Chav.)) also a newly recorded pest of the privet (*Ligustrum lucidum* Ait.). Journal of the Agricultural Association of China 1995;1995(172):93–108.
141. Wiech K, Pniak M. Parazytoidy Gasienic zwójek wystepujacych na żywopłotach ligustrowych. [Parasitoids of leaf roller caterpillars occurring on common privet hedgerows.] Progress in Plant Protection 2002;42(2):464–6. [In Polish with English summary].

142. Schaffner JV. Microlepidoptera and their parasites reared from field collections in the northeastern United States. USDA, Miscellaneous no 767 1959; 97 pp.
143. Brown JW, Passoa S. Larval foodplants of Euliini (Lepidoptera: Tortricidae): from *Abies* to *Vitis*. Pan-Pacific Entomologist 1998;74(1):1–11.
144. Bentancourt CM, Scatoni IB, Gonzalez A, Franco J. Biology of *Bonagota cranaodes* (Meyrick) (Lepidoptera: Tortricidae) on seven natural foods. Neotropical Entomology 2004;33 (3):299–306.
145. Rocca M, Brown JW. New host records for four species of tortricid moths (Lepidoptera: Tortricidae) on cultivated blueberries, *Vaccinium corymbosum* (Ericaceae), in Argentina. Proceedings of the Entomological Society of Washington 2013;115(2):167–72.
146. Bradley JD, Tremewan WG, Smith A. British Tortricoid Moths. Cochylidae and Tortricidae: Tortricinae. The Ray Society, London; 1973. p. viii + 251.
147. Powell JA. Records of the Palearctic tortricid, *Clepsis consimilana*, in the Pacific northwest: can an urban moth be overlooked for half a century? Pan-Pacific Entomologist 1988;64(1):98–9.
148. Yasuda T. The Tortricinae and Sparganothinae of Japan (Lepidoptera: Tortricidae) (Part II). Bulletin of the University of Osaka Prefecture (series B) 1975;27:79–251.
149. She DS, Feng FJ. Biological characteristics of blackhead fire-worm infesting *Ligustrum sinense*. Scientia Silvae Sinicae 2008;44(5):75–8. [In Chinese with English summary].
150. Lewis JA, Sohn JC. Lepidoptera: Yponomeutidea I (Argyresthiidae, Attevidae, Praydidae, Scythropiidae and Yponomeutidae). Brill, Leiden, The Netherlands; 2015. p. xxiv + 253.
151. Bellman H. A Field Guide to the Grasshoppers and Crickets of Britain and Northern Europe. Collins, London; 1985. p. 213.
152. Cary FH, DanaharGW, Morris RJ. An investigation of speckled bush crickets *Leptophyes punctatissima* (Bosc.) (Orthoptera: Tettigoniidae) found in privet bushes. Entomologist 1995;114:70–8.
153. Orosz S, Jenser G, Reider-Sally K. Környezeti tényezők hatása a fagyaltripsz (*Dendrothrips ornatus* Jablonowski) populációdinamikájára. [The effect of environmental factors on the population dynamics of privet thrips (*Dendrothrips ornatus* Jablonowski).] Növényvédelem 2002;38(2):61–6. [In Hungarian with English summary].
154. Ye JL, She DS. Biological characteristics of *Dendrothrips ornatus* and test on control by pesticide. Journal of Zhejiang Forestry Science and Technology 2010;30(4):76–8. [In Chinese with English summary].
155. Mead FW. Bureau of nematology. Tri-Ology Technical Report 1989;28(10):5–6.
156. Lehman P (editor). Bureau of nematology – detections of special interest. Tri-ology Technical Report 1990;29(3):1–2.
157. de Freina JJ. Eine neue population von *Brahmaea ledereri* Rogenhofer, 1873, (*Brahmaea ledereri zaba* ssp.n.) in Türkisch-Kurdistan, sowie Nachweis der Konspezifität von *Brahmaea ledereri* Rogenhofer, 1873, und *Brahmaea christophi* Staudinger, 1879 (Lepidoptera, Brahmaeidae). Entomofauna 1982;3(9):129–39.
158. Saccardo PA. 1888. Sylloge Fungorum 7, 806, 865.
159. Cummins GB. Uredinales of continental China collected by S.Y. Cheo II. Mycologia 1951;43:78–98.
160. Tranzschel W. Conspectus Uredinalium, p. 303. Academy of Sciences USSR, Moscow; 1939. p. 426.
161. Dietel P. Einige Uredineen aus Ostasien. Hedwigia 1898;37:212–43.
162. Ji J-X, Li Z, Wang Q, Li Y, Kakishima M. Life cycle of *Aecidium klugkistianum* on *Ligustrum* and its new combination, *Puccinia klugkistiana*. Mycoscience 2017;58:307–11.
163. Saccardo PA. 1898. Sylloge Fungorum 14, 311.
164. Morimoto Y. Notes on rust fungi collected from Amami-Ohshima. Japanese Journal of Botany 1956;31:37–41.
165. Petrak F. 1932–35. Petrak's List 7, 812, 870, 924, 972.
166. Martin PMD. Studies in the Xylariaceae: supplementary note. South African Journal of Botany 1976;42:71–83.
167. Kirschstein W. Über neue, seltene und kritische Kleinpilze. Hedwigia 1944;81:193–224.
168. Saccardo PA. 1891. Sylloge Fungorum 9, 469.
169. Nag Raj TR. Coelomycetous Anamorphs with Appendage-Bearing Conidia, p. 895. Mycologue Publications, Waterloo, Canada; 1993. p. 1101.
170. Saccardo PA. Fungi Veneti novi vel critici vel mycologiae, Venetae addendi. Micheli 1878;1:133–221.
171. Saccardo PA. 1886. Sylloge Fungorum 4, 330, 471, 722.
172. Saccardo PA. 1895. Sylloge Fungorum 11, 332, 339, 515, 527, 565.
173. Wei CT. Notes on *Corynespora*. Mycological Papers 1950;34:1–10.

174. Miller JN, Alfieri SA. Leaf spot of *Ligustrum sinense* caused by *Corynespora cassiicola* and its control. Phytopathology 1974;64:255–6.
175. Guo YL. Four new species of *Corynespora*. Acta Mycologica Sinica 1984;3:161–9.
176. Fabre JH. Essai sur les Sphériacées du Department de Vaucluse. Annales des Sciences Naturelles Botanique 1879;9:66–118.
177. Hennings P, Fungi S. Paulensis a cl. Puttemans collecti. Hedwigia 1902;41:295–311.
178. Zhang M, Zhang TY. Taxonomic studies of *Helminthosporium* from China 4. Six new species and a key to *Helminthosporium* from China. Mycotaxon 2009;109:399–413.
179. Merezhko TO. New taxonomic combinations in the family Sphaerioidaceae Sacc. Ukrainian Botanical Journal 1979;36:366–8.
180. Sawada K. 1943. Descriptive Catalogue of the Formosan Fungi, Part VIII. Report of the Department of Agriculture Government Research Institute of Formosa 85, p. 28, 130 pp.
181. Petrak F. 1936–39. Petrak's List 8, 42, 76.
182. Saccardo PA. 1884. Sylloge Fungorum 3, 116, 347, 497, 704.
183. Moller FH. Fungi of the Faröes, Part II. p. 152. Munksgaad, Copenhagen; 1958. p. 286.
184. Politis JC. 1935. Contribution à l'étude des champignons de l'Attique. Pragmateiai tés Akademie Athenon 4, p. 17, 44 pp.
185. Hawksworth DL. Ascospore sculpturing and generic concepts in the Testudinaceae (Zopfiaceae). Canadian Journal of Botany 1979;57:91–9.
186. Hosagoudar VB. Studies on foliicolous fungi – VII. Journal of Economic & Taxonomic Botany 2004;28:187–95.
187. Heald FD, Wolf FA. New species of Texas fungi. Mycologia 1911;3:5–22.
188. Chupp CD. A Monograph of the Fungus Genus *Cercospora*, p. 471. Privately Printed, New York, 1954. p. 667.
189. Chiddawar PP. Contribution to our knowledge of the Cercosporae of Bombay State II. Indian Phytopathology 1960;12:111–21.
190. Braun U. Taxonomic notes on species of the *Cercospora* complex. Mycotaxon 1993;48:275–98.
191. Deighton FC. Studies on *Cercospora* and allied genera. VI. *Pseudocercospora* Speg., *Pantospora* Cif., and *Cercoseptoria* Petr. Mycological Papers 1976;140(78):168.
192. Maublanc A. Sur quelques espèces nouvelles ou peu connués de champignons infériers. Bulletin de la Société Mycologique de France 1906;22:63–70.
193. Kickx JJ. Flore Cryptogamique des Flandres, p. 354. Librairie Générale, Ghent, Belgium; 1867. p. 521.
194. Saccardo PA. Notae mycologicae series XVIII. Annales Mycologici 1914;12, 282–314.
195. Saccardo PA. 1892. Sylloge Fungorum 10, 115, 607, 702.
196. Weir BS, Johnston PR, Damm U. The *Colletotrichum gloeosporioides* species complex. Studies in Mycology 2012;73:115–80.
197. Petrak F. 1922–28. Petrak's List 3, 65:213, 231.
198. Vajna L, Bagyinka T. Occurrence of privet anthracnose in Hungary caused by *Glomerella cingulata*. New Disease Reports 2004;4:22.
199. Shen J, Dong LK, Wang ZH, Yu JY, Zou FL. First report of anthracnose caused by *Colletotrichum gloeosporioides* on *Ligustrum japonicum* in China. Plant Disease 2017;101:1329.
200. Esfandiari E. Neue Iranische Pilze. Sydowia 1951;5:366–70.
201. Petrak F. Pilzflora von Mähren und Osterr.-Schlesien. Annales Mycologici 1916;14:159–76.
202. Saccardo PA. 1927. Sylloge Fungorum 26, 321, 979.
203. von Höhnel F. Über die *Cytospora-Arten* auf Salix in Europa. Mitteilungen des Botanischen Instituts der Technischen Hochschule Wien 1928;5:54–8.
204. Reid J, Booth C. Winterella, the correct name for *Cryptospora* and *Ophiovalsa*. Canadian Journal of Botany 1987;65:1320–42.
205. Grove WB. British Stem and Leaf Fungi (Coelomycetes) 1, p. 196. Cambridge University Press, Cambridge; 1935. p. 488.
206. Petrak F. Mykologische Notizen. Annales Mycologici 1921;19:17–128.
207. Saccardo PA. 1882. Sylloge Fungorum 1, 133.
208. Petrak F. Uber Englerula und die Englerulaceen. Annales Mycologici 1928;26:385–413.
209. Hughes SJ. Five species of *Sarcinella* from North America, with notes on *Questieriella* n. gen., *Mitteriella, Endophragmiopsis, Schiffnerula*, and *Clypeolella*. Canadian Journal of Botany 1983;61:1727–67.

210. Viswanathan TS, Ananthanarayanan S. *Cyclotheca kamatii* sp. nov. – a new record to India. Mycopathologia Mycologia Applicata 1966;29:373–6.
211. Aa HA, Vanev S. A Revision of the Species Described in *Phyllosticta*, p. 289. CBS, Utrecht; 2002. p. 510.
212. Saccardo PA. 1899. Sylloge Fungorum 15, 965.
213. Petrak F. 1930. Petrak's List 5, 203.
214. Diedicke H. Neue seltene Pilze aus Thüringen. Annales Mycologici 1904;2:511–4.
215. Saccardo PA. 1906. Sylloge Fungorum 18, 184, 327.
216. Petrak F. 1929. Petrak's List 4, 208.
217. Brunaud P. Miscellanées mycologiques. Actes de la Société Linnéenne de Bordeaux 1888;42:85–10.
218. Tassi F. I generi *Phyllosticta* Pers., Phoma Fr., *Macrophoma* (Sacc.) Berl. & Voglino e i loro generi analoghi, giusta la legge d analogia. Bollettino del Laboratorio de Orto Botanico Reale Universita Siena 1902;5:1–76.
219. Sawada K. 1931. Descriptive catalogue of the Formosan fungi V. Report of the Department of Agriculture Government Research Institute of Formosa 51, p. 23, 131 pp. 220. von Schweinitz LD. Synopsis fungorum in America boreali media degentium. Transactions of the American Philosophical Society 1832;4:141–316.
221. Sherwood MA. The Ostropalean fungi. Mycotaxon 1977;5:1–277.
222. Patil MS, Mahamulkar SH. Studies on Meliolaceae of India – IV. Indian Phytopathology 1999;52:245–53.
223. Hosagoudar VB, Goos RD. Meliolaceous fungi from the state of Kerala, India II. The genus *Meliola*. Mycotaxon 1990;37:217–72.
224. Braun U, Takamatsu S. Phylogeny of *Erysiphe, Microsphaera, Uncinula (Erysipheae)* and *Cystotheca, Podosphaera, Sphaerotheca* (Cystotheceae) inferred from rDNA ITS sequences – some taxonomic consequences. Schlechtendalia 2000;4:1–33.
225. Ma J, Ma LG, Zhang YD, Xia JW, Zhang XG. New species and a new record of *Solicorynespora* from southern China. Mycotaxon 2012;119:95–102.
226. Petrak F. Mykologische Notizen. VI. Annales Mycologici 1923;21:182–335.
227. Hughes SJ. Conidiophores, conidia and classification. Canadian Journal of Botany 1953;31:577–659.
228. Svrček M. A taxonomic revision of inoperculate discomycetes described by J. Velenovsky in the genus *Helotium*, preserved in National Museum, Prague. Acta Musei Nationalis Pragae 1984;40B:129–215.
229. Săvulescu T, Sandu-Ville C. Beitrag zur Kenntnis der Micromyceten Rumäniens. Hedwigia 1935;75:159–233.
230. Sutton BC. Some Hyphomycetes with holoblastic sympodial conidiogenous cells. Transactions of the British Mycological Society 1973;61:417–29.
231. Yoshikawa M, Yokoyama T. *Thedgonia ligustrina* on *Ligustrum japonicum* and *Cercospora kyotensis* sp. nov. on *Hydrangea serrate* var. *thunbergii*. Transactions of the Mycological Society of Japan 1992;33:177–84.
232. Braun U. A Monograph of *Cercosporella*, *Ramularia* and Allied Genera (Phytopathogenic Hyphomycetes), p. 290. IHW Verlag: Munich; 1995. p. 333.
233. Crous PW, Braun U, Wingfield MJ, Wood AR, Shin HD, Summerell BA, *et al.* Phylogeny and taxonomy of obscure genera of microfungi. Persoonia 2009;22:139–61.
234. Kowalik M, Bigaj K. Causal agents of leaf spot disease of common privet (*Ligustrum vulgare* L.). Progress in Plant Protection 2012;52:69–72.
235. Saccardo PA. 1889. Sylloge Fungorum 8, 580.
236. Kirk PM, Ansell AE. Authors of Fungal Names. CAB International, Wallingford, UK; 1992. p. 95.
237. Kaiser W, Crous PW. *Mycosphaerella lupini* sp. nov., a serious leaf spot disease of perennial lupin in Southcentral Idaho, USA. Mycologia 1998;90:726–31.

20 Status of *Opuntia* invasions in the arid and semi-arid lands of Kenya

Eunice W. Githae*

Department of Biological Sciences, Chuka University, P. O. Box 109-60400, Chuka, Kenya

Abstract

Invasive alien species (IAS) are among the leading threats to biodiversity, food security and human well-being. *Opuntia* (prickly pear cactus) is one of the most widespread and naturalized in the arid and semi-arid areas of Kenya, with extreme effects on rural livelihoods and the environment. However, comprehensive information on the status of invasion in the country is lacking, which is crucial for developing strategies for prevention and management. This paper, therefore, provides an overview of the interacting factors that influence its invasion in Kenya in terms of species diversity, pathways of introduction, negative impacts and the effectiveness of regulations and control measures. Seven invasive species of *Opuntia* are present in the country with the most abundant (*O. stricta*) being under biocontrol trial. These species have the same habitat preferences, physiological traits, negative impacts, introduction pathways and management options. Invasion is mainly attributed to changes in land use and consequent land degradation. Introduction pathways are largely intentional for ornamental purposes but unintentional at a local scale through escape from gardens and natural dispersal. The most applied methods of managing *Opuntia* are mechanical and chemical methods that are unsustainable and labour-intensive at large scale. Effective policies are therefore needed to prevent an increase in the significant negative impacts caused by IAS including those that have a limited distribution.

Keywords: Biological invasions, invasive alien species, *Opuntia*

Review Methodology

The approach employed was a review of published information from a variety of sources, including online scientific publications, books, field guides and experts' opinion on invasive alien species with a focus on *Opuntia* species in Kenya. Information was also gathered from relevant databases, including the Global Invasive Species Database, National Museums of Kenya, Centre for Agricultural Biosciences International (CABI), BioNET-EAFRINET, Germplasm Resources Information Network (GRIN), Flora of North America, Kenya Wildlife Service, Kenya Forestry Service and the Ministry of Environment and Natural Resources. Key priority pathways and impacts were categorized according to ecological and socio-economic impacts as follows: Minimal Concern (MC) – discernible impacts but no effects on individual fitness of native species; Minor (MN) fitness of individuals reduced but no impact on populations; Moderate (MO) – changes to populations but not to community composition; Major (MR) – community changes that are reversible; and Massive (MV) irreversible community changes and extinctions. They were also categorized based on the current framework of invasive species.

*Corresponding author: egithaeh@gmail.com, egithae@chuka.ac.ke

Introduction

Invasive alien species (IAS) are those species whose introduction threatens ecosystems, habitats or species. They are regarded as among the biggest threats to global ecosystems and biodiversity [1]. They have a high ecological and competitive ability as well as efficient reproduction and dispersal mechanisms. In a new environment, such species prosper in absence of predators or parasites that would regulate their relative abundance [2]. Besides, IAS are able to occupy vacant niches or take advantage of ecologically stressed or disturbed environments by changing resource availability [3]. This, in turn, increases the competitive ability of the IAS in the new environment. In the rangelands, impacts of IAS are mostly related to livestock production and include livestock deaths, reduced forage, interference with grazing practices, increased costs of management and reduced livestock yields [4]. They may also negatively impact wildlife habitats, deplete soil and water resources and reduce the diversity of plants and animals. Effects on rural livelihoods are linked to the extra cost of management, reduced ecosystem productivity, increased the threat to health and reduced utilities from natural resources [5]. These impacts not only reduce biodiversity but also threaten food security and human well-being.

Much review on the status and impacts of IAS around the world has been done to update knowledge on the occurrence and drivers of introduction [6–8]. Such studies are crucial for developing protocols for prevention and management. Nevertheless, more comparative studies on the target species, their pathways of introduction and spatial distribution are necessary to improve understanding of the interacting factors that influence their invasion. Pathways of introduction can either be intentionally for some benefits (e.g. use in biological production systems, landscaping and ornamental purposes) or unintentionally through transport, trade and travel [9, 10]. Identifying these pathways provide information to reduce the rates of unintentional introductions and for developing interventions for monitoring and management [9]. This not only informs the development policies and management strategies but also evaluates the effectiveness of control measures.

The cactus family (Cactaceae) is among the most widely introduced IAS across the world due to its popularity in the horticultural trade [6]. *Opuntia* Mill (Prickly pear cactus) which is native to America is the most common and naturalized in the arid and semi-arid lands of Kenya. The genus is also naturalized in many parts of the world, including Australia, Europe and many African countries [6]. In Kenya, *Opuntia* species were introduced in the 1940s as ornamental plants in the Northern drylands but their distribution is basically being changed by human activities. [11]. Some studies have proposed that *Opuntia* invasion after 50 years of introduction is as a response to changes in land use primarily settlement of pastoralists, followed by overgrazing and subsequent land degradation [11]. Despite the urgency in controlling the species, there are various gaps that hinder its management in East Africa. These include a poor description of the species because of variation in growth form, lack of information on proper management and poor monitoring and control measures [12, 13]. A broad national assessment of particular taxa of concern is, therefore, necessary for the formulation of appropriate control strategies. This may uncover new patterns, processes and risks that can improve the existing management strategies [6].

Description of *Opuntia*

Opuntia belongs to a group of plants known as a cactus (plural: cacti'). It is the largest genus in Family Cactaceae with more than 360 species and a range of geographical distribution. Cactus plants are of different shapes, sizes and growth forms and often with clusters of spines that arise from modified axillary buds called areoles [6]. The stems (pads) are also of varied lengths and widths and generally flattened [14]. This morphological variation is a trait that is strongly associated with invasiveness [15]. About 1600 species of *Opuntia* are recognized worldwide. Most have become invasive due to their ability to survive well in dry periods, compete well in disturbed conditions, have low levels of parasites and have a high reproductive potential [16]. However, description of cactus plants is generally difficult because of variation in phenotypes and presence of many hybrids [17].

Opuntia species form dense thickets through vegetative reproduction when detached cladodes become rooted in the ground [18]. They also have a specialized photosynthetic system known as Crussulacean Acid Metabolism with the ability to withstand prolonged drought and high temperatures [19]. The flowers range in colour depending on the species and bloom once a year [20]. Flowers open on a single day or 2 days consecutively hence well suited to reducing the risk of poor or no pollination [21]. The flowers are pollinated effectively by bees and beetles [22]. The fruits are red to reddish-purple and vary from pear-shaped to almost spherical. The species produce many seeds that are spread by various agents, including baboons, birds and large mammals [18, 23]. Germination occurs during the rainy season, although seeds can persist in the soil for at least 19 months [21]. Plants reach reproductive maturity at about nine years of age and can live longer than 30 years [21]. In addition, *Opuntia* species have relatively shallow and horizontally spreading roots that use water efficiently [24]. Such traits ease their spreading and colonization in dry areas.

Species types and distribution

Seven species of *Opuntia* are present in Kenya and have similar traits and habitat preferences (Table 20.1). These species are weeds in many parts of the world [25] and in most cases, they occur in association with each other [26]. The counties most invaded include Narok, Laikipia, Baringo, Nakuru and Taita Taveta (Figure 20.1) with small patches occurring in the neighbouring counties. *Opuntia stricta* var. *stricta* is one of the most widespread in Kenya [27]. The species was introduced in Central Kenya as an ornamental plant and has since spread across the rangelands and conservation areas [11]. It is widespread and abundant in Laikipia County and Taita Taveta [27]. In Ethiopia, the variety is widespread and abundant in the Eastern part where it is associated with *O. ficus-indica* [27]. Further studies show that the species grow in association with *Opuntia elatior* along the Kenyan coastline with a few infestations near Lake Baringo [27]. However, it has little benefits and is the most problematic in terms of impeding mobility, reducing fodder availability and impacting negatively on human and livestock health. On the other hand, *O. elatior* is less widespread but capable of forming dense thickets in the arid and semi-arid areas [26]. *Opuntia ficus-indica* is another serious invasive species that has destroyed grazing land in the Kenyan and Tanzanian drylands [28]. Some studies have discovered that the species is the most diverse in the Rift Valley and Eastern areas of Kenya [29]. Another species *Opuntia engelmannii* is naturalized and invasive in South Africa and Kenya [30]. The species is prohibited in South Africa and should be controlled in all situations [31] while in Kenya it is not yet declared as a weed but has invaded several ecosystems especially in Baringo and Laikipia [30]. The species starts as a minor weed in localized areas but later explodes in abundance causing invasions in wider areas [25]. In addition, the species is fast-growing and forms impenetrable spiny thickets that overwhelm native vegetation and prevent free movement of people and animals [30]. *Opuntia monacantha* is widespread in Kenya but is not particularly common. The species is scattered in Laikipia, Baringo, along the Coast and Narok It is also naturalized in other parts of the world, including Australia, South Africa and south-eastern USA. The species is reported as a problem by spreading rapidly especially on degraded rangelands, disturbed areas and agricultural land [30]. Large infestations of *Opuntia exaltata* are found in parts of Nakuru and Naivasha. The species displaces native plant species, lowers the value of pasture and prevents free movement of people, livestock and wildlife. *Opuntia exaltata* is listed as a noxious weed in South Africa [32]. All *Opuntia* species in Kenya are spiny except *Opuntia microdasys* which has no spines but many glochides. It rapidly establishes large and dense infestations that prevent access to water and other resources [26]. The species has a high frequency of vegetative multiplication that can develop dense populations that readily colonize new and wide areas [33].

Table 20.1. *Opuntia* species present in Kenya with their common names and the distinguishing features.

	Species	Common name	Description	Habitat
1.	*Opuntia elatior* Mill.	Red flower prickly pear	Forms dense branched clumps, cladodes are olive-green and egg-shaped with almost parallel sides, spines are needle-like and dark brown	Disturbed and open lands, roadsides, grasslands and dry gullies in arid and semi-arid areas
2.	*Opuntia engelmannii* Salm-Dyck	Cows tongue cactus	Erect shrub with many ascending to sprawling branches forming dense clumps, cladodes are broadly egg-shaped to diamond-shaped or tapering at the end, spines are white to yellow, red to dark brown at the base	Disturbed and open lands, sandy and gravelly soils and grasslands in arid and semi-arid areas
3.	*Opuntia exaltata* A. Berger	Long spine cactus	Much branched shrub and develops a trunk with age, curved and cylindrical cladodes with small rounded outgrowths, spines are straight and yellow-brown	Disturbed and open lands, grasslands and roadsides
4.	*Opuntia ficus-indica* (L.) Mill.	Sweet prickly pear	Shrub forming a trunk with age, cladodes are much longer than broad, spines are white, absent or 1–2 or more per areole	Most common in arid and semi arid areas, conservation areas, disturbed and open lands, savannah and roadsides.
5.	*Opuntia microdasys* (Lehm.) Pfeiff.	Teddy-bear cactus/Bunny ears	Forms thickets, cladodes are velvety, egg-shaped or almost round, no spines but many glochidea	Disturbed and open lands in arid and semi-arid areas
6.	*Opuntia monacantha* Haw.	Drooping prickly pear	Shrub with a large, much-branched top and drooping upper segments, cladodes are egg-shaped to somehow elongated and tapered towards the base, many spines on the trunks	Disturbed and open lands, roadsides, abandoned agricultural lands, savannah and coastal bush lands
7.	*Opuntia stricta* (Haw.) Haw.	Australian pest pear	Thicket-forming, cladodes are longer than broad; spines are straight, flattened and yellow	Most common in arid and semi arid areas, disturbed and open areas, rangelands, conservation areas, roadsides, wastelands, grasslands, coastal bush lands, dry gullies and river banks

Pathways and risks

Three main pathways of *Opuntia* introduction in Kenya are identified according to the CBD (Convention on Biological Diversity) [34]: (i) intentional introduction as ornamentals but escapes unintentionally, (ii) unintentional introduction through natural dispersal and (iii) unintentional introduction through trans-portation of habitat material from the infested

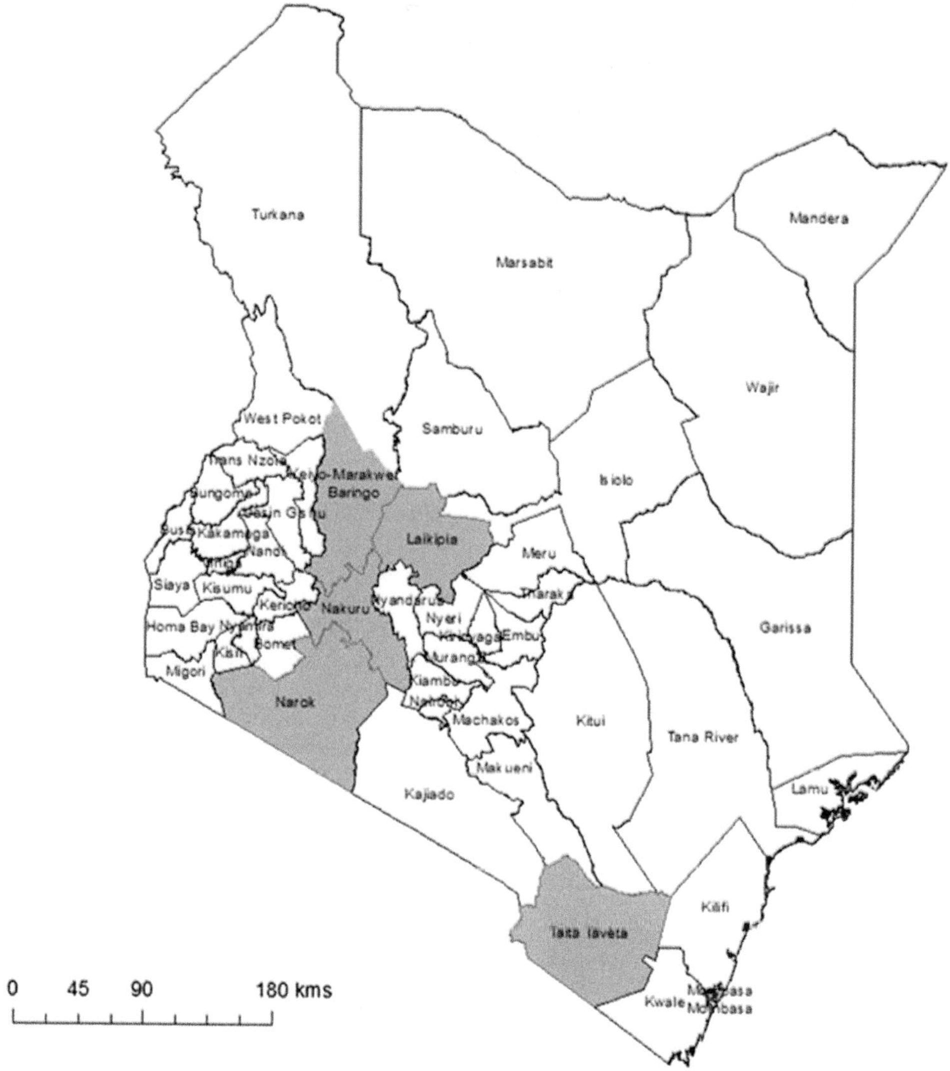

Fig. 20.1. Major areas of *Opuntia* invasion in Kenya.

areas to new areas. However, most introduction pathways in the arid and semi-arid areas of Kenya are due to intentional introductions for ornamental purposes. This is because cactus plants are very popular in the horticultural trade. This trade has developed over the years and is responsible for the spread of many IAS that have had devastating effects on people and livelihoods [8, 27]. Horticultural trade of *Opuntia* continues in Kenya and should be regulated and monitored to prevent further spread and future invasions. The escape is from cultivation especially through garden throw-outs [35] while dispersal is through vegetative reproduction [16], floods [36] and rapid dispersal of seeds by various vectors [18]. A study in the Serengeti–Mara ecosystem of East Africa revealed that more than 80% of alien plants were intentionally introduced with 24% having established naturalized populations and spreading rapidly in areas far away from human settlement, hence posing a major threat to biodiversity and tourism [37]. Such pathways and their associated vectors require proper management to limit the movement and subsequent invasions. Strategies for control and management include increased public awareness of the risks of invasions, screening of seed imports at boarders and promotion of safer non-invasive plant species for ornamental purposes.

Based on the impact category [34], four of the species; *O. stricta, O. ficus-indica, O. engelmannii* and *O. monacantha* were the most widely distributed and of major concern. *Opuntia elatior, O. exaltata* and *O. microdasys*, on the other hand, were of minor infestations and of moderate concern. All these species belong to category E of the unified framework for biological invasions where they are fully invasive with individuals dispersing, surviving and reproducing at different sites across many habitats [38]. But a significant correlation exists between population size and impact where cacti with larger population size are more likely to cause substantial impacts [39]. However, an invasive plant species with limited distribution can still have greater impacts than an abundant widespread invader if no urgent management actions are put into place. This can cause massive irreversible community changes and extinctions.

Negative impacts of Opuntia

Despite that *Opuntia* plants are appreciated for both direct and indirect uses by many livelihoods in the world [19], the species is hardly used in Kenya due to the presence of spines and glochids that cause injuries to both human and animals. They are only valued for their use as ornamental plants or as a live hedge. However, these uses cannot compensate for the overall negative impacts they portray. Once they escape into the environment, they usually adapt to a broad range of environmental conditions where they are widely established. *Opuntia* species mainly invade the natural pasture which forms about 80% of the Kenyan land mass. Pastoralists are the main inhabitants of these lands and contribute more than 50% to agricultural Gross Domestic Product (GDP). However, the loss of productivity, stability and ecological functioning of these lands by IAS hinder the economic development [40]. In many African countries, forage losses to unpalatable invasive plants have reduced rangeland capacities to three-folds [2]. The potential costs associated with *Opuntia* invasion in the rangelands are based on reduced grazing land, replacement of native forage species as well as negative impacts to human and livestock health [27]. Infestation of invasive plants also influences the availability of products by reducing the abundance of native plant species, which in turn affect the ecosystem services [41]. In addition, invasive weeds alone are responsible for reducing overall crop yields by not less than 25% with extreme cases of more than 75% [2]. Some studies have assessed the socio-ecological impacts of *Opuntia* in Eastern Africa and have reported that the species invade valuable grazing land, reduce native plant populations and rangeland condition, affect human health and impede the mobility of human and animals [27]. Although the full cost of the impacts of cactus invasions in Kenya has not yet been quantified, some studies show an average economic loss of US$ 500–1000 per rural household per year [27].

Very few households in the rural areas are actively involved in managing the species hence revealing the need for management interventions [27]. Nevertheless, many invasion opportunities follow disturbances that are largely caused by human activities and include intensive grazing and land clearance for agriculture [42]. These later degrade the lands and open opportunities for exploitation of invasive trees and shrubs [10]. *Opuntia*, in particular, invades areas that have experienced large land-cover transformations especially in woodlands and bushlands located near urban areas [43]. At the ecosystem level, invasive alien plants change community structure and composition and their eradication may open niches for the establishment of other undesirable plants unless restoration practices are incorporated [44].

Effective control measures

The most applied methods of managing *Opuntia* in Kenya are mechanical and manual methods such as chopping, burying and burning. This is because they are target specific and economical [40]. However, they can be labour-intensive in large-scale infestations and the disturbed areas may encourage re-invasion [4]. Herbicides, on the other hand, improve the effectiveness of chopping but their use alone is very ineffective since the plant regenerates after some time. This is a major drawback in the use of herbicide especially where populations are replenished from soil-stored seeds [45]. Besides, herbicides are very expensive, do not give long-term effects and can lead to the replacement of sensitive species as well as resistance after continuous use [4]. The control of *O. stricta* is almost entirely reliant on biological control with two effective natural enemies; *Cactoblastis cactorum* Berg and *Dactylopius opuntiae* Costa [46]. The latter was introduced in a pilot project in Laikipia, Kenya and showed a positive result. The main benefits of biocontrol are that they are cost-effective, permanent and include areas which are not accessible for chemical or mechanical control [47]. However, a study conducted after the biocontrol trial revealed that reduction of *O. stricta* populations affected the native baboon-plant mutualism where the baboons had to supplement their diets with other plant species [48]. Biocontrol species may also become invasive and negatively impact non-targeted species [49]. Besides, removal of invasive plant species by chemical, mechanical or biocontrol methods may open niches for other invasion unless followed by rehabilitation [11, 43]. Instead of relying on a single control method, successful management of IAS requires an integrated approach that uses multiple control methods to prevent re-invasion and improve ecosystems.

Regulations to prevent introductions

The threats of invasive cactus are part of many pressures and drivers of biodiversity loss and environmental change and therefore cannot be treated in isolation. The responses should, therefore, be long-term and addressed at various scales. Regardless of the many laws and institutions governing the management of invasive species in Kenya, invasion of plant species continues as a threat and key contributor to environmental degradation as a result of movement of plant materials [28]. Most policies addressing invasive species focus mainly on well-established invasive species despite the fact that prevention and eradication of invaders with limited distribution are more cost-effective [50]. Effective policies are therefore needed to prevent an increase in the significant negative impacts caused by invasive species including those that are not well established and with limited distribution [51]. Nevertheless, addressing the problem wholesomely can start by categorizing IAS into groups of invasive, prohibited and permitted lists [52]. Grouping species for management identifies not only common goals but also common stakeholders and allows simplification of decision-making processes [32]. Other regulations include constricting pathways, intercepting movements at borders and assessing risk for intentional imports [50]. Generally, unintentional introduction of any invasive plant into Kenya is quarantined based on risk analysis and existing scientific knowledge on the distribution, biology and pests of the plant [53]. Inspections are also carried out at the entry points; sea ports, international airports and borders [53]. For the intentional introduction of bioproducts, risk assessment is first carried out and those approved for introduction are referred to relevant research institutions for efficacy or registration. However, appropriate measures on the monitoring, control or elimination of these species are required. Furthermore, laws and regulations for prohibiting the introduction and use of such species should be enforced and maintained.

Conclusion

Seven species of *Opuntia* have been identified in Kenya and they share the same traits. Their introduction is largely intentional and attributed to changes in land use. The most applied methods of managing *Opuntia* are mechanical and chemical methods that are unsustainable. To properly manage *Opuntia* in the country, mapping and inventory of the species in all the counties are necessary together with community awareness on the effects of invasions and possible solutions. Furthermore, effective policies are required to prevent an increase in the significant negative impacts caused by IAS including those that are not well established and with limited distribution.

Acknowledgements

The author acknowledges the Stellenbosch Training Institute for Advanced Study (STIAS) and the Centre for Invasion Biology (CIB), Stellenbosch University for technological support.

References

1. Baillie JEM, Hilton-Taylor C, Stuart SN. IUCN Red List Of Threatened Species, A Global Species Assessment. IUCN, Gland, Switzerland and Cambridge, UK; 2004.
2. Boy G, Witt A. Invasive Alien Plants and Their Management in Africa. UNEP/GEF Removing Barriers to Invasive Plant Management Project. CABI, Africa; 2013.
3. Gordon DR. Effects of invasive, non-indigenous plant species on ecosystem processes: lessons from Florida. Ecological Applications 1998;8(4):975–89.
4. DiTomaso JM. Invasive weeds in rangelands: species, impacts, and management. Weed Science 2000;48:255–65.
5. Haile ZM. Invasion of *Prosopis juliflora* (SW.) DC and rural livelihoods: the case of Afar pastoralists at Middle Awash area of Ethiopia. MSc Thesis, Department of International Environmental and Development Studies, Norwegian University of Life Sciences, Ås, Norway. 2008.
6. Novoa A, Le-Roux JJ, Robertson MP, Richardson DM. Introduced and invasive cactus species: a global review. AoB Plants 2015;7:plu078.
7. Shackleton CM, McGarry D, Fourie S, Gambiza J, Shackleton SE, Fabricius C. Assessing the effect of invasive alien species on rural livelihoods: case examples and a framework from South Africa. Human Ecology 2007;35:113–27.
8. Richardson DM, Rejmánek M. Trees and shrubs as invasive alien species – a global review. Diversity and Distributions 2011;17:788–809.
9. Hulme PE, Bacher S, Kenis M, Klotz S, Kuhn I, Minchin D, *et al.* Grasping at the routes of biological invasions: a framework for integrating pathways into policy. Journal of Applied Ecology 2008;45:403–14.
10. Clare S, Nattley W, Lothar G. A Guide to Designing Legal and Institutional Frameworks on Alien Invasive Species. IUCN, Gland, Switzerland; 2000.
11. Strum SC, Stirling G, Mutunga SK. The perfect storm: land use change promotes *Opuntia stricta*'s invasion of pastoral rangelands in Kenya. Journal of Arid Environments 2015;118:37–47.
12. Gichua M, Njoroge G, Shitanda D, Ward D. Invasive species in East Africa: current status for informed policy decisions and management. JAGST 2013;15(1):45–55.
13. Richardson DM, Pysek P. What is an invasive species? 2004. Available from: URL: http://www.cabicompendium.org/cpc/library/invasiveplants/.
14. Pinkava DJ. Cactaceae cactus family, part six. *Opuntia* P. Miller Prickly-Pears. Journal of the Arizona-Nevada Academy of Science 2003;35(2):137–50.
15. Anderson EF. The Cactus Family. Timber Press, Portland; 2001.
16. Zimmermann HG, Moran VC, Hoffmann JH. Invasive cactus species (Cactaceae). In: Muniappan R, Reddy GVP, Raman A, editors. Biological Control of Tropical Weeds Using Arthropods. Cambridge University Press, Cambridge, UK; 2009. p. 108–29.
17. Ochoa J. Cactus pear (*Opuntia* spp.) varieties main characteristics at República Argentina. FAO Cactusnet Newslettera. Nùmero especial 2003;3:29.
18. Foxcroft LC, Pickett STA, Cadenasso ML. Expanding the conceptual frameworks of plant invasion ecology. Perspectives in Plant Ecology, Evolution and Systematics 2011;13:89–100.
19. Grünwaldt JM, Guevara JC, Grunwaldt EG, Martínez Carretero E. Cacti (*Opuntia* sps.) as forage in Argentina dry lands. Revista de la Facultad de Ciencias Agrarias. Universidad Nacional de Cuyo. Mendoza. Argentina 2015;47(1):263–82.

20. Turner RM, Bowers JE, Burgess TL. Sonoran Desert Plants: An Ecological Atlas. University of Arizona Press, Tucson, 1995.
21. Bowers JE. Temporal variation in longevity of *Opuntia engelmannii* (Cactaceae) flowers. Madroño 2004;51:280–86.
22. Grant V, Grant KA, Hurd PD. Pollination of *Opuntia lindheimeri* and related species. Plant Systematics and Evolution 1979;132:313–20.
23. Padrón B, Nogales M, Traveset A, Vilá M, Martínez-Abraín A, Padilla DP, *et al.* Integration of invasive *Opuntia* spp. by native and alien seed dispersers in the Mediterranean area and the Canary Islands. Biological Invasions 2011;13:831–44.
24. Snyman HA. Cactus pears *Opuntia ficus-indica* and *O. robusta* and their water utilization. Grassroots: Newsletter of the Grassland Society of Southern Africa 2007;7(4):41–8.
25. Walters M, Figueiredo E, Crouch NR, Winter PJD, Smith GF, Zimmermann HG, *et al.* Naturalised and Invasive Succulents of Southern Africa. Abc Taxa, Belgium; 2011.
26. Witt A. Guide to the Naturalized and Invasive Plants of Laikipia. CABI, Oxfordshire, UK, 2017.
27. Shackleton RT, Witt ABR, Piroris FM, van Wilgen BW. Distribution and socio-ecological impacts of the invasive alien cactus *Opuntia stricta* in Eastern Africa. Biological Invasions 2017;19(8):2427–441.
28. Obiri JF. Invasive plant species and their disaster-effects in dry tropical forests and rangelands of Kenya and Tanzania. Journal of Disaster Risk Studies 2011;3:417–28.
29. Omweri AH, Rimberia FK, Mwangi SG, Sila DN. Morphological characterization and distribution of cactus species (Cactaceae) in arid and semi-arid lands of Kenya. International Journal of Agronomy and Agricultural Research 2016;9:182–91.
30. Pasiecznik N. *Opuntia engelmannii* (cactus apple) invasive species compendium datasheets, maps, images, abstracts and full text on invasive species of the world. 2015. Available from: URL: https://www.cabi.org/isc/datasheet/115969.
31. Nel JL, Richardson DM, Rouget M, Mgidi TN, Mdzeke N, Le Maitre DC, *et al.* A proposed classification of invasive alien plant species in South Africa: towards prioritizing species and areas for management action. South African Journal of Science 2004;100:53–64.
32. Henderson L. Alien weeds and invasive plants. A complete guide to declare weeds and invaders in South Africa. Plant Protection Research Institute Handbook No. 12, 2001, 300pp.
33. Reyes-Aguero JA, Aguirre RJR, Valiente-Banuet A. Reproductive biology of *Opuntia*: a review. Journal of Arid Environments 2006;64:549–85.
34. Convention on Biological Diversity (CBD). Pathways of Introduction of Invasive Alien species, Their Prioritisation and Management. Secretariat of the Convention on Biological Diversity, Montréal. UNEP/CBD/SBSTTA/18/9/Add.1; 2014.
35. Hodkinson DJ, Thompson K. Plant dispersal: the role of man. Journal of Applied Ecology 1997;34:1484–96.
36. Parendes LA, Jones JA. Role of light availability and dispersal in exotic plant invasion along roads and streams in the H. J. Andrews experimental forest, Oregon. Conservation Biology 2000;14:64–75.
37. Witt ABR, Kiambi S, Beale T, Van Wilgen BW. A preliminary assessment of the extent and potential impacts of alien plant invasions in the Serengeti-Mara ecosystem. East Africa Koedoe 2017;59(1):1–16. doi: 10.4102/koedoe.v59i1.1426.
38. Blackburn TM, Pyšek P, Bacher S, Carlton JT, Duncan RP, Jarosık V, *et al.* A proposed unified framework for biological invasions. Trends in Ecology and Evolution 2011;26:333–9.
39. Novoa A, Kumschick S, Richardson DM, Rouget M, Wilson JRU. Native range size and growth form in Cactaceae predicts invasiveness and impact. NeoBiota 2016;30:75–90.
40. Taylor WR, Whitson TD. Plains Prickly Pear Cactus Control. University of Wyoming, Cooperative Extension Service; 1999. Bulletin No. B-1074.
41. Rai RK, Scarborough H, Subedi N, Lamichhane B. Invasive plants – Do they devastate or diversify rural livelihoods? Rural farmers' perception of three invasive plants in Nepal. Journal for Nature Conservation 2012;20:170–6.
42. Davis MA, Grime JP, Thompson K. Fluctuating resources in plant communities: a general theory of invasibility. Journal of Ecology 2000;88:528–34.
43. Vilá M, Burriel JA, Pino J, Chamizo J, Llach E, Porterias M, *et al.* Association between *Opuntia* species invasion and changes in land cover in the Mediterranean region. Global Change Biology 2003;9:1–6.
44. Masters RA, Sheley RL. Principles and practices for managing rangeland invasive plants. Journal of Range Management 2001;54:502–17.

45. Foxcroft LC, Rouget M, Richardson DM, MacFayden S. Reconstructing 50 years of *Opuntia stricta* invasion in the Kruger National Park, South Africa: environmental determinants and propagule pressure. Diversity and Distribution 2004;10:427–37.
46. Zimmermann HG, Moran VC. Biological control of prickly pear, *Opuntia ficus indica* (Cactaceae), in South Africa. Agriculture, Ecosystems and Environment 1991;37:29–35.
47. Witt A. *Opuntia* biological control in Kenya. *Ecology & Management of Alien Plant Invasions (EMAPI) Conference.* Hawaii, USA. 2015.
48. Dyck MA. Restoration of Native Baboon-Plant Mutualisms Following Biocontrol of the Invasive Prickly Pear Cactus (*Opuntia stricta*) in Kenya. Honors Theses, Nairobi; 2017.
49. McLeod R. Counting the Cost: Impact of Invasive Animals in Australia. Cooperative Research Centre for Pest Animal Control, Canberra; 2004.
50. Simberloff D. We can eliminate invasions or live with them. Successful management projects. Biological Invasions 2009;11(1):149–57.
51. Panetta FD. Weed eradication feasibility: lessons of the 21st century. Weed Research 2015;55:226–38.
52. Novoa A, Kaplan H, Kumschick S, Wilson JRU, Richardson DM. Soft touch or heavy hand? Legislative approaches for preventing invasions: insights from cacti in South Africa. Invasive Plant Science and Management 2015;8:307–16.
53. Kedera C, Kuria B. Invasive alien species in Kenya: status and management. IPPC secretariat. Identification of risks and management of invasive alien species using the IPPC framework. Proceedings of the Workshop on Invasive Alien Species and the International Plant Protection Convention, Braunschweig, Germany, 2003. p. 22–6.

Index

Note: Page numbers in **bold** type refer to **figures**; page numbers in *italic* type refer to *tables*

Printed and bound by CPI Group (UK) Ltd, Croydon, CR0 4YY

13/09/2026

14961088-0004